P9-DCW-646

BIOLOGY FOR A CHANGING WORLD

Michèle Shuster
New Mexico State University

Janet Vigna
Grand Valley State University

Gunjan Sinha

Matthew Tontonoz

W. H. Freeman and Company • New York

Publisher Kate Ahr Parker
Senior Acquisitions Editor Marc Mazzoni
Developmental Editors Andrea Gawrylewski and Susan Weisberg
Associate Director of Marketing Debbie Clare
Managing Editor for First Edition Elaine Palucki, PhD
Senior Media Editor Patrick Shriner
Supplements Editor Amanda Dunning
Assistant Editor Anna Bristow
Project Editors Leigh Renhard and Dana Kasowitz
Art Director Diana Blume
Text Designers Matthew Ball and Diana Blume
Senior Illustration Coordinator Bill Page
Artwork Precision Graphics
Photo Editors Christine Buese and Ted Szczepanski
Photo Researcher Elyse Rieder
Production Manager Ellen Cash
Composition MPS Limited, a Macmillan Company
Printing and Binding Quad Graphics–Versailles

Library of Congress Control Number: 2010941508

ISBN-13: 978-0-7167-7324-5
ISBN-10: 0-7167-7324-4

© 2012 by W. H. Freeman and Company. All rights reserved.

Printed in the United States of America

Second printing

W. H. Freeman and Company
41 Madison Avenue
New York, NY 10010
Houndmills, Basingstoke RG21 6XS, England
www.whfreeman.com

To our teachers and students: You are our inspiration

About the Authors

Michèle Shuster, Ph.D., is an assistant professor in the biology department at New Mexico State University in Las Cruces, New Mexico. She focuses on the scholarship of teaching and learning, studying introductory biology, microbiology, and cancer biology classes at the undergraduate level, as well as working on several K-12 science education programs. Michèle is an active participant in programs that provide mentoring in scientific teaching to postdoctoral fellows, preparing the next generation of undergraduate educators. She is the recipient of numerous teaching awards, including a Donald C. Roush Excellence in Teaching Award at NMSU. Michèle received her Ph.D. from the Sackler School of Graduate Biomedical Sciences at Tufts University School of Medicine, where she studied meiotic chromosome segregation in yeast.

Janet Vigna, Ph.D., is an associate professor in the biology department at Grand Valley State University in Allendale, Michigan. As a member of the Integrated Science Program, she teaches courses in genetics and science education for preservice teachers, and is active in a variety of K-12 science education programs. She has been teaching university-level biology for 14 years, with a special focus on effectively teaching biology to nonmajors. Her current research focuses on the environmental effects of the biological pesticide *Bacillus thuringiensis israelensis* on natural frog communities. She received her Ph.D. in microbiology from the University of Iowa.

Gunjan Sinha is a freelance science journalist who writes regularly for *Scientific American*, *Science*, and *Nature Medicine*. Her article on the biochemistry of love, "You Dirty Vole," was published in *The Best American Science Writing 2003*. She holds a graduate degree in molecular genetics from the University of Glasgow, Scotland, and currently lives in Berlin, Germany.

Matthew Tontonoz has been a developmental editor for textbooks in introductory biology, cell biology, evolution, and environmental science. He received his B.A. in biology from Wesleyan University, where he did research on the neurobiology of birdsong, and his M.A. in the history and sociology of science from the University of Pennsylvania, where he studied the history of the behavioral and life sciences. His writing has appeared in *Science as Culture*. He lives in Brooklyn, New York.

About the Publishers

ALL OF US INVOLVED in science education understand the importance of scientific literacy. How do we get the attention of a nonscientist? And if we can get it, how do we keep it—not only for the duration of the course, or the chapter in a textbook, but beyond?

How do we convey in our courses and our textbooks not just what we know but also how science is done? These are the challenges we hope to address with our new series of textbooks specifically for the nonscientist.

With this series, W. H. Freeman and *Scientific American* join forces not just to engage nonscientists but also to equip them with critical life tools.

W. H. FREEMAN

DISTINGUISHED by a discerning editorial vision and a long-standing commitment to superior quality, W. H. Freeman works closely with top researchers and educators to develop superior teaching and learning materials in the sciences. We know that a dedicated instructor and the right textbook have the power to change the world–one student at a time.

SCIENTIFIC AMERICAN

COMMITTED to bringing first-hand developments in modern science to its audience, *Scientific American* has long been the world's leading source for science and technology information, featuring more articles by Nobel laureates than any other consumer magazine. The oldest continuously published magazine in the United States, *Scientific American* has been independently ranked among the top 10 U.S. consumer media outlets as "Most Credible" and "Most Objective."*

*Erdos & Morgan 2008-2009 Opinion Leader Survey

From the Authors

The development of this book has taken us all on an extremely long and winding road, on which we have met fascinating people and had incredible experiences. The authors would like to thank Elizabeth Widdicombe, Kate Parker, and the folks at W. H. Freeman and Company and *Scientific American* for supporting this vision for biology education. They recognized our diverse strengths and brought us together to make this vision a reality. We have learned so much from one another on this challenging and rewarding professional journey, and none of us has likely worked so hard and so passionately on a project as we all have on this one.

We would like to thank all of the people who were interviewed and generously contributed information for these chapters. Their stories are central to the impact that this book will have on the students we teach. They are authentic examples of biology in a changing world, and they bring this book to life.

A special thank you is required for our Senior Acquisitions Editor, Marc Mazzoni, for his unwavering encouragement and ability to bring stable direction and support to the project. Developmental Editors Andrea Gawrylewski and Susan Weisberg and Assistant Editor Anna Bristow have spent many hours in the pages of this book, editing the details, managing our chaos, and smoothing our rough edges. We thank them for their dedication, patience, experience, and expertise. Thanks go to Patrick Shriner and Amanda Dunning for their tireless work on our media and supplements program. And we must thank Elaine Palucki, who has been with us from the very beginning, bringing enthusiasm and a fresh voice to our discussions. Elaine has recruited an outstanding pool of reviewers for this project, to whom we owe a debt of gratitude.

Many thanks to the production team, Leigh Renhard, Dana Kasowitz, Philip McCaffrey, Nancy Brooks, Matthew Ball, Diana Blume, Bill Page, Christine Buese, Ted Szczepanski, Elyse Rieder, Ellen Cash, and all the people behind the scenes at W. H. Freeman for translating our ideas into a beautiful, cohesive product. We would like to thank Rachel Rogge and Jan Troutt at Precision Graphics for their outstanding work on the Infographics. We appreciate their patience with the many edits and quick timelines throughout the project. They do amazing work.

We'd like to thank Debbie Clare for her enthusiasm and hard work in promoting this book in the biology education community. We thank the enthusiastic group of salespeople who connect with biology educators across the country and do a wonderful job representing this book.

The authors would like to thank our families and friends who have been close to us during this process. They have been our consultants, served as sounding boards about challenges, celebrated our successes, shared our passions, and supported the extended time and energy we often diverted away from them to this project. We are grateful for their patience and unending support.

And finally, a sincere thank you to our many teachers, mentors, and students over the years who have shaped our views of biology and the world, and how best to teach about one in the context of the other. You are our inspiration.

Brief Contents

Contents

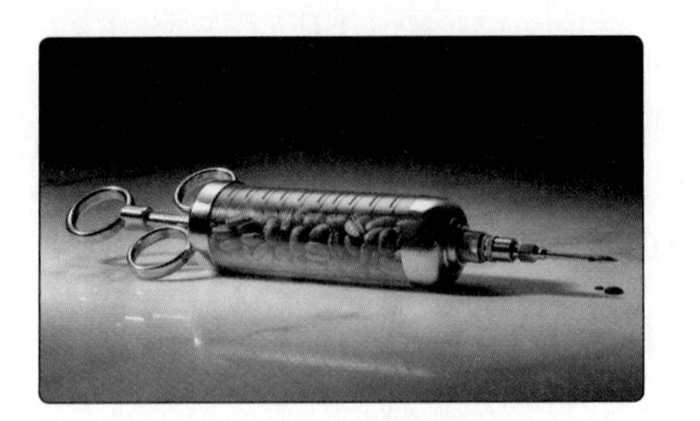

Survivors

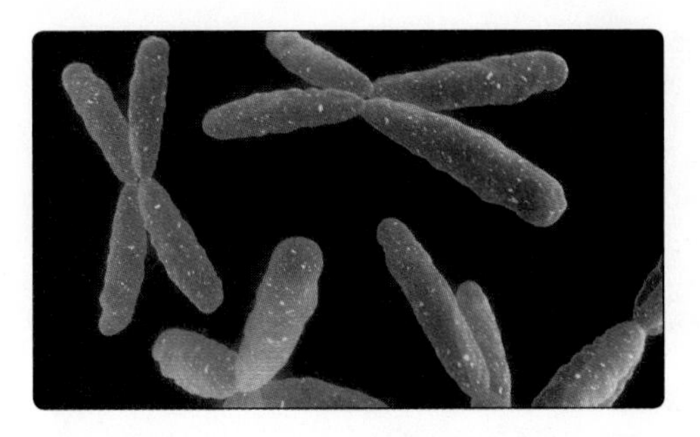

12. Complex Inheritance 229

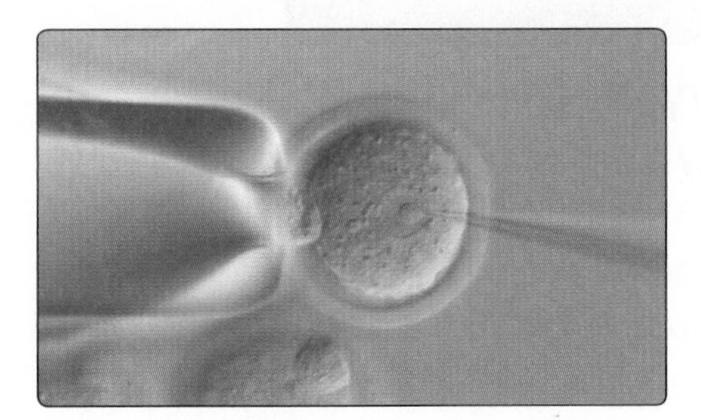

13. Stem Cells and Cell Differentiation 253

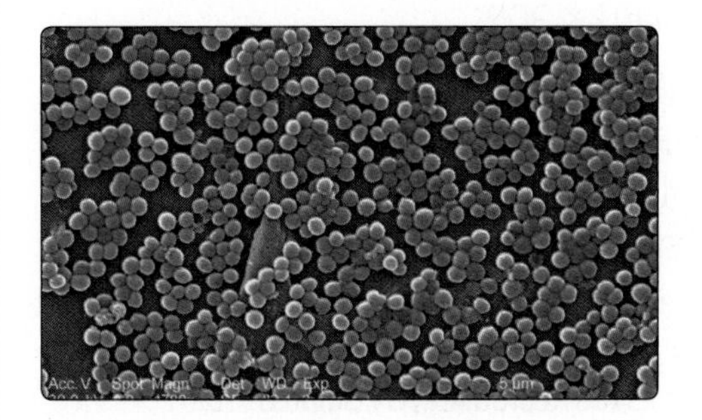

14. Natural Selection and Adaptation 271

Biology for a Changing World at a Glance

Biology for a Changing World was written by a team of two full-time college biology instructors and two science writers, with extensive input from nonmajors biology instructors across the United States and Canada. The authors identified newsworthy stories to convey key concepts, then gathered feedback from instructors to ensure that these stories are relevant, useful, and (most important) interesting to students.

Key Features

- Engaging stories carry students through each chapter, demonstrating how biology relates to their daily lives.
- Magazine-style design balances words and images while providing students with the learning tools they need.

> "This format does exactly what I would like to do—it takes a real-life example as an application of the material and uses the example in the form of an unfolding story to *both* teach the material to the student and at the same time demonstrate why and how the material is important to society and the student today."
> —Chris Haynes, Shelton State Community College

Chapter 8 Genes to Proteins

Medicine from milk

Scientists genetically modify animals to make medicine

In a Massachusetts barn nestled among willow and oak trees, rows of juglike machines drone in a constant hum. Goats, dozens of them, are being milked. But this is no ordinary dairy operation. This farm is among several worldwide practicing the art of "pharming"–using genetically modified animals to churn out therapeutic drugs.

The first drug produced from such **transgenic** animals is already available, manufactured by GTC Biotherapeutics, a firm based in Framingham, Massachusetts. The drug consists of a human protein called antithrombin that was extracted from transgenic goats' milk. Antithrombin is most commonly used to treat patients who either inherit or acquire a deficiency of the antithrombin protein, which puts them at risk of developing dangerous blood clots.

For decades, scientists had extracted antithrombin from human blood donations. But blood contains only small amounts of antithrombin, and the supply depends on the number of blood donors. Transgenic goats, however, can produce massive amounts of the drug in a relatively short period of time. Moreover, relying on a herd of goats instead of human volunteers ensures a consistent supply. And because the animals live in a controlled environment, there is less risk of transmitting infections such as HIV and hepatitis to healthy people through contaminated donor blood.

"This is very exciting, it is novel and has great potential for where we can go with this new technology." —Bernadette Dunham

Because of all these advantages, some people are predicting that transgenic animals may one day replace human donors as the source for therapeutic agents extracted from blood. "This is very exciting, it is novel and has great potential for where we can go with this new technology," Bernadette Dunham, director of the FDA's Center for Veterinary Medicine, told

144 UNIT 2: HOW IS LIFE PERPETUATED? CELL DIVISION AND INHERITANCE

SEBASTIANKNIGHT/DREAMSTIME.COM

TRANSGENIC Refers to an organism that carries one or more genes from a different species.

PROTEIN A macromolecule made up of repeating subunits known as amino acids, which determine the shape and function of a protein. Proteins play many critical roles in living organisms.

AMINO ACIDS The building blocks of proteins. There are 20 different amino acids.

the *Washington Post* in February 2009, when the company's drug for antithrombin deficiency was approved for market.

Antithrombin: From Gene to Protein

Antithrombin is a **protein.** Recall from Chapter 2 that proteins are one of the four main macromolecules that make up cells. Proteins have myriad functions in the body: they allow our muscles to contract, give our hair and skin its texture, and facilitate the thousands of chemical reactions that occur in our cells. In fact, proteins play a huge role in all basic cellular functions. Proteins can perform such a variety of different tasks because they come in many shapes and sizes.

All proteins are made of the same building blocks called **amino acids.** There are 20 different amino acids in all. All amino acids have the same basic core structure, but each also has a unique chemical side group that distinguishes the amino acids from one another. Amino acids bond together to form linear chains. The human antithrombin protein is a chain of 432 amino acids. Many human proteins are in this size range, but chain lengths vary from just a few to thousands of amino acids. The longest human protein, titin, is a single chain of 34,350 amino acids.

The sequence of amino acids in any given chain makes each chain unique, and also determines how that chain ultimately folds into a

CHAPTER 8: GENES TO PROTEINS 145

INFOGRAPHIC 7.1

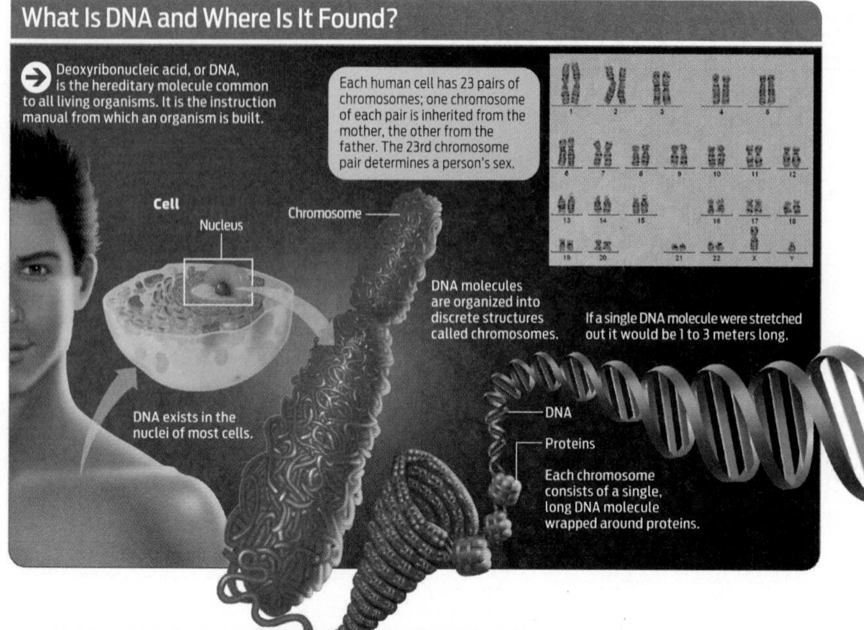

What Is DNA and Where Is It Found?
Deoxyribonucleic acid, or DNA, is the hereditary molecule common to all living organisms. It is the instruction manual from which an organism is built.
Each human cell has 23 pairs of chromosomes; one chromosome of each pair is inherited from the mother, the other from the father. The 23rd chromosome pair determines a person's sex.
Cell
Nucleus
Chromosome
DNA molecules are organized into discrete structures called chromosomes.
If a single DNA molecule were stretched out it would be 1 to 3 meters long.
DNA exists in the nuclei of most cells.
DNA
Proteins
Each chromosome consists of a single, long DNA molecule wrapped around proteins.

"The graphics are head and shoulders above anything in other texts that target a nonscience audience."
—Mark Bucheim, University of Tulsa

- **Infographics** *Scientific American*-style illustrations teach core biological concepts by combining easy-to-follow images with straightforward explanations. Each graphic provides a complete picture of a fundamental scientific principle.
- End-of-chapter pedagogy provides a point-by-point review of each chapter's key concepts:
  - **Test Your Knowledge** self-tests are aligned with each chapter's key concepts
  - **Know It** questions assess general comprehension
  - **Use It** questions assess whether students can apply what they've learned

"This is a great way to reinforce the 'science of the story.' The Know It and Use It segments. reinforce scientific information and allow the student to apply concepts to everyday situations."
—Pamela Anderson Cole, Shelton State Community College

Chapter 8 Test Your Knowledge

PROTEIN STRUCTURE AND FUNCTION
Proteins have a unique three-dimensional structure that specifies their function. The structure of a protein is determined by its corresponding gene sequence.

HINT See Infographics 8.1–8.4.

KNOW IT
1. What determines a protein's function?
2. The final product of gene expression is
a. a DNA molecule.
b. an RNA molecule.
c. a protein.
d. a ribosome.

USE IT
3. Heating can cause a protein to denature, or unfold. What do you think would happen to a protein's function in this case? Explain your answer.
4. Insulin is a protein that is used therapeutically to treat people with diabetes. In your own words, describe the relationship between the insulin gene and the insulin protein.

GENE STRUCTURE
All genes have two key parts: a regulatory sequence and a coding sequence. To review gene structure, refer to Infographics 8.4 and 8.5.

KNOW IT
5. The difference between two alleles of a gene is best ascertained by
a. examining the amount of protein produced from each allele.
b. examining the structure of the protein produced from each allele.
c. examining the amount of mRNA produced from each allele.
d. examining the nucleotide sequence of each allele.
6. If a functional allele of antithrombin is expressed,
a. blood clots will be more likely to form in the wrong place.
b. blood clots will be less likely to form in the wrong place.
c. functional antithrombin protein will be present in blood.
d. a and c
e. b and c

USE IT
7. You are a doctor. Your patient has reduced levels of normal functioning antithrombin. Would you suspect a problem in the regulatory or in the coding sequence of the antithrombin gene? Why?
8. If you wanted to use genetic engineering to increase the amount of antithrombin this patient produces, would you modify the regulatory sequence or the coding sequence? Explain your answer.

MAKING TRANSGENIC ORGANISMS
Transgenic organisms are becomingly increasingly important in agriculture and medicine.

HINT See Infographics 8.5 and 8.6.

KNOW IT
9. Melanin is expressed in skin cells and gives skin its color. If you wanted to express a different gene in skin cells, which part of the melanin gene would you use? Why? If you wanted to produce melanin in yeast cells, what part of the melanin gene would you use? Why?

USE IT
10. Explain why scientists used the beta casein regulatory sequence to express human antithrombin in goats' milk.

GENE EXPRESSION
Gene expression is the multistep process of converting the information of DNA into proteins.

HINT See Infographics 8.7–8.10.

KNOW IT
11. For each structure or enzyme listed, indicate by N (nucleus) or C (cytoplasm) its active location in eukaryotic cells:
RNA polymerase _____
Ribosome _____
tRNA _____
mRNA _____

158 UNIT 2: HOW IS LIFE PERPETUATED? CELL DIVISION AND INHERITANCE

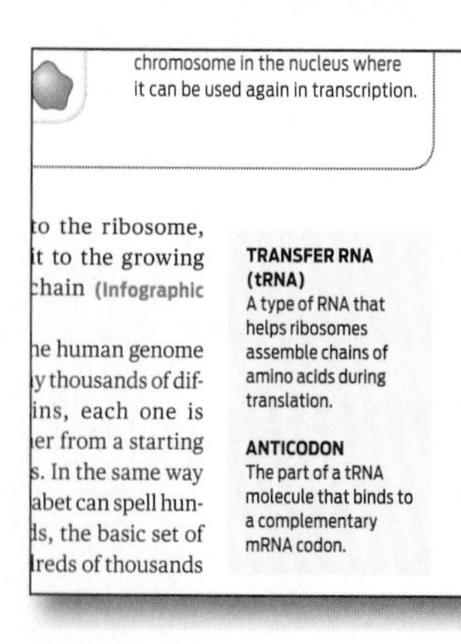

- Running Glossary provides immediate, concise definitions for key terms.

Media and Supplements

Biology for a Changing World is supported by a robust set of study and teaching resources and products. These support materials have been written by a team of experienced nonmajors educators and are tied together by peer-reviewed Learning Objectives for each chapter. These objectives allow instructors to identify the core concepts that most challenge their students and enable them to target student needs earlier and more effectively. In addition, they provide instructors with a way to demonstrate that their students have mastered specific chapter goals.

Our program is outlined below, please ask your sales representative to see our supplement sampler for more clarification.

- **Instructor Resources**
  - **Story Abstracts** The abstracts offer a brief story synopsis, providing interesting details relevant to the chapter and to the online resources not found in the book.
  - **Active Learning Activities** Our activities aim to enhance the student's natural curiosity and to inspire critical thinking about the topics. These will also provide alternative examples to the stories in the text.
  - **Clicker Questions** Designed to be used by students working in teams as well as in large lectures.
  - **Optimized Figure JPEGS and PowerPoints** Infographics are optimized and split apart to be used for projection in large lecture halls.
  - **Stepped Art Sequences and Animations** Every piece of art in the text is interactive in some way, either through an art sequence or an animation.
  - **Lecture PowerPoints**–Prebuilt lectures to help with the transition to a new textbook.
  - **Test Questions/Quizzes** All assessment is organized into the textbook's "Know it" and "Use it" categories.

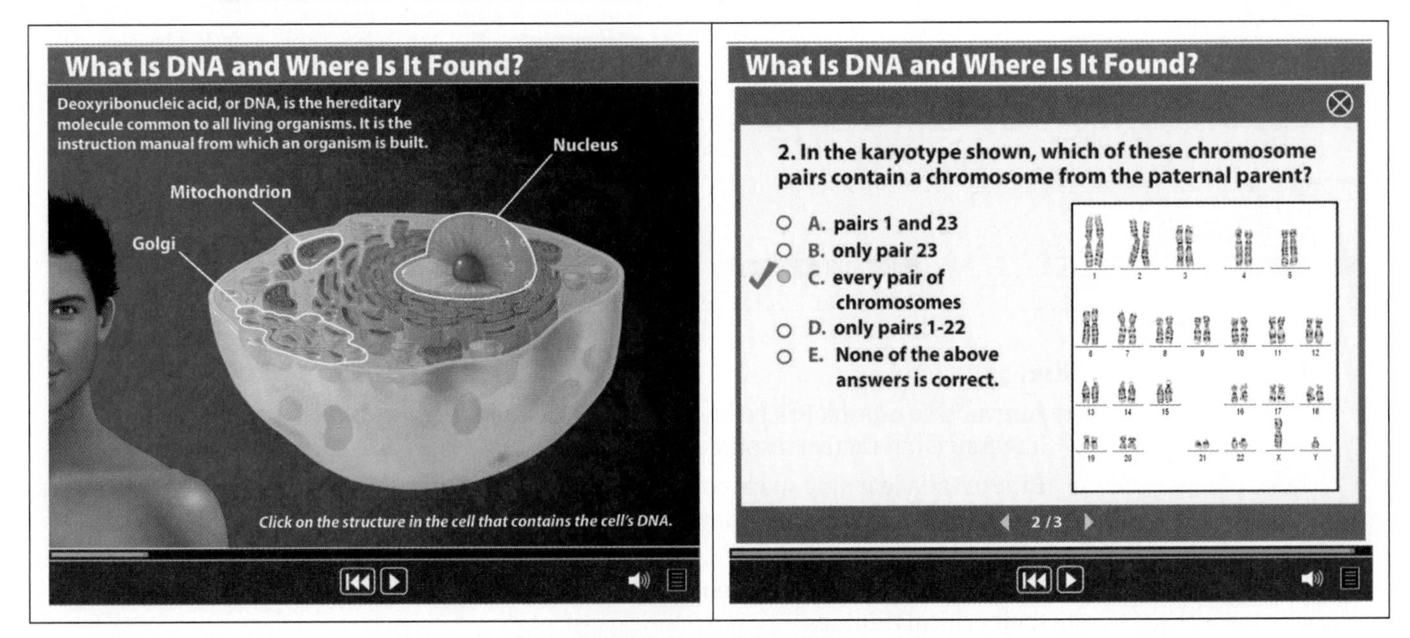

Interactive Infographic Tutorial

- **Instructor Products**
  - **Test Bank/Computerized Test Bank** More than 100 questions per chapter presented in a sortable, searchable platform.
  - **Interactive e-book** Priced lower than the printed textbook and featuring seamlessly integrated interactive resources and study tools.
  - **BioPortal** A learning space to help instructors administer their courses by combining our fully customizable e-book, instructor resources, student resources, news feeds, and homework management tools.
  - **Course Management System e-packs** Available for Blackboard, WebCT, and other course management platforms.
  - **Faculty Lounge** The only publisher-provided Web site linking the nonmajors biology community, where instructors can share lecture ideas, videos, animations, and other resources.
  - **Instructors Resource DVD**
- **Student Resources**
  - **Key Term Flashcards** Students can drill and learn the most important terms in each chapter using interactive flashcards.
  - **Lecture Companion Art** The Infographics for each chapter are available as PDF files that students can download and print before lectures.
  - **Quizzing with Feedback** Response-specific feedback helps explain concepts and correct student misunderstandings.
  - **Interactive Infographics** All Infographics in the text include an animated interactive tutorial or an infographic activity.
  - **LearningCurve** A new learning tool that evaluates what students know and don't know and provides them with a personalized study plan to guide their study of each chapter.

Focused Quizzing
7.1 How DNA helped free an innocent man
Which of the following contains DNA?
chromosomes
blood
hair
both blood and hair
All of these contain DNA.
Topic Breakdown
Show Personalized Study Plan
20% DNA Replication
50% DNA Structure
69% Understanding PCR
Take a Break
Next Question
You're currently 31% of the way towards completing the activity, as shown by the dark bar above.

Focused Quizzing
Great! You've answered enough questions to complete the activity!
Don't stop now, though: you can keep answering questions as long as you wish for practice.
Your grade will not be affected if you continue answering questions.
Personalized Study Plan
Topic Breakdown: These need some work
29% DNA Replication
63% DNA Structure
You scored better on these:
89% Understanding PCR
Click topics above to view relevant eBook sections and additional resources.
More Questions

- **Student Products**
  - **Interactive e-book** Priced lower than the printed textbook and featuring seamlessly integrated interactive resources and study tools.
  - **BioPortal** A learning space combining our fully customizable e-book, student resources, news feeds, and homework management tools.
  - **Printed Study Guide** Covers all the topics in each chapter, breaking down each infographic by offering students clear learning objectives and providing questions to test their critical thinking.
  - **Free Book Companion Web Site** Featuring most student resources in an online format.

Acknowledgments

We would like to thank the many reviewers who have helped with the development of this text.

Stephanie Aamodt, *Louisiana State University-Shreveport*
Marilyn Abbott, *Lindenwood University*
Julie Adams, *Ohio Northern University*
Tadesse A. Addisu, *Northern Virginia Community College-Annandale*
Adjoa Ahedor, *Rose State College*
Ann Aguanno, *Marymount Manhattan College*
Zulfiqar Ahmad, *East Tennessee State University*
Mark Ainsworth, *Seattle Central Community College*
Carol Allen, *Montgomery College*
Pamela Anderson Cole, *Shelton State Community College*
Ken Andrews, *East Central University*
Corrie Andries, *Central New Mexico Community College*
Josephine Arogyasami, *Southern Virginia University*
Joseph A. Arruda, *Pittsburg State University*
Tami Asplin, *North Dakota State University*
Kim Atwood, *Cumberland University*
Felicitas Avendano, *Grandview University*
James Backer, *Daytona State College*
David Bailey, *Delta College*
Andy Baldwin, *Mesa Community College*
Mary Ball, *Carson-Newman College*
Verona Barr, *Heartland Community College*
Tina Beams Jones, *Shelton State Community College*
Lynne Berdainer, *Gainesville State College*
Christine Bezotte, *Elmira College*
Bill Rogers, *Ball State University*
Curtis Blankenspoor, *Calvin College*
Lisa Boggs, *Southwestern Oklahoma State University*
Cheryl Boice, *Lake City Community College*
Larry Boots, *University of Montavello*
Barbara Boss, *Keiser University*
Brenda Bourns, *Seattle University*
Bradley Bowden, *University of Connecticut*
Mark Boyland, *Union University*
Dean Bratis, *Villanova University*
Mimi Bres, *Prince George's Community College*
Randy Brewton, *University of Tennesee-Knoxville*
Marguerite (Peggy) Brickman, *University of Georgia*
Clay Britton, *Methodist University*
Gregory Brown, *McGill University*
Carole Browne, *Wake Forest University*
Sara Browning, *Palm Beach Atlantic University*
Joseph Bruseo, *Holyoke Community College*
Mark Buchheim, *University of Tulsa*
Anne Bunnell, *East Carolina University*
Jamie Burchill, *Troy University*
Greg Butcher, *Centenary College of Louisiana*
David Byres, *Florida State College at Jacksonville-South Campus*
Carolee Caffrey, *Hofstra University*
Jane Caldwell, *Washington & Jefferson College*
Jamie Campbell, *Truckee Meadows Community College*
Shillington Cara, *Eastern Michigan University*
Michael Carr, *Oakton Community College*
Dale Casamata, *University of North Florida*
Deborah A. Cato, *Wheaton College*
Jeannie Chapman, *University of South Carolina-Upstate*
Xiaomei Cheng, *Mount St. Mary's College*
Steven D. Christenson, *Brigham Young University-Idaho*
Kimberly Cline-Brown, *University of Northern Iowa*
Yvonne Cole, *Lindenwood University*
Claudia Cooperman, *Philadelphia University*
Erica Corbett, *Southeastern Oklahoma State University*
David Corey, *Midlands Technical College-Beltline Campus*
Cathy Cornett, *University of Wisconsin-Platteville*
Frank Coro, *Miami-Dade College-InterAmerican Campus*
Angela Costanzo, *Hawaii Pacific University*
Richard Cowart, *University of Dubuque*
Jan Crook-Hill, *North Georgia College and State University*
Peter Cumbie, *Winthrop University*
Kathleen L. Curran, *Wesley College*
Jennifer Cymbola, *Grand Valley State University*
Gregory Dahlem, *Northern Kentucky University*
Don Dailey, *Austin Peay State University*
Michael S. Dann, *Pennsylvania State University*
Farahad Dastoor, *University of Maine*
Cara L. Davies, *Ohio Northern University*
Renne Dawson, *University of Utah*
Nishantha de Silva, *Lock Haven University*
Jodi Denuyl, *Grand Valley State University*
Elizabeth DeStasio, *Lawrence University*
Chris Dobson, *Grand Valley State University*
Therese Dudek, *Kiswaukee College*
Denise Due-Goodwin, *Vanderbilt University*
Jacquelyn Duke, *Baylor University*
Michael Edgehouse, *Cabrillo College*
Susan S. Epperson, *University of Colorado at Colorado Springs*
Paul Farnsworth, *University of New Mexico*
Steven Fields, *Winthrop University*
Lynn Firestone, *Brigham Young University-Idaho*
Teresa Fischer, *Indian River Community College*
Carey Fox, *Brookdale Community College*
Karen Francl, *Radford University*
Barbara S. Frank, *Idaho State University*
Diane Fritz, *Northern Kentucky University*
Richard Gardner, *Southern Virginia University*
Shelley Garrett, *Guilford Technical Community College*
Phil Gibson, *University of Oklahoma*
Julie L. Glenn, *Gainesville State College-Oconee Campus*
Inna Goldenberg, *Oakton Community College*
Stephen Gomez, *Central New Mexico Community College*
Brad Goodbar, *College of the Sequoias*
Kate Goodrich, *Widener University*
Sherri Graves, *Sacramento City College*
Madoka Gray-Mitsumune, *Concordia University*
Bradley Griggs, *Piedmont Technical College*
Cheryl Hackworth, *West Valley College*
Janelle Hare, *Morehead State University*
Katherine Harris, *Hartnell College*
Joe Harsh, *Butler University*
Roberta Hayes, *St. Johns College of Liberal Arts and Sciences*
Chris Haynes, *Shelton State Community College*
Steve Heard, *University of New Brunswick*
Jason Heaton, *Samford University*
Susan Hengeveld, *Indiana University*
Kelly Hogan, *University of North Carolina-Chapel Hill*
Andrew Holmgren, *Heartland Community College*
Ann Marie Hoskinson, *Minnesota State University-Mankato*
Tim Hoving, *Grand Rapids Community College*
Tonya Huff, *Riverside Community College*
Evelyn Jackson, *University of Mississippi*
Laurie Johnson, *Bay College*
Tanganika K. Johnson, *Southern University and A&M College*

David Jones, *Dixie State College*
Jackie Jordan, *Clayton State University*
Marian Kaehler, *Luther College*
John Kell, *Radford University*
Michael Kennedy, *Missouri Southern State University*
Janine Kido, *Mt. San Antonio College*
Kerry Kilburn, *Old Dominion University*
Dennis J. Kitz, *Southern Illinois University-Edwardsville*
Cindy Klevickis, *James Madison University*
Jeannifer Kneafsey, *Tulsa Community College*
Brenda Knotts, *Eastern Illinois University*
Olga Kopp, *Utah Valley College*
Ari Krakowski, *Laney College*
Dan Krane, *Wright State*
Wendy A. Kuntz, *Kapiolani Community College*
Holly Kupfer, *Central Piedmont Community College*
Dale Lambert, *Tarrant County College*
Kirkwood Land, *University of the Pacific*
Elaine Larsen, *Skidmore College*
Mary Lehman, *Longwood University*
Beth Leuck, *Centenary College of Louisiana*
Robert Levine, *McGill University*
Patrick Lewis, *Sam Houston State University*
Tammy Liles, *Bluegrass Community College*
Susanne Lindgren, *California State University-Sacramento*
Matthew Linton, *University of Utah*
Cynthia Littlejohn, *University of Southern Mississippi*
Madelyn Logan, *North Shore Community College*
Ann S. Lumsden, *Florida State University*
Will Mackin, *Elon University*
Paul H. Marshall, *Northern Essex Community College*
Mary Martin, *Northern Michigan University*
Ron Mason, *Mt. San Jacinto College-Menifee*
Helen Mastrobuoni, *County College of Morris*
Amie Mazzoni, *Fresno City College*
Rob McCandless, *Methodist*
Brett McMillan, *McDaniel College*
Malinda McMurry, *Morehead State University*
Michael McVay, *Green River Community College*
Scott Medler, *State University of New York-Buffalo*
Judith Megaw, *Indian River State College*
Diane L. Melroy, *University of North Carolina-Wilmington*
Paige Mettler-Cherry, *Lindenwood University*
Jim Mickle, *North Carolina State University*
Hugh Miller, *East Tennessee State University*
Scott Moody, *Ohio University*
John Moore, *Taylor University*
Lia Muller, *San Diego Mesa College*
Ann Murkowski, *North Seattle Community College*
Shawn Nordell, *University of St. Louis*
Peter Oelkers, *University of Michigan-Dearborn*
Margaret Oliver, *Carthage College*
Joanna Padolina, *Virginia Commonwealth University*
Karen Pasko, *Emmanuel College*
Forrest E. Payne, *University of Arkansas at Little Rock*
Joseph Peabody, *Brigham Young University-Independent Study*
Linda Peters, *Holyoke Community College*
Stephanie Toering Peters, *Wartburg College*
William Pietraface, *State University of New York-Oneonta*
Joel Piperberg, *Millersville University*
Gregory J. Podgorski, *Utah State University*
Jeff Podos, *University of Massachusetts-Amherst*
Therese Poole, *Georgia State University*
Michelle Priest, *Irvine Valley College*
Kenneth Pruitt, *University of Texas at Brownsville*
Dianne Purves, *Crafton Hills College*
Scott Quinton, *Johnson County Community College*
Logan Randolph, *Polk State College*
Nick Reeves, *Mt. San Jacinto College-Menifee*
Kim Regier, *University of Colorado-Denver*
Nancy Rice, *Western Kentucky University*
Stanley Rice, *Southeastern Oklahoma State University*
Brendan Rickards, *Gloucester County College*
Jennifer Robbins, *Xavier University*
Laurel Roberts, *University of Pittsburg*
Peggy Rolfsen, *Cincinnati State Technical and Community College*
Amy Rollins, *Clayton State University*
Deanne Roquet, *Lake Superior College*
Karen Rose, *Shelton State Community College*
Barbara Salvo, *Carthage College*
Ken Saville, *Albion College*
Michael Sawey, *Texas Christian University*
Karen Schaffer, *Northwest Missouri State University*
Daniel Scheirer, *Northeastern University*
Bronwyn Scott, *Bellevue College*
David Serrano, *Broward College-Central Campus*
Marilyn Shopper, *Johnson County Community College*
Laurie Shornicle, *University of Missouri-St. Louis*
Brad Shuster, *New Mexico State University*
Tamara Sluss, *Kentucky State University*
Patricia Smith, *Valencia Community College-East Campus*
Sharon Smth, *Florida State College at Jacksonville-Deerwood Center*
Adrienne Smyth, *Worcester State University*
James Sniezek, *Montgomery College*
Andrea Solis, *Mount St. Mary's University*
Anna Bess Sorin, *University of Memphis*
Carol St. Angelo, *Hofstra University*
Wendy Stankovich, *University of Wisconsin-Platteville*
Rob Stark, *California State University-Bakersfield*
Amanda Starnes, *Emory University*
Alicia Steinhardt, *Hartnell College*
Bethany Stone, *University of Missouri*
Christine Stracey, *Westminster College*
Sheila Strawn, *University of Oklahoma*
Steve Taber, *Saginaw Valley State University*
John R. Taylor, *Southern Utah University*
Sonia Taylor, *Lake City Community College*
Don Terpening, *State University of New York-Ulster*
Pamela Thineson, *Century College*
Janice Thomas, *Montclair State University*
Paula Thompson, *Florida State College at Jacksonville-North Campus*
Heather Throop, *New Mexico State University*
Sanjay Tiwary, *Hinds Community College-Raymond Campus*
Jeff Travis, *State University of New York-Albany*
Eileen Underwood, *Bowling Green University*
Craig Van Boskirk, *Florida State College at Jacksonville-Deerwood Center*
Bina Vanmali, *University Missouri-Columbia*
José Vázquez, *New York University*
R. Steve Wagner, *Central Washington University*
Rebekah Waikel, *Eastern Kentucky University*
Timothy Wakefield, *John Brown University*
Helen Walter, *Mills College*
Paul Wanda, *Southern Illinois University-Edwardsville*
Katherine Warpeha, *University of Illinois at Chicago*
Arthur C. Washington, *Florida Agricultural and Mechanical University*
Amanda Waterstrat, *Eastern Kentucky University*
Kathy Webb, *Bucks Country Community College*
Karen Wellner, *Arizona State University*
Mike Wenzel, *California State University-Sacramento*
Brad Wetherbee, *University of Rhode Island*
Alicia Whatley, *Troy University*
Robert S. Whyte, *California University of Pennsylvania*
Tara Williams-Hart, *Lousiana State University-Shreveport*
Christina Wills, *Rockhurst University*
Carol Wymer, *Morehead State University*
Lan Xu, *South Dakota State University*
Rick Zechman, *California State University-Fresno*
Michelle Zjhra, *Georgia Southern University*
Elena Zoubina, *Bridgewater College*
Jeff Zuiderveen, *Columbus State University*

Java Report

What You Will Be Learning

- 1.1 Conflicting Conclusions
- 1.2 Science Is a Process: Narrowing Down the Possibilities
- 1.3 Anatomy of an Experiment
- 1.4 Sample Size Matters
- 1.5 Everyday Theory vs. Scientific Theory
- 1.6 Caffeine Side Effects
- 1.7 Correlation Does Not Equal Causation
- 1.8 From the Lab to the Media: Lost in Translation

Java Report

Making sense of the latest buzz in health-related news

In 1981, a study in the *New England Journal of Medicine* made headlines when it reported that drinking two cups of coffee a day doubled a person's risk of getting pancreatic cancer; five or more cups a day supposedly tripled the risk. "Study Links Coffee Use to Pancreas Cancer," trumpeted the *New York Times*. "Is there cancer in the cup?" asked *Time* magazine. The lead author of the study, Dr. Brian MacMahon of the Harvard School of Public Health, appeared on the *Today* show to warn of the dangers of coffee. "I will tell you that I myself have stopped drinking coffee," said MacMahon, who had previously drunk three cups a day.

Just five years later, MacMahon's research group was back in the news reporting in the same journal that a second study had found *no* link between coffee and pancreatic cancer. Subsequent studies, by other authors, also failed to reproduce the original findings.

A sometime health villain, coffee's reputation seems to be on the rise. Recent studies have suggested that, far from causing disease, the beverage may actually help *prevent* a number of conditions—everything from Parkinson disease and diabetes to cancer and tooth decay. A 2010 CBS News headline announced, "Java Junkies Less Likely to Get Tumors," and a blog proclaimed, "Morning Joe Fights Prostate Cancer." The September 2010 issue of *Prevention* magazine ran an article titled "Four Ways Coffee Cures."

Not everyone is buying the coffee cure, however. Public health officials are increasingly alarmed by our love affair with—some might say, addiction to—caffeine. Emergency rooms are reporting more caffeine-related admissions, and poison control centers are receiving more calls related to caffeine "overdoses." In response, the state of California is even considering forcing manufacturers to put warning

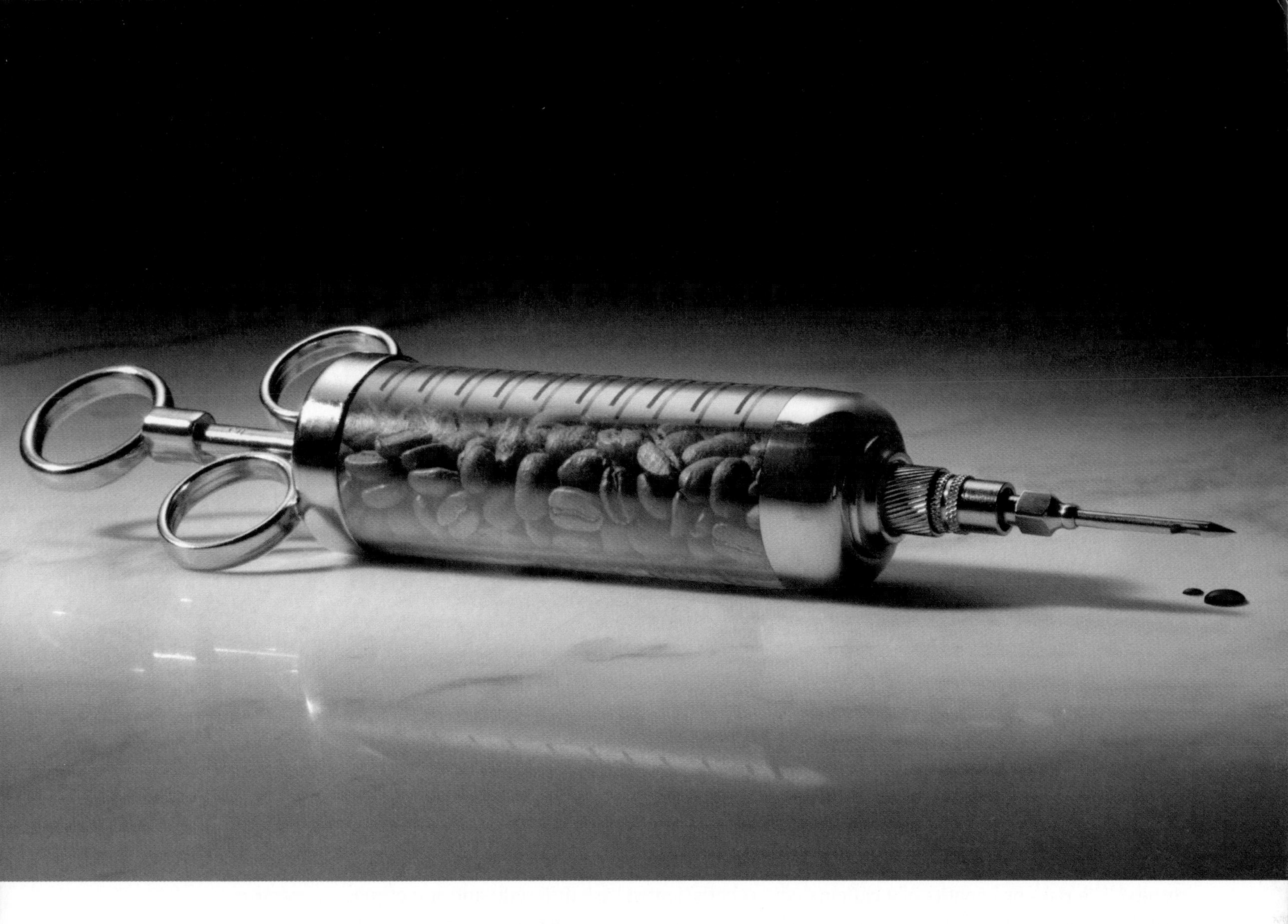

labels on energy drinks. Nevertheless, caffeine's "energizing" effect is advertised on nearly every street corner, where, increasingly, you're also likely to find a coffee shop; as of 2010, there were 222 Starbucks within a five-mile radius of a Manhattan zip code according to Foodio54.com; nationally, the average within the same radius is 10.

Consumers are flooded with a firehose of health information every day.
–Gary Schwitzer

Conflicting messages like these are all too common in the news. From the latest cancer therapies to the ecological effects of global warming, a steady but often contradictory stream of scientific information vies for our increasingly Twitter-size attention spans.

Why the mixed messages? Are researchers making mistakes? Are journalists getting their facts wrong? While both of these possibilities may be true at times, the bigger problem is widespread confusion over the nature of science and the meaning of scientific evidence.

"Consumers are flooded with a firehose of health information every day from various media sources," says Gary Schwitzer, publisher of the consumer watchdog blog HealthNewsReview.org and former director of health journalism at the University of Minnesota. "It can be–and often is–an ugly picture: a bazaar of disinformation." Too often, he says, the results of studies are reported in incomplete or misleading ways.

Consider the grande cup of coffee or the Red Bull you may have had with breakfast this morning. Why might consuming coffee or caffeine be

The national average number of Starbucks within a five-mile radius of a single zip code is 10.

associated with such dramatically different results? The risks or benefits of a caffeinated beverage may depend on the amount a person drinks–one cup versus a whole pot. Or maybe it matters *who* is drinking the beverage. The *New England Journal of Medicine* study, for example, looked at hospitalized patients only. Would the same results have been seen in people who weren't already sick? Sometimes, to properly evaluate a scientific claim, we need to look more closely at how the science was done (**Infographic 1.1**).

Science is less a body of established facts than a way of knowing.

Science Is a Process

Science is less a body of established facts than a way of knowing–a method of seeking answers to questions on the basis of observation and experiment. Scientists draw conclusions from the best evidence they have at any one time, but the process is not always easy or straightforward. Conclusions based on today's evidence may be modified in the future as other scientists ask different–and sometimes better–questions. Moreover, with improved technology, researchers may uncover better data; new information can cast old conclusions in a new light. Science is a never-ending process.

Let's say you want to investigate the "energizing" effects of coffee scientifically–how might you go about it? A logical place to start would be your own personal experience. You may notice that you feel more awake when you drink coffee. It seems to help you concentrate as you pull an all-nighter to finish a paper. Such informal, personal observations are called **anecdotal evidence.** It's a type of evidence that may be interesting but is often unreliable, since it wasn't based on systematic study. You could perhaps poll your classmates to find out if they experience coffee in the same way.

SCIENCE
The process of using observations and experiments to draw evidence-based conclusions.

ANECDOTAL EVIDENCE
An informal observation that has not been systematically tested.

PEER REVIEW
A process in which independent scientific experts read scientific studies before their publication to ensure that the authors have appropriately designed and interpreted their study.

HYPOTHESIS
A testable and falsifiable explanation for a scientific observation or question.

INFOGRAPHIC 1.1

Conflicting Conclusions

A variety of studies published in peer-reviewed scientific journals report different conclusions about the risks and benefits of coffee. In order for the public to understand and use these outcomes to its advantage, a closer look at the scientific process and the factors that surround coffee drinking is necessary.

Scientific studies report that drinking coffee...

- May cause pancreatic cancer
- Is linked to infertility and low infant birth weight
- Lowers the risk of Parkinson disease
- Does not cause pancreatic cancer
- Reduces risk of ovarian cancer

So, is it really the coffee?
Or other factors associated with drinking coffee?

- Chemicals naturally present in coffee, including caffeine
- The climate and soil in which different coffee plants are grown (which in turn influences the chemicals in coffee)
- How the beans are roasted and processed
- How much coffee a person drinks
- The gender, age, and general health of a coffee drinker
- Other social factors, such as whether coffee is consumed with a meal or with a cigarette, or with other foods and beverages that may interact in some way with coffee
- Other unknown factors that just happen to correlate with coffee drinking

TESTABLE
A hypothesis is testable if it can be supported or rejected by carefully designed experiments or nonexperimental studies.

FALSIFIABLE
Describes a hypothesis that can be ruled out by data that show that the hypothesis does not explain the observation.

EXPERIMENT
A carefully designed test, the results of which will either support or rule out a hypothesis.

Nevertheless, this anecdotal evidence might lead you to formulate a question: Does coffee improve mental performance? To get a sense of what information currently exists on the subject, you could read relevant coffee studies that have already been conducted, available in online databases of journal articles or in university libraries. Generally, you can trust the information in scientific journals because it has been subject to **peer review,** meaning that independent and unbiased experts have critiqued the soundness of a study before it was published. The aim of peer review is to weed out sloppy research, as well as overstated claims, and thus to ensure the integrity of the journal and its scientific findings. To further reduce the chance of bias, authors must declare any possible conflicts of interest and name all funding sources (for example, pharmaceutical or biotechnology companies). With this information, reviewers and readers can view the study with a more critical eye.

Based on what you learn from reading journal articles, you could formulate a **hypothesis** to explain how coffee improves mental performance. A hypothesis is a narrowly focused statement that is **testable** and **falsifiable,** that is, it can be proved wrong. A hypothesis represents one possible answer to the question under investigation. One hypothesis to explain coffee's effects, for example, is that drinking coffee improves memory. Another might be: high levels of caffeine increase concentration. Not all explanations will be *scientific* hypotheses, though. Statements of opinion, and hypotheses that use supernatural or mystical explanations that cannot be tested or refuted, fall outside the realm of scientific explanation. (Some call such explanations "pseudoscience"; astrology is a good example.)

With a clear scientific hypothesis in hand—"coffee improves memory"—the next step is to test it, generating evidence for or against the idea. If a hypothesis is shown to be false—"coffee does not improve memory"—it can be rejected and removed from the list of possible answers to the original question. On the other hand, if data support the hypothesis, then it will be accepted, at least until further testing and data show otherwise. Because it is impossible to test whether a hypothesis is true in every possible situation, a hypothesis can never be proved true once and for all. The best we can do is support the hypothesis with an exhaustive amount of evidence **(Infographic 1.2)**.

There are multiple ways to test a hypothesis. One is to design a controlled **experiment** in which you measure the effects of coffee drinking on a group of subjects. In 2002, Lee Ryan, a psychologist at the University of Arizona, decided to do just that. Ryan noticed that memory is often optimal early in the morning in adults over age 65 but tends to decline as the day goes on. She also noticed that many adults report feeling more alert after drinking caffeinated coffee. She therefore hypothesized that drinking coffee might prevent this decline in memory, and devised an experiment to test her hypothesis.

First she collected a group of participants—40 men and women over age 65, who were active, healthy, and who reported consuming some form of caffeine daily. She then randomly divided these people into two groups: one that would get caffeinated coffee, and one that would receive decaf. The caffeine group is known as the **experimental group,** since caffeine is what's being tested in the experiment. The decaf group is known as the **control group**—it serves as the basis of comparison. Both groups were given memory tests at 8 A.M. and again at 4 P.M. on two nonconsecutive days. The experimental group received a 12-ounce cup of regular coffee containing approximately 220–270 mg of caffeine 30 minutes before each test. The control group received a **placebo:** a 12-ounce cup of decaffeinated coffee containing no more than 5 to 10 mg of caffeine per serving.

By administering a placebo, Ryan could ensure that any change observed in the experimental group was a result of consuming caffeine and not just any hot beverage. Moreover, participants did not know whether they were drinking regular or decaf, so a **placebo effect** was also ruled out. In addition, all participants were forbidden to eat or drink any other caffeine-containing foods or drinks—like chocolate, soda, or coffee—for at least four hours before

EXPERIMENTAL GROUP
The group in an experiment that experiences the experimental intervention or manipulation.

CONTROL GROUP
The group in an experiment that experiences no experimental intervention or manipulation.

PLACEBO
A fake treatment given to control groups to mimic the experience of the experimental groups.

INFOGRAPHIC 1.2

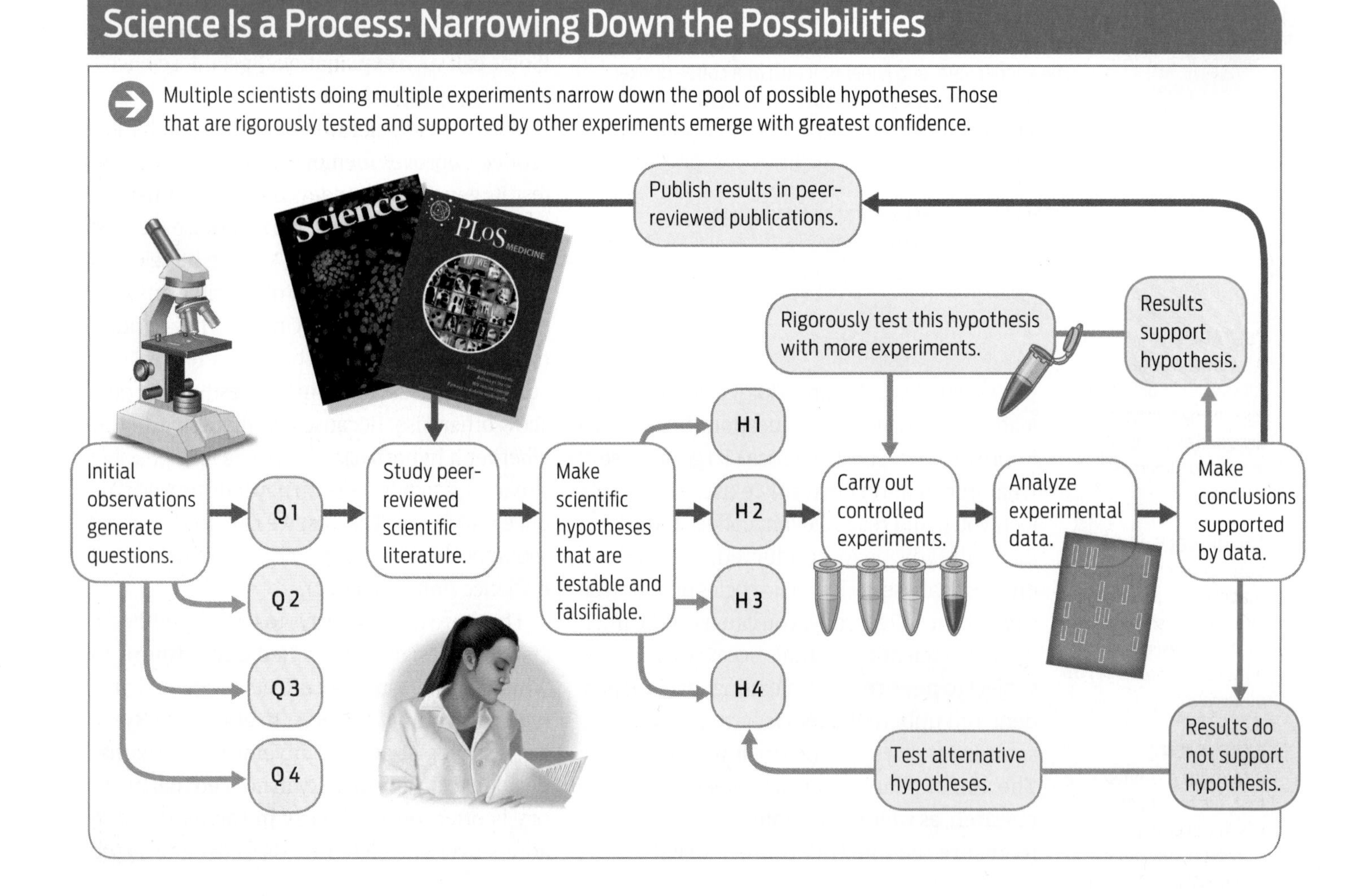

The studies in scientific journals are reviewed by experts before publication to ensure accuracy.

each test. Thus, the control group was identical to the experimental group in every way except for the consumption of caffeine.

In this experiment, caffeine consumption was the **independent variable**—the factor that is being changed in a deliberate way. The tests of memory are the **dependent variable**—the outcome that may "depend" on caffeine consumption.

Ryan found that people who drank decaffeinated coffee did worse on tests of memory function in the afternoon compared to the morning. By contrast, the experimental group who drank caffeinated coffee performed equally well on morning and afternoon memory tests. The results, which were reported in the journal *Psychological Science*, support the hypothesis that caffeine, delivered in the form of coffee, improves memory—at least in certain people **(Infographic 1.3)**.

Because other factors might, in theory, explain the link between coffee and mental performance (perhaps coffee drinkers are more active, and their physical activity rather than their coffee consumption explains their mental performance), it's too soon to see these results as proof of coffee's memory-boosting powers. To win our confidence, the experiment must be repeated by other scientists and, if possible, the methodology refined.

Size Matters

Consider the size of Ryan's experiment—40 people, tested on two different days. That's not a very big study. Could the results have simply been due to chance? What if the 20 people who drank caffeinated coffee just happened to have better memory?

One thing that can strengthen our confidence in the results of a scientific study is **sample size.** Sample size is the number of individuals participating in a study, or the number of times an experiment or set of observations is

PLACEBO EFFECT
The effect observed when members of a control group display a measurable response to a placebo because they think that they are receiving a "real" treatment.

INDEPENDENT VARIABLE
The variable, or factor, being deliberately changed in the experimental group.

DEPENDENT VARIABLE
The measured result of an experiment, analyzed in both the experimental and control groups

SAMPLE SIZE
The number of experimental subjects or the number of times an experiment is repeated. In human studies, sample size is the number of subjects.

INFOGRAPHIC 1.3

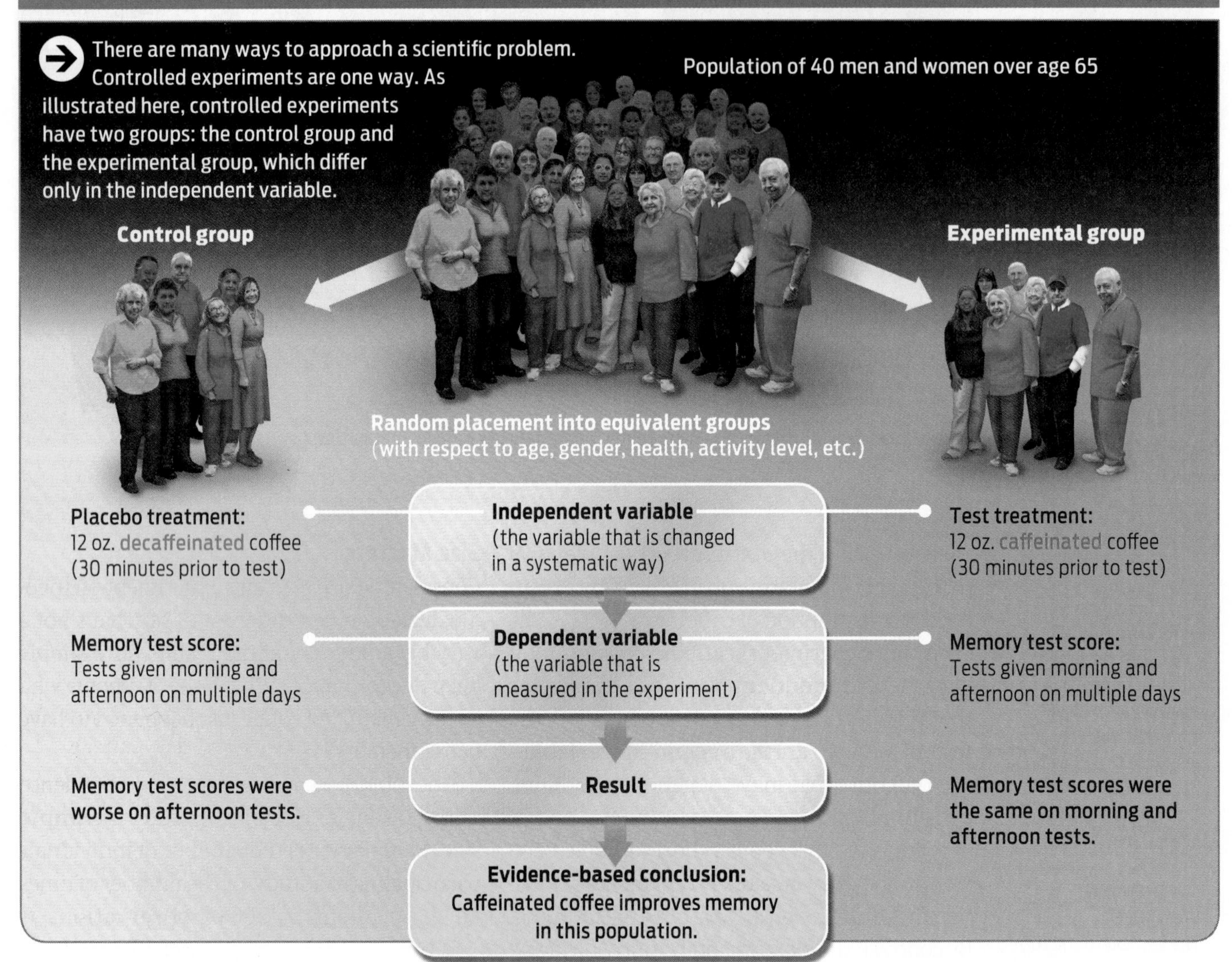
Anatomy of an Experiment
There are many ways to approach a scientific problem. Controlled experiments are one way. As illustrated here, controlled experiments have two groups: the control group and the experimental group, which differ only in the independent variable.
Population of 40 men and women over age 65
Control group
Experimental group
Random placement into equivalent groups
(with respect to age, gender, health, activity level, etc.)
Placebo treatment:
12 oz. decaffeinated coffee
(30 minutes prior to test)
Independent variable
(the variable that is changed in a systematic way)
Test treatment:
12 oz. caffeinated coffee
(30 minutes prior to test)
Memory test score:
Tests given morning and afternoon on multiple days
Dependent variable
(the variable that is measured in the experiment)
Memory test score:
Tests given morning and afternoon on multiple days
Memory test scores were worse on afternoon tests.
Result
Memory test scores were the same on morning and afternoon tests.
Evidence-based conclusion:
Caffeinated coffee improves memory in this population.

repeated. The larger the sample size, the more likely the results will have **statistical significance**–that is, they will not be due to random chance (**Infographic 1.4**).

News reports are full of statistics. On any given day, you might hear that 75% of the American public opposes a piece of legislation. Or that 15% of a group of people taking a medication experienced a certain unpleasant side effect–like nausea or suicidal thoughts–compared to, say, 8% of people taking a placebo. Are these differences significant or important? Whenever you hear such numbers being cited, it's important to keep in mind the total sample size. In the case of the side effects, was this a group of 20 patients (15% of 20 patients is 3 people), or was it 2,000? Only with a large enough sample size can we be confident that the results of a given study are statistically significant and represent something more than chance. Moreover, it's important to consider the population being studied. For example, do the people reporting their views on a piece of legislation represent a broad cross section of the public, or are most of them watchers of the same television network, whose views lie at one extreme? Likewise, in Ryan's study, are the 65-year-old self-described "morning people"

STATISTICAL SIGNIFICANCE
A measure of confidence that the results obtained are "real," rather than due to random chance.

INFOGRAPHIC 1.4

Sample Size Matters

The more data collected in an experiment, the more you can trust the conclusions.

Data from only eight participants:

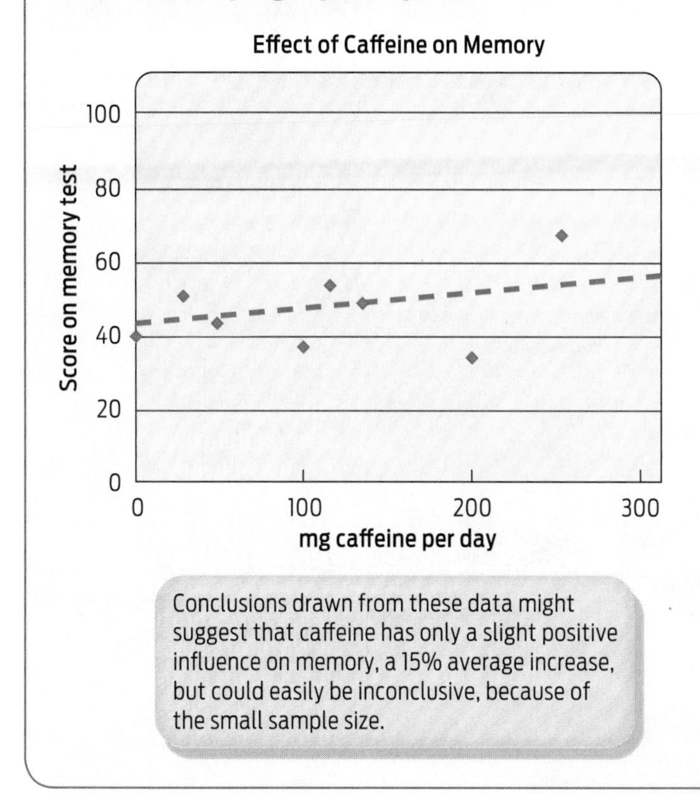

Conclusions drawn from these data might suggest that caffeine has only a slight positive influence on memory, a 15% average increase, but could easily be inconclusive, because of the small sample size.

Data from dozens of participants:

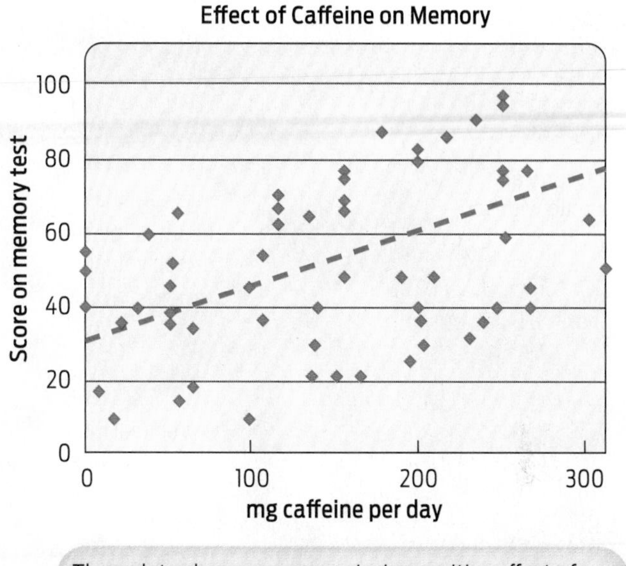

These data show a more convincing positive effect of caffeine on memory, a 45% average increase, because it is supported by more data. A statistical analysis would show that this positive influence is significant – in other words, it is not due to chance.

who regularly consume coffee representative of the wider population?

If you search for "caffeine and memory" on PubMed.gov (a database of medical research papers), you'll see that the memory-enhancing properties of caffeine is a well-researched topic. Many studies have been conducted, at least some of which tend to support Ryan's results. Generally, the more experiments that support a hypothesis, the more confident we can be that it is true. A hypothesis that continues to hold up after many years of rigorous testing may eventually be considered a **scientific theory.** Note that the word "theory" in science means something very different from its colloquial meaning. In everyday life we may say something is "just a theory," meaning it isn't proved. But in science, a theory is an explanation that is supported by a large body of evidence compiled over time by numerous researchers, and which remains the best explanation we have for an observed phenomenon (**Infographic 1.5**).

In science, a theory is the best explanation we have for an observed phenomenon.

SCIENTIFIC THEORY
A hypothesis that is supported by many years of rigorous testing and thousands of experiments.

This Is Your Brain on Caffeine

Caffeine is a stimulant. It is in the same class of psychoactive drugs as cocaine, amphetamines, and heroin (although less potent than these, and acting through different chemical pathways). Caffeine boosts not just memory and mental activity but physical activity as well. One study, in 2004, found that 33% of 193 track and field

INFOGRAPHIC 1.5

Everyday Theory vs. Scientific Theory

In everyday life, people use the word "theory" to refer to an idea that they would like to follow up.
In science, a theory is a hypothesis that has never been disproved, even after many years of rigorous testing.

Everyday theory:
Great idea based on a person's experience and knowledge

If you carry an umbrella with you, it won't rain.

The freezer is the safest place to keep valuables.

You feel more cheerful when you wear bright clothing.

Scientific theory:
Important hypotheses supported by thousands of scientific experiments

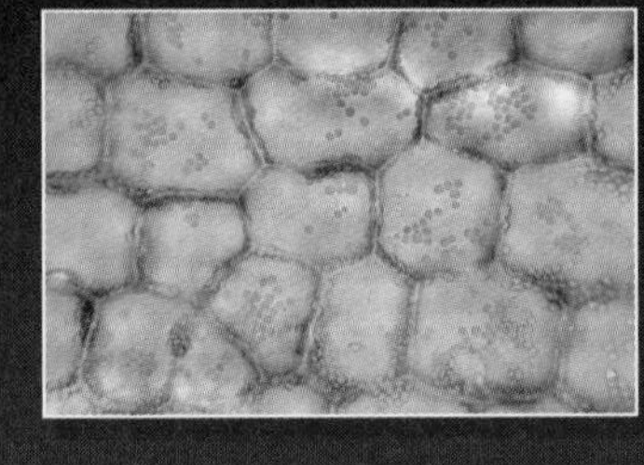

Cell Theory:
All living things are made of cells.

Theory of General Relativity:
Gravity influences time and space.

Theory of Evolution by Natural Selection:
Populations of organisms change over time, adapting to their environment.

athletes and 60% of 287 cyclists said they consumed caffeine to enhance their performance. Recognizing caffeine's reputation as a performance-enhancing drug, the International Olympic Committee prohibited athletes from using it until 2004 (when it decided to allow it, presumably because it had become too common a substance to regulate).

While the exact mechanisms are not fully understood, scientists think that caffeine exerts its energizing effect by counteracting the actions of a chemical in the brain called adenosine. Adenosine is the body's natural sleeping pill—its concentration increases in the brain while you are awake and by the end of the day promotes drowsiness. Caffeine blocks the effect of adenosine in the brain and keeps us from falling asleep.

Some researchers contend that coffee's mind-boosting effects are an indirect result of the cycle of dependency.

Though our understanding of the chemistry is relatively new, humans have enjoyed coffee's kick for more than a thousand years. It's said that an Ethiopian goatherd found his goats acting unusually frisky one afternoon after munching the leaves of a small bush. Chewing a few of the shrub's berries himself, he got a caffeine buzz, and the rest was history. Today, caffeine is the most wildly used stimulant on the planet **(Table 1.1)**.

In fact, consumption of caffeinated beverages has skyrocketed in the past 25 years; for example, young people now drink far more soda than milk. A 2009 study in the journal *Pediatrics*

INFOGRAPHIC 1.6

Caffeine Side Effects

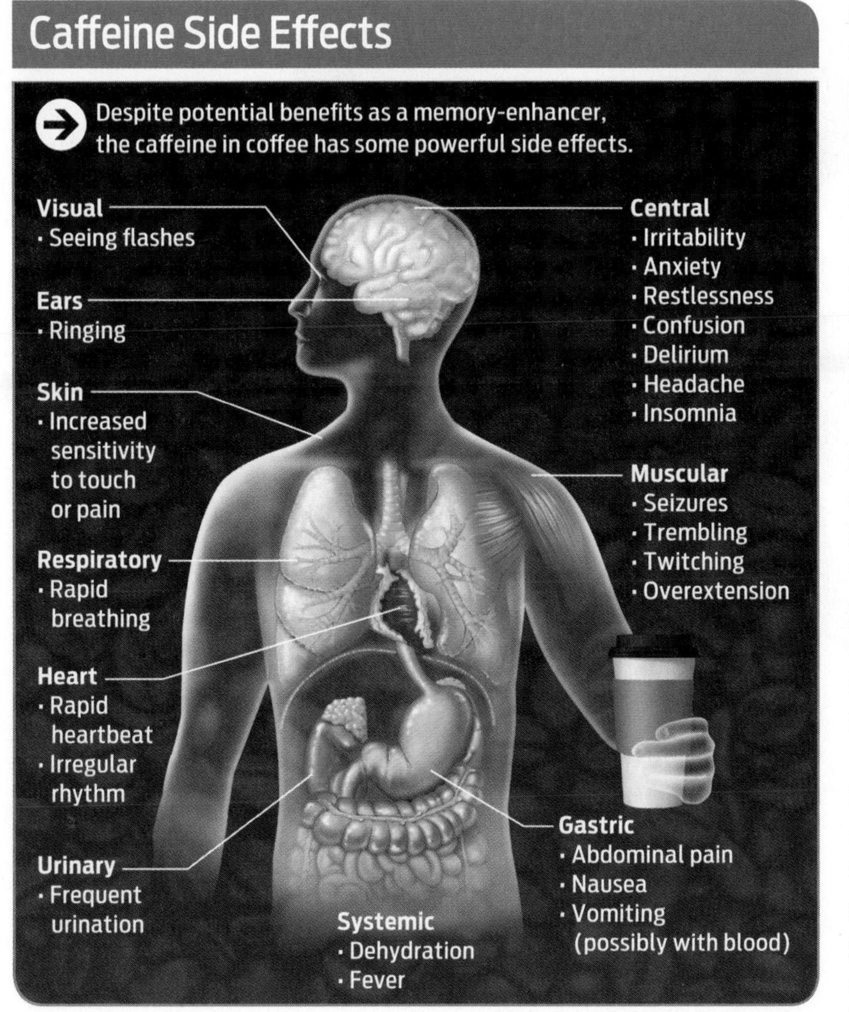

found that teenagers consume up to 1,458 mg of caffeine a day–nearly five times the recommended maximum adult dose of 300 mg. Caffeine can cause anxiety, jitters, heart palpitations, trouble sleeping, dehydration, and more serious symptoms–especially in people who are sensitive to it. In 2007, two high school students in Colorado Springs, Colorado, were hospitalized with stomach pain, nausea, and vomiting after drinking one 8-ounce can of Spike Shooter, a potent beverage that packs a walloping 300 mg of caffeine–the equivalent of almost four Red Bulls **(Infographic 1.6)**.

For regular coffee drinkers who crave their morning buzz, such symptoms are unlikely to convince them to kick the habit. This may be because, like many other psychoactive substances, caffeine is addictive. Those who drink a significant amount of coffee every day may notice that they don't feel quite right if they skip a day; they may be cranky or get a headache. These are symptoms of withdrawal. In fact, some researchers contend that coffee's mind-boosting effects are an indirect result of the cycle of dependency. Improvement in mood or performance following a cup of coffee, they say, may simply represent relief from withdrawal symptoms rather than any specific beneficial property of coffee.

To test this dependency hypothesis, scientists could conduct an experiment. They could compare the effects of drinking coffee in two groups: one group of regular coffee drinkers who had abstained from coffee for a short period, and another group of non–coffee drinkers. Does coffee give both groups a boost, or only the regular coffee drinkers looking for their fix?

In fact, this very experiment was done in 2010 by a group of researchers at the University of Bristol in England. Their study, published in the journal *Neuropsychopharmacology*, looked at caffeine's effect on alertness. Researchers gave caffeine or a placebo to 379 participants and asked them to take a test that rated their level of alertness. The study found that caffeine did not boost alertness in non–coffee drinkers compared to those drinking a placebo (although it did boost their level of anxiety and headache). Heavy coffee drinkers, on the other hand, experienced a steep drop in alertness when given the placebo.

"What this study does is provide very strong evidence for the idea that we don't gain a benefit in alertness from consuming caffeine," the study author, Peter Rogers, said. "Although we feel alert, that's just caffeine bringing us back to our normal state of alertness." Of course, this doesn't really explain why people get hooked on coffee in the first place.

Finding Patterns

Performing controlled laboratory experiments like those discussed above is one way that scientists try to answer questions. Another approach is to make careful observations or comparisons

of phenomena that exist in nature. This is the approach taken by scientists who study **epidemiology**—the incidence of disease in populations—or some other area, like the movement of stars or the nature of prehistoric life, that cannot be directly manipulated.

For example, if an epidemiologist wanted to learn about the relationship between cigarette smoking and lung cancer, he could compare the rates of lung cancer in smokers and nonsmokers, but he could not actually perform an experiment in which he made people smoke cigarettes and waited to see whether or not they got cancer. Such an experiment would be highly unethical.

Although epidemiological studies do not provide the immediate gratification of a laboratory experiment, they do have certain advantages. For one thing, they can be relatively inexpensive to conduct, since often the only procedure involved is a participant questionnaire. And you can study factors that are considered harmful, such as excess alcohol or smoking, that you would be unable to test experimentally. Finally, epidemiological studies have the power of numbers and time. The Framingham Heart Study, for example, is a famous epidemiological study that has tracked rates of cardiovascular disease in a group of people and their descendants in Framingham, Massachusetts, in order to identify common risk factors. Begun in 1948, the study has been going on for decades and has provided mountains of data for researchers in many fields, from cardiology to neuroscience.

Most of the health studies featured in the news are epidemiological studies. Consider a study on coffee and Parkinson disease published in the *Journal of the American Medical Association (JAMA)* in 2000. Researchers examined the relationship between coffee drinking and the incidence of Parkinson disease, a condition that afflicts more than 1 million people in the United States, including men and women of all ethnic groups. There is no known cure, only palliative treatments to help lessen symptoms, which include trembling limbs and difficulty coordinating speech and movement.

EPIDEMIOLOGY
The study of patterns of disease in populations, including risk factors.

TABLE 1.1

How Much Caffeine Is in Our Beverages?

The FDA Recommends No More than 65 mg of Caffeine in 12 oz.

BEVERAGE	SERVING SIZE	QUANTITY OF CAFFEINE
Coffee	8 oz	95 mg and up
Red Bull	8.3 oz (1 can)	76 mg
Rockstar	8 oz (half can)	80 mg
Amp	8.4 oz (1 can)	74 mg
Coke Classic	12 oz (1 can)	35 mg
Mountain Dew	12 oz (1 can)	54 mg
Barq's Root Beer	12 oz (1 can)	23 mg
Sprite	12 oz (1 can)	0 mg

Source: Mayo Clinic

For more than 30 years, researchers at the Veterans Affairs Medical Center in Honolulu followed more than 8,000 Japanese-American men, gathering all sorts of information about them: their age, diet, health, smoking habits, and other characteristics. Of these men, 102 developed Parkinson disease. What did these 102 men have in common? Epidemiologists found that most of them did not drink caffeinated beverages—no coffee, soda, or caffeinated tea.

By contrast, coffee drinkers had a lower incidence of Parkinson disease. In fact, those who drank the most coffee were the least likely to get the disease. Men who drank more than two 12-ounce cups of coffee each day had one-fifth the risk of getting the disease compared to non–coffee drinkers.

So does coffee prevent Parkinson disease? The occurrence and progression of many diseases are affected by a complex range of factors, including age, sex, diet, genetics, and exposure to bacteria and environmental chemicals, as well as lifestyle factors like drinking, smoking, and exercise. Although the study discussed here suggests a link—or **correlation**—between caffeine and lower incidence of Parkinson disease, it does not necessarily show that caffeine prevents the disease. In other words, correlation is not causation. Perhaps the people who like to drink coffee have different brain chemistry, and it's this different brain chemistry that explains the differing incidence of Parkinson disease among coffee drinkers (**Infographic 1.7**).

CORRELATION
A consistent relationship between two variables.

Indeed, other studies have found that cigarette smoking also correlates with a lower risk of Parkinson disease. Both coffee drinking and smoking could be considered types of thrill seeking, behavior observed in people who enjoy the "high" they get from stimulants such as caffeine or nicotine. The lower risk of Parkinson disease among coffee drinkers might therefore result from thrill-seeking brain chemistry that also happens to resist disease—rather than being caused by either smoking or drinking coffee per se.

Moreover, the study followed Japanese-American men. Would the same relationship of caffeine and Parkinson disease be seen in other ethnic groups or in women? Several

INFOGRAPHIC 1.7

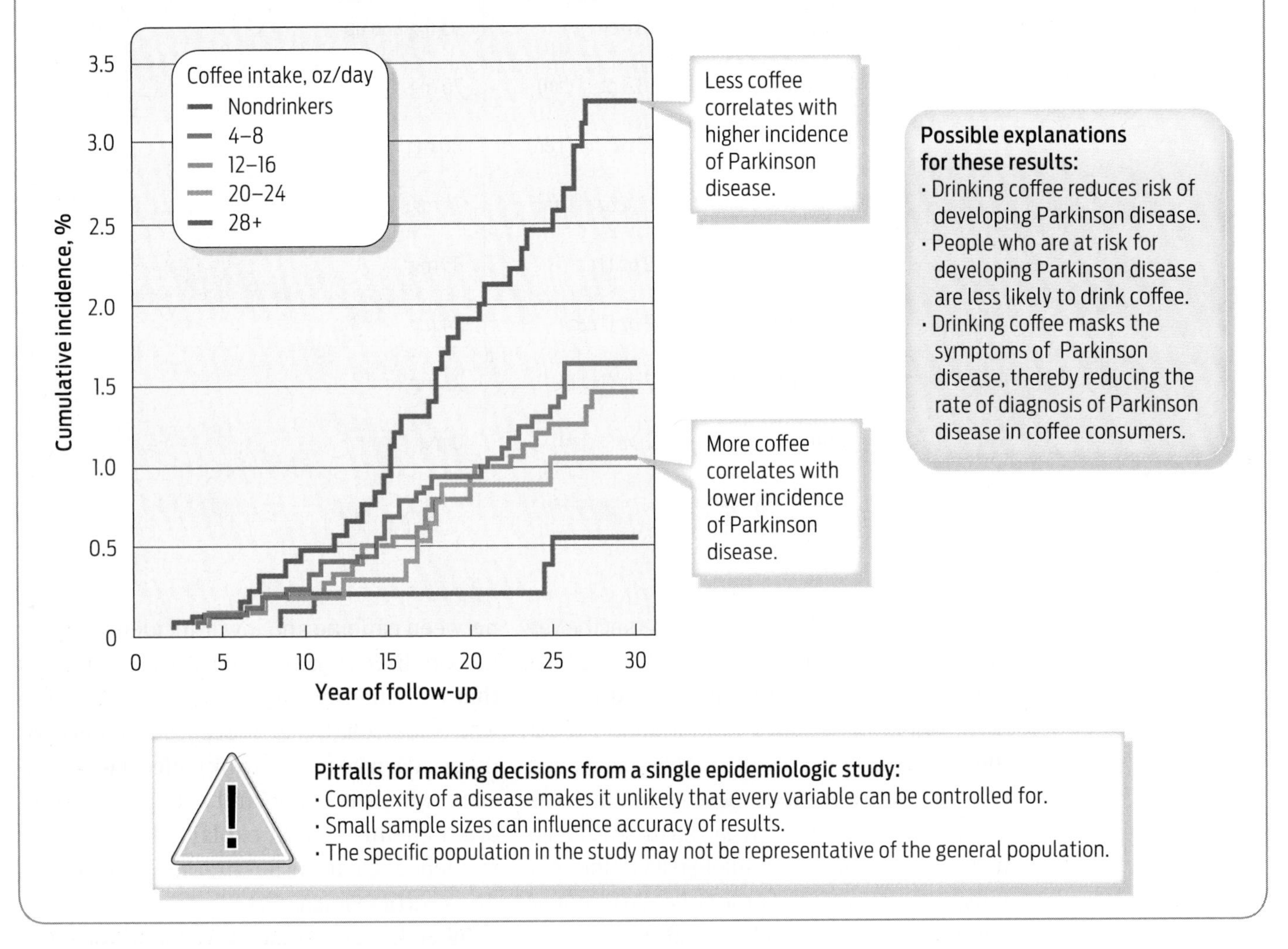

Correlation Does Not Equal Causation
While the data shown below show a convincing correlation between reduced caffeine intake and an increased risk of Parkinson disease, it is impossible to state that less coffee causes Parkinson disease. Other factors that were not tested or controlled for could be causing the reduced risk.
Coffee intake, oz/day
Nondrinkers
4–8
12–16
20–24
28+
Cumulative incidence, %
3.5
3.0
2.5
2.0
1.5
1.0
0.5
0
0
5
10
15
20
25
30
Year of follow-up
Less coffee correlates with higher incidence of Parkinson disease.
More coffee correlates with lower incidence of Parkinson disease.
Possible explanations for these results:
· Drinking coffee reduces risk of developing Parkinson disease.
· People who are at risk for developing Parkinson disease are less likely to drink coffee.
· Drinking coffee masks the symptoms of Parkinson disease, thereby reducing the rate of diagnosis of Parkinson disease in coffee consumers.
Pitfalls for making decisions from a single epidemiologic study:
· Complexity of a disease makes it unlikely that every variable can be controlled for.
· Small sample sizes can influence accuracy of results.
· The specific population in the study may not be representative of the general population.

SOURCE: ROSS ET AL., JAMA 2000; 283:2671–2679

other epidemiological studies have found a correlation between caffeine consumption and a lower incidence of Parkinson disease in men of other ethnicities. But in women the results have been inconclusive. All in all, there's still no direct evidence that caffeine actually prevents the disease in either men or women.

"While our study found a strong correlation between coffee drinkers and low rates of Parkinson's disease," said the study's lead author, G. Webster Ross in a press release issued by the U.S. Department of Veterans Affairs, "we have not identified the exact cause of this effect. I'd like to see these findings used as a basis to help other scientists unravel the mechanisms that underlie Parkinson's onset."

To get a clearer picture of caffeine's role in Parkinson disease, researchers could conduct a type of experiment known as a **randomized clinical trial,** in which the effects of coffee are measured directly under controlled conditions. One could divide a population into two groups, put one group on coffee and the other on decaf, and then follow both groups for a number of years to see which one had the higher incidence of disease. The problem with such a study is that it is often very expensive to conduct, and it can be difficult to get

RANDOMIZED CLINICAL TRIAL
A controlled medical experiment in which subjects are randomly chosen to receive either an experimental treatment or a standard treatment (or placebo).

INFOGRAPHIC 1.8

From the Lab to the Media: Lost in Translation

The data as reported in peer-reviewed journals are often very complex. Scientists interpret these data in lengthy discussions, but the public receives them as isolated media headlines.

Data from scientific studies provide a large amount of information.

Unadjusted and Age-Adjusted incidence of Parkinson Disease (PD) According to Amounts of Coffee Consumed per Day

Based on 30 Years of Follow-Up After the 1965 to 1968 Examinations:

Coffee Intake (oz/day)	No. Cases of PD/No. Subjects at Risk	Incidence Rate/10,000 Person-Years: Unadjusted	Incidence Rate/10,000 Person-Years: Adjusted for Age	Adjusted Relative Hazard (95% Confidence) Compared with Top Category of Coffee Intake*
Nondrinker	32/1286	10.5	10.4	5.1 (1.8 – 14.4)§
4 to 8	33/2576	5.5†	5.3‡	2.7 (1.0 – 7.8)
12 to 16	24/2149	4.7†	4.7†	2.5 (0.9 – 7.3)
20 to 24	9/1034	3.6†	3.7†	2.0 (0.6 – 6.4)
≥28	4/959	1.7¦	1.9¦	Reference
Test for Trend		p<0.001	p<0.001	p<0.001
Nondrinkers vs. Drinkers				2.2 (1.4 – 3.3)¶

* Adjusted for age and pack-years of cigarette smoking.
† Significantly different from nondrinkers, p<0.01
‡ Significantly different from nondrinkers, p<0.05
¦ Significantly different from nondrinkers, p<0.001
§ Significant excess risk of PD, p <0.01
¶ Significant excess risk of PD, p <0.001

ADAPTED FROM: ROSS ET AL., JAMA 2000; 283:2671–2679

Translation of complex data into media headline

TODAY'S NEWS

Coffee Cure!

Drinking Coffee Prevents Parkinsons

But media reports don't have the time and space to explain all the information.

So the general public may not receive important details and potential limitations of the single study.

- As shown in the data table, even some coffee drinkers develop Parkinson disease, so not everyone will benefit.
- The results are reflecting a correlation, not a causation. This is not direct evidence that coffee is a cure.
- This study was carried out with a particular male population, so we cannot generalize the results to other populations (e.g., women).

people to stick to the regimen for the length of the study. (And such studies are unethical if the experimental treatment is likely to cause harm.)

Getting Beyond the Buzz

While a lower risk of Parkinson disease represents a potential boon to coffee drinkers, the news for caffeine addicts isn't all good. Over the years, epidemiological studies have linked caffeine consumption to *higher* rates of various diseases, including osteoporosis, fibrocystic breast disease, and bladder cancer. As with the link to Parkinson disease, however, such correlations do not necessarily prove that caffeine causes any of these diseases. Nevertheless, such studies are often quite influential and newsworthy—like the supposed link between coffee and pancreatic cancer that made headlines in 1981. That study was based on a single epidemiological study, which was later discounted by further research.

Journalists face unique challenges in covering health news, says Gary Schwitzer of HealthNewsReview.org: "They must cover complex topics, do it quickly, creatively, accurately, completely and with balance—and then be sure they don't 'dumb it down' too much for a general news audience. . . . If they can't do it right, they must realize the *harm* they can do by reporting inaccurately, incompletely, and in an imbalanced way" **(Infographic 1.8)**.

Journalists and scientists aren't the only ones who bear the responsibility of determining what information is trustworthy. As consumers and citizens, we can become more knowledgeable about how science is done and which studies deserve to influence our behavior. Whether it's the latest media report linking cell phones to brain tumors or vaccines to autism, the only way to really judge the value of a study is to sift through the evidence ourselves. Of course, to do that, we might first need a cup of coffee. ■

Summary

- Science is an ongoing process in which scientists conduct carefully designed studies to answer questions or test hypotheses.
- Scientific hypotheses are tested in controlled experiments or in nonexperimental studies, the results of which can support or rule out a hypothesis.
- Scientific hypotheses can be supported by experimental data but cannot be proved absolutely, as future experiments or technologies may provide new findings.
- The strength of the conclusions of a scientific study depends on, among other factors, the type of study carried out and the sample size.
- Every experiment should have a control—a group that is identical in every way to the experimental group except for one factor: the independent variable.
- The independent variable in an experiment is the one being deliberately changed in the experimental group (e.g., coffee intake). The dependent variable is the measured result of the experiment (e.g., effect of coffee on memory).
- Often a control group takes a placebo, a fake treatment that mimics the experience of the experimental group.
- In epidemiological studies, a relationship between an independent variable (such as caffeine intake) and a dependent variable (such as development of Parkinson disease) does not necessarily mean one caused the other; in other words, correlation does not equal causation.
- A randomized clinical trial is one in which test subjects are randomly chosen to receive either a standard treatment (or placebo) or an experimental treatment (e.g., caffeine).
- Scientists rely on peer-reviewed scientific reports to learn about new advances in the field. Peer review helps to ensure that the scientific results are valid as well as accurately and fairly presented.
- Most of the general public relies on media reports for their scientific information. Media reports are not always completely accurate in how they portray the conclusions of the scientific studies.
- Scientific theories are different from everyday theories. A scientific theory has withstood the test of time and extensive testing and is supported by a significant body of evidence.

Chapter 1 Test Your Knowledge

PROCESS OF SCIENCE

Science is a method of seeking answers to questions on the basis of observation and experiment.

HINT **See Infographics 1.1. and 1.2.**

KNOW IT

1. When scientists carry out an experiment, they are testing a

- **a.** theory.
- **b.** question.
- **c.** hypothesis.
- **d.** control.
- **e.** variable.

2. Of the following, which is the earliest step in the scientific process?

- **a.** generate a hypothesis
- **b.** analyze data
- **c.** conduct an experiment
- **d.** draw a conclusion
- **e.** ask a question about an observation

USE IT

3. When a scientist reads a scientific article in a scientific or medical journal, he or she is confident that the report has been peer reviewed. What does this mean? Why is peer review important?

DESIGNING EXPERIMENTS

Many considerations go into the design and implementation of a scientific experiment.

HINT **See Infographics 1.3–1.4.**

KNOW IT

4. In a controlled experiment, which group receives the placebo?

- **a.** the experimental group
- **b.** the control group
- **c.** the scientist group
- **d.** the independent group
- **e.** all groups

5. In the studies of coffee and memory discussed, the independent variable was _________ and the dependent variable was ______________.

- **a.** caffeinated coffee; decaffeinated coffee
- **b.** memory; caffeinated coffee
- **c.** caffeine; memory
- **d.** memory; caffeine
- **e.** decaffeinated coffee; caffeinated coffee

USE IT

6. You are working on an experiment to test the effect of a specific drug on reducing the risk of breast cancer in postmenopausal women. Describe your control and experimental groups with respect to age, gender, and breast cancer status.

7. Design a randomized clinical trial to test the effects of caffeinated coffee on brain activity. Design your study so that the results will be as broadly applicable as possible.

EVALUATING EVIDENCE

Many factors can influence the strength of a scientific claim.

HINT **See Infographics 1.4–1.8.**

KNOW IT

8. From what you have read in this chapter, would you say a 21-year-old Caucasian female can count on caffeinated coffee to reduce her risk of Parkinson disease?

- **a.** yes, because the results of a peer-reviewed study showed that drinking caffeinated beverages reduced the risk of Parkinson disease
- **b.** no, because subjects in that peer-reviewed study were Japanese-American males; it cannot be inferred that the same results would hold for Caucasian females
- **c.** no; she would have to restrict her consumption of coffee to decaffeinated coffee to reduce her risk of Parkinson disease
- **d.** yes; coffee is known to reverse the symptoms of Parkinson disease
- **e.** There is no data on the relationship between drinking caffeinated beverages and Parkinson disease because it would be unethical to conduct such an epidemiological study.

9. In which type of study would you have the most confidence?

- **a.** a randomized clinical trial with 10,000 subjects
- **b.** a randomized clinical trial with 5,000 subjects
- **c.** an epidemiological study with 15,000 subjects
- **d.** an endorsement of a product by a movie star
- **e.** a report on a study presented by a new organization

➔ USE IT

10. Your friend's mother has always been a coffee addict. She recently received a diagnosis of Parkinson disease. Does her experience negate the results of the *JAMA* study described in this chapter? Why or why not?

11. Depending on the television station that you watch, you may have seen advertisements that show beautiful people with clear skin who claim that a specific skin care product is "scientifically proven" to reduce acne. The product reportedly gave these people their glowing, clear skin.

a. Is their testimony itself strong enough evidence for you to act on? Why or why not?

b. What kind of scientific evidence would convince you to spend money on this product? Explain your answer.

SCIENCE AND ETHICS

12. You know that scientific reports are subject to peer review before being published in scientific journals. Do you think that scientists should also review media reports about their studies and work to correct any misleading statements? Why or why not? Who is ultimately responsible for what is reported in the popular press?

13. Your grandmother has told you about the changes she is making to her diet because of stories she has read in the news. Make a checklist of things she should consider before changing her behavior.

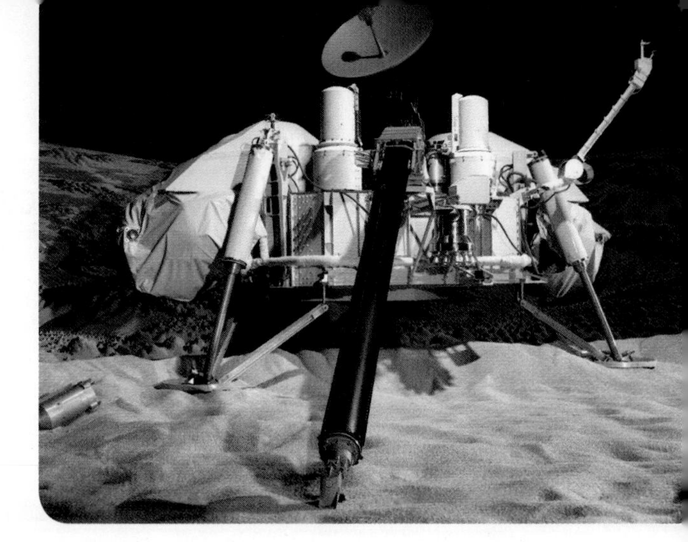

Chapter 2 Chemistry and Molecules of Life

What Is Life?

What You Will Be Learning

What Is Life?

Evidence from space heats up an age-old debate

With a flash of fiery light, a shooting star streaks across the night sky. Thirteen thousand years later, on December 27, 1984, geologist Roberta Score picks up that shooting star and holds it in her mittens. It is a grapefruit-size rock, weighing just over 4 pounds, whose dark gray-green color stands out sharply against the brilliant white of the frozen Antarctic ice cap.

Score is one of a team of six researchers with ANSMET, the Antarctic Search for Meteorites program, who for six weeks every year crisscross the mile-thick ice in snowmobiles, searching for booty from space.

Score knew the rock was special as soon as she saw it. Coated in a layer of molten glass, or fusion crust, it had the telltale sign of having blazed through the atmosphere, but was otherwise unique in color and texture. "Yowza-Yowza," wrote the team in their field report. The first meteorite to be catalogued that year, it was named ALH84001, after Allan Hills, the patch of ice where it was discovered.

Each year, tens of thousands of such meteorites, often called shooting stars, fall to earth. Most are commonplace chunks of interstellar debris left over from the dawn of the solar system. But this one was special. At 4.5 billion years old, it is by far the oldest of only a handful of meteorites known to have come from Mars. NASA scientists believe the rock was kicked off the surface of Mars and jettisoned into space when a comet or meteorite struck that planet some 16 million years ago. It then floated in space until nudged again, this time toward earth.

In 1996, just 12 years after its discovery, the rock was catapulted into international fame when a team of NASA researchers claimed to have found evidence of Martian life inside it. Presenting their findings in the journal *Science*, lead author David McKay, a planetary scientist at NASA's Johnson Space Center,

The surface of Mars.

described what he said was convincing evidence of "primitive life on early Mars" found within the ancient rock.

"If this discovery is confirmed, it will surely be one of the most stunning insights into our universe that science has ever uncovered." —Bill Clinton

The report sent shock waves through the press: "Life on Mars: Official," proclaimed the UK *Daily Mirror*. "We're Not Alone," echoed the Montreal *Gazette*. "E.T., phone Mars," requested the *Boston Globe*. President Bill Clinton held a press conference to mark the occasion, declaring, "Today, rock 84001 speaks to us across all those billions of years and millions of miles. It speaks of the possibility of life. If this discovery is confirmed, it will surely be one of the most stunning insights into our universe that science has ever uncovered."

And yet what began with excitement and fanfare quickly took a decidedly sour turn when other researchers stepped up to cast doubt on the evidence. The microscopic findings in the meteorite could have been produced without life, skeptics argued. NASA scientists had overblown the significance of their findings, critics said.

The *Viking Lander 1* spacecraft.

Two years after NASA's historic announcement, biologist Andrew Knoll of Harvard University told *Science,* "You would have a hard time finding even a small number of people who are enthused by the idea of life being recorded in this meteorite."

But what would definitive evidence of Martian life look like? Would we even recognize it? These are not just idle questions mulled over by imaginative *Star Trek* fans. They go to the heart of a fundamental debate in biology, one that has been raging since Aristotle: What is life?

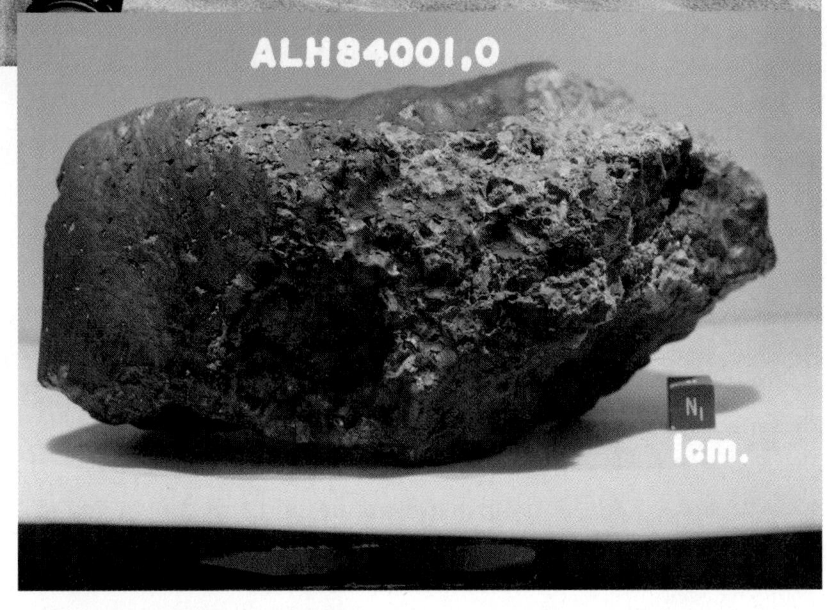

The Martian meteorite ALH84001.

The Search for Alien Life

NASA's search for life on Mars began in 1964, when the *Mariner 4* spacecraft photographed the planet during a deep-space flyby, providing us with the first up-close pictures of the red planet. The photographs revealed a dry, rocky landscape, more reminiscent of our lifeless moon than the lush, blue marble we call home. But looks can be deceiving, so NASA followed up its *Mariner* missions with *Viking Lander 1*—the first spacecraft to land on the Martian surface, touching down on July 20, 1976.

Equipped with mechanical arms that could grab and test Martian soil, *Viking Lander 1* was designed to look for signs of life. NASA scientists hypothesized that if life were present in the soil, then they should be able to measure its chemical signature. Was anything emitting

HOMEOSTASIS
The maintenance of a relatively constant internal environment.

ENERGY
The ability to do work. Living organisms obtain energy from food, which they either make using the energy of sunlight or consume from the environment.

INFOGRAPHIC 2.1

Five Functional Traits of Life

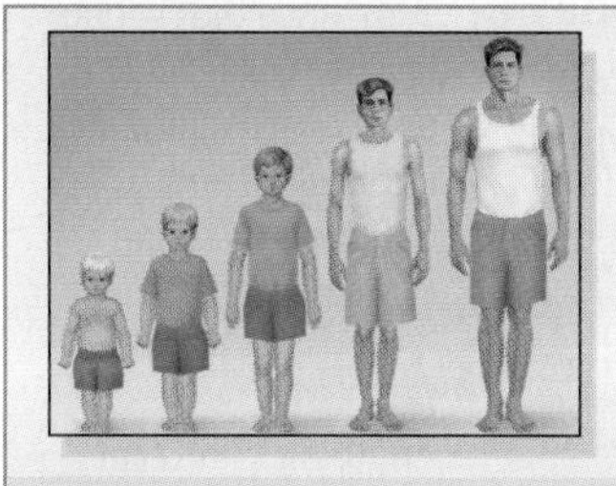

Growth:
For unicellular (one-cell) organisms, this is an increase in cell size prior to reproduction. For multicellular organisms, growth refers to an increase in an organism's size, as the number of cells making up the organism increases.

Reproduction:
The process of producing new organisms. Offspring are similar, but not necessarily identical, to their parents in general structure, function, and properties.

Homeostasis:
Organisms maintain a stable internal environment, even when the external environment changes.

Sense and Respond to Stimuli:
Organisms respond to stimuli in many ways. For example, they may move toward a food source or move away from a threatening predator.

Obtain and Use Energy:
All living organisms require an input of energy to power their activities. Organisms obtain energy from food (which they either produce themselves or consume from the environment). Chemical reactions convert that energy into usable forms. The sum total of all these reactions is metabolism.

METABOLISM
All the chemical reactions taking place in the cells of a living organism that allow it to obtain and use energy.

carbon dioxide, for example, as many organisms on earth do? Researchers put Martian soil in a sterile container filled with nutrients and waited to see what would happen.

Initially, the results seemed promising: something in the Martian soil did indeed seem to be breaking down the nutrients and producing carbon dioxide gas. More intriguing, when the experiment was repeated after the soil was heated to a very high temperature (a temperature that would kill most life), no carbon dioxide was measured. The researchers interpreted this experiment as evidence for Martian life.

But the results were far from definitive. Subsequent analyses revealed that the Martian soil could have produced carbon dioxide through strictly abiotic (that is, nonliving) means by a chemical reaction similar to combustion. Whether or not living organisms were responsible for the carbon dioxide remained unclear.

In looking for a specific chemical reaction, the NASA scientists were employing a definition of life based on what living things *do*. Biologists generally agree that–on earth at least–all living things have in common five functional traits, traits that rocks and sand will never have and robots don't yet have. Specifically, living things (1) grow and (2) reproduce: they increase in size and produce offspring that are similar but not necessarily identical to their parents. Living things also (3) maintain a relatively stable internal environment in the face of changing external circumstances–producing heat when they're cold, for example–a phenomenon known as **homeostasis.** To maintain homeostasis, (4) living things sense and respond to their environment, as when a plant grows toward sunlight. And to carry out these and other life-defining activities, (5) all living organisms obtain and use **energy,** the power to do work. Energy comes from sunlight or food, which living things break down through a series of chemical reactions, the sum total of which is called **metabolism** **(Infographic 2.1)**.

In looking for carbon dioxide, the NASA scientists were looking for evidence of chemical metabolism. But the inconclusive results of the experiment demonstrate why it is risky to rely on any one functional trait as the defining feature of life: it's always possible to come up with an exception to the rule.

For example, the ability to reproduce would seem to be a fundamental principle of life–and it is. Yet this definition alone would exclude some entities that are clearly alive, such as mules, which are sterile and thus cannot reproduce. Similarly, if the sole definition is that living things consume energy and grow, we could claim that fire is alive, and yet that doesn't seem right.

Carol Cleland is a philosopher at the University of Colorado and a member of NASA's Astrobiology Institute who has spent a lot of time thinking about the problem of defining life. At a NASA-sponsored conference on astrobiology held in 2006 she said, "There's a serious problem with trying to answer the question 'What is life?' and designing a search for life based upon definitions. Yet this is something that's been commonly done." It's what NASA's *Viking* mission did, for example. The problem with this strategy, she explained, is that a functional "definition" of life will match only our current beliefs about life; it leaves no room for life that functions or behaves differently from the way it does on earth.

Another approach, which NASA scientists have also used to look for life on Mars, is to search for the distinctive chemical building blocks of life. Regardless of how it functions, at its most basic level all life is a chemical concoction, a chemical soup. We can therefore analyze life, in part, by analyzing that soup's ingredients.

ELEMENT
A chemically pure substance that cannot be chemically broken down; each element is made up of and defined by a single type of atom.

MATTER
Anything that takes up space and has mass.

ATOM
The smallest unit of an element that cannot be chemically broken down into smaller units.

INFOGRAPHIC 2.2

All Matter on Earth Is Made of Elements

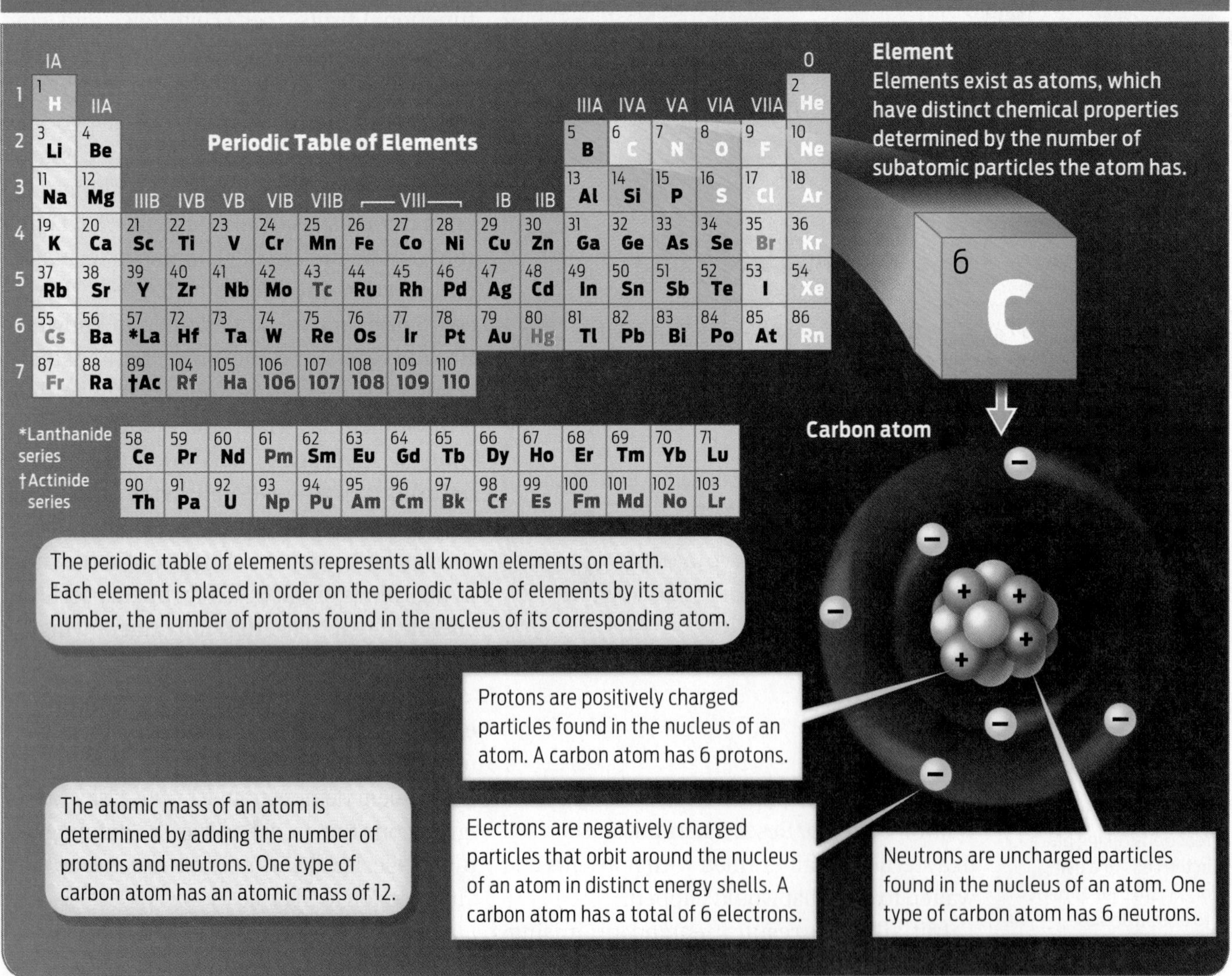

Periodic Table of Elements
*Lanthanide series
†Actinide series
The periodic table of elements represents all known elements on earth. Each element is placed in order on the periodic table of elements by its atomic number, the number of protons found in the nucleus of its corresponding atom.
Element
Elements exist as atoms, which have distinct chemical properties determined by the number of subatomic particles the atom has.
6
C
Carbon atom
Protons are positively charged particles found in the nucleus of an atom. A carbon atom has 6 protons.
The atomic mass of an atom is determined by adding the number of protons and neutrons. One type of carbon atom has an atomic mass of 12.
Electrons are negatively charged particles that orbit around the nucleus of an atom in distinct energy shells. A carbon atom has a total of 6 electrons.
Neutrons are uncharged particles found in the nucleus of an atom. One type of carbon atom has 6 neutrons.

INFOGRAPHIC 2.3

Carbon Is a Versatile Component of Life's Molecules

Molecules are chains of atoms linked by covalent bonds. The element carbon is a key component of the molecules of living organisms because it can form multiple covalent bonds.

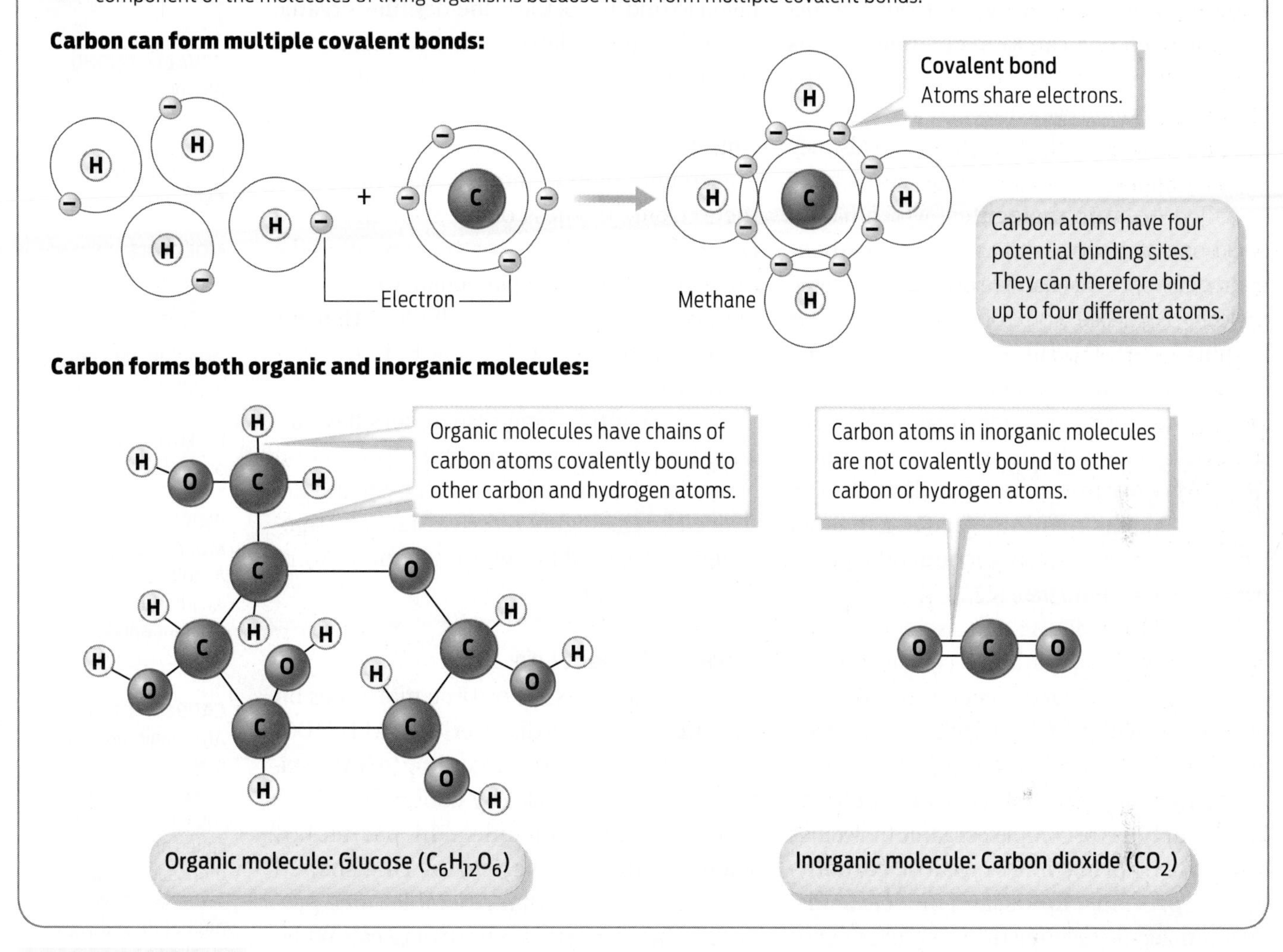

Carbon can form multiple covalent bonds:
H
H
H
H
+
C
Electron
Methane
Covalent bond
Atoms share electrons.
Carbon atoms have four potential binding sites. They can therefore bind up to four different atoms.
Carbon forms both organic and inorganic molecules:
Organic molecules have chains of carbon atoms covalently bound to other carbon and hydrogen atoms.
Carbon atoms in inorganic molecules are not covalently bound to other carbon or hydrogen atoms.
O
C
O

Organic molecule: Glucose ($C_6H_{12}O_6$)

Inorganic molecule: Carbon dioxide (CO_2)

PROTON
A positively charged subatomic particle found in the nucleus of an atom.

ELECTRON
A negatively charged subatomic particle with negligible mass.

NEUTRON
An electrically uncharged subatomic particle found in the nucleus of an atom.

NUCLEUS
The dense core of an atom.

Life's Recipe

So far, all life we know of–from amoeba to leaf to zebra–uses the same basic chemical recipe: a stew of carbon-based ingredients floating in a broth of water. Carbon is one of approximately 100 different **elements** found on earth. Elements are substances that cannot be broken down by chemical means into smaller substances. They are considered the fundamental components of anything that takes up space or has mass–the **matter** in the universe.

The smallest unit of an element that still retains the property of an element is an **atom.** What gives each atom its identity is the specific number of positively charged **protons,** negatively charged **electrons,** and neutral **neutrons** each atom has.

The element carbon, for example, is made up of atoms with six protons, six electrons, and six neutrons. The relatively heavy protons and neutrons are packed into the atom's dense core, or **nucleus,** while the tiny electrons orbit around it (Infographic 2.2).

Carbon is the fourth most common element in the universe and the second most common element in your body. In fact, just six elements make up the bulk of you: oxygen (65%), carbon (18.5%), hydrogen (9.5%), nitrogen (3.3%), and phosphorus and sulfur (2%).

Carbon has unique properties that make it an ideal backbone for life. Most important, it easily forms long chains and rings. You can think of carbon atoms as having four attachment, or bonding, sites. By sharing electrons with its neighbors, carbon can form **covalent bonds** with two, three, or four other atoms, giving the element enormous versatility.

When atoms are linked by covalent bonds, they form **molecules.** Living things are made up of so-called **organic molecules,** which have a backbone of carbon with at least one carbon-hydrogen bond. An example of a simple organic molecule is glucose, a type of sugar. Its molecular formula is $C_6H_{12}O_6$. This means that each molecule of glucose has 6 carbon atoms, 12 hydrogen atoms, and 6 oxygen atoms. Glucose is a ring-shaped molecule, with the carbon atoms forming the backbone of the ring. Carbon dioxide (CO_2), however, is an **inorganic molecule**–it does not have a carbon-carbon backbone and a carbon-hydrogen bond **(Infographic 2.3)**.

When astrobiologists (and science fiction writers) talk about "carbon-based life forms," they are talking about our chemical makeup of organic molecules. The particular organic molecules that NASA scientists hoped to find in Martian soil during the *Viking* mission were any of the four types of complex organic molecules that make up living things on earth: **carbohydrates, proteins, lipids,** and **nucleic acids.** Every molecule forming the structure of your body can be classified as one of these organic molecules. Your skin, for example, is composed of the proteins collagen and elastin, and the padding in your soft spots is composed of lipids, also known as fats.

Carbohydrates, proteins, and nucleic acids can be quite large and are therefore considered **macromolecules.** Macromolecules share a similar organization in that they are composed of subunits called **monomers** linked together in a chain. When two or more monomers join together they form a **polymer.** Carbohydrates, for example, are polymers made up of linked monomers called **monosaccharides;** similarly, proteins are made up of subunits called **amino acids** that are bonded together; and nucleic acids are polymers composed of **nucleotides** that form long chains (see **Up Close: Molecules of Life**).

Despite careful efforts, NASA's *Viking* probe failed to find any of these life-defining organic molecules in Martian soil. At the time, NASA's conclusion was that the Martian surface is self-sterilizing, meaning that no living organisms could survive in the harsh conditions. The combination of intense solar radiation, the extreme dryness of the soil, and a soil chemistry resembling combustion all make the Martian surface a particularly inhospitable place. Not to mention the fact that it's extremely cold: −120°C (−184°F) in the predawn winter.

More recently, some researchers have argued that any organic molecules present on Mars would be quickly broken down and destroyed by the highly reactive Martian atmosphere. This could explain why none were detected by *Viking.*

Traces of Ancient Life

Viking's failure to find organic molecules on Mars made the 1996 discoveries in ALH84001 all the more surprising. According to NASA scientists, ALH84001 clearly contains carbon-based organic molecules. In particular, scientists found a variety of ring-shaped organic molecules that resemble ones produced when living things burn or decay. While the presence of such organic molecules does not in itself prove the presence of life–they can be produced without life–NASA scientists argued that their location within the meteorite, near other potential markers of life, strengthened the case for life on Mars.

Where did these organic molecules come from if *Viking* did not detect them in Martian soil but they are clearly present in ALH84001? The meteorite definitely came from Mars. Scientists know this because the trapped gases in the rock perfectly match the profile of gases recorded by *Viking Lander 1.* Scientists believe that ALH84001 is a piece of volcanic rock that was churned up from deep within the Martian surface. Since ALH84001 likely came from sub-

COVALENT BOND
A strong chemical bond resulting from the sharing of a pair of electrons between two atoms.

MOLECULE
Atoms linked by covalent bonds.

ORGANIC MOLECULE
A molecule with a carbon-based backbone and at least one C–H bond.

INORGANIC MOLECULE
A molecule that lacks a carbon-based backbone and C–H bonds.

CARBOHYDRATE
An organic molecule made up of one or more sugars. A one-sugar carbohydrate is called a monosaccharide; a carbohydrate with multiple linked sugars is called a polysaccharide.

PROTEIN
An organic molecule made up of linked amino acid subunits.

LIPIDS
Organic molecules that generally repel water.

NUCLEIC ACIDS
Organic molecules made up of linked nucleotide subunits; DNA and RNA are examples of nucleic acids.

terranean Mars, could there be life beneath the surface of the red planet?

It's a distinct possibility. NASA hopes one day to be able to answer this question definitively by drilling deep into the Martian crust and hauling Martian soil back to earth for analysis—but that mission is a long way off. In the shorter term, NASA plans to send a rover to Mars in 2011 to explore Martian soil more closely than was possible with *Viking*. Known as the *Mars Science Laboratory*, or *Curiosity*, the rover will be able to perform a variety of extremely sensitive chemical tests on the soil, including ones designed to detect minute quantities of amino acids, carbohydrates, lipids, and nucleic acids.

"We shouldn't lock ourselves into a definition that might blind us to the presence of unfamiliar forms of life."
—Carol Cleland

Given the problems of defining life—knowing what to look for—what does philosopher Cleland think of NASA's plan to search for life using organic molecules as an indicator? "I think it is a good idea," she says, noting that they're what life on earth is made of. But she warns that we shouldn't turn the detection of organic molecules into an absolute requirement for life, or make it our definition. "Because our experience of life is limited to a single example—familiar earth life," she explains, "we shouldn't lock ourselves into a definition that might blind us to the presence of unfamiliar forms of life should we be so fortunate to encounter them."

Martian Bacteria?

Besides organic molecules, the really tantalizing find in meteorite ALH84001 was the presence of what looked like the fossilized remains of microscopic organisms. A widely publicized photo that has since become famous shows what looks to be a wormlike creature inching its way through a sample of the meteorite. Other pictures of the meteorite show jelly bean-shaped structures resembling bacteria.

The tiny fossilized "beans" found in ALH84001 resemble a type of bacteria on earth

MACROMOLECULES
Large organic molecules that make up living organisms; they include carbohydrates, proteins, and nucleic acids.

MONOMER
One chemical subunit of a polymer.

POLYMER
A molecule made up of individual subunits, called monomers, linked together in a chain.

MONOSACCHARIDE
The building block, or monomer, of a carbohydrate.

AMINO ACID
The building block, or monomer, of a protein.

NUCLEOTIDE
The building block, or monomer, of a nucleic acid.

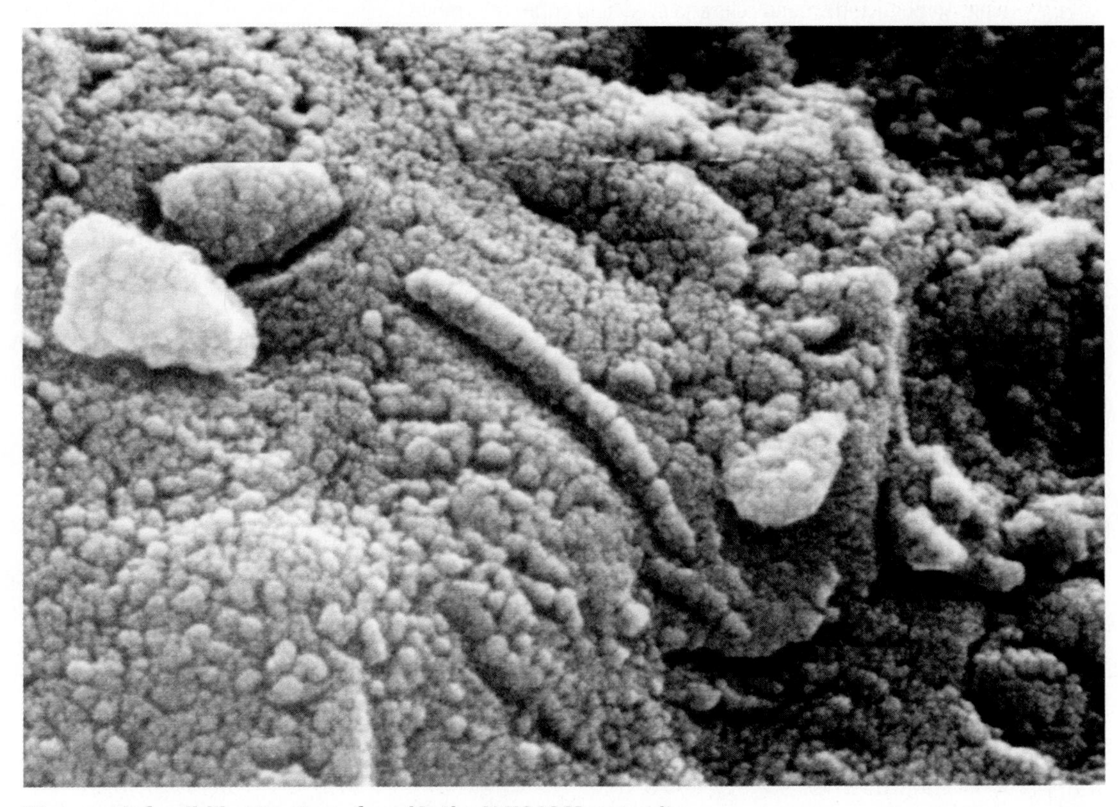

Microscopic fossil-like structures found in the ALH84001 meteorite.

UP CLOSE Molecules of Life: Carbohydrates, Proteins, Lipids, Nucleic Acids

a. Carbohydrates Are Made of Monosaccharides

Carbohydrates are made up of repeating subunits known as monosaccharides, or simple sugars. Carbohydrates act as energy-storing molecules in many organisms. Other carbohydrates provide structural support for cells.

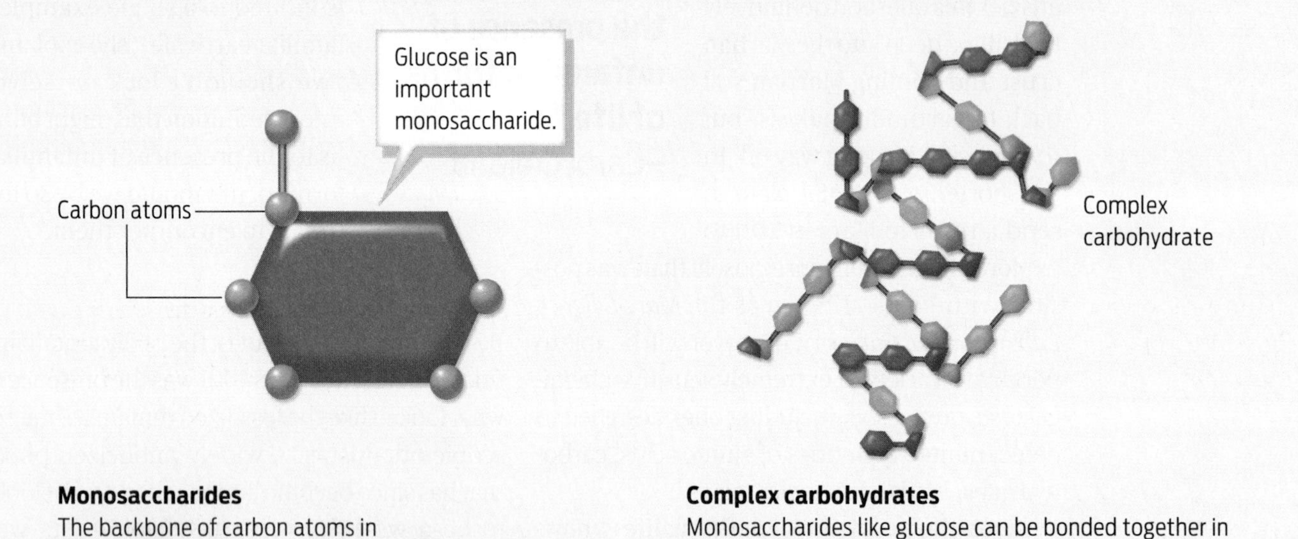

Monosaccharides
The backbone of carbon atoms in monosaccharides is most often arranged in a ring.

Complex carbohydrates
Monosaccharides like glucose can be bonded together in straight or branching chains called complex carbohydrates.

b. Proteins Are Made of Amino Acids

Proteins are polymers of different small repeating units called amino acids joined together by peptide bonds. Proteins carry out many functions in cells. They help speed up the rate of chemical reactions. They also move things through and around cells and even help entire cells move.

Amino Acid
There are 20 different amino acids found in proteins. Each amino acid shares a common "core" structure (shown in green).

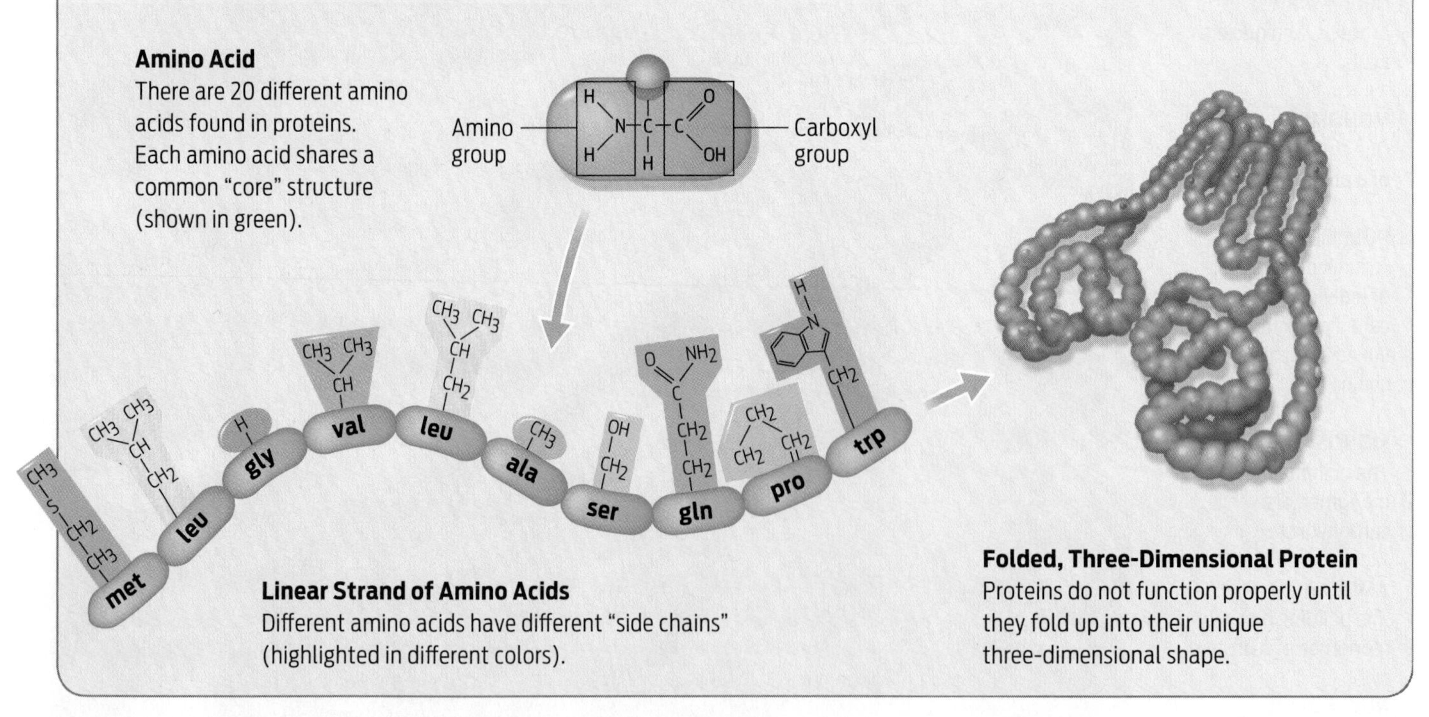

Linear Strand of Amino Acids
Different amino acids have different "side chains" (highlighted in different colors).

Folded, Three-Dimensional Protein
Proteins do not function properly until they fold up into their unique three-dimensional shape.

c. Lipids Are Hydrophobic Molecules

There are different types of lipids, each with a distinct structure and function. Lipids are not made up of repeating subunits or building blocks, but they are all hydrophobic molecules, meaning they don't mix with water.

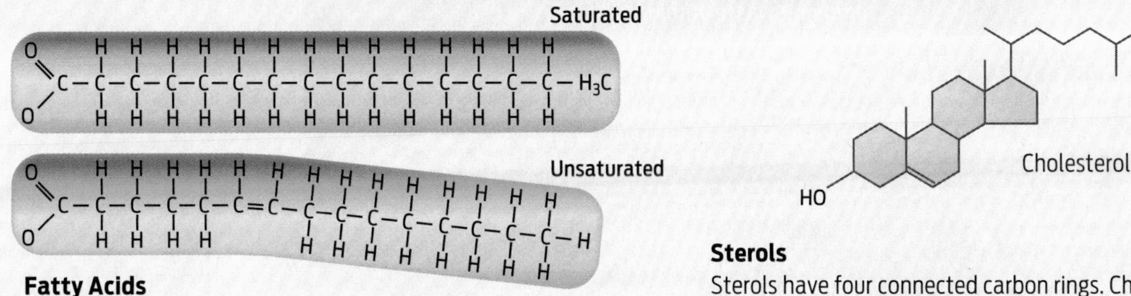

Fatty Acids

Fatty acids contain long chains of carbon atoms bonded to one another and to hydrogen atoms.

Sterols

Sterols have four connected carbon rings. Cholesterol is a sterol that's an important component of cell membranes. Other sterols may be hormones or color-inducing pigments.

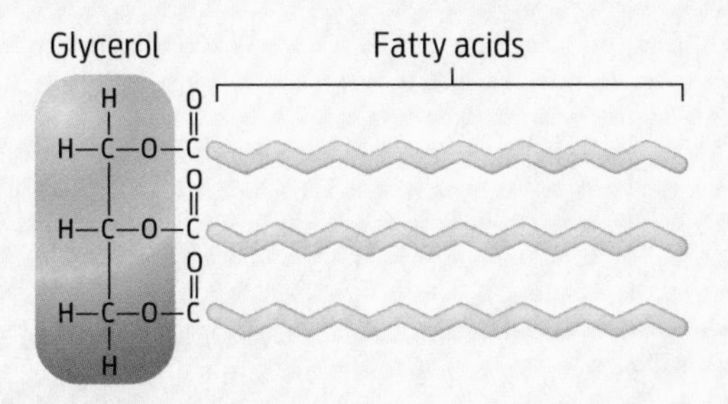

Triglycerides

Triglycerides, also known as fat, have three fatty acid chains attached to a glycerol molecule. Fats store large amounts of energy and also provide padding and thermal insulation.

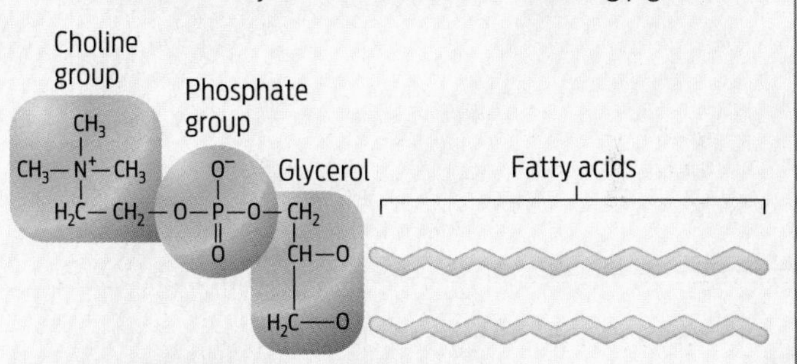

Phospholipids

Phospholipids have two fatty acid chains and a phosphate group attached to a glycerol molecule. Phospholipids are an important component of cell membranes.

d. Nucleic Acids Are Made of Nucleotides

Nucleic acids are polymers of repeating subunits known as nucleotides. There are two types of nucleic acids, DNA and RNA, each of which is made up of slightly different types of nucleotides. DNA and RNA are critical for the storage, transmission, and execution of genetic instructions.

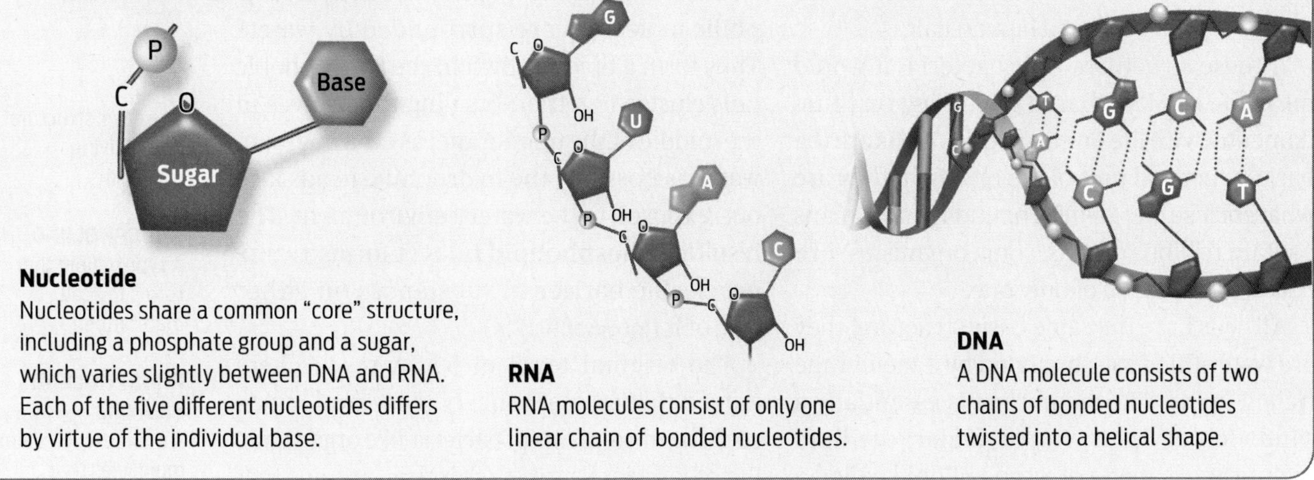

Nucleotide

Nucleotides share a common "core" structure, including a phosphate group and a sugar, which varies slightly between DNA and RNA. Each of the five different nucleotides differs by virtue of the individual base.

RNA

RNA molecules consist of only one linear chain of bonded nucleotides.

DNA

A DNA molecule consists of two chains of bonded nucleotides twisted into a helical shape.

INFOGRAPHIC 2.4

A Layer Rich in Phospholipids Defines Cell Boundaries

Cells have an aqueous interior that is separated from a chemically distinct aqueous exterior by a lipid bilayer.

Water outside
Water outside the cell dissolves molecules, bringing the cell important external signals and removing waste eliminated from the cell.

A simple cell

Water inside
Water inside the cell dissolves molecules and supports their chemical interaction required for cell functions.

Lipid bilayer
A lipid bilayer separates the aqueous inside from the aqueous outside of the cell.

Phospholipid
Water-loving head (hydrophilic)
Water-hating tails (hydrophobic)

Phospholipids assemble into bilayers in water. Water-hating tails congregate between water-loving heads, forming a lipid sandwich.

Phospholipid bilayer

known as nanobacteria—"nano" for their exceptionally tiny size. The largest of the fossilized beans are 100 nanometers in diameter—less than 1/100 the width of a human hair.

If these structures *were* bacteria, it would mean life on Mars shared something very fundamental with life on earth: cells. **Cells** are the basic structural unit of life on earth; they are what enclose life, giving it boundaries. Humans contain trillions of cells; some organisms, like bacteria, are made of only one.

All cells have the same basic structure: they are water-filled sacs bounded by a membrane rich in lipids. The membrane is essentially a sandwich of lipids. In particular, the lipid membrane is made of a type of lipid called a **phospholipid.** Each phospholipid has one **hydrophobic** ("water-fearing") end that repels water and a **hydrophilic** ("water-loving") end that attracts it. What happens when a bunch of partly hydrophobic, partly hydrophilic molecules are surrounded by water? They form a lipid sandwich: the hydrophobic tails cluster together, burying themselves in the middle of the membrane, as far away from water as possible; the hydrophilic heads face out, exposed to the watery environment. The resulting **phospholipid bilayer** forms a semipermeable barrier to substances on either side of it **(Infographic 2.4)**.

The original team of NASA researchers argued that at least some of these oval lumps could be the remains of bacteria-like organisms. But other scientists were skeptical, arguing that the lumps were far too small to house the necessary components of living cells. More likely,

CELL
The basic structural unit of living organisms.

PHOSPHOLIPID
A type of lipid that forms the cell membrane.

PHOSPHOLIPID BILAYER
A double layer of lipid molecules that characterizes all biological membranes.

HYDROPHOBIC
"Water-fearing"; hydrophobic molecules will not dissolve in water.

HYDROPHILIC
"Water-loving"; hydrophilic molecules dissolve in water.

SOLVENT
A substance in which other substances can dissolve; for example, water.

SOLUTE
A dissolved substance.

SOLUTION
A mixture of solutes dissolved in a solvent.

POLAR MOLECULE
A molecule in which electrons are not shared equally between atoms, causing a partial negative charge at one end and a partial positive charge at the other; for example, water.

IONIC BOND
A strong electrical attraction between oppositely charged ions.

ION
An electrically charged atom, the charge resulting from the loss or gain of electrons.

HYDROGEN BOND
A weak electrical attraction between a partially positive hydrogen atom and another atom with a partial negative charge.

they said, the structures were formed from nonliving chemical processes that just happened to form oval shapes.

Chris McKay, an astrophysicist with NASA's Ames Research Center who is generally sympathetic to the quest to find life on Mars, is skeptical of the famous "nano-worm." There is "no evidence that these shapes had anything to do with biology," says McKay.

Though the case for cellular life in ALH84001 has been weakened, it has not been completely ruled out. According to many researchers, the strongest evidence for life in the meteorite is the presence of so-called magnetite grains—tiny magnetic particles composed of iron that are found alongside the bacteria-like beans. On earth, similar magnetite particles are used by certain bacteria as a kind of navigation device, like a magnet in a compass. In the bacterial compass, the magnetite grains are arranged end to end, like beads on a string.

"Liquid water is the key requirement in the search for life." —Chris McKay

"The magnetite grains remain intriguing enough that I am sure that this will be one of the first things investigated on a Mars sample return," says Chris McKay. "If we find the magnetite grains aligned in 'string-of-pearls' fashion this would be good evidence of a biological origin." It other words, it would mean Martian bacteria, and therefore Martian cells.

Follow the Water

In their search for extraterrestrial life, astrobiologists often say, "Follow the water." Water is viewed as a proxy for life because it is so crucial to life on earth. Water makes up 75% to 85% of a cell's weight. All of life's chemical reactions take place in water, and many living things can survive only a few days without it.

A simple Mickey Mouse-shaped molecule consisting of one oxygen atom bound to two hydrogens, water comes pretty close to being a miracle substance. It is a universal **solvent,** capable of dissolving just about any substance—even gold. Water transports all of life's dissolved molecules, or **solutes,** from place to place—whether through a cell, a body, or an ecosystem. Life, in essence, is a water-based **solution.**

But water is more than just a stage on which the chemical reactions in question take place. It is a principal actor. Many biological molecules, like proteins and DNA, have the necessary shapes they do only because of the surrounding water that they interact with.

What makes water such a good solvent? Because the electrons in a water molecule are shared unequally between the oxygen and hydrogen atoms, water is considered a **polar molecule.** With a partial negative and a partial positive charge on either end, water is an excellent solvent for other polar molecules with partial charges and substances like salt that contain **ionic bonds.** Ionic bonds are strong bonds formed between oppositely charged **ions.** By surrounding each charged ion, water dissolves the bond between them **(Infographic 2.5)**.

When astrobiologists speak about the importance of water for life, they make an important qualification: *liquid* water. Frozen water is found throughout the universe; there are abundant quantities on Mars, for example. But only on earth does water exist primarily in its liquid form at ambient temperature and pressure.

"Liquid water is the key requirement in the search for life," says astrophysicist Chris McKay. "The other worlds of the solar system have enough light, enough carbon, and enough of the other key elements for life. Water in the liquid form is rare."

Why is water liquid at room temperature? Essentially, it's because water molecules are "sticky." Each water molecule has a partial charge on each end and can therefore form electrostatic attractions, known as **hydrogen bonds,** with one another and with other molecules. These hydrogen bonds act as a kind of glue holding water molecules together and keeping them liquid at room temperature. You can see water's stickiness wherever you look:

a drop of water clinging to a leaf despite the downward pull of gravity, for example, or an insect able to land on the surface of a pond **(Infographic 2.6)**.

Compared to other molecules its size, water also has a large liquid range–freezing at 0°C (32°F) and boiling at 100°C (212°F). That's because water molecules can absorb a lot of energy before they get hot and vaporize (that is, turn into a gas)–again because of their tenacious hydrogen bonds. Because of these bonds, water gets hot more slowly than do other liquids and also holds onto heat longer. And water's liquid range can be extended even further: add salt to water and you can lower the freezing point to −46°C (−50°F); increase the pressure and you can bump up the boiling point to over 343°C (650°F). It's because there is so much salt in seawater that most oceans don't freeze in winter.

Finally, unlike most substances on earth, water has the unusual property of being less dense as a solid than as a liquid: ice floats. And because it does, fish can live beneath frozen lakes in winter and not turn into ice cubes–which is good for both the fish and us.

Given its amazing properties, scientists want to find out whether liquid water exists on Mars. From the *Viking* missions, scientists know that frozen water exists in the form of large ice caps on the surface of Mars and also as a layer of permafrost just beneath the surface. In 2008, NASA's *Phoenix Lander* provided further evidence of frozen water in Martian soil. But so far no liquid water has been found. Scientists suspect that the Martian atmosphere is so thin and so cold that any liquid water would rapidly evaporate or freeze.

Though liquid water is not present on the surface of Mars today, many scientists suspect that liquid water–lots of it–once covered the planet. Clues to this ancient water can be seen all over the Martian surface, which in many places is carved out like sections of the Grand Canyon. The *Phoenix Lander* also found telltale signs of liquid water's past on the surface of Mars in the form of salt deposits like those you can see when seawater evaporates.

INFOGRAPHIC 2.5

Water Is a Good Solvent Because It Is Polar

Water is a polar molecule because electrons are not shared equally between the oxygen and hydrogen atoms. Electrons are pulled closer to the oxygen atom than to the hydrogen atoms, creating a slightly negative oxygen atom and slightly positive hydrogen atoms. The partial charges on each water molecule can interact with charged ions or other molecules, allowing water to "coat" or dissolve the hydrophilic solutes.

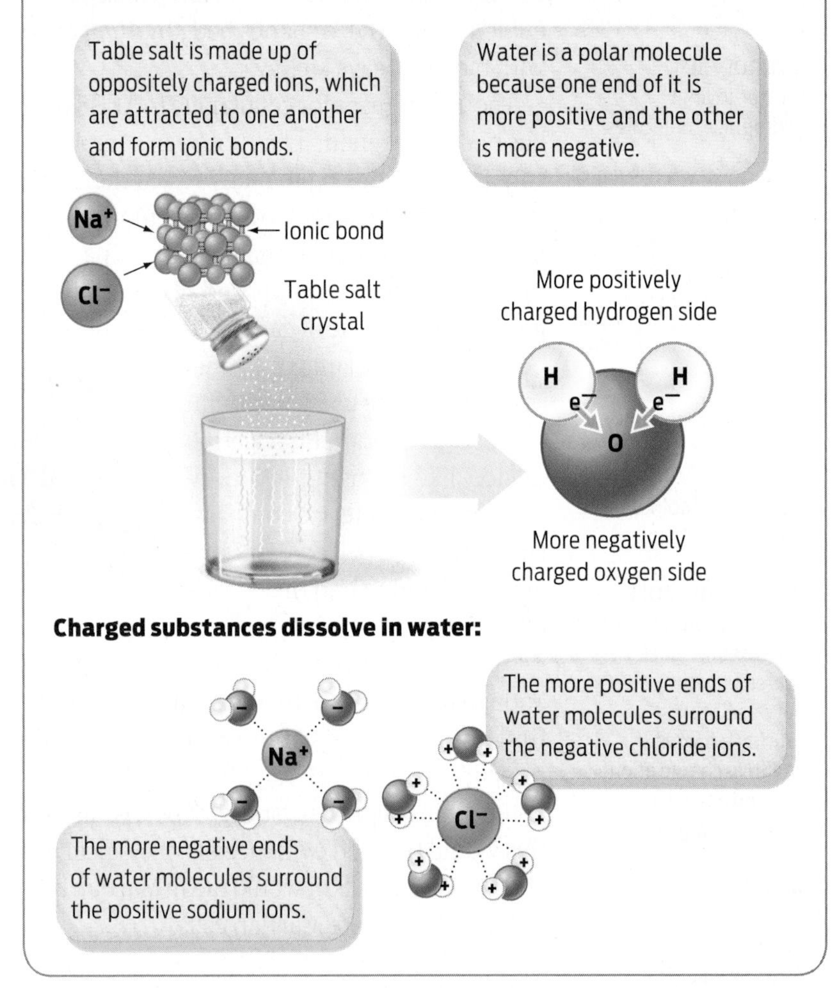

Table salt is made up of oppositely charged ions, which are attracted to one another and form ionic bonds.
Water is a polar molecule because one end of it is more positive and the other is more negative.
Na+
Cl−
Ionic bond
Table salt crystal
More positively charged hydrogen side
H
H
e−
e−
O
More negatively charged oxygen side
Charged substances dissolve in water:
Na+
Cl−
The more positive ends of water molecules surround the negative chloride ions.
The more negative ends of water molecules surround the positive sodium ions.

Additional support for the presence of ancient water comes from meteorite ALH84001. Crevices of the meteorite are filled with carbon-rich globules that resemble those produced by bacteria on earth. Scientists believe these globules could have formed only if liquid water had once percolated through the meteorite, carrying CO_2 from the Martian atmosphere into the rock.

Where all this water went, no one knows. But some scientists suspect that liquid water

pH
A measure of the concentration of H^+ in a solution.

ACID
A substance that increases the hydrogen ion concentration of solutions, making them more acidic.

BASE
A substance that reduces the hydrogen ion concentration of solutions, making them more basic.

INFOGRAPHIC 2.6

Water Is "Sticky" Because It Forms Hydrogen Bonds

When many water molecules are near one another, the partially positive hydrogen atoms of some molecules are attracted to the partially negative oxygen atoms of neighboring water molecules. These attractions are hydrogen bonds, weak electrical attractions.

Hydrogen Bonds — two polar molecules are attracted to each other

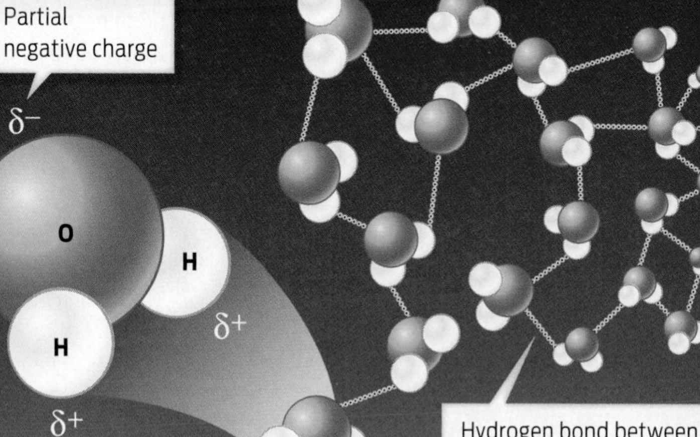

Cohesion
Hydrogen bonding between water molecules is strong enough to defy gravity, allowing water to flow up stems of even the tallest plants. This cohesive property supports life, for example by providing surface tension on lakes for insects to land on.

Adhesion
The partial charges on water molecules allow them to readily bind to many surfaces, making them wet. Leaves can collect water for the organisms that live on them.

may still exist beneath the surface of the planet, and may even bubble to the surface periodically, as is suggested by photographs of apparent water flows taken in 2004 and 2005 by NASA's *Mars Global Surveyor* satellite. The possibility that water existed on Mars in the past, and may still exist today beneath the surface, raises the question of whether a belowground habitat exists that is conducive to life.

If there is water within Mars, would it have the properties of earth water? Depending on what's dissolved in it, water can have a large range of characteristics–from caustic drain cleaner and ammonia to tart lemon juice and cavity-causing soda. The different chemical properties of water-based solutions reflect their **pH,** the concentration of hydrogen ions (H^+) in a solution, which is defined as ranging from 0 to 14. Here's the background of measurement by pH: water molecules (H_2O) can split briefly into separate hydrogen (H^+) and hydroxide (OH^-) ions. In pure water, the number of separated H^+ ions is by definition exactly equal to the number of separated OH^- ions, and the pH is therefore 7, or neutral. Acidic solutions, or **acids,** have a higher concentration of hydrogen ions (H^+) and a pH closer to 0. When acids are added to water, they increase the concentration of hydrogen ions and make the solution more acidic. Basic solutions, or **bases,** on the other hand, have a lower concentration of H^+ ions and a pH closer to 14. Bases remove H^+ ions from a solution, thereby increasing the proportion of OH^- ions.

Strong acids and bases are highly reactive with other substances, which makes them destructive to the molecules in a cell. Also, many biochemical reactions take place only at a certain pH. Living things are thus extremely sensitive to changes in pH, and most function best when their pH stays within a specific range. The pH of human blood ranges from about 7.35 to 7.4. If that pH were to fall even slightly, to 7, our biochemistry would malfunction and we would die.

The *Viking* experiments determined that the pH of Martian soil is roughly 7.2. The *Phoenix Lander* recently calculated it at 7.7–mild enough to grow asparagus, as the mission's chief chemist put it. **(Infographic 2.7)**

"Weird Life"

So far, NASA's search for life on Mars has hewn very closely to our understanding of life on earth, where living things seem to share certain chemical and structural properties, like carbon-based molecules and cells. Nevertheless, there are a few exceptions, or boundary cases, that seem to bend the rules of life on earth. **Viruses** are an example. Viruses reproduce and pass their genetic information on to new viruses, but they are not made of cells at all. Instead, they are infectious particles consisting of a protein shell that encloses genetic information. Viruses reproduce by infecting a host cell and hijacking its cellular machinery to make copies of itself. Other noncellular, self-reproducing entities include **prions,** infectious proteins that are responsible for mad cow disease and related human and animal illnesses. Whether or not viruses and prions are truly alive is hotly debated among scientists.

If viruses and prions bend the rules, then might not Martian life as well? In 2008, the National Academy of Sciences issued a "weird life" report suggesting that NASA not be so narrowly focused on water and organic molecules in its search for life on other planets. True, water may be crucial to life on earth, but that doesn't mean that other solvents–ammonia or methane, for example–could not support life elsewhere, the report noted. The report also urged the space agency to avoid being "fixated on carbon," even though carbon forms the scaffold of life on earth. Other elements, like silicon, for example, could in theory provide a functional scaffold for life on other planets.

INFOGRAPHIC 2.7

Solutions Have a Characteristic pH

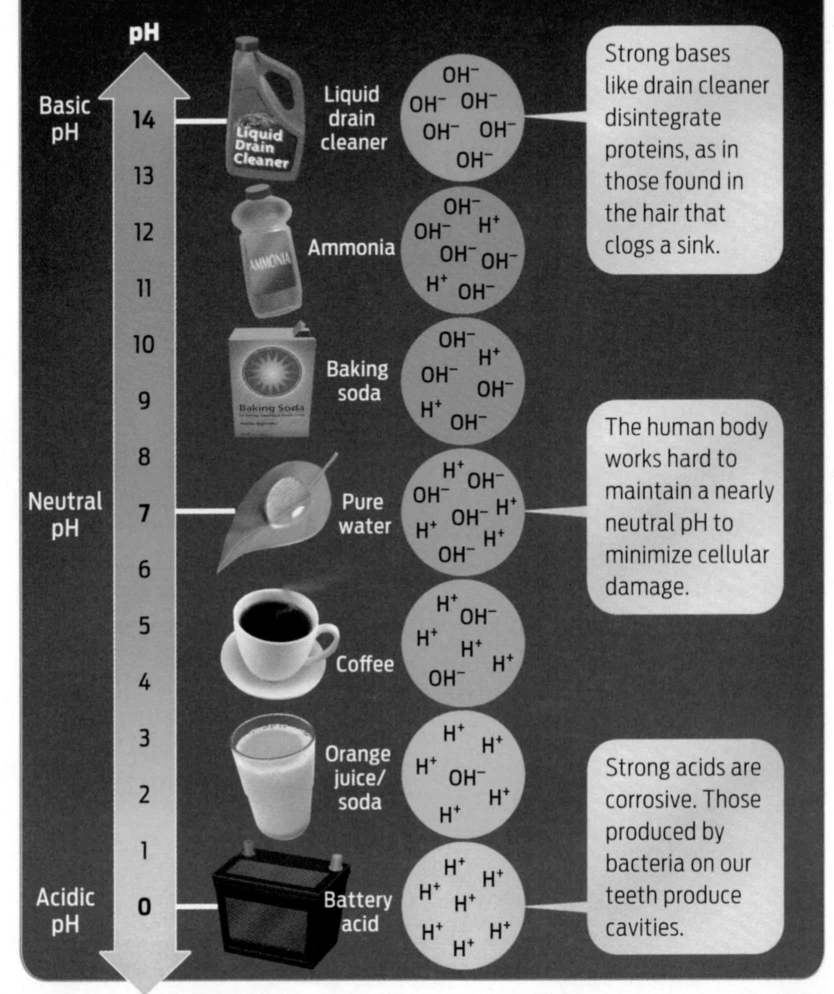

The pH of a solution is a measure of the concentration of hydrogen ions (H⁺) in it. Solutions with a low concentration of H⁺ ions have a basic pH (greater than pH 7). Solutions with a high concentration of H⁺ ions have an acidic pH (a pH of less than 7). Both acids and bases can be damaging because they are highly reactive with other substances. A neutral solution has a pH of 7.
pH
Basic pH
Neutral pH
Acidic pH
14
13
12
11
10
9
8
7
6
5
4
3
2
1
0
Liquid drain cleaner
Liquid Drain Cleaner
Ammonia
Baking soda
Pure water
Coffee
Orange juice/ soda
Battery acid
Strong bases like drain cleaner disintegrate proteins, as in those found in the hair that clogs a sink.
The human body works hard to maintain a nearly neutral pH to minimize cellular damage.
Strong acids are corrosive. Those produced by bacteria on our teeth produce cavities.

A binocular microscopic view of carbonate globules in ALH84001.

VIRUS
An infectious agent made up of a protein shell that encloses genetic information.

PRION
A protein-only infectious agent.

Our first panoramic view of the surface of Mars.

"I spend my time and energy in the search for evidence of life on Mars." –Chris McKay

Recent discoveries by biologists who study microorganisms are also challenging our notions of what life looks like and where it can survive. Microscopic organisms have been found growing just about anywhere, from radioactive waste and boiling geysers to sunless deep-sea vents and Arctic springs made almost entirely of salt. Such extreme-loving organisms reveal that life is nothing if not adaptive. Could similarly adaptive organisms have once inhabited Mars? Might they still? At least some astrobiologists are cautiously optimistic.

"I spend my time and energy in the search for evidence of life on Mars," says Chris McKay. "Obviously, this is because I think there must have been life there and we have a good chance of finding evidence of it." ■

Summary

- "Life" is difficult to define in universal terms because we have only a single example of it to consider: life on earth.
- On earth, living organisms share a number of functional characteristics: they grow and reproduce, maintain homeostasis, sense and respond to their environment, and rely on energy to carry out their functions.
- All matter is composed of elements, of which there are about 100 on earth. Each element has a unique atomic structure, with a particular number of protons, neutrons, and electrons.
- When atoms share pairs of electrons, they form covalent bonds, making molecules.
- On earth, living organisms are made up of organic molecules, those containing a backbone of the element carbon.
- Four types of carbon-based organic molecules make up living things: proteins, carbohydrates, nucleic acids, and lipids.
- Living organisms on earth are made of cells, which contain water and are surrounded by a membrane of lipids; cells are the smallest unit of life.
- Water is a polar molecule, with a partial positive and a partial negative charge.
- Water has many properties that make it a crucial component of life on earth: it is a good solvent, it is "sticky," it regulates heat well, and it floats when frozen.
- Substances, like salt, that easily dissolve in water are considered hydrophilic; substances, like lipids, that do not dissolve in water are hydrophobic.
- The concentration of H^+ ions in a solution determines its pH. Most chemical reactions in cells take place at a nearly neutral pH.
- If life is found on other planets, it may or may not use the chemical framework used by life on earth.

Chapter 2 Test Your Knowledge

PROPERTIES OF LIFE

Life on earth is marked by certain functional, structural, and chemical properties.

HINT See Infographics 2.1–2.4 and Up Close: Molecules of Life.

KNOW IT

1. Which of the following is *not* a generally recognized characteristic of most (if not all) living organisms?

- **a.** the ability to reproduce
- **b.** the ability to maintain homeostasis
- **c.** the ability to obtain energy directly from sunlight
- **d.** the ability to sense and respond to the environment
- **e.** the ability to grow

2. What is homeostasis? Why it is important to living organisms?

3. The basic building blocks of life are

- **a.** DNA molecules.
- **b.** cells.
- **c.** proteins.
- **d.** phospholipids.
- **e.** inorganic molecules

4. What subatomic particles are located in the nucleus of an atom?

- **a.** protons
- **b.** neutrons
- **c.** electrons
- **d.** protons, neutrons, and electrons
- **e.** protons and neutrons

5. When an atom loses an electron, what happens?

- **a.** It becomes positively charged.
- **b.** It becomes negatively charged.
- **c.** It becomes neutral.
- **d.** Nothing happens.
- **e.** atoms cannot lose an electron because atoms have a defined number of electrons

6. What does it mean to say a macromolecule is a polymer? Give an example.

7. A collection of amino acids could be used to build a

- **a.** protein.
- **b.** complex carbohydrate.
- **c.** triglyceride.
- **d.** nucleic acid.
- **e.** cell

USE IT

8. How would you assess whether or not a possibly living organism from another planet were truly alive?

9. Which of the characteristics of living organisms (if any) allow you to distinguish between living and formerly living (that is, dead) organisms? Explain your answer.

10. What are the arguments for and against viruses being considered living organisms?

11. If, in a mound of dirt, you had evidence that carbon dioxide was being consumed and converted to glucose, what could you conclude about the presence of a living organism? Explain your answer.

12. How does a sterol, such as cholesterol, differ from a triglyceride? Structurally, what do triglycerides and phospholipids have in common?

WATER: THE SOLVENT OF LIFE

Water has many properties that make it a suitable medium for living things and their chemical reactions.

HINT See Infographics 2.5–2.7.

KNOW IT

13. Is olive oil hydrophobic or hydrophilic? What about salt? Explain your answer.

14. Two water molecules can bond to each other by ___________ bonding; this is an example of ______________.

- **a.** hydrogen; adhesion
- **b.** covalent; adhesion
- **c.** non-covalent; cohesion
- **d.** covalent; cohesion
- **e.** hydrogen; cohesion

15. Coffee, tea, or any water-based beverage with sugar in it is an example of a(n) _________ solution.
   a. What is the solvent in such a beverage?
   b. What is the solute in such a beverage?

16. As an acidic compound dissolves in water, the pH of the water _________.
   a. becomes higher
   b. remains neutral
   c. becomes lower
   d. doesn't change
   e. becomes basic

17. The bond between the oxygen atom and a hydrogen atom in a water molecule is a(n) _________ bond.
   a. covalent
   b. hydrogen
   c. ionic
   d. hydrophobic
   e. noncovalent

18. How do ionic bonds compare to hydrogen bonds? What are the similarities and differences?

USE IT

19. Why do olive oil and aqueous vinegar tend to separate in salad dressing? Will added salt dissolve in the oil or the vinegar? Explain your answer.

20. Why do deserts cool off more at night than do seaside towns?

21. Which of the following would be most likely to dissolve in olive oil?
   a. a polar molecule
   b. a nonpolar molecule
   c. a hydrophilic molecule
   d. a and c
   e. b and c

SCIENCE AND ETHICS

22. One approach to finding out if there is life on Mars is to bring Martian dirt samples to earth for analysis. What are possible considerations for science and society if a Martian life form is released on earth? If an earth life form is introduced onto Mars?

Wonder Drug

What You Will Be Learning

3.1 How Penicillin Was Discovered

3.2 Cell Theory: All Living Things Are Made of Cells

3.3 Membranes: All Cells Have Them

3.4 Prokaryotic and Eukaryotic Cells Have Different Structures

3.5 Some Antibiotics Target Bacterial Cell Walls

3.6 Some Antibiotics Inhibit Prokaryotic Ribosomes

3.7 Molecules Move across the Cell Membrane

3.8 Eukaryotic Cells Have Organelles

UP CLOSE Eukaryotic Organelles

Wonder Drug

How a chance discovery in a London laboratory revolutionized medicine

On a September morning in 1928, biologist Alexander Fleming returned to his laboratory at St. Mary's Hospital in London after a short summer vacation. As usual, the place was a mess—his bench piled high with the petri dishes on which he was growing bacteria. On this day, as Fleming sorted through the plates, he noticed that one was growing a patch of fluffy white mold. It had been contaminated, likely by a rogue mold spore that had drifted in from a neighboring laboratory.

Fleming was about to toss the plate in the sink when he noticed something unusual: wherever mold was growing, there was a zone around the mold where the bacteria did not seem to grow. Curious, he looked under a microscope and saw that the bacterial cells near the mold had burst, or lysed. Something in the mold was killing the bacteria.

Experiments confirmed that the mold was capable of killing many kinds of bacteria, including *Streptococcus, Staphylococcus,* and *Pneumococcus.* Fleming published his results in 1929 in the *British Journal of Experimental Pathology.* He named the antibacterial substance "penicillin," after the fungus producing it, *Penicillium notatum*. It was the birth of the first **antibiotic**.

Fleming was not the first to notice the bacteria-killing property of *Penicillium*, but he was the first to study it scientifically and publish the results. In fact, Fleming had been looking for bacteria-killing substances for a number of years, ever since he had served as a medical officer in World War I and witnessed soldiers dying from bacteria-caused infections. He had already discovered one such antimicrobial agent—the chemical lysozyme—which he detected in his own tears and nasal mucus, so he knew what bacteria-killing signs to look for.

If you've ever seen a piece of moldy bread or rotting fruit, then you've met the *Penicillium* fungus. It doesn't look very impressive, but the

ANTIBIOTIC
A chemical that can slow or stop the growth of bacteria; many antibiotics are produced by living organisms.

Fleming in his lab.

chemical it produces ushered in a whole new age of medicine. For the first time, doctors had a way to treat such deadly illnesses as bacterial pneumonia, syphilis, and meningitis. As physician Lewis Thomas, former president of Memorial Sloan-Kettering Cancer Center in New York City, wrote in his 1992 memoir *Fragile Species,* "We could hardly believe our eyes on seeing that bacteria could be killed off without at the same time killing the patient. It was not just amazement, it was a revolution" **(Infographic 3.1)**.

CELL THEORY
The concept that all living organisms are made of cells and that cells are formed by the reproduction of existing cells.

Bug Bullet

What makes antibiotics special is not just their ability to kill bacteria. After all, cyanide kills bacteria just fine. The important thing about antibiotics is that they exert their destructive effects on bacteria without (typically) harming their human or animal host, even if taken internally.

"We could hardly believe our eyes on seeing that bacteria could be killed off without at the same time killing the patient. It was not just amazement, it was a revolution." –Lewis Thomas

Although Fleming didn't know it at the time, penicillin and other antibiotics preferentially kill bacteria because they target what is unique about bacterial cells. According to the **cell theory,** all living things are made of cells, and

INFOGRAPHIC 3.1

How Penicillin Was Discovered

A fortuitous observation by Fleming led to the discovery of the first antibiotic. He realized that the fungus on his culture plate was somehow inhibiting the reproduction of bacteria.

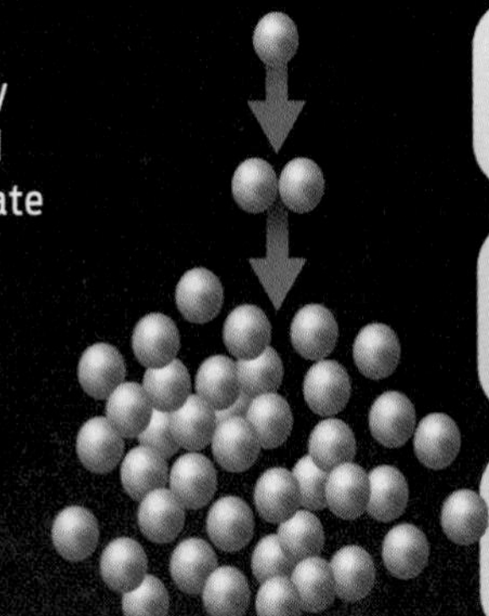

A single bacterial cell lands on a culture plate far away from the mold.

Nutrients in the plate support the growth and division of the bacterial cells.

After many rounds of cell division, enough cells accumulate in this spot to be visualized as a colony on the plate.

Staphylococcus bacterial colonies form at locations far away from the mold.

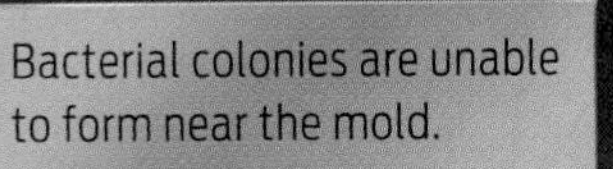

Bacterial colonies are unable to form near the mold.

Penicillium mold

Penicillium under a microscope and on an orange

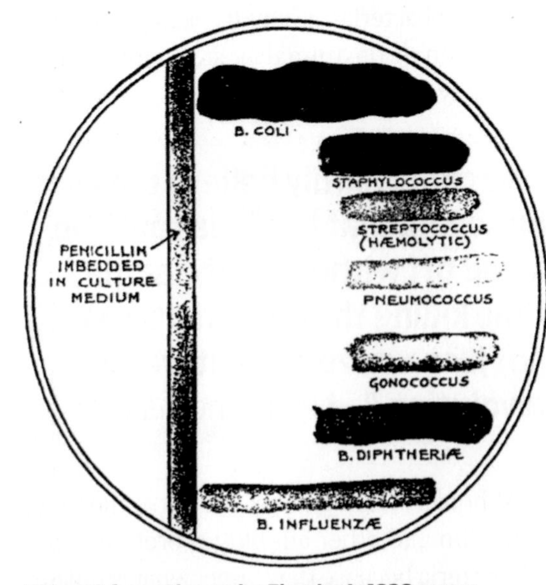

Figure 2 from Alexander Fleming's 1929 paper, showing the response of different bacteria to penicillin.

every new cell comes from the division of a pre-existing one. But not all cells are alike. Cells come in many shapes and sizes and perform various functions, depending on where they are found **(Infographic 3.2)**. Moreover, they fall into two fundamentally different categories: **prokaryotic** or **eukaryotic.** Prokaryotic cells are relatively small and lack internal membrane-bound compartments, called **organelles**. Eukaryotic cells, by contrast, are much larger and contain many such organelles. Penicillin and other antibiotics target structures that are unique to prokaryotic cells.

To understand why antibiotics affect prokaryotic and eukaryotic cells differently, it helps

PROKARYOTIC CELLS
Cells that lack internal membrane-bound organelles.

EUKARYOTIC CELLS
Cells that contain membrane-bound organelles, including a central nucleus.

ORGANELLES
The membrane-bound compartments of eukaryotic cells that carry out specific functions.

INFOGRAPHIC 3.2

Cell Theory: All Living Things Are Made of Cells

All living organisms are composed of cells. These cells arise from the reproduction of existing cells. Different cells have different structures and functions.

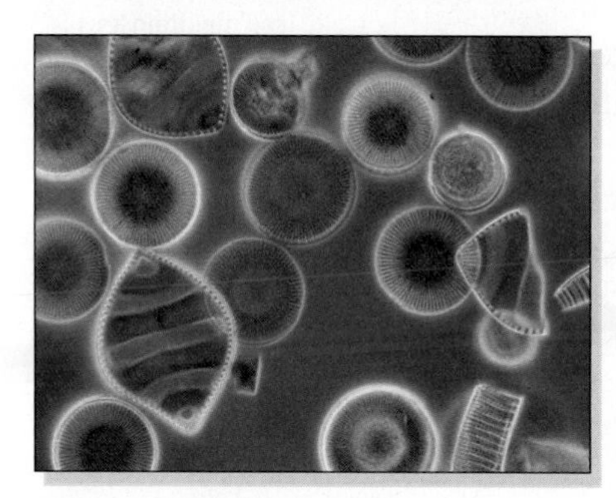
Diatoms: single-cell eukaryotes

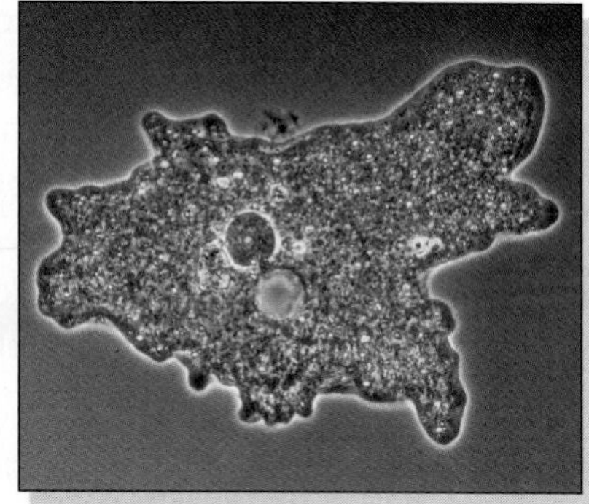
Amoeba (a protozoan): a single-cell eukaryote

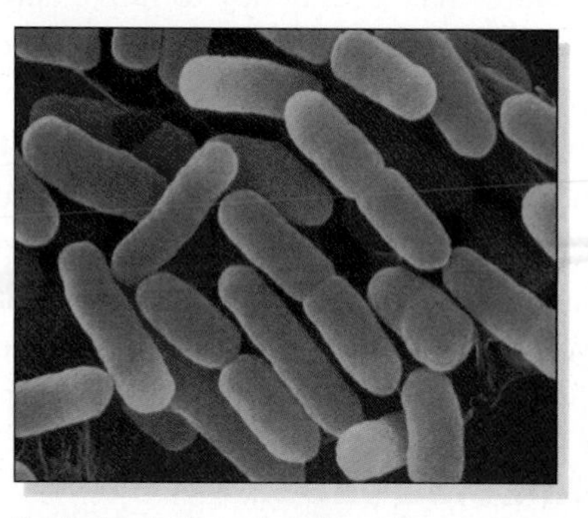
Bacteria: single-cell prokaryotes

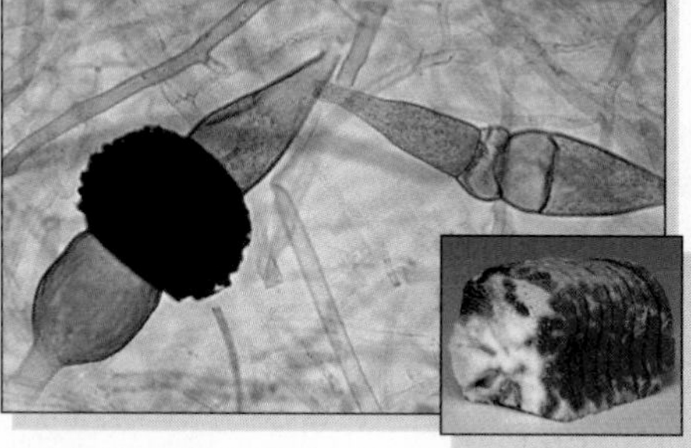
Molds (fungi): single and multicellular eukaryotic cells

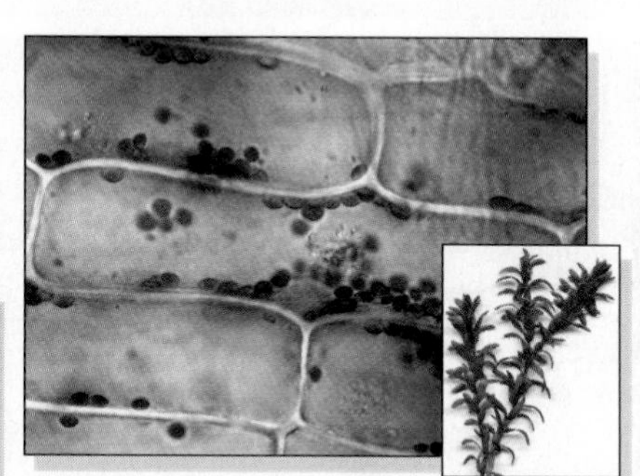
Elodea (an aquatic plant): a multicellular eukaryote

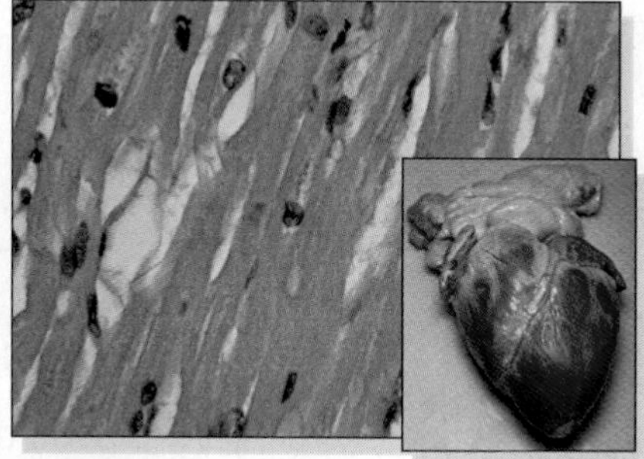
Humans (these are heart cells): multicellular eukaryotes

CELL MEMBRANE
A phospholipid bilayer with embedded proteins that forms the boundary of all cells.

CYTOPLASM
The gelatinous, aqueous interior of all cells.

RIBOSOME
A complex of RNA and protein that carries out protein synthesis in all cells.

NUCLEUS
The organelle in eukaryotic cells that contains the genetic material.

to understand first what the two cell types have in common. All cells, both prokaryotic and eukaryotic, are surrounded by a **cell membrane.** This flexible yet sturdy structure forms a boundary between the external environment and the cell's watery **cytoplasm** and literally holds the cell together. Partly hydrophobic, partly hydrophilic molecules known as phospholipids make up the bulk of the cell membrane, and proteins embedded in the membrane perform particular functions, such as transporting nutrients in and wastes out. The cell membrane forms a semipermeable barrier to substances on either side of it **(Infographic 3.3).**

In addition to a flexible cell membrane, both prokaryotic and eukaryotic cells have two other elements in common: **ribosomes,** which synthesize the proteins that are crucial to cell function; and DNA, the molecule of heredity.

Beyond these three features, however—cell membrane, ribosomes, and DNA—the two cell types are structurally quite different. In a prokaryotic cell, for instance, the DNA floats freely within the cell's cytoplasm, while in a eukaryotic cell it is housed within a central command center called the **nucleus.** The nucleus is one of many organelles found within eukaryotic cells, but not in their simpler prokaryotic cousins **(Infographic 3.4).**

Penicillin kills bacteria because of one important difference between prokaryotic and eukaryotic cells. Unlike human and other animal cells, most bacteria are surrounded by a **cell wall.** This rigid structure is what allows bacteria to survive in watery environments–say, your intestines or a pond.

Water has a tendency to move across cell membranes from lower to higher solute concentration, a process called **osmosis.** In a low-solute environment, water will tend to rush into the solute-rich cytoplasm of a cell, causing it to swell. This swelling is potentially fatal to bacteria. Without a cell wall, bacterial cells would fill up with water and burst. Their sturdy cell wall, however, counteracts this osmotic pressure, keeping too much water from rushing in. (Eukaryotic cells are protected from osmotic pressure by the cholesterol in their cell membrane.)

What makes the bacterial cell wall rigid is the molecule **peptidoglycan,** a polymer made of sugars and amino acids that link to form a chainlike sheath around the cell. Different bacterial walls can have different structures, but all have peptidoglycan, which is found only in bacteria. By interfering with the synthesis of

CELL WALL
A rigid structure enclosing the cell membrane of some cells that helps the cell maintain its shape.

INFOGRAPHIC 3.3

Membranes: All Cells Have Them

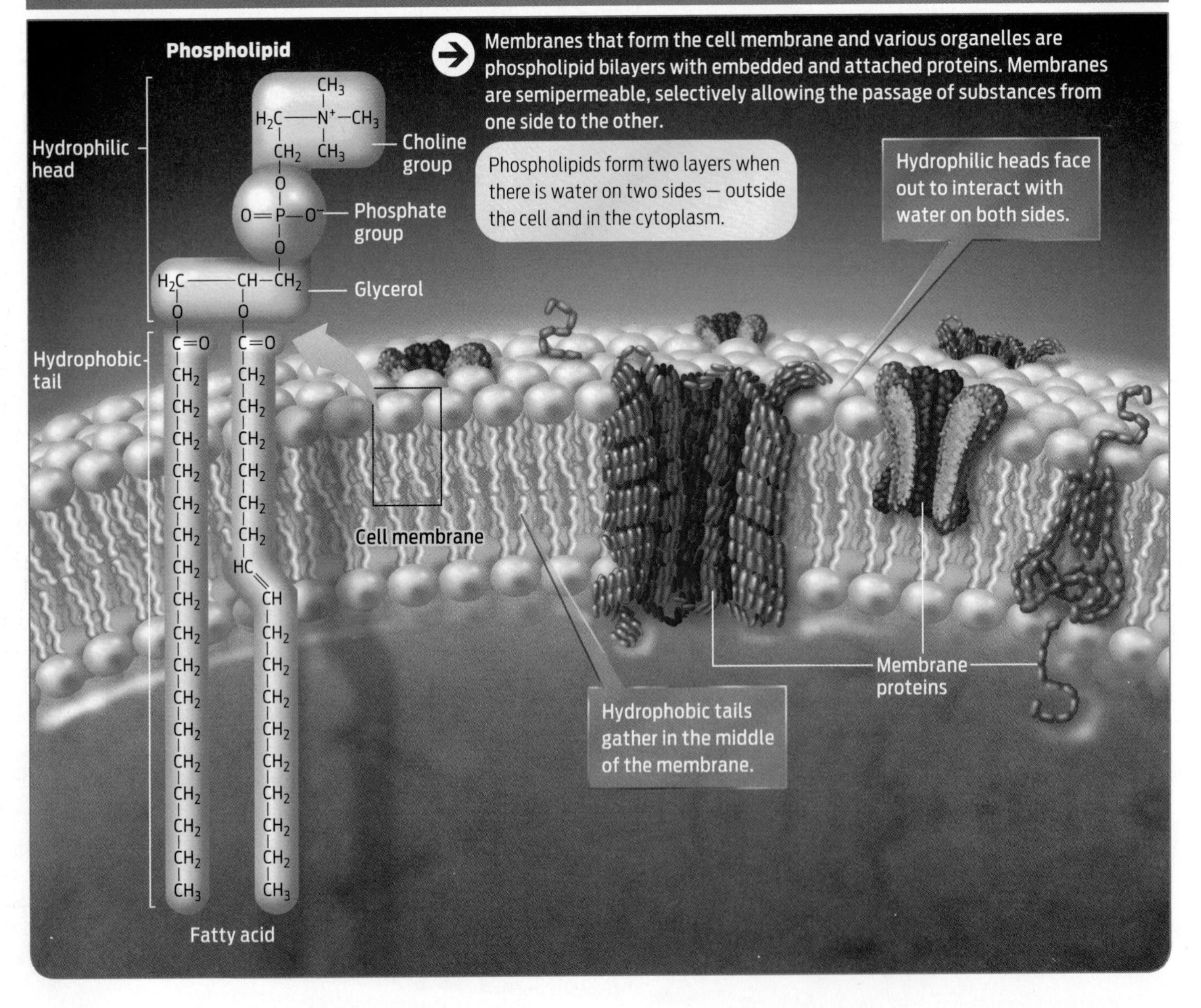

Phospholipid
Hydrophilic head
Choline group
Phosphate group
Glycerol
Hydrophobic tail
Fatty acid
Membranes that form the cell membrane and various organelles are phospholipid bilayers with embedded and attached proteins. Membranes are semipermeable, selectively allowing the passage of substances from one side to the other.
Phospholipids form two layers when there is water on two sides — outside the cell and in the cytoplasm.
Hydrophilic heads face out to interact with water on both sides.
Cell membrane
Hydrophobic tails gather in the middle of the membrane.
Membrane proteins

INFOGRAPHIC 3.4

Prokaryotic and Eukaryotic Cells Have Different Structures

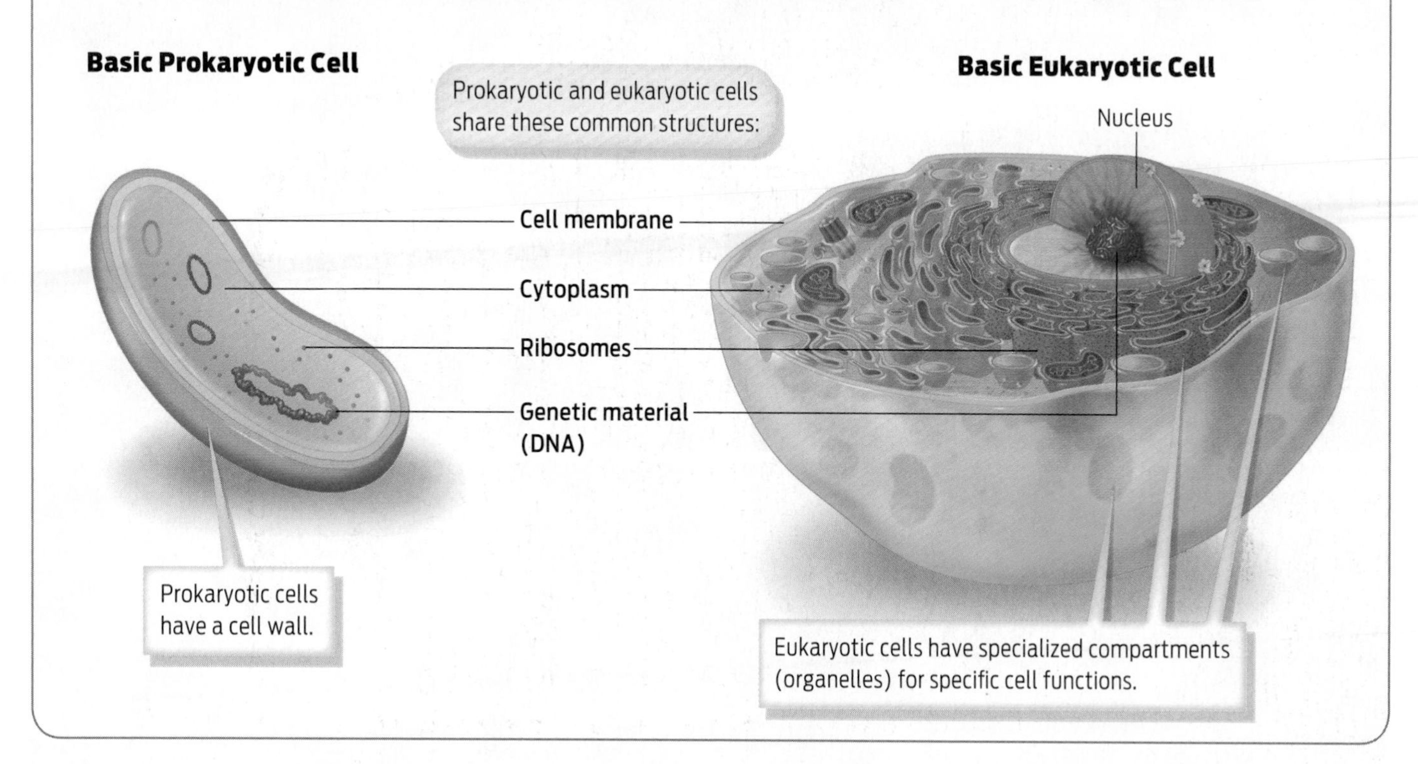

While all cells have a cell membrane, cytoplasm, ribosomes and DNA, there are specific structural differences between prokaryotic and eukaryotic cells. Eukaryotic cells contain a variety of membrane-enclosed organelles while prokaryotic cells do not.
Basic Prokaryotic Cell
Prokaryotic and eukaryotic cells share these common structures:
Basic Eukaryotic Cell
Nucleus
Cell membrane
Cytoplasm
Ribosomes
Genetic material (DNA)
Prokaryotic cells have a cell wall.
Eukaryotic cells have specialized compartments (organelles) for specific cell functions.

OSMOSIS
The diffusion of water across a semipermeable membrane from an area of lower solute concentration to an area of higher solute concentration.

PEPTIDOGLYCAN
A macromolecule that forms all bacterial cell walls and provides rigidity to the cell wall.

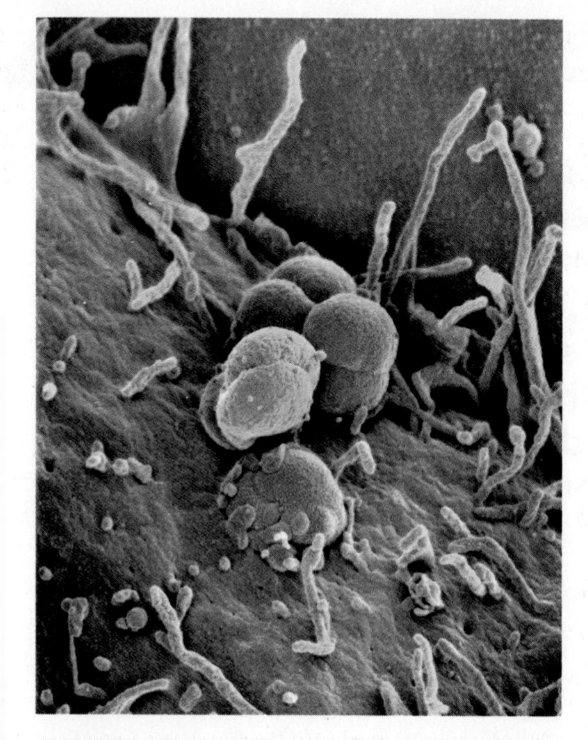

Scanning electron micrograph of the bacteria that cause gonorrhea.

peptidoglycan, penicillin weakens the cell wall, which is then no longer able to counteract osmotic water pressure. Eventually, the cell bursts (**Infographic 3.5**).

Bacteria are not the only organisms with a cell wall (plant cells and fungi have them, too), but they are the only ones that have a cell wall made of peptidoglycan–which is why penicillin is such a selective bacteria killer.

Ironically, despite its remarkable killing powers, penicillin was not immediately recognized as a medical breakthrough when it was first discovered. In fact, Fleming didn't think his mold had much of a future in medicine. At the time, the idea that an antiseptic agent could kill bacteria without at the same time harming the patient was unheard of, so Fleming never considered that penicillin might be taken internally. Nor was he a chemist, so he lacked the expertise to isolate and purify the active ingredient from the mold. While he found that his

INFOGRAPHIC 3.5

Some Antibiotics Target Bacterial Cell Walls

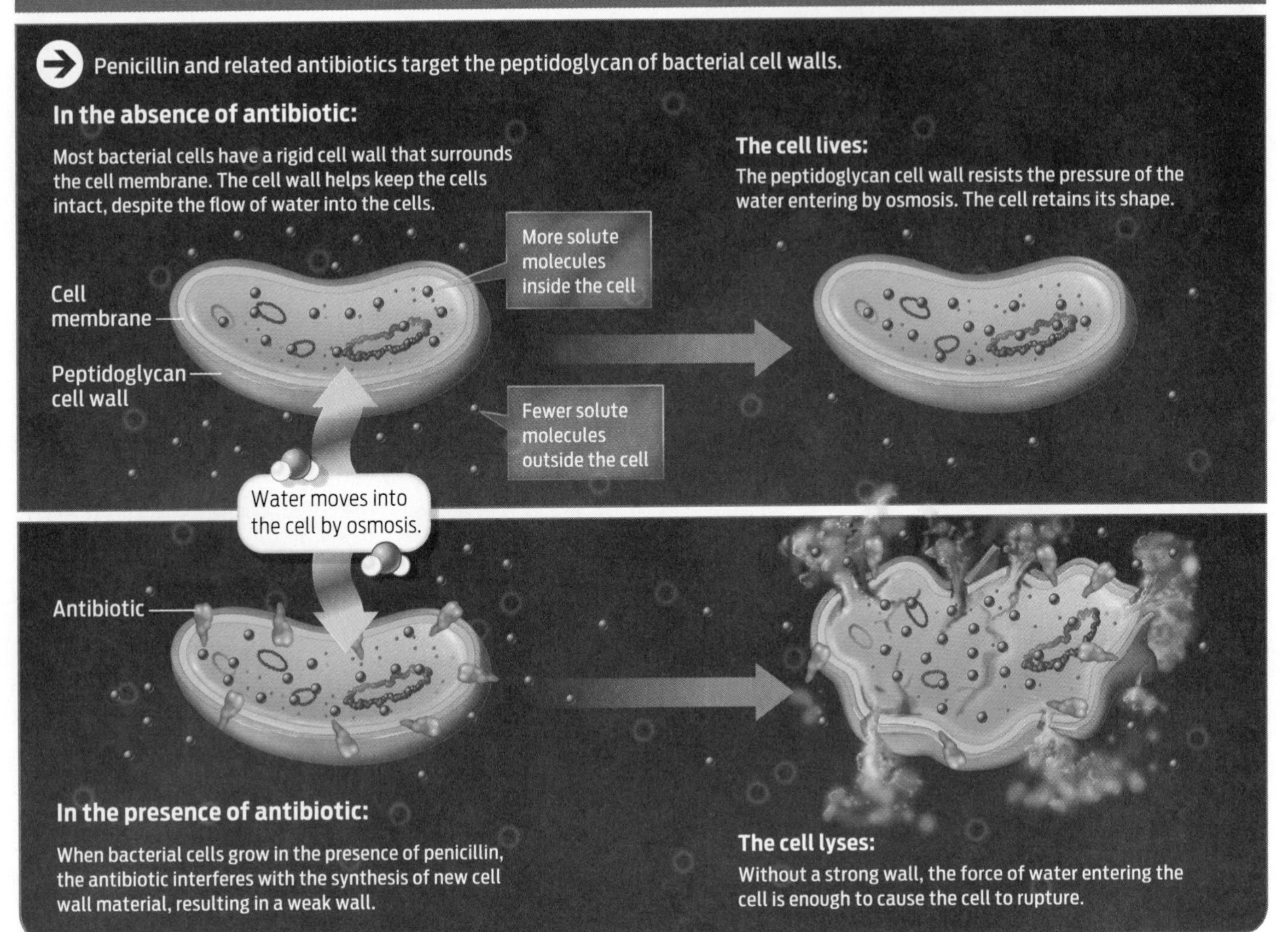

Penicillin and related antibiotics target the peptidoglycan of bacterial cell walls.
In the absence of antibiotic:
Most bacterial cells have a rigid cell wall that surrounds the cell membrane. The cell wall helps keep the cells intact, despite the flow of water into the cells.
Cell membrane
Peptidoglycan cell wall
More solute molecules inside the cell
Fewer solute molecules outside the cell
The cell lives:
The peptidoglycan cell wall resists the pressure of the water entering by osmosis. The cell retains its shape.
Water moves into the cell by osmosis.
Antibiotic
In the presence of antibiotic:
When bacterial cells grow in the presence of penicillin, the antibiotic interferes with the synthesis of new cell wall material, resulting in a weak wall.
The cell lyses:
Without a strong wall, the force of water entering the cell is enough to cause the cell to rupture.

mold juice made a "reasonably good" topical antiseptic, he noted in a 1940 paper that "the trouble of making it seemed not worth while," and largely gave up working on it.

Ten years would pass before anyone reconsidered Fleming's mold. By then, history had intervened and given new urgency to the search for antibacterial medicines.

With few other antibacterial medicines available, penicillin suddenly became the focus of research during World War II.

From Fungus to Pharmaceutical

On September 1, 1939, Germany invaded Poland, plunging the world into war for the second time in a generation. With the horrors of World War I still seared into memory, many feared the death toll that would result from the hostilities. Millions of soldiers and civilians had died in World War I, many not as a result of direct combat injuries but from infections resulting from surgeries meant to treat those injuries. With few other antibacterial medicines available, penicillin suddenly became the focus of research during World War II.

In 1938, Ernst Chain, a German-Jewish biochemist, was working in the pathology department at Oxford University, having fled Germany for England in 1933 when the Nazis came to power. Both Chain and his supervisor, Howard Florey, were interested in the biochemistry of antibacterial substances. Chain stumbled across

Manufacturing penicillin in 1943: culture flasks are filled with the nutrient solution in which penicillin mold is grown.

Fleming's 1929 paper on penicillin and set about trying to isolate and concentrate the active ingredient from the mold, which he succeeded in doing by 1940. Chain's breakthrough allowed Florey's group to begin testing the drug's clinical efficacy. They injected the purified chemical into bacteria-infected mice and found that the mice were quickly rid of their infection. Human trials followed next, in 1941, with the same remarkable result.

As encouraging as these results were, there was one nagging problem: it took up to 2,000 liters of mold fluid to obtain enough pure penicillin to treat one person. The Oxford doctors used almost their whole supply of the drug treating their first patient, a policeman ravaged by a staphylococcal infection. The team stepped up their purification efforts—even culturing the mold in patients' bedpans and re-purifying the drug from patients' urine—but there was no way they could keep up with demand.

The turning point came in 1941, when Oxford scientists approached the U.S. government and asked for help in growing penicillin on a large scale. The method they devised took advantage of something the United States had in abundance: corn. Using a by-product of large-scale corn processing as a culture medium in which to grow the fungus, the scientists were able to produce penicillin in much greater quantities.

At first, all the penicillin harvested from U.S. production plants came from Fleming's original strain of *Penicillium notatum*. But researchers continued to look for more potent strains to improve yields. In 1943, they got lucky: researcher Mary Hunt discovered one such strain growing on a ripe cantaloupe in a Peoria, Illinois, supermarket. This new strain, called *Penicillium chrysogenum*, produced more than 200 times the amount of penicillin as the origi-

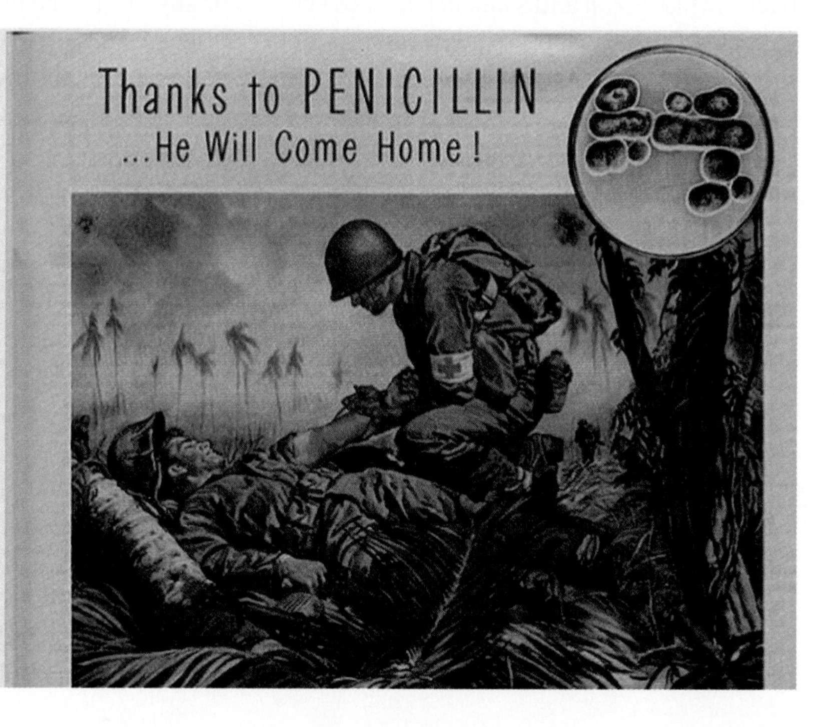
Thanks to PENICILLIN
...He Will Come Home!

"For the first time in human history, most people felt that infectious disease was ceasing to be a threat."

nal strain. With it, production of the drug soared. By the time the Allies invaded France on June 6, 1944—D-day—they had enough penicillin to treat every soldier that needed it. By the following year, penicillin was widely available to the general public.

"Penicillin seemed to justify a carefree attitude to infection," says medical historian Robert Bud, principal curator of the Science Museum in London. "In western countries, for the first time in human history, most people felt that infectious disease was ceasing to be a threat, and sexually infectious disease had already been conquered. For many it seemed cure would be easier than prevention."

Yet, as effective as penicillin was, it was effective only against certain types of bacteria; against others, it was powerless.

Stockpiling the Antibiotic Arsenal

As Fleming knew, most of the bacterial world falls into one of two categories, **Gram-positive** and **Gram-negative;** these names reflect the way bacterial cell walls trap a dye known as Gram stain (after its discoverer, the Danish scientist Hans Christian Gram). Fleming found that while penicillin easily killed Gram-positive bacteria like *Staphylococcus* and *Streptococcus*, it had little effect on Gram-negative bacteria like *E. coli* and *Salmonella*, whose cell walls have an extra layer of lipids surrounding them. This extra lipid layer prevents penicillin from reaching the peptidoglycan beneath it.

The discovery that penicillin was effective only on Gram-positive bacteria led other researchers in the 1940s to look for other antibiotics that could kill Gram-negative bacteria. The first such broad-spectrum antibiotic was streptomycin, discovered in 1943 by Albert Schatz and Selman Waksman at Rutgers University. In addition to killing Gram-negative bacteria, streptomycin was the first effective treatment for the deadly bacterial disease tuberculosis.

Like other antibiotics in the class known as aminoglycosides, streptomycin works by interfering with protein synthesis on bacterial ribosomes. Ribosomes are the molecular machines that assemble a cell's proteins. While both eukaryotic and prokaryotic cells have ribosomes, their ribosomes are different sizes and have different structures. Because streptomycin targets features specific to bacterial ribosomes, it doesn't harm the human who is taking it (Infographic 3.6).

Antibiotics can also target bacteria by inhibiting a bacterium's ability to make a critical vitamin or to copy its DNA before dividing. When this happens, the bacterium dies instead of reproducing.

"Penicillin seemed to justify a carefree attitude to infection. . . . For many it seemed cure would be easier than prevention." —Robert Bud

Why can broad-spectrum antibiotics, like streptomycin or gentamicin, kill Gram-negative bacteria when penicillin cannot? It's because these drugs have a chemical structure that allows them to pass more easily through the outer lipid layer of the Gram-negative bacterial cell wall. Although natural penicillin cannot pass this layer, many modern synthetic varieties of penicillin, known collectively as beta-lactams, can.

Crossing Enemy Lines

For any drug to be effective, it has to reach its designated target. In the case of many antibiotics, that means getting inside the cell to do their work. How do antibiotics penetrate a cell's outer defenses?

In all cells, the cell membrane acts as a barrier to transport, allowing only certain substances to pass through it.

With its densely packed collection of hydrophobic phospholipid tails, the cell membrane prevents many large molecules, like glucose, and hydrophilic substances, like sodium ions, from wandering across the cell membrane. In fact, the only things that do cross the membrane easily are small, uncharged molecules like oxygen (O_2), which can travel relatively easily across by a process known as **simple diffusion.**

Simple diffusion takes advantage of the natural tendency of dissolved substances to spread

GRAM-POSITIVE
Refers to bacteria with a cell wall that includes a thick layer of peptidoglycan that retains the Gram stain.

GRAM-NEGATIVE
Refers to bacteria with a cell wall that includes a thin layer of peptidoglycan surrounded by an outer lipid membrane that does not retain the Gram stain.

SIMPLE DIFFUSION
The movement of small, hydrophobic molecules across a membrane from an area of higher concentration to an area of lower concentration; simple diffusion does not require energy.

INFOGRAPHIC 3.6

Some Antibiotics Inhibit Prokaryotic Ribosomes

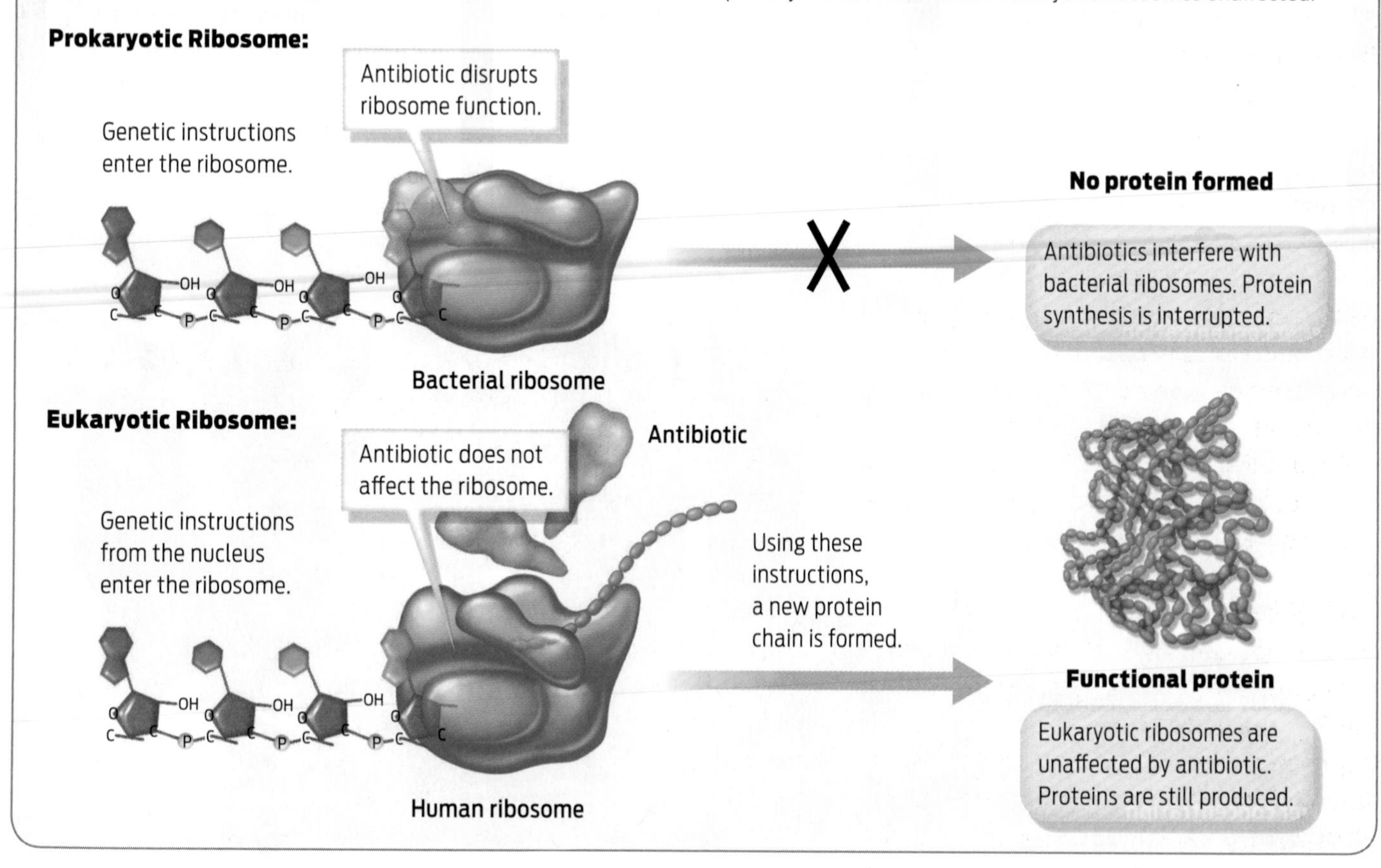

Ribosomes are responsible for the synthesis of proteins in both prokaryotic and eukaryotic cells, but their structure is slightly different in the two types of cells. Antibiotics that interfere with prokaryotic ribosomes leave eukaryotic ribosomes unaffected.
Prokaryotic Ribosome:
Antibiotic disrupts ribosome function.
Genetic instructions enter the ribosome.
No protein formed
Antibiotics interfere with bacterial ribosomes. Protein synthesis is interrupted.
Bacterial ribosome
Eukaryotic Ribosome:
Antibiotic
Antibiotic does not affect the ribosome.
Genetic instructions from the nucleus enter the ribosome.
Using these instructions, a new protein chain is formed.
Functional protein
Eukaryotic ribosomes are unaffected by antibiotic. Proteins are still produced.
Human ribosome

out from an area of higher concentration to one of lower concentration–think of food coloring diffusing in a glass of water. Because the substance is moving from the side of the membrane with a higher concentration to the side with a lower concentration, no energy is required to move substances across the membrane. Take oxygen, for example. The concentration of oxygen molecules, which are small and uncharged, is often higher outside the cell and lower inside. This concentration difference, or gradient, allows oxygen to diffuse easily into the cell–a good thing, because the cell needs oxygen in order to survive.

But the cell also needs some large or hydrophilic molecules in order to survive–one of them is glucose, the cell's energy source. To move such molecules across the membrane the cell makes use of **transport proteins.** Transport proteins sit in the membrane bilayer with one of their ends outside the cell and the other inside. By acting as a kind of channel, carrier, or pump, transport proteins provide a passageway for those large or hydrophilic molecules to cross the membrane. They are also very specific: a protein that transports glucose will not transport calcium ions, for example. The cells of your body contain hundreds of types of transport proteins.

TRANSPORT PROTEINS
Proteins involved in the movement of molecules across the cell membrane.

FACILITATED DIFFUSION
The process by which large or hydrophilic solutes move across a membrane from an area of higher concentration to an area of lower concentration with the help of transport proteins.

Some antibiotics are small hydrophobic molecules that can cross the cell membrane directly by simple diffusion–tetracycline, for example. Others, including penicillin and streptomycin, require the assistance of transport proteins. Transport proteins can move substances either up or down a concentration gradient. When a substance uses a transport protein to move down a concentration gradient, the process is called **facilitated diffusion.** Like simple diffusion, facilitated diffusion requires no energy

In Europe, however, the landscape is quite different. Since 2007, food manufacturers that want to market products with health claims must apply to the European Food Safety Authority for approval before introducing the product to the market. Scientific evidence documenting the claim must be submitted with the application.

Tighter regulation in Europe is one reason that large food manufacturers such as Nestlé have been pumping more money into research to support product claims. And because functional foods are by their very nature intended to provide health benefits beyond basic nutrition, companies that produce them are more eager to get government approval to be able to market them with specific health claims.

Digestion, Enzymes, and Metabolism

Nestlé's Boost Glucose Control drink, for example, contains fiber, a type of carbohydrate known to slow down digestion. Most functional foods, in fact, are built on the way our bodies naturally digest and use specific types of food molecules. Digestion is the process of breaking down the huge food molecules into smaller pieces so that our bodies can use them. It is a series of chemical reactions that take place throughout the digestive system. In our mouths, stomachs, and small intestines, chemical reactions break the bonds that hold food molecules together.

For most of us, carbohydrates constitute the largest portion of our diet. Bread, pasta, and rice are rich in carbohydrates. When we eat a plate of pasta, for example, our digestive system breaks down the carbohydrates into smaller sugar molecules. These sugar molecules are absorbed from the small intestine into the bloodstream, in which they are transported to cells in the rest of the body for use in building cell structures and carrying out cell functions.

To break down any macromolecule into its constituent parts, however, the chemical reactions that take place during digestion require the help of chemical facilitators called **enzymes,** which are specialized proteins that speed up the rate of a chemical reaction.

Enzymes work by lowering the amount of energy required to nudge a chemical reaction into motion. Enzymes substantially reduce **activation energy,** and so the reaction occurs more easily.

Enzymes bind to molecules called **substrates.** The part of the enzyme that binds to substrates is called its **active site.** Each enzyme is made so that its active site fits only one particular substrate molecule or group of highly similar substrate molecules. For example, each type of digestive enzyme in our mouth, stomach, and intestine binds to one specific type of organic molecule present in food.

Because of enzymes our bodies are able to break apart the chemical bonds in food molecules to release their component building blocks. Reactions that break molecules into smaller units–those that digest food, for instance–are called **catabolic reactions.** Reactions that build organic molecules from their simpler building blocks–such as those that build new muscle–are known as **anabolic reactions.** All the chemical reactions that take place inside our bodies are collectively called **metabolism** (Infographic 4.3).

Digesting Carbs

In some diseases, normal metabolism goes awry. People with diabetes, for example, have trouble controlling sugar levels in their blood. Healthy people have ways of regulating sugar, the end product of carbohydrate digestion. Cells in the pancreas, a small organ located near the stomach, secrete a hormone called **insulin** in response to high blood-sugar levels. Insulin is a protein–it is a chain of amino acids produced by the pancreas. When insulin binds to most cells in the body, it enables them to absorb sugar from the blood. People with type 1 diabetes, however, cannot make insulin; in people with type 2 diabetes, the receptors on their cells respond poorly to insulin, triggering the pancreas to release more insulin, and eventually the pancreas can "burn out," leading to insulin deficiency and elevated blood sugar. Over time, high blood sugar can lead to serious impairments,

Nestlé's Boost Glucose Control drink contains fiber, a type of carbohydrate known to slow down digestion.

ENZYME
A protein that speeds up the rate of a chemical reaction.

ACTIVATION ENERGY
The energy required for a chemical reaction to proceed. Enzymes accelerate reactions by reducing their activation energy.

SUBSTRATE
A compound or molecule that an enzyme binds to and on which it acts.

ACTIVE SITE
The part of the enzyme that binds to substrates.

CATABOLIC REACTION
Any chemical reaction that breaks down complex molecules into simpler molecules.

INFOGRAPHIC 4.3

Enzymes Facilitate Chemical Reactions

Cells require enzymes to break down and build up macromolecules. Enzymes are proteins that speed up chemical reactions by reducing the amount of activation energy required to set them in motion.

a. Catabolic Reaction: Bonds are broken

Substrates

Enzyme

Active site

Bond linking substrates

1. Substrates bind to the active site of a specific enzyme.
2. The active site of the enzyme changes shape, stressing the bond, reducing the activation energy, thereby making it easier to break.
3. The substrate bond breaks and the resulting products are released from the enzyme. The enzyme is ready to be used again.

b. Anabolic Reaction: Bonds are created

Substrates

Active site

Enzyme

Bond linking substrate subunits

1. Substrate binds to the active site of a specific enzyme.
2. The active site of the enzyme changes shape, which orients substrates so they can bind. The enzyme also reduces the amount of activation energy required, which facilitates bonding.
3. The enzyme releases the resulting products. The enzyme is ready to be used again.

ANABOLIC REACTION
Any chemical reaction that combines simple molecules to build more-complex molecules.

METABOLISM
All biochemical reactions occurring in an organism, including reactions that break down food molecules and reactions that build new cell structures.

INSULIN
A hormone secreted by the pancreas that regulates blood sugar.

including cardiovascular disease, kidney failure, and blindness.

To combat high blood sugar, many diabetics inject themselves with insulin. But as insulin causes sugar in the blood to rush into cells, blood sugar can plummet quickly. Low blood sugar is equally dangerous: it can cause sweating, shakiness, hunger, dizziness, and nausea. To stave off these highs and lows, diabetics are advised to regulate the amount of sugar in their diet.

Because the body breaks down most carbohydrates into sugars, carbohydrates present the most trouble for diabetics–patients must keep track of the amount of carbs in their diets. Not all carbohydrates are the same, however, nor do all types cause spikes in blood sugar.

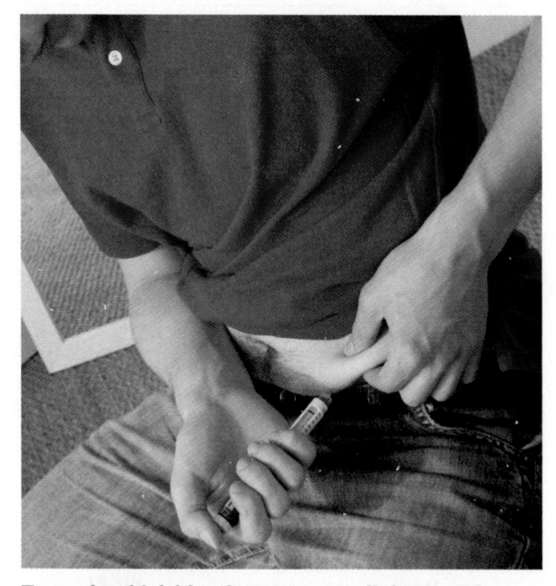

To combat high blood sugar, many diabetics inject themselves with insulin.

Most of the carbohydrates we eat are **complex carbohydrates**–large molecules with branch-like extensions found in plant and meat products. Because they are made of many smaller sugar molecules bound together, complex carbohydrates are also called **polysaccharides. Starch** is a complex carbohydrate found in plant products such as rice and potatoes; **glycogen** is a complex carbohydrate found in chicken and steak.

Enzymes in our digestive tract break complex carbohydrates into their component subunits. The smallest carbohydrate subunit is called a **simple sugar**, or **monosaccharide**. The most common simple sugar released from food is glucose, one of two sugars found in table sugar (**Infographic 4.4**).

But there are some carbohydrates that humans can't digest; one of them is fiber. **Fiber** is a type of indigestible complex carbohydrate found in fruits and vegetables. Humans lack the necessary enzyme to break down fiber, so most fiber passes undigested through the digestive system and out in feces. Although not technically a nutrient because it is not absorbed by the body, fiber is an important part of a healthful diet–it can lower cholesterol and decrease our risk of various cancers. It also

COMPLEX CARBOHYDRATE (POLYSACCHARIDE)
A carbohydrate made of many simple sugars linked together, that is, a polymer of monosaccharides; examples are starch and glycogen.

STARCH
A complex plant carbohydrate made of linked chains of glucose molecules; a source of stored energy.

INFOGRAPHIC 4.4

Complex Carbohydrates Are Broken Down into Simple Sugars

Meats, vegetables, fruits, and grains are rich sources of complex carbohydrates, also called polysaccharides. Digestion breaks down complex carbohydrates into their monosaccharide subunits, also called simple sugars. Not all complex carbohydrates are digestible by humans.

Complex Carbohydrates
Simple Sugars
a. Glycogen is the energy-storing carbohydrate in animal cells.
Glycogen
Human enzymes can easily break the bonds between each glucose molecule.
CH_2OH CH_2OH CH_2OH
O O
OH OH OH
Human enzyme
Glucose is absorbed into the bloodstream and delivered to cells.
b. Starch is the energy-storing carbohydrate in plant cells.
Starch
Human enzymes can easily break the bonds between glucose molecules.
CH_2OH CH_2OH CH_2OH
O O
OH OH OH
Human enzyme
Glucose is absorbed into the bloodstream and delivered to cells.
c. Fibers act as structural carbohydrates in plants.
Fiber
Humans do not have an enzyme that can break the bonds between the glucoses in fiber.
OH CH_2OH OH
O O
CH_2OH OH CH_2OH
No human enzyme has the shape for binding fiber.
Fiber passes undigested through the digestive system and is eliminated from the body in feces.

helps regulate digestion. One of the primary therapeutic ingredients in Boost Glucose Control is fiber.

Because diabetics have trouble balancing their blood-sugar levels, doctors recommend that they avoid foods made of simple sugars–candy, for example–and instead eat complex carbohydrates such as potatoes and oats (in addition to other foods required for a healthful diet). That's because complex carbohydrates are digested more slowly than simple sugars, which means that blood-sugar levels rise more slowly as well. But as anyone who has ever dieted knows, sticking to a restrictive diet is hard. And it's hard even for diabetics. What's more, diabetics suffer the same consequences of dieting that normal people do: deprivation can lead to bingeing on foods that they should avoid.

"Some food additives, like dietary fiber, do have solid evidence behind them." –Jeya Henry

To stave off dangerous fluctuations in blood sugar, Nestlé's Boost Glucose Control drink is an alternative to other snacks. It is essentially a mix of protein, digestible complex carbohydrates, and fiber. Combining the two types of carbohydrates makes the digestible carbohydrates less accessible to enzymes, thus slowing the release of sugar even more. Protein also takes longer to digest than does carbohydrate. Slowing down digestion reduces the risk of surges in blood sugar that are dangerous for diabetics (Infographic 4.5).

Does it work? "Some food additives, like dietary fiber, do have solid evidence behind them," says Jeya Henry, professor of human nutrition at Oxford Brookes University, in Oxford, England, who also heads the university's Functional Foods Centre. In addition to helping regulate blood sugar, fiber appears to cut the risk of many other diseases, too, including heart disease and diseases of the gastrointestinal tract.

In 2009, Henry co-authored a paper that showed that beta-glucan–a type of fiber normally found in oats and barley–when added in certain doses to a type of flat bread significantly slowed down rises in blood-sugar levels relative to blood-sugar levels in people who ate flatbread without beta-glucan. This is important, he says, because some studies have shown that processing or cooking foods with beta-glucan can degrade it. The study showed that mild cooking does not necessarily affect beta-glucan's effectiveness.

Of course, anyone can always eat foods that naturally contain high amounts of fiber–whole-grain breads and lentils, for example–but such foods aren't palatable to everyone, says Henry. In fact, the average American eats only about 15 grams of fiber a day; the recommended amount is 20 to 40 grams, depending on sex, age, and other factors.

Nestlé has shown in one clinical study that their Boost Glucose Control drink does not cause spikes in blood sugar (whereas most snack foods with a high simple-sugar content cause abnormally high blood-sugar levels in diabetics). However, the company is primarily relying on evidence already published in the scientific literature that has shown that each individual active ingredient–the specific types of protein, carbohydrates, and fiber–in its drink can help diabetics control swings in blood-sugar levels.

Fighting Chronic Disease

Developing foods with health benefits isn't Nestlé's only goal. Nestlé is, after all, a company, and companies are in the business of earning money. For Nestlé, foods marketed to diabetics presented a lucrative opportunity–according to the World Health Organization (WHO), about 171 million people around the world suffer from diabetes. And their numbers are likely to more than double by 2030.

Many chronic diseases, however, can be either prevented or slowed down with diet and lifestyle changes–which is why they make a good target for food manufacturers. Take osteoporosis, for example. In America alone about 10 million people already have osteoporosis and

GLYCOGEN
A complex animal carbohydrate made of linked chains of glucose molecules; a source of stored energy.

SIMPLE SUGAR (MONOSACCHARIDE)
A carbohydrate made up of a single sugar subunit; an example is glucose.

FIBER
A complex plant carbohydrate that is not digestible by humans.

INFOGRAPHIC 5.8

Photosynthesis: A Closer Look

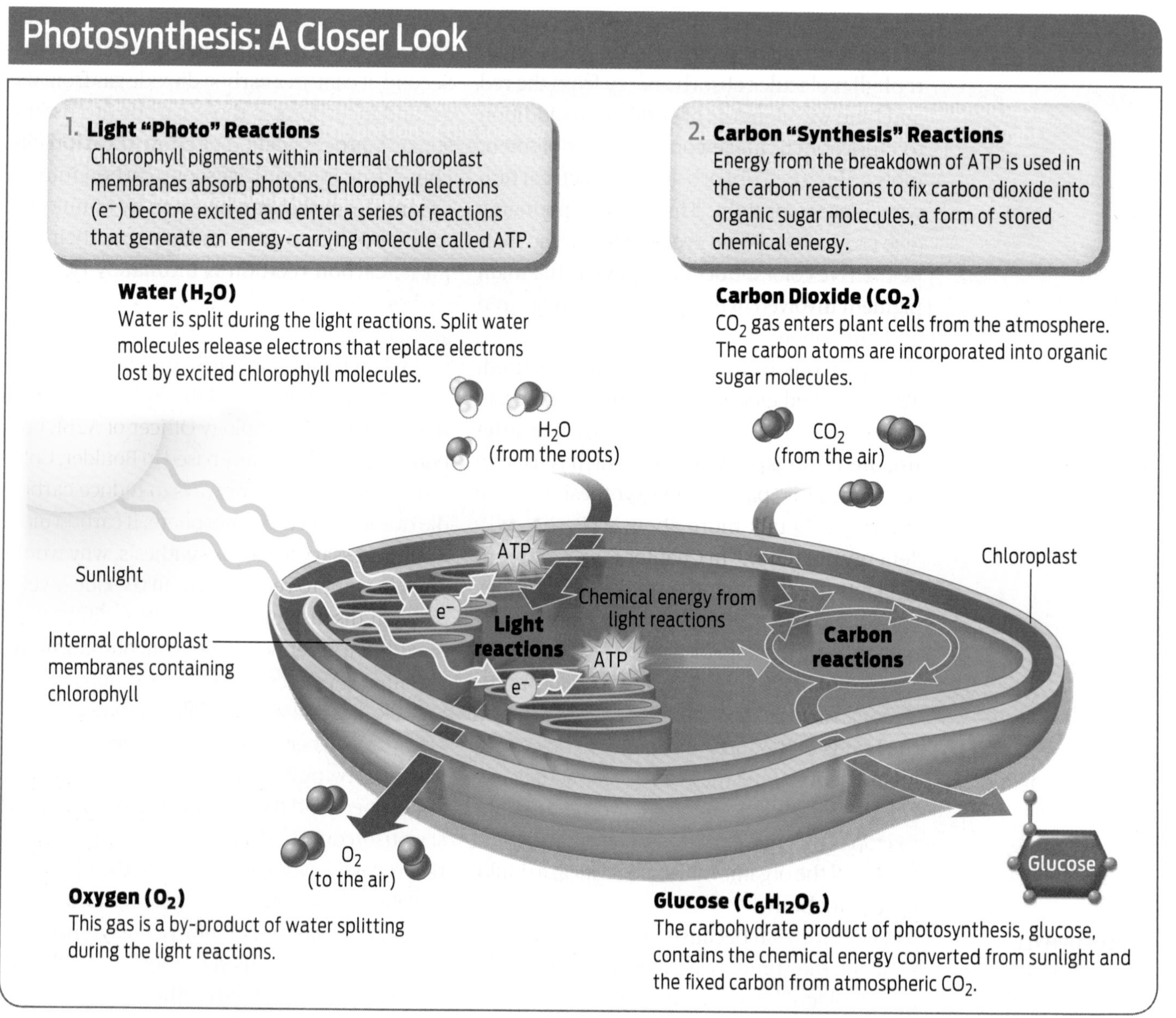

microorganisms to regenerate after being destroyed–and in the meantime, the damaged soil is unable to remove carbon dioxide from the atmosphere.

1. Light "Photo" Reactions
Chlorophyll pigments within internal chloroplast membranes absorb photons. Chlorophyll electrons (e⁻) become excited and enter a series of reactions that generate an energy-carrying molecule called ATP.
2. Carbon "Synthesis" Reactions
Energy from the breakdown of ATP is used in the carbon reactions to fix carbon dioxide into organic sugar molecules, a form of stored chemical energy.
Water (H_2O)
Water is split during the light reactions. Split water molecules release electrons that replace electrons lost by excited chlorophyll molecules.
Carbon Dioxide (CO_2)
CO_2 gas enters plant cells from the atmosphere. The carbon atoms are incorporated into organic sugar molecules.
H_2O (from the roots)
CO_2 (from the air)
Sunlight
Chloroplast
ATP
e⁻
Light reactions
Chemical energy from light reactions
Carbon reactions
Internal chloroplast membranes containing chlorophyll
O_2 (to the air)
Glucose
Oxygen (O_2)
This gas is a by-product of water splitting during the light reactions.
Glucose ($C_6H_{12}O_6$)
The carbohydrate product of photosynthesis, glucose, contains the chemical energy converted from sunlight and the fixed carbon from atmospheric CO_2.

Sears, however, has a solution. His new company takes small samples of microorganisms from healthy soil, grows them in a contained facility, and then transplants them to damaged soil, where they spread out and thrive. He estimates that if 1 billion hectares of land were restored in this way, one-seventh of the world's greenhouse gas problem would be solved because of the vast amounts of carbon dioxide that would be pulled out of the atmosphere by the photosynthetic cyanobacteria in the regenerated soil.

> **"Algae truly are the foundations of our entire planet."**
> **–Jim Sears**

When you think about it, it's amazing that organisms like algae and cyanobacteria that seem so simple could be so vital to life on earth. But they are. Not only did they provide the

Jim Sears examines a restored soil sample.

planet with the first breaths of atmospheric oxygen millions of years ago, but soon they could become the world's most important fuel source as well as a potential solution to climate change. All this from single-cell organisms that have just one major claim to fame: they can convert the energy of sunlight into energy-rich organic molecules. "Algae truly are the foundations of our entire planet," says Sears. ■

Summary

- All living organisms require energy to live and grow. The ultimate source of energy on earth is the sun.
- Photosynthesis is a series of chemical reactions that captures the energy of sunlight and converts it into chemical energy in the form of sugar and other energy-rich molecules. This energy is used by all living organisms to fuel cellular processes.
- Photosynthesis can be divided into two main parts: a "photo" part, during which the pigment chlorophyll captures light energy, and a "synthesis" part, during which captured energy is used to fix carbon dioxide into glucose.
- Photosynthetic organisms are known as autotrophs; they include plants, algae, and some bacteria. Animals do not photosynthesize; they are known as heterotrophs.
- Energy is neither created nor destroyed, but is converted from one form into another, a principle known as the conservation of energy.
- Kinetic energy is the energy of motion and includes heat and light energy. Potential energy is stored energy and includes chemical energy.
- Energy flows from the sun, is captured and transferred through living organisms, and then flows back into the environment as heat.
- Energy conversions are inefficient. Some energy is lost as heat with every conversion of energy.
- The energy-rich molecules produced by some photosynthetic algae include oils that can be used as an energy source to power automobiles and aircraft. These alternative fuels show great promise in terms of sustainable consumption.

Chapter 5 Test Your Knowledge

PHOTOSYNTHESIS

Photosynthesis is the process by which the energy of sunlight is captured and stored as chemical energy.

HINT See Infographics 5.2, 5.5–5.8.

➲ KNOW IT

1. The energy of sunlight exists in the form of
- **a.** glucose.
- **b.** photons.
- **c.** gamma rays.
- **d.** ions.
- **e.** particles.

2. Which photon wavelength contains the greatest amount of energy?
- **a.** violet
- **b.** red
- **c.** green
- **d.** yellow
- **e.** blue

3. Why does algae appear green?

4. Glucose is a product of photosynthesis. Where do the carbons in glucose come from?
- **a.** starch
- **b.** cow manure
- **c.** carbon dioxide
- **d.** water
- **e.** soil

5. Compare and contrast the ways photosynthetic algae and animals obtain energy.

6. Mark each of the following as an INPUT (I) or an OUTPUT (O) of photosynthesis.

Oxygen _____
Carbon dioxide_____
Photons _____
Glucose _____
Water _____

➲ USE IT

7. Global warming is linked to elevated atmospheric carbon dioxide levels. How might this affect photosynthesis? If global warming should cause ocean levels to rise, in turn causing forests to be immersed in water, how would photosynthesis be affected?

8. Why are energy-rich lipids from algae more "useful" as a fuel than energy-rich sugars and other carbohydrates produced by photosynthetic organisms like corn and wheat?

ENERGY FLOW

Energy is initially captured by autotrophs and flows through other organisms and machines. As energy flows, some of it is lost as heat.

HINT See Infographics 5.2, 5.3, and 5.4.

➲ KNOW IT

9. The energy in an energy bar is __________ energy. The energy of a cyclist pedaling is _____________ energy.
- **a.** light; chemical
- **b.** potential; chemical
- **c.** chemical; kinetic
- **d.** potential; potential
- **e.** kinetic; potential

10. Kinetic energy is best described as
- **a.** stored energy.
- **b.** light energy.
- **c.** the energy of movement.
- **d.** heat energy
- **e.** any of the above, depending on the situation.

➲ USE IT

11. If you wanted to get the most possible energy from photosynthetic algae, should you eat algae directly or feed algae to a cow, and then eat a burger made from that cow? Explain your answer.

ALGAE AND BIOFUELS

Algae that produce large amounts of lipids are being developed as new and sustainable fuel sources.

HINT See Infographics 5.1, 5.2, and 5.5.

➲ KNOW IT

12. Photosynthetic algae are
- **a.** eukaryotic autotrophs.
- **b.** prokaryotic autotrophs.
- **c.** eukaryotic heterotrophs.
- **d.** prokaryotic heterotrophs.

13. Which of the following is/are necessary for biofuel production by algae?

a. sunlight
b. sugar
c. CO_2
d. soil
e. all of the above
f. a & b
g. a & c

USE IT

14. Many types of algae can divert the sugars they make by photosynthesis into pathways to make biodiesel, a fuel that essentially consists of lipids. Biodiesel is a promising replacement for fossil fuels. Describe the energy transfers required to make biodiesel and explain why biodiesel might be a more promising fuel than lipids extracted from animals.

15. What do you think are some of the advantages and disadvantages of growing algae in enclosed tubes or bags compared to growing them in open vats? Be sure to explain your reasoning.

SCIENCE AND ETHICS

16. Many biofuels require arable land as part of their production process. Discuss competing needs for arable lands in the context of human needs for food and fuel, and how algae may alleviate this tension.

Chapter 6 Dietary Energy and Cellular Respiration

Supersize Me?

What You Will Be Learning

Supersize me?

Changing our culture of eating

For years Paul Rozin, a professor of psychology at the University of Pennsylvania, was baffled by this question: How are the French able to eat rich cheeses, butter-laden sauces, fatty meat, and still stay slimmer than Americans?

As of 2008, a whopping 72% of American men and 64% of American women were overweight or obese. Compare these figures to the corresponding ones in France: about 56% of men and 40% of women.

How are the French able to eat rich cheeses, butter-laden sauces, fatty meat, and still stay slimmer than Americans?

Americans aren't the only ones tipping the scales in greater numbers. People around the world are getting heavier. As of 2005, approximately 1.6 billion adults over the age of 15 were overweight and at least 400 million were obese, according to the World Health Organization (WHO). By 2020, WHO expects those numbers to double, a figure that will amount to almost half of today's global population. Our increasing girth alone wouldn't be a problem were it not for what comes with it: obesity has ushered in increased rates of heart disease, diabetes, and other related illnesses. In fact, by 2020 chronic illnesses resulting from obesity will likely kill more people than infectious diseases.

Why are people getting heavier? Biologists argue that humans are predisposed to gain weight. Throughout human evolution, famine was the norm, and people had to work hard to grow or hunt and gather their food. Our bodies have adapted by storing extra food as fat for times when food is scarce.

How heavy is too heavy? Even with a few extra pounds here and there, most people still fall within a healthy weight range. Only when our total body fat passes a certain point do the scales tip toward unhealthy. That point depends on a number of factors, including gender, body type, and frame size. To get a rough estimate, some health care professionals rely on the **body mass index (BMI).** The BMI estimates body fat from

BODY MASS INDEX (BMI)
An estimate of body fat based on height and weight.

OBESE
Having 20% more body fat than is recommended for one's height, as measured by a body mass index greater than 30.

the indirect measures of height and weight. People with a BMI between 25 and 29.9 are considered overweight; people with a BMI of 30 and above are considered **obese**. BMI can be misleading, however. Athletes and people with more muscle mass will sometimes register as overweight or obese when in fact they are perfectly healthy **(Infographic 6.1)**.

But sociologists peg another weight-gain culprit: eating behavior. Culture, they argue, is as much to blame for the obesity epidemic as biology. Some societies have remained relatively thin, they point out, despite similar biology. But when these societies adopt American eating habits, which include fast-food, snacks, and soda, they tend to put on pounds, too, which is why Rozin wanted to study the eating habits of the French.

Rozin has found that culture affects not only *how* we eat but also *how much*. In studies conducted in America and in France, Rozin found that portion sizes in America are often bigger than those in France. In other words, one reason that a greater percentage of Americans are heavier than the French is simply because in America, more food is spooned onto our plates. Eating more food than our bodies need means more food-energy stored as fat.

"Culture is underrated as a contributing factor to unhealthy eating," says Rozin. "The idea of what a proper meal is and your own habits are largely instituted by the culture in which you live."

Clearly there's little we can do about our biology. But culture is another matter. There is a growing movement in America to reign in what has been the cultural norm of unhealthful eating. For example, public health experts have been lobbying the government to pass legislation that would improve people's access to more-heathful foods–fresh fruits and vegetables, for example, that are nutritious and also low in Calories–to counteract our larger serving sizes. Many local governments have already banned restaurants from using an unhealthful type of fat to fry foods. And Congress has passed legislation that limits what types of food can be sold in schools so that children aren't filling themselves up with Calorie-dense food with little nutritional value. Many are pressing the U.S. government to do even more to change the way Americans eat (**Infographic 6.2**).

INFOGRAPHIC 6.1

Body Mass Index (BMI)

A BMI chart provides an indirect measure of body fat, based on the ratio of body height to weight. Because it does not take muscle mass or frame size into account, BMI is only an estimate of body fat. Some people may register as overweight even though they are a healthy weight.

Height in Feet and Inches	Weight in Pounds: 120	130	140	150	160	170	180	190	200	210	220	230	240	250
4'6"	29	31	34	36	39	41	43	46	48	51	53	56	58	60
4'8"	27	29	31	34	36	38	40	43	45	47	49	52	54	56
4'10"	25	27	29	31	34	36	38	40	42	44	46	48	50	52
5'0"	23	25	27	29	31	33	35	37	39	41	43	45	47	49
5'2"	22	24	26	27	29	31	33	35	37	38	40	42	44	45
5'4"	21	22	24	26	28	29	[illegible]	[illegible]	[illegible]	[illegible]	[illegible]	[illegible]	[illegible]	43
5'6"	19	21	23	24	[illegible]	[illegible]	[illegible]	[illegible]	[illegible]	[illegible]	[illegible]	[illegible]	[illegible]	[illegible]
5'8"	18	20	21	23	24	26	[illegible]	[illegible]	[illegible]	[illegible]	[illegible]	[illegible]	[illegible]	[illegible]
5'10"	17	19	20	22	23	24	26	27	29	30	31	33	35	36
6'0"	16	18	19	20	22	23	24	26	27	28	30	31	33	34
6'2"	15	17	18	19	21	22	23	24	26	27	28	30	31	32
6'4"	15	16	17	18	20	21	22	23	24	26	27	28	29	30
6'6"	14	15	16	17	19	20	21	22	23	24	25	27	28	29
6'8"	13	14	15	17	18	19	20	21	22	23	24	25	26	28

A person 5'6" tall weighing 150 lbs is within the healthy weight range.

Underweight | Healthy Weight | Overweight | Obese

"Unhealthy food choices have become the default food choice," says Kelly Brownell, director of the Rudd Center for Food Policy and Obesity at Yale University. "The question is what can we do about it?"

"Culture is underrated as a contributing factor to unhealthy eating." –Paul Rozin

What's in a Meal?

A few years ago, Rozin and colleagues at the Centre National de la Recherche Scientifique in France set out to see just how much more Americans eat than their French counterparts. They compared portion sizes at eleven restaurants in Philadelphia and in Paris. Their results, published in the journal *Psychological Science*, didn't surprise them. The average portion size in the Paris restaurants weighed 277 grams. By contrast, the average portion size in the Philadelphia restaurants weighed 346 grams–25% more! Even restaurant chains like McDonald's served smaller portions of certain foods: in

INFOGRAPHIC 6.2

Obesity Is Influenced by Biology and Culture

Biological History
Famine was common. Our bodies have evolved to hoard energy in the form of body fat to get them through times when food was scarce.

+

Cultural Influence
An abundance of high-fat, processed food is increasingly common in developed countries.

=

Modern Obesity
Today people consume many more Calories than during any other time in history because food is abundant. Our bodies store extra Calories as fat, as they have been evolutionarily programmed to do.

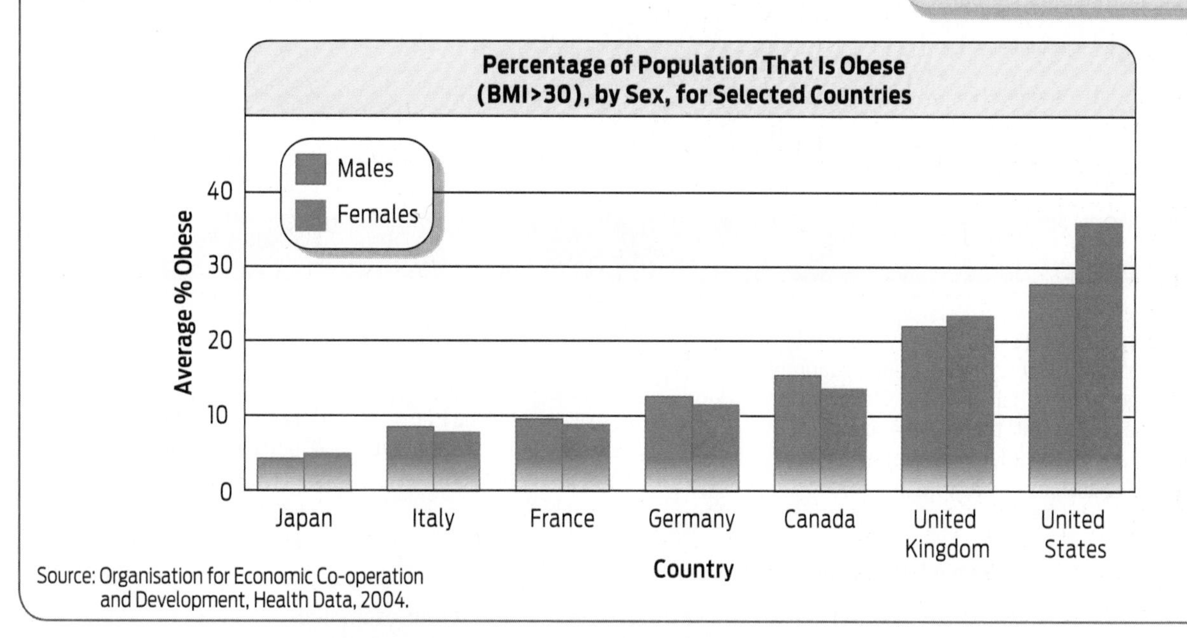

Source: Organisation for Economic Co-operation and Development, Health Data, 2004.

Paul Rozin, University of Pennsylvania psychology professor, is examining the differences between American and French eating culture.

Paris, medium fries weighed 90 grams, large fries 135 grams; in Philadelphia, medium fries weighed 155 grams and large fries 200 grams.

Once they had surveyed the restaurant scene, Rozin's team went further. They compared the sizes of packaged food in American and in French supermarkets, and they found the same trend: the portions of the majority of food items they tested–from ice cream to chewing gum to yogurt–were smaller in France. Even portion sizes for ingredients in the most commonly used French cookbook were smaller

For a number of the items McDonald's serves, portion sizes are larger in Philadelphia than in Paris.

than those in the celebrated American favorite, *Joy of Cooking* **(Infographic 6.3)**.

How does portion size contribute to weight gain? The answer lies in the energy content of the food we eat. We've already seen, in Chapter 4, that all food consists of mixtures of carbohydrates, fats, proteins, and nucleic acids, and that the relative proportion of each macronutrient varies in different types of food. Meat, for example, contains more protein per unit weight than do potatoes; potatoes have more carbohydrates than meat does. To nourish our bodies, we must eat a balanced diet that includes appropriate amounts of all macronutrients. But food is not only a source of nutrition; it is also a source of chemical energy that powers our activities. Food is fuel.

INFOGRAPHIC 6.3

Americans Eat Large Portions

Researchers compared portion sizes in restaurants in Philadelphia to those in Paris. In all but one restaurant, U.S. portions were larger at least half the time. The average portion size in Paris was 277 grams; the average size in Philadelphia was 346 grams. Philadelphians eat an average of 25% more food than Parisians at every meal.

For sampled menu items, U.S. restaurants consistently serve larger portion sizes.

Numbers >1 reflect larger portion sizes in the U.S., compared to France

Table 1. Restaurant portion sizes

Restaurant in Paris	Restaurant in Philadelphia	No. of items sampled/ No. larger in U.S.	Mean size ratio (U.S./France)
McDonald's	McDonald's	6/4	1.28
Hard Rock Cafe	Hard Rock Cafe	2/0	0.92
Pizza Hut	Pizza Hut	2/2	1.32
Häagen-Dazs	Häagen-Dazs	2/2	1.42
French: local bistro	French: local bistro	1/1	1.17
Quick	Burger King	5/4	1.36
Local Chinese	Local Chinese	6/4	1.72
Italian: Bistro Romain	Olive Garden	3/2	1.02
Crêpes: local	Crêpes: local	4/2	1.04
Ice cream: local	Ice cream: local	2/2	1.24
Pizza: local	Pizza: local	2/2	1.32

ROZIN, P ET AL. PSYCHOLOGICAL SCIENCE. 203, 14:450–454.

INFOGRAPHIC 6.4

Food Powers Cellular Work

Cells use these subunits either as building blocks to make new macromolecules (Chapter 4) or as energy to fuel cellular work. Fats are the most energy-dense molecules.
Carbohydrate
Simple sugars
Energy 4 Calories/gram
Protein
Amino acids
Energy 4 Calories/gram
Fat
Fatty acids and glycerol
Energy 9 Calories/gram
Nucleic acid
Nucleotides
Not a significant source of energy for cells
The body breaks down ingested food into subunits, which then pass into the bloodstream, which delivers them to the body's cells.

calorie
The amount of energy required to raise the temperature of 1 gram of water by 1° Celsius.

CALORIE
1,000 calories or 1 kilocalorie (kcal); the capital "C" in Calorie indicates "kilocalorie." The Calorie is the common unit of energy used in food nutrition labels.

Scientists measure energy in units called calories. A **calorie** (in lower case) is the amount of energy required to raise the temperature of 1 gram of water by 1° Celsius. In essence, a calorie is a measurement of energy—the capacity to perform a certain amount of work. On most food labels, the amount of energy stored is listed in kilocalories, which are also referred to as kcals or as Calories (the capital "C" indicates that kilocalories, not calories, are meant). One **Calorie** is equal to 1,000 calories, or 1 kcal.

Of all the organic molecules, fats are the most energy dense: each gram of fat stores approximately 9 Calories in its chemical bonds. Proteins and carbohydrates are about half as energy dense: each gram stores about 4 Calories. Clearly, a 200-gram serving of fatty bacon contains many more Calories than does a 200-gram serving of asparagus **(Infographic 6.4)**.

All our activities—everything from thinking and digesting to sleeping and running—require energy. So all bodies expend some Calories each day just to stay alive. A person's daily energy needs largely depend on gender, age, body type, and activity levels. A sedentary college-age average-size male, for example, would need to ingest anywhere between 2,200 and 2,400 Calories per day to power his activities and maintain his weight, whereas a football player would need more than 3,200 Calories a day to power and maintain his. Exercise or other physical activities require additional energy beyond the basic life-sustaining energy needs of the body. Consequently, athletes, or those who exercise a great deal, generally need to eat more to fuel their activities than do their less-active peers.

Exactly how much more should an athlete eat? Consider the college football player, who must consume 800 to 1,000 Calories more than his sedentary roommate. An average cheeseburger contains anywhere between 400 and 600 Calories depending on its size, so the football player would need about two extra cheeseburgers per day. That's probably less than you thought. Now suppose that same athlete ate a cheeseburger off-season. It would take 1.5 hours of slow swimming, or 2.5 hours of walking, or 3 hours of cycling at 5.5 miles per hour, to use up that extra energy **(Table 6.1)**.

Surprisingly, not everyone burns energy at the same rate. There are people who seem to be able to eat to their heart's content and hardly gain an ounce. And there are those who seem to gain weight just by looking at food. Genetics plays a large role in how much food each one of us actually needs, but there are other factors, too–gender, for one. Men, because their bodies naturally produce more muscle-building hormones than do women's, generally have more muscle mass and therefore need to eat more than women do. Since muscle cells burn more calories than do fat cells, the ratio of muscle mass compared to fat content in our bodies is another factor.

Putting on Pounds

Our bodies are fairly efficient at extracting energy from food, but we humans eat not only for sustenance but also for pleasure–which is where problems can arise. Many of us eat more food than our bodies need. We also have a natural preference for fatty and sugary foods because such foods are energy dense. For our ancestors, it was likely important to load up on those foods to store energy for times when food was scarce. Today, this ancient taste preference has become a vice that snack food companies have become very good at exploiting. For the large majority of us, when we eat Calories beyond what our bodies require, the extra energy is stored in one of two places: as **glycogen** in muscle and liver cells, or as **triglycerides** in fat cells.

Glycogen is the energy-storing carbohydrate found in animal cells. You can think of glycogen as a short-term storage system. When we require

TABLE 6.1

Calories In, Calories Out

CALORIES IN SELECT FOODS

FOOD	CALORIES
8 oz. unsweetened green tea	2
1 large slice whole wheat bread	79
½ cup cooked white rice	102
12 oz. nonfat milk	120
12 oz. cola	140
1 glazed doughnut	200
1 slice thick-crust cheese pizza	256
1 Starbucks grande mocha frappucino with whipped cream	380
1 McDonald's Big Mac	540
1 Burger King Whopper	670

CALORIES BURNED DURING SELECT ACTIVITIES*

ACTIVITY	CALORIES/HOUR
Sleeping	55
Sitting	85
Standing	100
Office work	140
Golf (walking)	240
Gardening (planting)	250
Walking (3 mph)	280+
Tennis	350+
Biking (moderate)	450+
Jogging (5 mph)	500+
Swimming (active)	500+
Hiking	500+
Power walking	600+
Cycling (stationary)	650
Squash	650+
Running	700+

*Approximate number of Calories burned per hour by a 150-pound woman.

INFOGRAPHIC 6.5

Glycogen and Fat Store Excess Calories

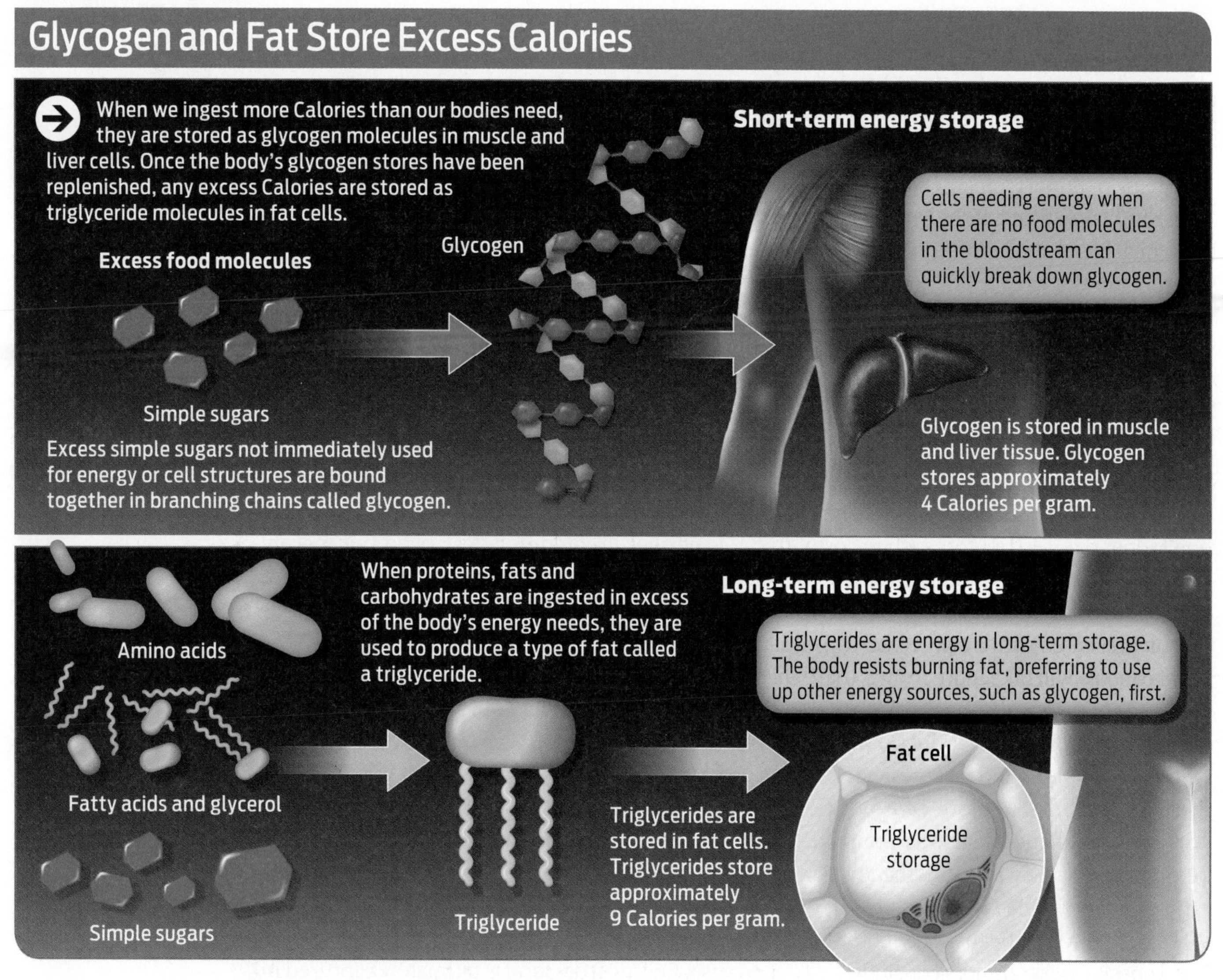

When we ingest more Calories than our bodies need, they are stored as glycogen molecules in muscle and liver cells. Once the body's glycogen stores have been replenished, any excess Calories are stored as triglyceride molecules in fat cells.
Excess food molecules
Glycogen
Short-term energy storage
Cells needing energy when there are no food molecules in the bloodstream can quickly break down glycogen.
Simple sugars
Excess simple sugars not immediately used for energy or cell structures are bound together in branching chains called glycogen.
Glycogen is stored in muscle and liver tissue. Glycogen stores approximately 4 Calories per gram.
Amino acids
When proteins, fats and carbohydrates are ingested in excess of the body's energy needs, they are used to produce a type of fat called a triglyceride.
Long-term energy storage
Triglycerides are energy in long-term storage. The body resists burning fat, preferring to use up other energy sources, such as glycogen, first.
Fat cell
Fatty acids and glycerol
Triglycerides are stored in fat cells. Triglycerides store approximately 9 Calories per gram.
Triglyceride storage
Simple sugars
Triglyceride

short bursts of energy—as in a sprint, for example—the body breaks down glycogen to obtain energy. However, because a gram of glycogen stores only half as many Calories as a gram of fat (about 4 Calories per gram versus 9), our bodies would have to carry around twice as much glycogen to store the same amount of Calories. So our bodies store most excess Calories as fat, which actually allows us to carry around less weight overall. The downside, however, is that it takes sustained activity to burn fat. The body burns fats only after it has already used up food molecules in the bloodstream and in stored glycogen.

For our ancestors who lived during times of frequent famine, this system of storing Calories as fat would have come in handy. Their bodies could burn fat for energy to carry them through times of food scarcity. Today, people in most of the developed world have plenty of food. But they are largely sedentary and eat more Calories than they need—which is why they've started to pack on, and keep on, the pounds **(Infographic 6.5)**.

Because each type of energy-rich organic molecule that we ingest—whether protein, carbohydrate, or fat—stores a different amount of energy, it's not only how much we eat but also what we eat that contributes to weight gain. We are more likely to gain weight from a portion of ice cream than an equivalent portion of broccoli, for example, because ice cream contains more fat—and therefore more Calories—than broccoli.

GLYCOGEN
A complex animal carbohydrate, made up of linked chains of glucose molecules, that stores energy for short-term use.

TRIGLYCERIDE
A type of lipid found in fat cells that stores excess energy for long-term use.

Some scientists have used this fact to argue that there must be some factor other than small portion size that explains why the French have remained relatively thin. The French may eat less, but they are also world renowned for their penchant for buttery, creamy sauces, dense desserts, and fatty meats. If the French load up on fat, which has more Calories, how do they manage to keep the weight off? Research such as Rozin's has shown that even though the French may eat more fat-laden or fat-heavy foods than Americans, it still has to do with portion size: they still consume fewer total Calories.

How do the French manage to eat only small portions? In France and many other European countries, small is the cultural norm. The French don't super size. Distributors such as Costco that sell bulk items don't exist in France, at least not yet. Research by several groups suggest that a person presented with a bigger package of, say, M&M candies will take more from it than when presented with a smaller package. A 2007 study led by Jennifer Fischer at Baylor College of Medicine, for example, found that preschool-age children consumed 33% more energy when the portion size of the meal was doubled. This behavior combined with meals made with a high ratio of energy-dense ingredients–oil, butter, and sugar–is a significant contributor to childhood obesity, these authors concluded.

Or it may have to do with the *way* the French eat; not only do they eat smaller portions at each meal, they don't snack, they don't run for second helpings, and they don't skip meals. Mireille Guiliano in her book *French Women Don't Get Fat* described how she gained 20 pounds during her 5-month stay in America. She snacked, she drank a lot of soda, and she ate standing up, she wrote. She found that she had forgotten how to enjoy the taste of food she was used to in France and compensated by eating larger portions. Part of her diet plan when she returned to France, she wrote, was quitting in-between-meals snacking and reacquainting herself with the French culture of eating–part of which involves eating only a few bites of any dish, just enough to satisfy the taste buds, and then pushing the plate away.

The French also spend more time eating. In his study, Rozin compared the average time people spent eating at McDonald's in Paris and in Philadelphia. In Paris, the average time of the meal was 22 minutes; in Philadelphia, only 14. While scientists don't know for sure how longer meals help people eat less, they speculate that taking it slow may help people enjoy their food more and recognize when they are full.

However it is that the French are able to curb their appetites, one thing is clear: they weigh less because they either eat less or are more active and burn more energy. The only way to gain weight is by taking in more Calories than we expend. In other words, our waistlines obey the principle of conservation of energy: energy is neither created nor destroyed but merely converted from one form into another. If more food energy is taken in than is used to power cellular reactions and physical movement, the excess (minus what is released as heat with every energy conversion) is stored as fat.

Extracting Energy from Food

Getting energy from food seems simple enough: we eat food and we have energy. To provide us with fuel, food goes through a series of complex biochemical reactions that convert the chemical energy stored in food into a form of fuel we can use. Energy from food is ultimately captured in

Research shows that large portion sizes are behind weight gain.

INFOGRAPHIC 6.6

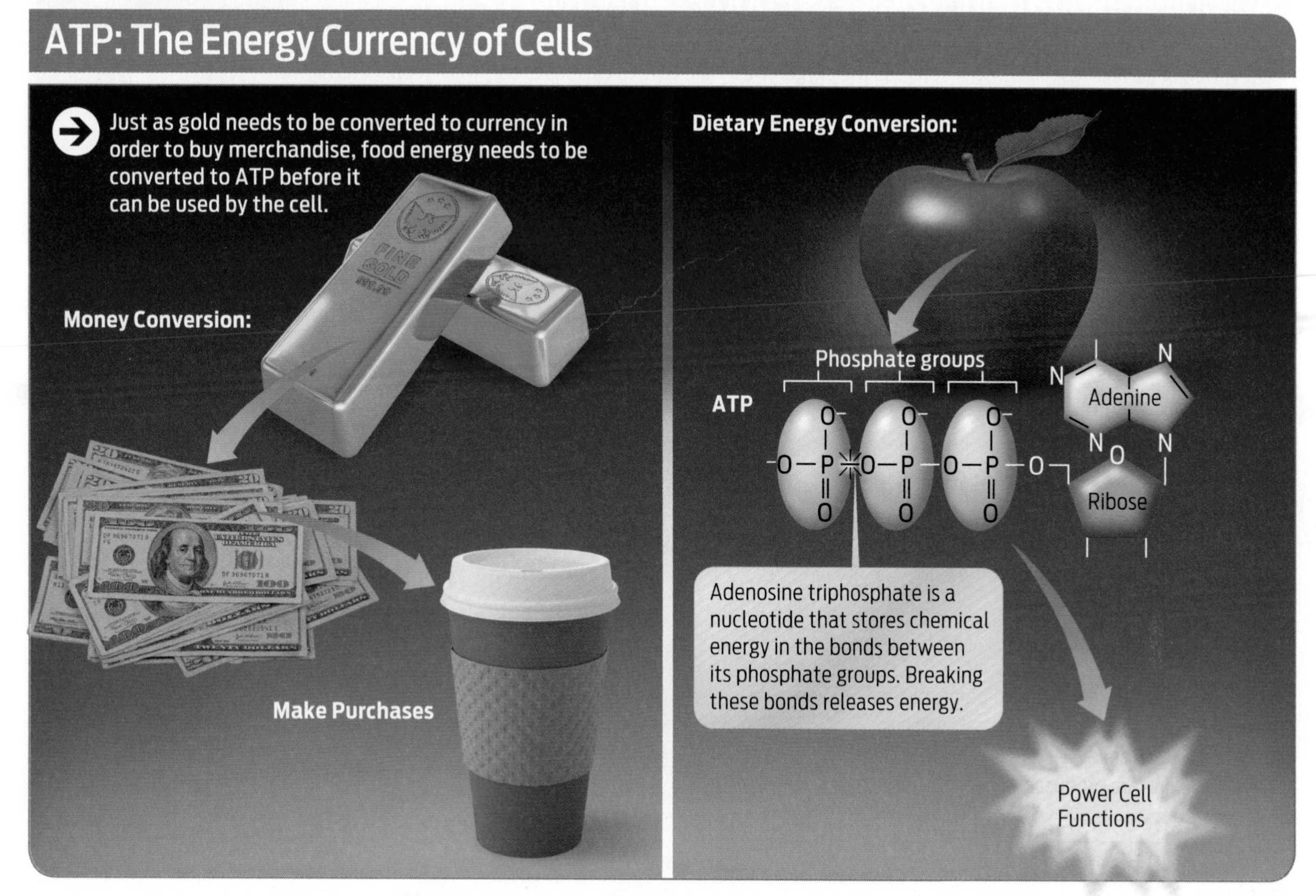

ATP: The Energy Currency of Cells
Just as gold needs to be converted to currency in order to buy merchandise, food energy needs to be converted to ATP before it can be used by the cell.
Money Conversion:
FINE GOLD
Make Purchases
Dietary Energy Conversion:
Phosphate groups
ATP
Adenine
Ribose
Adenosine triphosphate is a nucleotide that stores chemical energy in the bonds between its phosphate groups. Breaking these bonds releases energy.
Power Cell Functions

a molecule called **adenosine triphosphate (ATP)** that our cells use to carry out energy-requiring functions.

You can think of food as a bar of gold: it has a great deal of value, but if you carried that gold bar to your local convenience store, you wouldn't be able to buy even a cup of coffee with it. You would first have to convert your gold bar into bills and coins. ATP is the energetic equivalent of bills and coins; it's currency that your body can actually spend **(Infographic 6.6)**.

To make ATP, our bodies first break down food molecules into their smaller subunits through the process of digestion. Once released from food, such subunits as fatty acids, glycerol, amino acids, and sugars leave the small intestine and enter the bloodstream, which transports them to the body's cells. Inside the cells, enzymes break apart the bonds holding these subunits together. The energy stored in those bonds is then captured and converted into the molecular bonds that make up ATP. When ATP bonds are broken, energy is released, allowing cells to "spend" their ATP currency and carry out normal cellular functions.

The primary process that all eukaryotic organisms, including plants, use to convert energy stored in food molecules to form ATP is called **aerobic respiration.** Of the subunits released from food, sugar–in the form of glucose–is the most common source of energy for all organisms, from bacteria to humans. The aerobic respiration of glucose can be summarized by this equation:

$$\text{Glucose} + \text{Oxygen} \longrightarrow \text{Carbon dioxide} + \text{Water} + \text{Energy (+ heat)}$$

That is, the bonds holding the glucose molecule together are broken. Oxygen from the air we

ADENOSINE TRIPHOSPHATE (ATP)
The molecule that cells use to power energy-requiring functions; the cell's energy "currency."

AEROBIC RESPIRATION
A series of reactions that occurs in the presence of oxygen and converts energy stored in food into ATP.

INFOGRAPHIC 6.7

Aerobic Respiration Transfers Food Energy to ATP

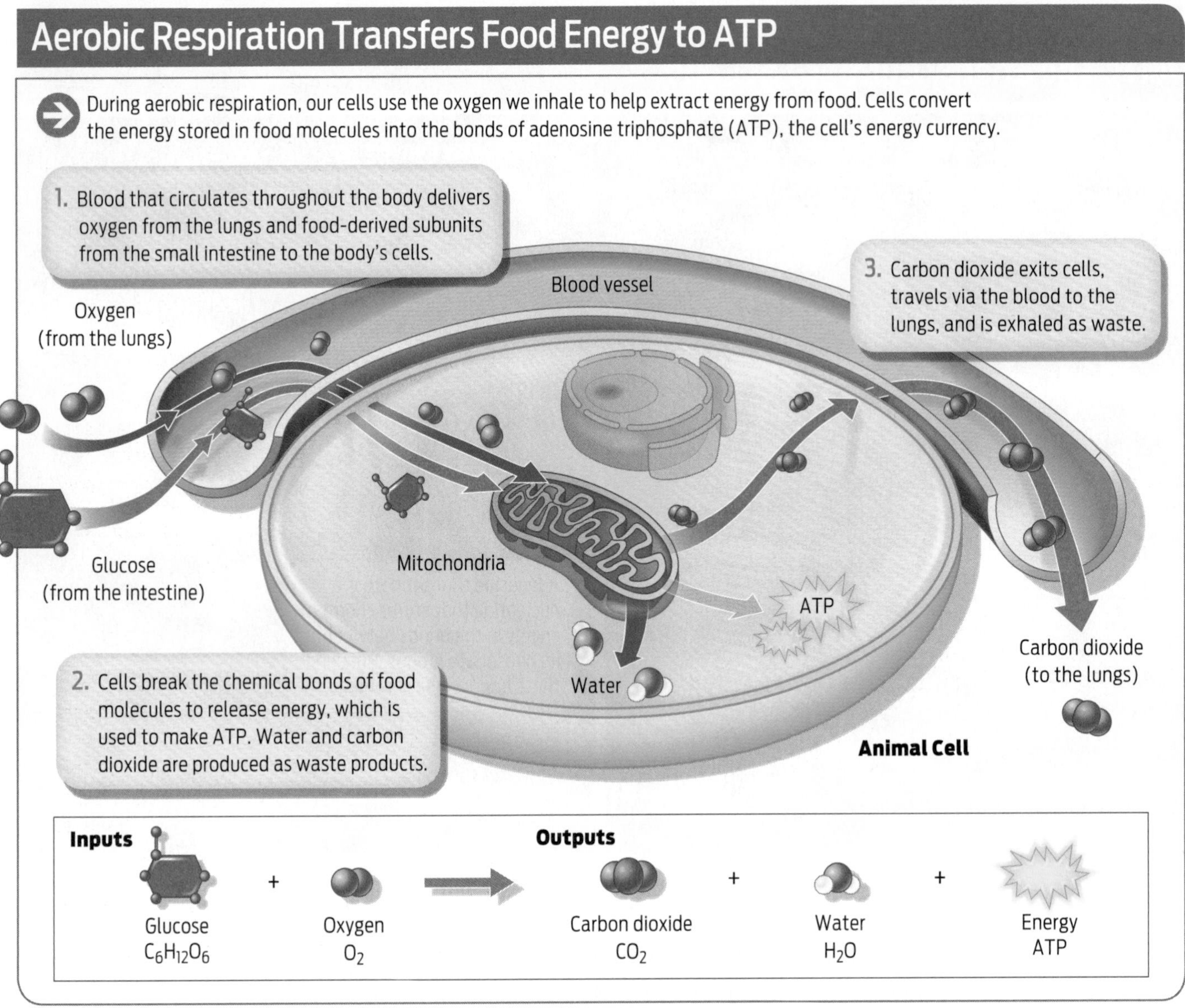

breathe is consumed in the process. When the bonds of glucose are broken, the energy released is used to form ATP and heat. Water and carbon dioxide are given off as waste products of the process **(Infographic 6.7)**.

During aerobic respiration, our cells use the oxygen we inhale to help extract energy from food. Cells convert the energy stored in food molecules into the bonds of adenosine triphosphate (ATP), the cell's energy currency.
1. Blood that circulates throughout the body delivers oxygen from the lungs and food-derived subunits from the small intestine to the body's cells.
Blood vessel
3. Carbon dioxide exits cells, travels via the blood to the lungs, and is exhaled as waste.
Oxygen (from the lungs)
Glucose (from the intestine)
Mitochondria
ATP
Carbon dioxide (to the lungs)
Water
2. Cells break the chemical bonds of food molecules to release energy, which is used to make ATP. Water and carbon dioxide are produced as waste products.
Animal Cell
Inputs
Outputs
Glucose $C_6H_{12}O_6$
+
Oxygen O_2
Carbon dioxide CO_2
+
Water H_2O
+
Energy ATP

Aerobic respiration is a multistep process that takes place in different parts of the cell. The initial steps of this process, known as **glycolysis,** take place in the cell's cytoplasm. Glycolysis is a series of chemical reactions that splits glucose into two smaller molecules. The products of glycolysis then enter the cell's mitochondria, where the last two steps of aerobic respiration occur.

During the second step of aerobic respiration, the **citric acid cycle,** a series of reactions strips electrons from the bonds between carbon atoms. The process releases CO_2, which is ultimately exhaled from an organism's lungs.

The electrons stripped from the carbon bonds are carried to the inner membranes of the mitochondria, where they go through the last step of aerobic respiration: the **electron transport chain.** Electrons stripped from the bonds in glucose contain a lot of potential energy. During electron transport, these energetic electrons are passed like hot potatoes down a chain of molecules, mostly proteins, in the inner mitochondrial membrane. Eventually the electrons pass to oxygen molecules, which accept the electrons and combine with hydrogen atoms to produce water. As electrons pass down the chain, they supply the energy needed

GLYCOLYSIS
A series of reactions that breaks down sugar into smaller units; glycolysis takes place in the cytoplasm and is the first step of both aerobic respiration and fermentation.

CITRIC ACID CYCLE
A set of reactions that takes place in mitochondria and helps extract energy (in the form of high-energy electrons) from food; the second step of aerobic respiration.

INFOGRAPHIC 6.8

Aerobic Respiration: A Closer Look

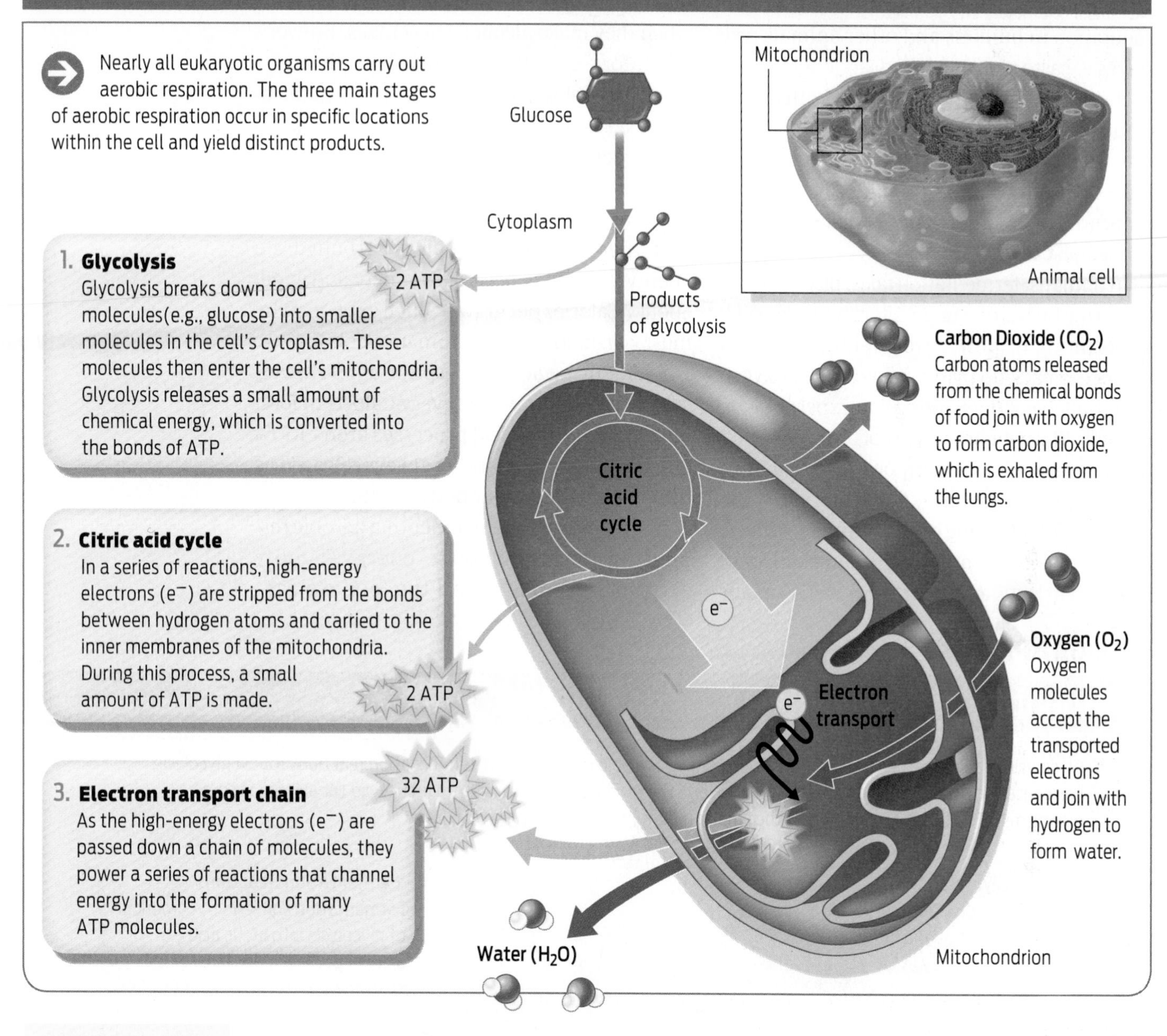

Nearly all eukaryotic organisms carry out aerobic respiration. The three main stages of aerobic respiration occur in specific locations within the cell and yield distinct products.
Glucose
Mitochondrion
Animal cell
Cytoplasm
1. Glycolysis
Glycolysis breaks down food molecules(e.g., glucose) into smaller molecules in the cell's cytoplasm. These molecules then enter the cell's mitochondria. Glycolysis releases a small amount of chemical energy, which is converted into the bonds of ATP.
2 ATP
Products of glycolysis
Carbon Dioxide (CO_2)
Carbon atoms released from the chemical bonds of food join with oxygen to form carbon dioxide, which is exhaled from the lungs.
Citric acid cycle
2. Citric acid cycle
In a series of reactions, high-energy electrons (e^-) are stripped from the bonds between hydrogen atoms and carried to the inner membranes of the mitochondria. During this process, a small amount of ATP is made.
2 ATP
e^-
e^-
Electron transport
Oxygen (O_2)
Oxygen molecules accept the transported electrons and join with hydrogen to form water.
3. Electron transport chain
As the high-energy electrons (e^-) are passed down a chain of molecules, they power a series of reactions that channel energy into the formation of many ATP molecules.
32 ATP
Water (H_2O)
Mitochondrion

ELECTRON TRANSPORT CHAIN
A process that takes place in mitochondria and produces the bulk of ATP during aerobic respiration; the third step of aerobic respiration.

FERMENTATION
A series of chemical reactions that takes place in the absence of oxygen and converts some of the energy stored in food into ATP. Fermentation produces far less ATP than does aerobic respiration.

to form ATP. This electron transport chain produces the bulk of ATP (**Infographic 6.8**).

We've focused on glucose, but cells can also burn fats and amino acids for fuel during aerobic respiration. Because fats generally have more carbon-hydrogen bonds than do sugars and amino acids, they have more electrons to be stripped in the citric acid cycle. More electrons stripped means that more ATP molecules are produced during electron transport (which also explains why a gram of fat contains more Calories than a gram of sugar or protein).

Feel the Burn

Aerobic respiration requires a continual source of oxygen ("aerobic" means "in the presence of oxygen"). If the rate at which cells consume oxygen exceeds the rate at which they take it in when we breathe, aerobic respiration comes to a halt; the electron transport chain has no oxygen to which it can deliver electrons. While glycolysis still occurs in the absence of oxygen, its products are shunted into a different process, called **fermentation,** which takes place in the cell's cytoplasm (as opposed to the mitochondria). Instead of carbon dioxide, fermentation in humans and other animals produces a waste product called lactic acid.

Because fermentation bypasses both the citric acid cycle and the electron transport chain, much less ATP is produced—only about 2 molecules of ATP from each molecule of glucose compared to 36 ATP produced by aerobic respiration **(Infographic 6.9)**.

In humans, fermentation takes place primarily during bursts of energy-intensive tasks, such as sprinting or power weight-lifting. It is, in essence, a back-up plan for times when oxygen isn't available. (The panting you experience on a treadmill is your body's way of trying to obtain more oxygen.) But for many organisms, like certain fungi and bacteria, fermentation is the main way of obtaining energy. In some of these organisms, fermentation produces alcohol rather than lactic acid as a by-product. Humans take advantage of these fermentation reactions when they make alcoholic beverages. Brewer's yeast, for example, is a fungus that ferments sugar in the absence of oxygen, producing alcohol as a result. Humans use brewer's yeast to make beer and wine.

Since fermentation does not break glucose down as completely as does aerobic respiration, there is still quite a bit of carbohydrate energy left in such beverages as beer and wine, about 7 Calories per gram—which explains why most weight-loss diets eliminate alcohol.

Even during aerobic respiration, however, our bodies don't convert every Calorie in food into ATP. The chemical processes aren't 100% efficient, so some energy is always released as heat, which keeps your body warm.

It's important to remember that aerobic respiration does not create energy—it only extracts it from food. All the food we eat—

INFOGRAPHIC 6.9

Fermentation Occurs When Oxygen Is Scarce

Glycolysis occurs whether or not oxygen is present. In the absence of oxygen, fermentation reactions follow glycolysis. Fermentation occurs in the cytoplasm and converts the products of glycolysis into lactic acid (or alcohol in some organisms). The only ATP produced is the small amount produced during glycolysis.

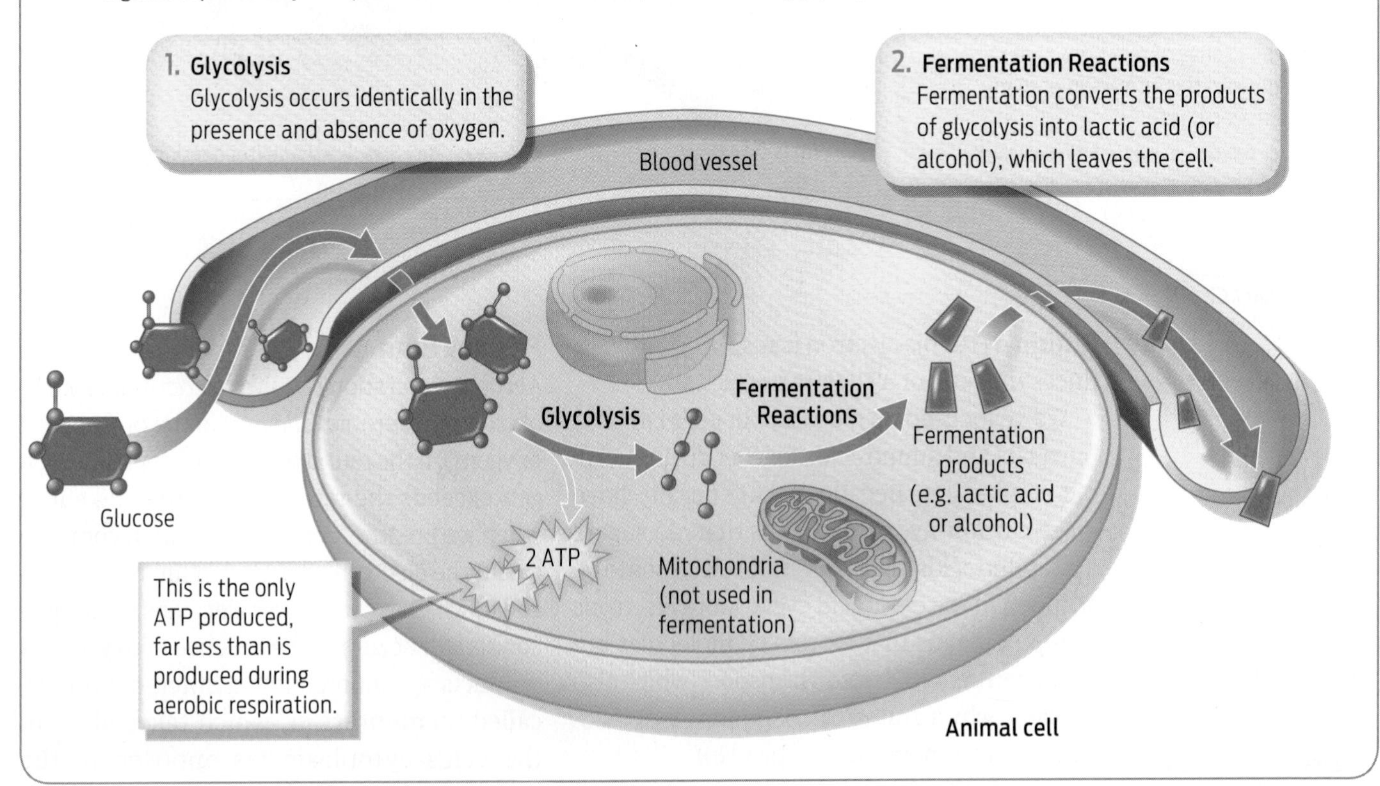

1. Glycolysis
Glycolysis occurs identically in the presence and absence of oxygen.
2. Fermentation Reactions
Fermentation converts the products of glycolysis into lactic acid (or alcohol), which leaves the cell.
Blood vessel
Glucose
Glycolysis
Fermentation Reactions
Fermentation products (e.g. lactic acid or alcohol)
2 ATP
Mitochondria (not used in fermentation)
This is the only ATP produced, far less than is produced during aerobic respiration.
Animal cell

In humans, fermentation takes place primarily during energy-intensive tasks.

whether burger, chicken leg, or Caesar salad—originally gets its energy from the sun, by way of photosynthesis. Photosynthesizers such as plants and algae capture the energy of sunlight and convert it into chemical energy stored in sugar. We then eat this sugar (or eat animals that have eaten this sugar), and that stored energy becomes available to us. Plants benefit from the relationship, too: plants use our carbon dioxide waste as raw material for making sugar during photosynthesis. In this way, photosynthesis and respiration form a continual cycle, with the outputs of one process serving as the inputs of the other (**Infographic 6.10**).

The Culture of Eating

Some scientists are interested in understanding how humans metabolize food in order to find a means of blocking some Calorie-dense molecules from being absorbed by the body. Food

INFOGRAPHIC 6.10

Photosynthesis and Aerobic Respiration Form a Cycle

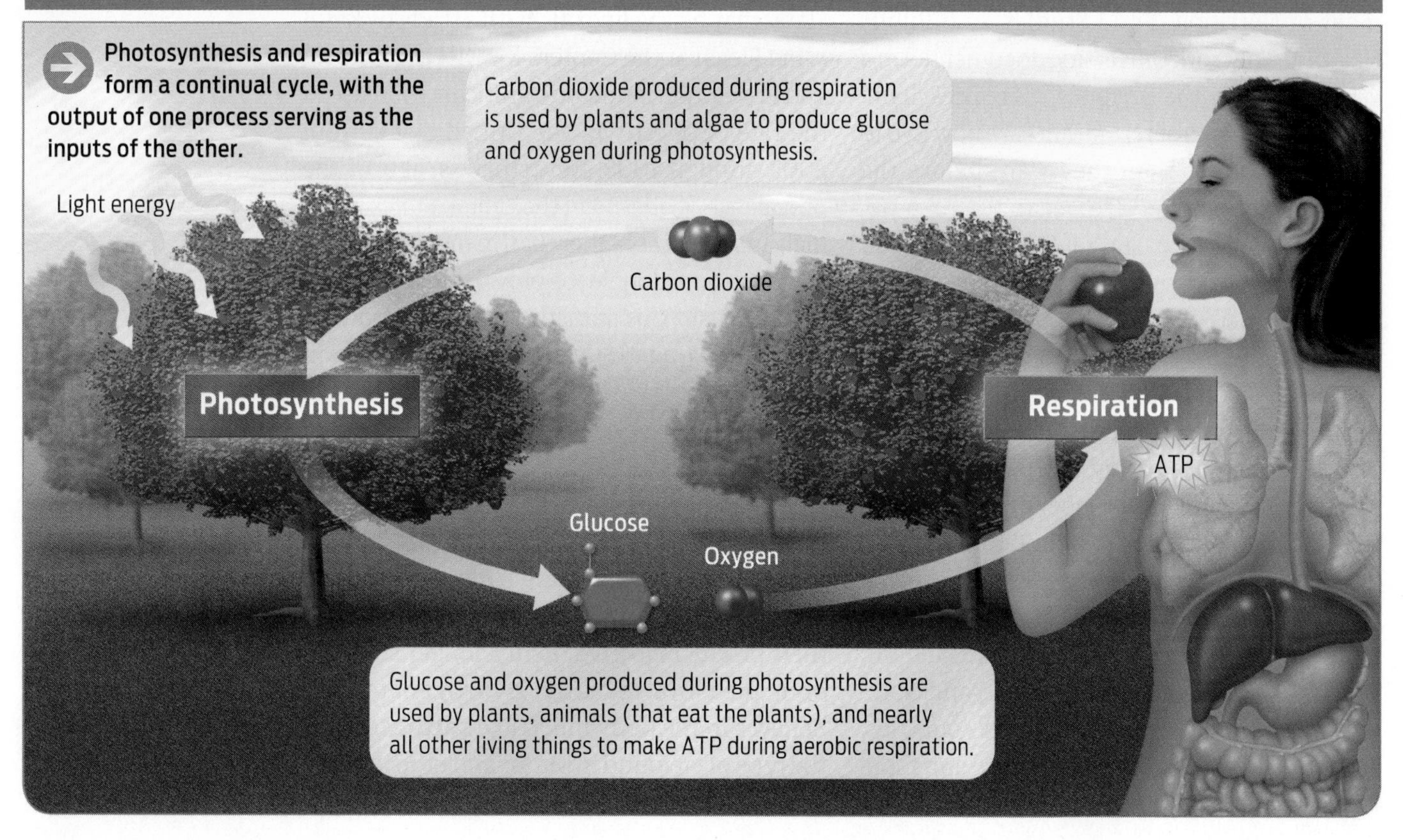

Photosynthesis and respiration form a continual cycle, with the output of one process serving as the inputs of the other.
Carbon dioxide produced during respiration is used by plants and algae to produce glucose and oxygen during photosynthesis.
Light energy
Carbon dioxide
Photosynthesis
Respiration
ATP
Glucose
Oxygen
Glucose and oxygen produced during photosynthesis are used by plants, animals (that eat the plants), and nearly all other living things to make ATP during aerobic respiration.

industry scientists, for example, have developed diet drugs that keep the body from absorbing some of the fat molecules in food. Food manufacturers would surely increase their markets if they could develop foods that pass through the intestines without being absorbed. (You may have heard of potato chips made with olestra, a fat substitute that is not absorbed by the intestines, and so passes through the body as waste, with famously unpleasant side effects.)

Of course, many food manufacturers specialize in low-fat and fat-free foods, which are less Calorie dense. But these efforts at tackling the problem of obesity offer temporary solutions at best. Studies have shown that 90% of dieters who avoid specific foods to lose weight over the short term regain most of their lost weight over time. That's because most people find it difficult to stick to the food restrictions prescribed by diets and tend to revert to their former eating habits.

A more permanent change in eating culture, on the other hand, offers a more sustainable fix because it emphasizes a change in eating habits, not the type of food itself. In fact, Americans didn't always consume large portion sizes. Lisa Young and Marion Nestle, both of the Department of Nutrition, Food Studies, and Public Health at New York University, documented in a 2002 study that the current sizes of fries, hamburgers, and soda at restaurants are now two to five times larger than they were before the 1970s, when portion size began to creep up. Single servings of pasta, muffins, steaks, and bagels today now exceed the government-recommended serving size by 480%, 333%, 224%, and 195%, respectively. And cookies exceed the standard by a factor of 8, according to Nestle's research. In America, bigger is better.

The French are eating more, too, and the results are evident: the number of obese French people has grown from 8.6% in 1997 to 13.1% in 2006, according to a 2008 study published by Marie-Aline Charles at the Institut National de la Santé et la Recherche Médicale in Villejuif, France. As American music, movies, and clothing have become pervasive in other countries, so, too, have our eating habits. More and more French people are eating large amounts of nutritionally poor, energy-dense foods. And as jobs increasingly place people in front of computers, they have become less physically active. More French people are eating on the go, eating fast food, and spending less time enjoying formal meals. Much to the dismay of public health experts, French eating culture is tipping toward unhealthful.

Senator Kel Seliger displays a box of candy in the Texas Senate during debate on a bill that would ban trans fats in restaurants—he opposed the bill.

But not all obesity news is bad. Recent studies suggest that the rate of obesity in the United States may be leveling off. A 2010 study by Katherine Flegal at the Centers for Disease Control and Prevention found no significant increase in the rate of obesity in the United States from 2003 through 2008. Nevertheless, at 30% the prevalence of obesity in the United States is still high and remains higher than in most European countries (and double the prevalence in France).

Because of this statistic, there is a growing trend in the United States to legislate changes in the foods people eat. Some cities, for example, have banned the use of **trans fat,** a type of hydrogenated vegetable fat that studies have shown contributes to heart disease. Commercially prepared foods such as cookies, French fries, doughnuts, and margarine often contain trans fat, which food manufacturers add to their products to give them a longer shelf life or a pleasing texture. Hydrogenated fat behaves in the body much like **saturated fat,** the type of fat found in butter and other animal products. Studies have shown that eating large

TRANS FAT
A type of vegetable fat which has been hydrogenated, that is, hydrogen atoms have been added, making it solid at room temperature.

SATURATED FAT
An animal fat, such as butter; saturated fats are solid at room temperature.

UNSATURATED FAT
A plant fat, such as olive oil; unsaturated fats are liquid at room temperature.

amounts of saturated fat can clog arteries. By contrast, **unsaturated fats**, which come from plants and are liquid at room temperature, are considered more healthful (although they are still high in Calories).

"Actions by governments are the only way conditions will change enough to have a major public health impact," says Kelly Brownell. For America's obesity woes, Brownell blames the food industry for their relentless marketing of unhealthful foods, agricultural and trade policies that promote unhealthful diets, and economic policies that make unheathful foods cheaper than healthful ones.

Not everyone agrees, however, that it is the government's job to restrict our food choices. Americans equate choice in food with democracy, argues Paul Rozin. "We could also overrespond to what many perceive as an obesity epidemic and that could be dangerous. It would restrict individual freedom."

Besides, while such government legislation would restrict *what* we eat, it wouldn't really affect *how* we eat. Rozin hopes people change their behavior voluntarily. For example, people could fit more exercise into their daily routines, climbing stairs instead of using an elevator or parking the car farther away from the entrance at the mall. To combat large portion sizes in restaurants, Rozin advocates ordering less, sharing dishes, or as Mireille Guiliano recommends, simply not eating everything that is on our plates.

The U.S. eating culture can change, Rozin says, but not overnight. Look at cigarette smoking: "It took 50 years to get cigarette smoking to decline and they [cigarettes] are much more harmful to health." ■

Summary

- The macronutrients in our food (proteins, carbohydrates, and fats) are sources of dietary energy.
- Fats are the most energy-rich organic molecules in our diet. Fats contain twice as many Calories per gram as carbohydrates and proteins.
- When we consume more Calories than we use, our bodies store the excess energy in the bonds of glycogen and body fat. Fats store more energy than does glycogen.
- Cells carry out chemical reactions that break down food to obtain usable energy in the form of ATP.
- In the presence of oxygen, aerobic respiration produces large amounts of ATP from the energy stored in food.
- Aerobic respiration occurs in three stages: (1) glycolysis, (2) the citric acid cycle, and (3) electron transport. The first stage occurs in the cytoplasm, the latter two in mitochondria. Electron transport produces the bulk of ATP.
- In the absence of oxygen, fermentation follows glycolysis and produces lactic acid in animals (or, in some organisms, alcohol). Fermentation produces far less ATP than does aerobic respiration.
- Exercise helps burn stored Calories. A combination of eating fewer Calories and exercising more will result in weight loss, although hereditary factors play a large role in determining a person's weight.
- During exercise, glycogen is used first. Stored fats are tapped only when glycogen stores have been depleted, as might occur during long periods of exercise.
- Photosynthesis and respiration form a cycle: the carbon dioxide given off by animals, plants, and all organisms that perform aerobic respiration is used by photosynthesizers to make glucose and oxygen during photosynthesis.

FOOD IS ENERGY

Each type of organic molecule found in food stores a different amount of energy. Both what we eat and how much we eat contribute to weight gain.

HINT **See Infographics 6.1–6.5.**

KNOW IT

1. Which type of organic molecule stores the most energy per gram?
- **a.** proteins
- **b.** starch
- **c.** nucleic acid
- **d.** fats (triglycerides)
- **e.** glycogen

2. A moderately active 21-year-old female has a choice of eating a 2,500-Calorie meal primarily of protein or a 2,500-Calorie meal primarily of sugar. What would be the result, in terms of energy, of choosing one over the other?
- **a.** Nothing; she would burn all these Calories, given her age, gender, and activity level.
- **b.** She would store the excess Calories as protein, regardless which meal she ate.
- **c.** She would store the excess Calories as protein if she ate the protein meal, and as glycogen if she ate the sugar meal.
- **d.** In either case, once her glycogen stores are replenished, she will store the excess Calories as fat.
- **e.** Regardless of the number of Calories, she will get more energy from the sugar meal.

3. A 5′6″ female weighs 167 pounds. Use Infographic 6.1 to determine her BMI. Would she be considered underweight, normal weight, overweight, or obese?

4. If you exercise for an extended period of time, you will use energy first from ________, then from ________.
- **a.** fats; glycogen
- **b.** proteins; fats
- **c.** glycogen; proteins
- **d.** fats; proteins
- **e.** glycogen; fats

5. Preparing foods with trans fats specifically increases the risk of which of the following conditions?
- **a.** cardiovascular disease
- **b.** obesity
- **c.** excessive weight gain
- **d.** arthritis
- **e.** colon cancer

USE IT

6. If you frequently crave French fries (that is, starchy potatoes fried in fat), how could you modify your lifestyle to eat fries without gaining weight? Explain your answer.

7. Consider a well-trained 130-pound female marathon runner. She has just loaded up on a carbohydrate meal and has the maximum amount of stored glycogen (6.8 grams of glycogen per pound of body weight).
- **a.** How many grams of glycogen is she storing?
- **b.** How many Calories does she have stored as glycogen?
- **c.** If this same number of Calories were stored as fat, how much would it weigh?
- **d.** Suppose she decides to go for a run at a pace of 9 miles per hour (she will be running 6.5-minute miles). Given her weight, she will burn 885 Calories per hour at this pace. How long will it take her to deplete her glycogen stores? How many miles can she run before her glycogen supplies run out? Will she be able to complete a 26.2-mile marathon?
- **e.** Once her glycogen supplies run out, what has to happen if she wants to keep running?

8. If the French eat meals with a higher fat content, why are the French on average not fatter than Americans?

AEROBIC RESPIRATION AND FERMENTATION

These reactions convert energy stored in food into usable forms.

HINT **See Infographics 6.6–6.10.**

KNOW IT

9. Which step is not correctly matched with its cellular location?
- **a.** glycolysis—cytoplasm
- **b.** citric acid cycle—mitochondria
- **c.** fermentation—mitochondria
- **d.** electron transport—mitochondria
- **e.** none of the above—they are all correctly matched

10. Compared to aerobic respiration, fermentation produces _______ ATP.

a. much more
b. the same amount of
c. a little less
d. much less
e. no

11. We obtain carbohydrates by eating them. How do plants obtain their carbohydrates?

a. by eating them
b. by cellular respiration
c. by fermentation
d. from the soil
e. by photosynthesis

12. In the presence of oxygen we use __________ to fuel ATP production. What process do plants use to fuel ATP production from food?

a. aerobic respiration; photosynthesis
b. aerobic respiration; aerobic respiration
c. fermentation; aerobic respiration
d. fermentation; photosynthesis
e. glycolysis; photosynthesis

USE IT

13. Draw a carbon atom that is part of a CO_2 molecule that you just exhaled. Using a written description or a diagram, trace what happens to that carbon atom as it is absorbed by the leaf of a spinach plant and then what happens to the carbon atom when you eat that leaf in a salad.

14. Given 1 gram of each, which of the following would yield the greatest amount of ATP by aerobic respiration?

a. fat
b. protein
c. carbohydrate
d. nucleic acid
e. alcohol

15. Explain how the presence or absence of oxygen affects ATP production. (The terms *aerobic respiration* and *fermentation* should appear in your answer.)

16. If you ingest carbon in the form of sugar, how is that carbon released from your body?

a. as sugar
b. as fat
c. as CO_2
d. as protein
e. in urine

SCIENCE AND ETHICS

17. Why do you think that longer meal times translate into fewer Calories consumed?

18. Do you think the government has a responsibility to regulate the information provided on nutrition labels? Explain your answer.

19. If the government were to issue tax incentives to reduce obesity in the United States, which of the following do you think would be most effective? Explain your choice.

a. taxing foods high in fat
b. giving tax breaks for people who join gyms or health clubs
c. giving rebates for purchasing fresh fruits and vegetables
d. paying enhanced salaries for teachers in elementary and middle schools to provide education about diet and nutrition

Biologically Unique

What You Will Be Learning

Biologically unique

How DNA helped free an innocent man

Roy Brown thought the police were just checking up on him when an officer knocked on his door one day in May 1991. Brown, a self-professed hard drinker who earned a living selling magazine subscriptions, had only a week before been released after serving an 8-month prison term. His crime: threatening to kill the director of the Cayuga County Department of Social Services in upstate New York. A caseworker had deemed Brown unfit to care for his 7-year-old daughter. Furious, Brown had threatened to kill the director and other workers. But he had served his time. What could the officer want from him now?

INNOCENCE PROJECT

Three days earlier, police had found the battered body of a woman lying in the grass about 300 feet from the farmhouse where she lived. Someone had burned the place to the ground. The body was identified as that of Sabina Kulakowski, a social worker at the Cayuga County Department of Social Services. The crime was horrific. The murderer had beaten the 49-year-old Kulakowski, bit her several times, dragged her outside, and then stabbed and strangled her to death. It was obvious that Kulakowski had struggled; her body was covered with defensive wounds.

Although Kulakowski was not involved in Brown's case, officers arrested Brown that day on suspicion of murder. Eight months later, a jury found Brown guilty of homicide and sentenced him to prison for 25 years to life. The prosecution argued that Brown's motive was revenge against the Department of Social Services. But what really nailed the case was testimony from an expert who stated that bite marks on the victim's body matched Brown's teeth.

Brown, however, maintained his innocence. "I never knew Ms. Kulakowski, and I had nothing to do with that woman's death . . . I am truly innocent," he told the court and onlookers after the verdict had been announced.

Even from jail Brown never stopped trying to prove his innocence. He repeatedly petitioned, in vain, for a retrial. Then something unexpected happened. Brown uncovered additional evidence that strongly suggested he was not the perpetrator. The evidence was so compelling, in

DEOXYRIBONUCLEIC ACID (DNA)
The molecule of heredity, common to all life forms, that is passed from parents to offspring.

CHROMOSOME
A single, large DNA molecule wrapped around proteins. Chromosomes are located in the nuclei of most eukaryotic cells.

fact, that in late 2004, after Brown had spent 12 years in prison, his lawyers decided to contact the Innocence Project–a nonprofit organization founded in 1992 by Peter Neufeld and Barry Scheck of the Benjamin N. Cardozo School of Law in New York City. Their mission: to use DNA evidence to free people wrongly convicted of crimes.

When the jury convicted Brown in 1992, analyzing crime scene evidence for traces of DNA wasn't established practice yet, so DNA was rarely used as evidence in criminal cases. But about a decade later, using DNA evidence in court cases became standard practice as science increasingly showed that it was an extremely accurate way to match crime scene evidence to perpetrators.

DNA as Evidence

How can scientists use DNA to identify a person? The answer lies in the chemical makeup of this molecule, often referred to as the "blueprint of life." **Deoxyribonucleic acid,** or **DNA,** is the hereditary molecule that is common to all life forms–from plants to bacteria to fungi–and that is passed from parents to offspring. DNA serves as the instruction manual from which we are built; it's the reason why you look like your parents, an aunt, or perhaps even a grandparent.

Where can you find DNA? The molecule exists inside the nucleus of almost every cell in our body in the form of **chromosomes,** strands of DNA wound around proteins. Humans have 23 pairs of chromosomes in the vast majority

Machines milk rows of goats.

three-dimensional structure–the protein itself. A protein's three-dimensional structure is important because it determines a protein's function. Some proteins, such as the antibody molecules that are a critical part of our defenses, or the hemoglobin that carries oxygen in our red blood cells, are made up of more than one folded amino acid chain. A protein's final overall shape–which is dictated by the placement of amino acid side groups– determines its specific function **(Infographic 8.1)**. Antithrombin folds into a compact globular structure. This structure is maintained through numerous chemical interactions between amino acid side groups. The position of every amino acid in the chain is important, contributing to the protein's overall shape and therefore its optimal function.

INFOGRAPHIC 8.1

Amino Acid Sequence Determines Protein Shape and Function

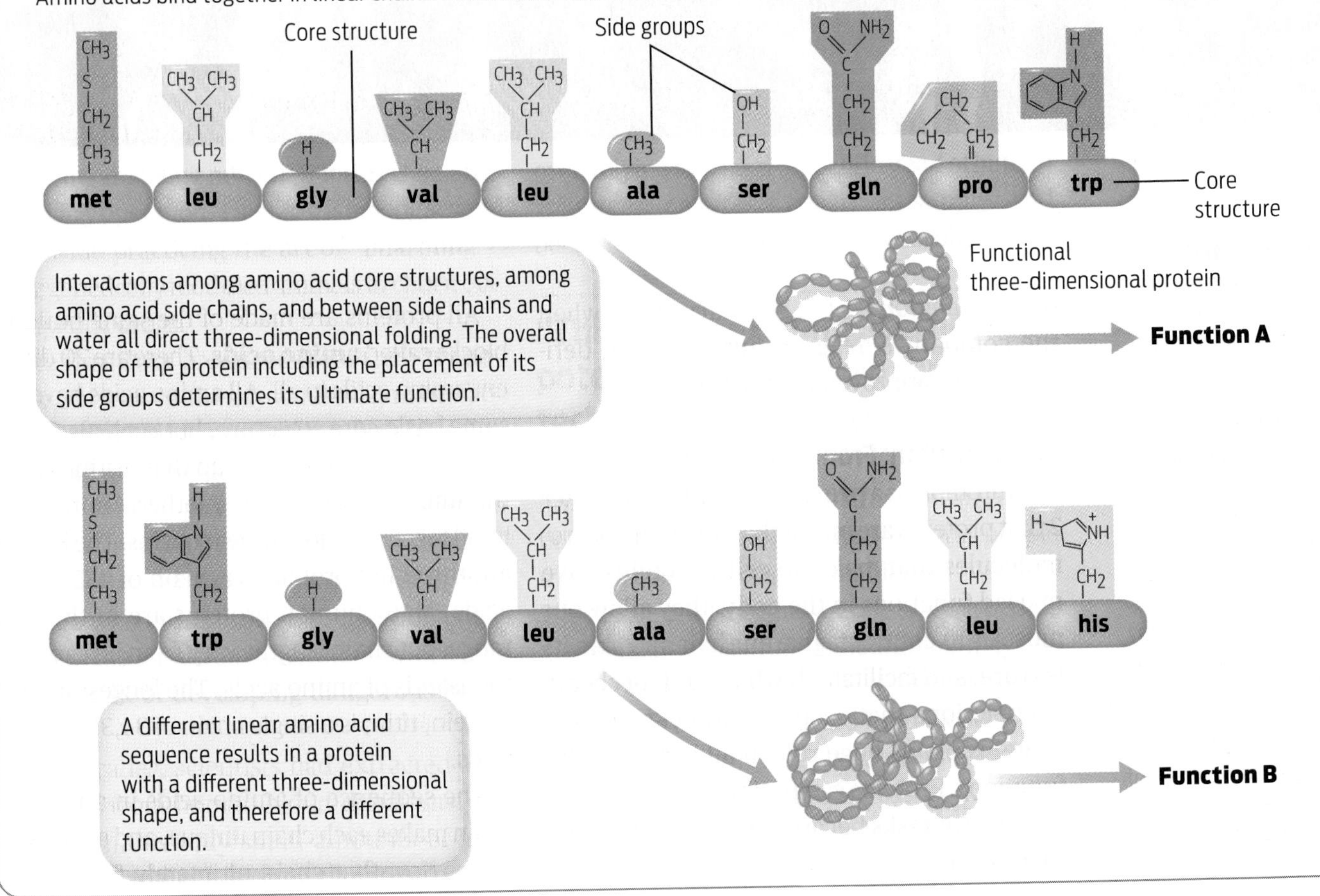

INFOGRAPHIC 8.2

Chromosomes Include Gene Sequences That Code for Proteins

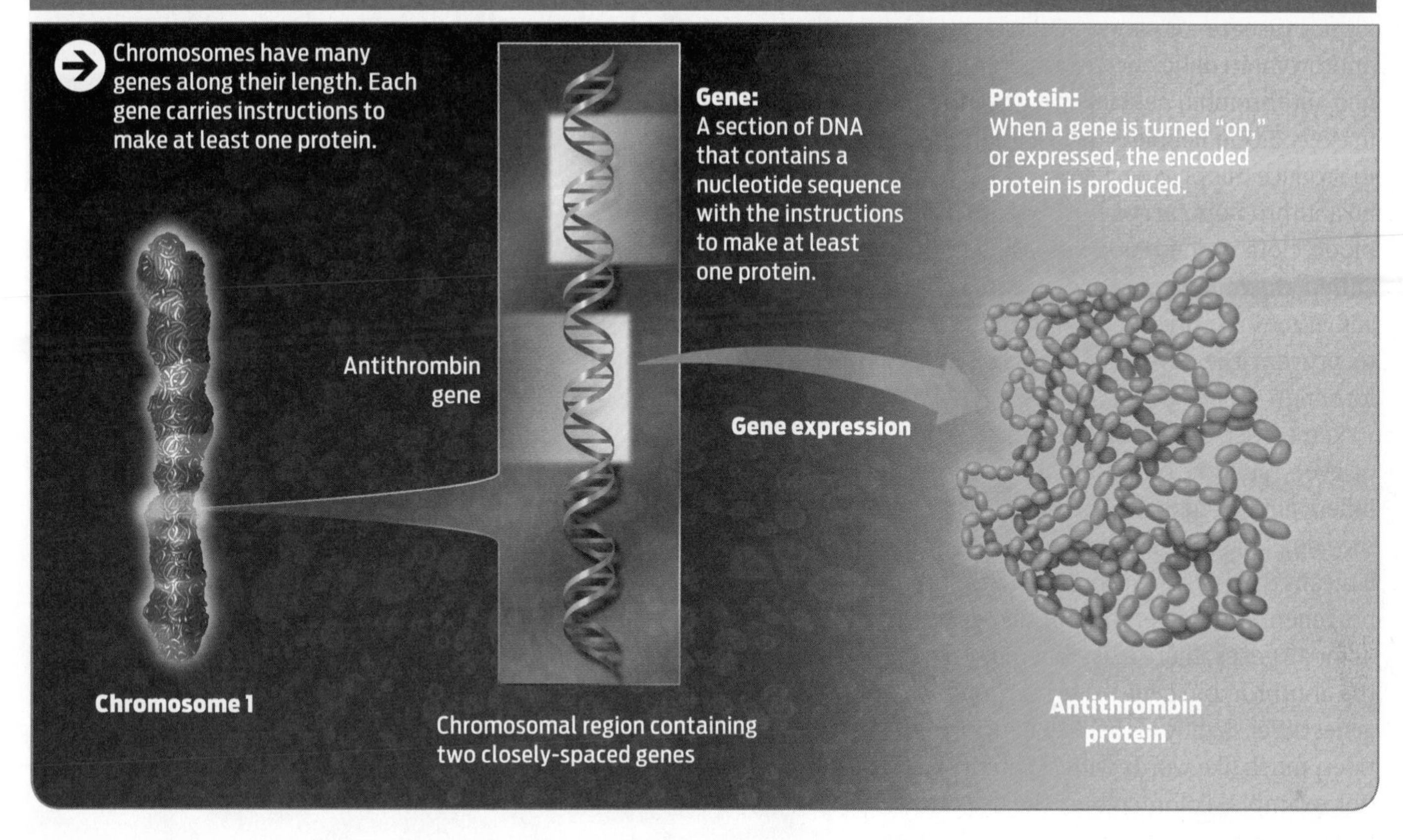

Chromosomes have many genes along their length. Each gene carries instructions to make at least one protein.
Gene:
A section of DNA that contains a nucleotide sequence with the instructions to make at least one protein.
Protein:
When a gene is turned "on," or expressed, the encoded protein is produced.
Antithrombin gene
Gene expression
Chromosome 1
Chromosomal region containing two closely-spaced genes
Antithrombin protein

GENE
A sequence of DNA that contains the information to make at least one protein.

GENE EXPRESSION
The process of using DNA instructions to make proteins.

GENOTYPE
The genetic make-up of an organism.

PHENOTYPE
The physical attributes of an organism, including both observable and internal or non-observable traits.

Because proteins play such important roles in the body, most drugs are either chemicals that interact with specific proteins, or, like antithrombin, are themselves protein molecules.

Where do proteins come from? Just as cells make DNA out of building blocks that we, in part, obtain from food, cells also make proteins using amino acid building blocks from our diet. But DNA and protein are not equals when it comes to their function in cells. Rather, the relationship between the two is hierarchical, with one directing the production of the other.

The instructions to make proteins are encoded in our DNA—our genes. A **gene** is a segment of DNA that contains instructions for making at least one protein. Genes are particular nucleotide sequences organized along the length of chromosomes, with each chromosome carrying a unique set of genes. The process of synthesizing a protein from the information encoded in a gene is called **gene expression** (Infographic 8.2). When we say a gene inside a cell is "expressed," we mean that the cell is making the protein encoded by that gene. In other words, our genes are the master instruction manual of our bodies; they dictate what proteins are made, when, and how many. Another way of saying this is that genes provide our **genotype,** but it is the proteins specified by those genes that, to a large extent, determine our physical traits, or **phenotype.**

The antithrombin gene, for example, sits on chromosome 1 and holds instructions to make a chain of 432 amino acids that folds into the antithrombin protein. When cells express the antithrombin gene, it means they produce antithrombin protein.

In the body, antithrombin protein prevents blood from clotting. The protein plays a regulatory role by inactivating enzymes that promote blood clotting. In this way, antithrombin prevents blood from clotting in the wrong place and causing a stroke or a heart attack.

Some people, however, can become antithrombin deficient because of heart, liver, or kidney disease, or cancer. Others can inherit antithrombin deficiency from a parent. Inherited antithrombin deficiency isn't rare—about 1 in every 5,000 people is born with the inability to produce this protein. People with too little or no antithrombin carry a high risk of developing blood clots inside blood vessels, a condition called thrombosis, and consequently they sometimes require antithrombin transfusions to prevent these vessel-blocking clots from forming **(Infographic 8.3)**.

Remember, each human cell has two copies of every gene in our genome. When people inherit antithrombin deficiency, it doesn't mean they don't have the antithrombin gene. Rather, it means that both their copies of the antithrombin gene are defective. This can happen because, as for all genes, there exist different versions of the antithrombin gene. Alternative versions of genes differ slightly in their sequence of nucleotides, much like words with different spellings (for example, color, colour; theater, theatre). Different versions of a gene with such alternative nucleotide "spellings" are called **alleles.** Sometimes, the allele of a gene encodes a protein with an abnormal shape. If a protein's shape is too contorted, it may not be able to do its job and the protein is nonfunctional. Having a nonfunctional protein is as harmful as not having one at all. If a person's two alleles of the

ALLELES
Alternative versions of the same gene that have different nucleotide sequences.

INFOGRAPHIC 8.3

Antithrombin Deficiency Can Cause Blood Clots

Antithrombin is an important protein that helps prevent blood clots (thrombosis). The antithrombin gene is expressed by cells in the liver, which then release antithrombin protein into the bloodstream.

Deep veins of the leg

Abundant antithrombin prevents blood clots. Blood flows freely through blood vessels.

Normal blood flow

Deep-vein thrombosis

Antithrombin deficiency may result in thrombosis, a blood clot that restricts blood flow. Thrombosis in blood vessels within the heart or lungs can kill.

INFOGRAPHIC 8.4

Genes to Proteins: Different Alleles Influence Phenotype

Versions of a gene with different nucleotide sequences are called alleles. Alternate nucleotide sequences change the type of protein coded for by a gene.

Expression of normal antithrombin allele:

Antithrombin gene allele 1

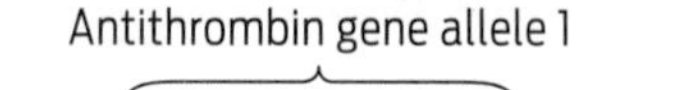

Gene expression

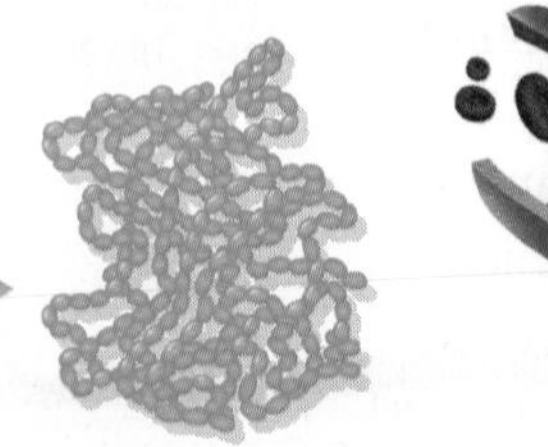

GGCGACCTGAGC
Normal nucleotide sequence

Normal Antithrombin Protein:
Protein has normal shape and therefore normal function.

Phenotype:
Blood flow is normal when blood contains only functional antithrombin.

Expression of abnormal antithrombin allele:

Antithrombin gene allele 2

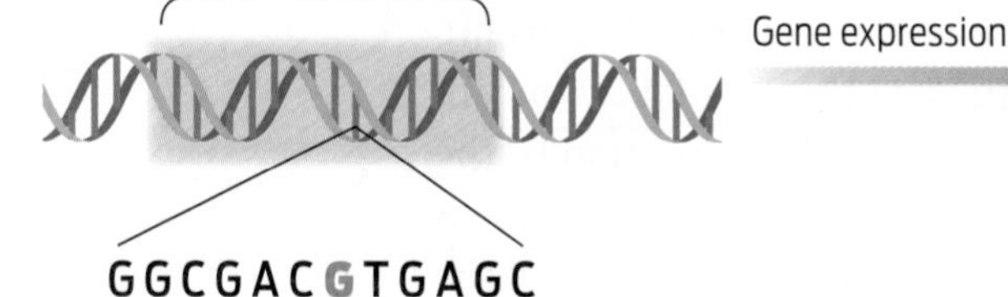

Gene expression

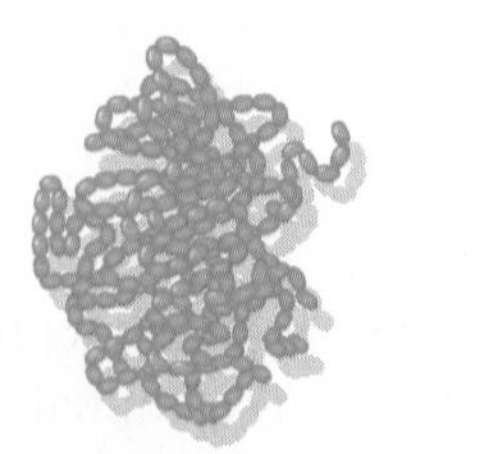

GGCGACGTGAGC
Alternate nucleotide sequence

Abnormal Antithrombin Protein:
Protein has abnormal shape and therefore is nonfunctional.

Phenotype:
Thrombosis prevents normal blood flow when blood contains only abnormal protein.

antithrombin gene both code for defective proteins, antithrombin deficiency results **(Infographic 8.4)**.

People with inherited antithrombin deficiencies usually take medication to thin their blood and prevent clots. At times when the risk of clots is high–during surgery, for example–they receive antithrombin treatment.

But it takes 50,000 blood donors to produce 1 kilogram of antithrombin. A single transgenic goat can produce the same amount in her milk in just one year, according to GTC Biotherapeutics.

Making Transgenic Goats

More than 20 years ago, when Harry Meade was working as a research scientist at a company called Biogen, it occurred to him that producing drugs in a mammal's milk might be more efficient than existing methods of large-scale protein production. The mammary gland is a natural protein factory, he reasoned. To nourish their young, all mammals produce proteins and secrete them into their milk.

Meade had been experimenting with getting genes from various organisms into hamster cells, which could be grown in large numbers in the laboratory, allowing scientists to purify significant amounts of protein from the cells. This method was effective, and in fact is still used today to express and harvest proteins of interest. But wouldn't it be more efficient, Meade wondered, to transfer a gene into a large mammal, such as a goat, so that the gene is expressed by the mammary gland? That way, when the goat lactates, the protein will collect in the goat's milk. This method would produce much greater quantities of a protein than could be produced by hamster cells in stainless steel vats.

To work on this project, Meade co-founded GTC Biotherapeutics. At his new company he helped devise a technique to create transgenic animals, the first of which was a transgenic goat. The basic idea is simple: isolate the gene of interest from a human chromosome and then insert it into the genome of a goat embryo. But in order to make sure the human gene is expressed properly in goat mammary glands, Meade and his colleagues had to create a hybrid gene that was part human and part goat.

The technique that Meade used takes advantage of the fact that every gene has two parts: a **regulatory sequence** and a **coding sequence.** Regulatory sequences are like on/off switches for genes; they determine when, where, and how much protein a gene makes. Coding sequences determine the identity of a protein—they specify the amino acid makeup (Infographic 8.5).

Meade realized that if he could attach the coding sequence of the human antithrombin gene to the regulatory sequence of a goat gene that is expressed only in the animal's mammary cells, he could get the goat's mammary cells to make the human protein. In other words, he could dupe the goat's mammary glands into making the human antithrombin protein and secreting it as part of the goat's milk.

About 10 years ago, Meade and his team successfully attached the coding sequence of the human antithrombin gene to the regulatory sequence of a goat gene expressed only by the mammary glands (the beta casein gene). Using the regulatory sequence of the beta casein gene ensures that the gene of interest, antithrombin, is expressed only in the mammary cells, and not in any other tissues. The antithrombin protein will be found solely in the goat's milk, and nowhere else. We'll look at the details of how genes are expressed later. For now, it's important to note that scientists took advantage of a natural phenomenon: since beta casein protein is normally found only in goat's milk, the antithrombin protein would also be found only in the animal's milk. With the exception of the mammary glands, no other tissue would express the antithrombin gene, so as not to disrupt the tissue's normal function and harm the animal.

INFOGRAPHIC 8.5

The Two Parts of a Gene

Genes are organized into two parts. Regulatory sequences determine when and how much protein a gene makes. Coding sequences determine the amino acid sequence of the encoded protein, which determines its shape and function.

Gene

Regulatory sequence: Controls the timing, location, and amount of gene expression.

Coding sequence: Determines the sequence of amino acids in the protein.

With the coding sequence of the human antithrombin gene fused to the regulatory sequence of the goat beta casein gene, the researchers could begin the process of putting the transgene inside a goat. Using a long, thin needle, a GTC scientist injected the gene construct into a fertilized single-cell goat embryo. He then implanted this transgenic embryo into a surrogate mother. As the embryo grew and the cells divided, the inserted gene replicated and was passed on to every cell in the developing goat (Infographic 8.6).

This is the technique the company used to create the first transgenic goats; today, GTC uses newer and more efficient methods to get gene constructs into animals.

Animals aren't the only organisms that have been genetically modified by humans. Much of the corn you eat today is transgenic—it contains genes from a soil bacterium. There are strains of transgenic soybeans, transgenic tomatoes, and transgenic insects. Trans-

REGULATORY SEQUENCE
The part of a gene that determines the timing, amount, and location of protein produced.

CODING SEQUENCE
The part of a gene that specifies the amino acid sequence of a protein. Coding sequences determine the identity, shape, and function of proteins.

INFOGRAPHIC 8.6

Making a Transgenic Goat

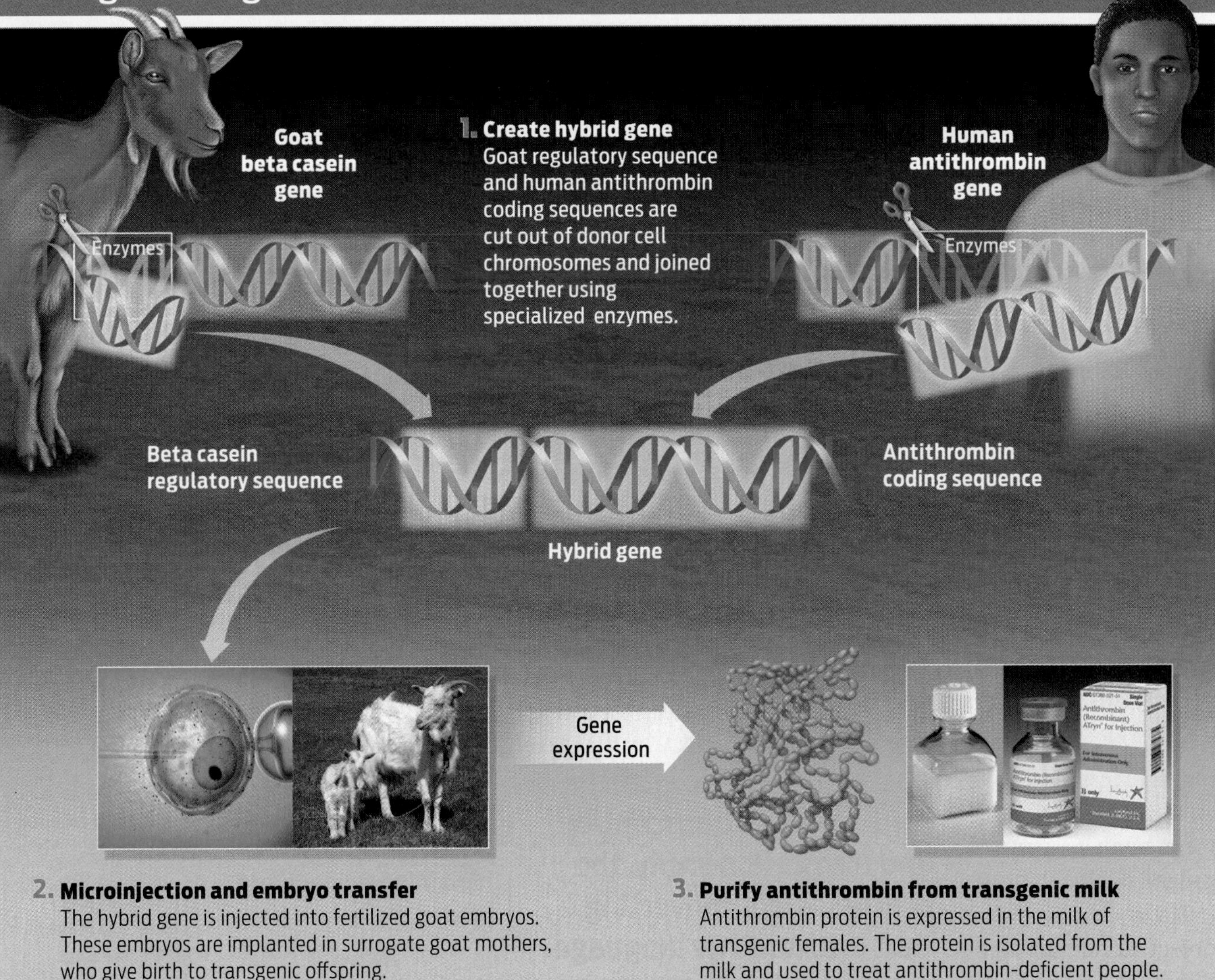
Goat beta casein gene
Enzymes
1. Create hybrid gene
Goat regulatory sequence and human antithrombin coding sequences are cut out of donor cell chromosomes and joined together using specialized enzymes.
Human antithrombin gene
Enzymes
Beta casein regulatory sequence
Antithrombin coding sequence
Hybrid gene
Gene expression
2. Microinjection and embryo transfer
The hybrid gene is injected into fertilized goat embryos. These embryos are implanted in surrogate goat mothers, who give birth to transgenic offspring.
3. Purify antithrombin from transgenic milk
Antithrombin protein is expressed in the milk of transgenic females. The protein is isolated from the milk and used to treat antithrombin-deficient people.

genic organisms are also called **genetically modified organisms (GMOs).** Transgenic crops such as corn and soybean usually contain genes for natural pesticides, which help the plants fight pests and reduce the amount of pesticide a farmer must use. Transgenic animals serve many purposes; sometimes they are used in research to study a gene's function, other times they can be used for a specific commercial purpose, such as producing medicines or other marketable products. Spider silk, for example, is a very strong, resilient fiber that can be produced in transgenic animals or in plants that carry spider genes.

GENETICALLY MODIFIED ORGANISM (GMO)
An organism that has been genetically altered by humans.

GENE THERAPY
A type of treatment that aims to cure disease by replacing defective genes with functional ones.

Such gene-swapping technology also has an important application in medicine: in **gene therapy** scientists attempt to replace a person's defective gene with a healthy one, an approach that can already treat, and in some cases cure, debilitating diseases such as severe combined immunodeficiency syndrome (SCID)—a disorder in which babies are born with deficient immune systems. Researchers hope that gene therapy might one day help treat several types of disorders caused by defective genes, such as cystic fibrosis, Huntington disease, and hemophilia.

Despite the many actual and potential benefits of genetic engineering, mixing and match-

Goats raised for medicine are kept in controlled environments.

ing genes inspires debate among scientists, environmentalists, and the general public.

Many people also object to human meddling with the biology of organisms that have evolved naturally. There are also other concerns, such as what might happen to a natural population of organisms if their genetically modified cousins were to escape into the environment and mate with the natural population; the consequences are unpredictable.

Although the idea of genetically engineering animals may be disquieting to some, humans have been tampering with the natural evolution of farm animals for centuries by selectively breeding them to have desirable traits. Moreover, from an animal-rights point of view, transgenic goats are treated no differently from goats farmed for their milk and meat.

That said, the prospect of being able to genetically modify—even clone whole organisms—for human purposes raises legitimate questions about how to conduct genetic engineering safely and humanely. For example, many people who find nothing ethically troubling about using gene therapy to treat human diseases would nonetheless find the prospect of cloning humans abhorrent. In this case, however, goats are being modified to save human lives—a much less controversial use of the technology.

Making Proteins, or How Genes Are Expressed

Transgenic or not, all organisms make proteins from genes in the same way. So far, we've been discussing gene expression in abstract terms: a gene provides instructions to make proteins. But what are those instructions? How is the antithrombin protein actually made by goat cells?

In order to get from a gene to a protein, cells carry out two major steps: **transcription** and **translation.** Briefly, transcription is the process of using DNA to make a **messenger RNA (mRNA)** copy of the gene. Translation is the process of using this mRNA copy as a set of instructions to assemble amino acids into a protein (Infographic 8.7).

As the names *transcription* and *translation* imply, the process is like converting a text into another language.

Why two separate steps? As the names *transcription* and *translation* imply, the process is like converting a text into another language. In this case, the text to be translated is a valuable, one-of-a-kind document: DNA. Just as you would be forbidden to borrow a rare manuscript from the library, and would instead have to rewrite the characters in it onto another sheet, the cell cannot take DNA out of its library—the nucleus. It must first make a copy—the mRNA. The cell can then take this mRNA copy into the cytoplasm, where it is translated into a protein.

Let's take a closer look at both steps. Transcription begins when an enzyme called **RNA polymerase** binds to the regulatory sequence of DNA just ahead of a gene's coding sequence. At that site, cellular machinery unwinds the

TRANSCRIPTION
The first stage of gene expression, during which cells produce molecules of messenger RNA (mRNA) from the instructions encoded within genes.

TRANSLATION
The second stage of gene expression. Translation "reads" mRNA sequences and assembles the corresponding amino acids to make a protein.

MESSENGER RNA (mRNA)
The RNA copy of an original DNA sequence made during transcription.

RNA POLYMERASE
The enzyme that accomplishes transcription. RNA polymerase copies a strand of DNA into a complementary strand of mRNA.

INFOGRAPHIC 8.7

Gene Expression: An Overview

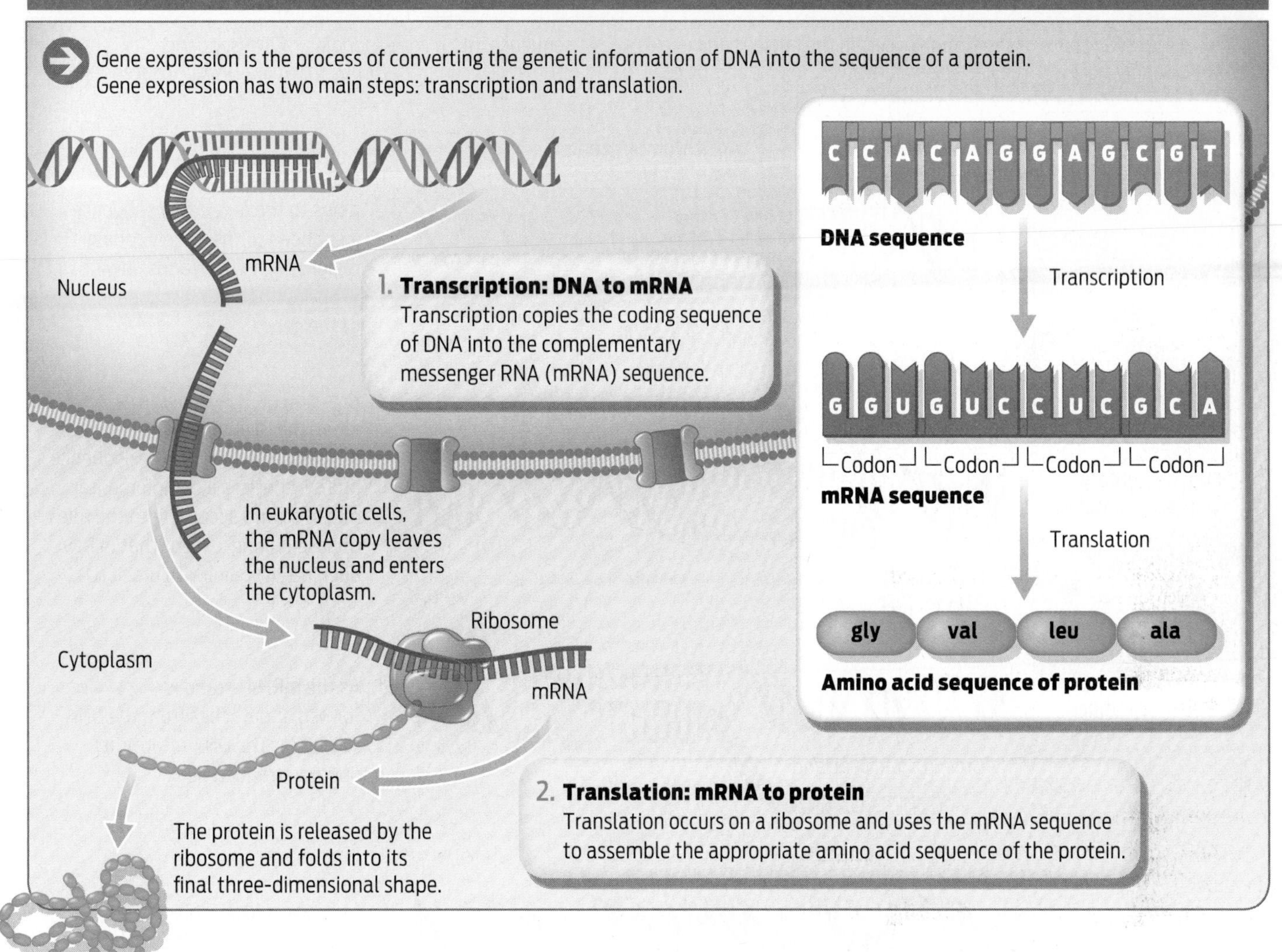

Gene expression is the process of converting the genetic information of DNA into the sequence of a protein.
Gene expression has two main steps: transcription and translation.
Nucleus
mRNA
1. Transcription: DNA to mRNA
Transcription copies the coding sequence of DNA into the complementary messenger RNA (mRNA) sequence.
In eukaryotic cells, the mRNA copy leaves the nucleus and enters the cytoplasm.
Cytoplasm
Ribosome
mRNA
Protein
The protein is released by the ribosome and folds into its final three-dimensional shape.
2. Translation: mRNA to protein
Translation occurs on a ribosome and uses the mRNA sequence to assemble the appropriate amino acid sequence of the protein.
C C A C A G G A G C G T
DNA sequence
Transcription
G G U G U C C U C G C A
Codon Codon Codon Codon
mRNA sequence
Translation
gly val leu ala
Amino acid sequence of protein

RIBOSOME
The cellular machinery that assembles proteins during the process of translation.

CODON
A sequence of three mRNA nucleotides that specifies a particular amino acid.

DNA double helix and RNA polymerase begins moving along one DNA strand. As it moves, the RNA polymerase "reads" the DNA sequence and synthesizes a complementary mRNA strand according to the rules of base pairing. The same rules of base pairing we discussed in the context of DNA apply here, with one difference: RNA nucleotides are made with the base uracil (U) instead of thymine (T). So the complementary base pairs are C with G and A with U **(Infographic 8.8)**.

As its moniker states, messenger RNA serves to relay information. Once the mRNA copy is made, it leaves the nucleus and attaches to a complex cellular machinery called the **ribosome.** This is the start of translation.

During translation, the ribosome "reads" the mRNA message and assembles a chain of amino acids. The ribosome acts like a factory in which mRNA serves as the instruction manual that specifies which amino acids should be joined together to form chains. Amino acids are specified by groups of three nucleotides called **codons.** Each codon is like a word: its letters name a particular amino acid (for example, the codon GGU specifies the amino acid glycine).

Although ribosomes are protein-assembling factories, they don't house all the parts needed to make proteins. Rather, they rely on a delivery system to bring the appropriate amino acids to the assembly site. The delivery system is

INFOGRAPHIC 8.8

Transcription: A Closer Look

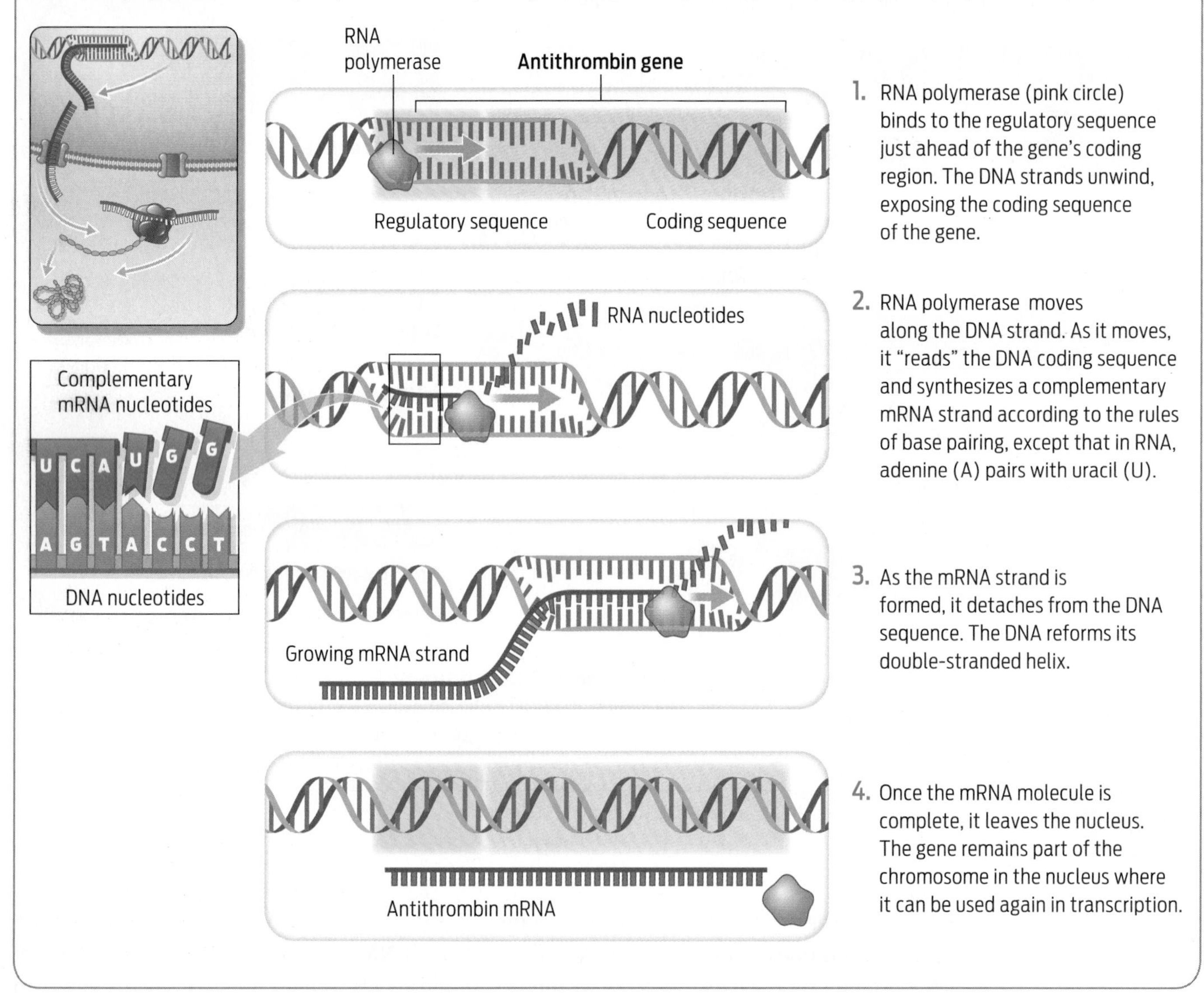

In eukaryotic cells, transcription occurs in the nucleus and copies a DNA sequence into a corresponding mRNA sequence. RNA polymerase is the key enzyme involved. In prokaryotic cells, transcription occurs in the cytoplasm, where DNA is located.
RNA polymerase
Antithrombin gene
Regulatory sequence
Coding sequence
1. RNA polymerase (pink circle) binds to the regulatory sequence just ahead of the gene's coding region. The DNA strands unwind, exposing the coding sequence of the gene.
RNA nucleotides
Complementary mRNA nucleotides
U C A U G G
A G T A C C T
DNA nucleotides
2. RNA polymerase moves along the DNA strand. As it moves, it "reads" the DNA coding sequence and synthesizes a complementary mRNA strand according to the rules of base pairing, except that in RNA, adenine (A) pairs with uracil (U).
Growing mRNA strand
3. As the mRNA strand is formed, it detaches from the DNA sequence. The DNA reforms its double-stranded helix.
Antithrombin mRNA
4. Once the mRNA molecule is complete, it leaves the nucleus. The gene remains part of the chromosome in the nucleus where it can be used again in transcription.

another type of RNA called **transfer RNA (tRNA),** which physically transports amino acids to the ribosome. Each tRNA is structured like an adaptor: one end binds to an amino acid, the other end binds to mRNA. The part that binds mRNA is called the **anticodon** because it base-pairs in a complementary fashion with an mRNA codon. When the amino acid-toting tRNA finds its codon match, it releases the amino acid to the ribosome, which adds it to the growing amino acid chain **(Infographic 8.9).**

The genetic code is universal, meaning that it is the same in all living organisms.

Although the human genome codes for many thousands of different proteins, each one is pieced together from a starting set of a mere 20 amino acids. In the same way that the 26 letters in the alphabet can spell hundreds of thousands of words, the basic set of amino acids can make hundreds of thousands

TRANSFER RNA (tRNA)
A type of RNA that helps ribosomes assemble chains of amino acids during translation.

ANTICODON
The part of a tRNA molecule that binds to a complementary mRNA codon.

INFOGRAPHIC 8.9

Translation: A Closer Look

In the cytoplasm, the ribosome reads the mRNA sequence and "translates" it into a chain of amino acids to make a protein.

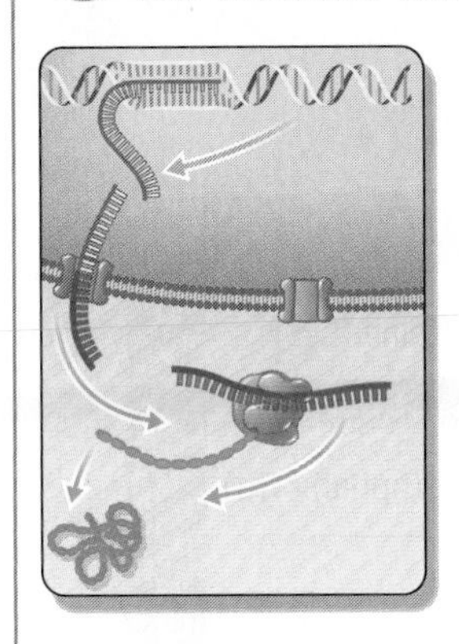

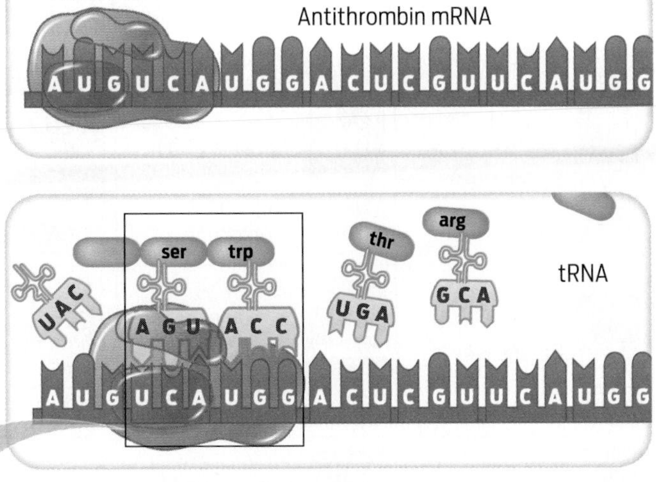

1. The newly transcribed mRNA associates with a ribosome.

2. As the ribosome moves along the mRNA, it "reads" the mRNA sequence in groups of three nucleotides called codons. Each codon specifies a particular amino acid, which is brought to the ribosome by tRNA, which uses its anticodon sequence to find a matching mRNA codon.

Amino acid: Corresponds to the mRNA codon

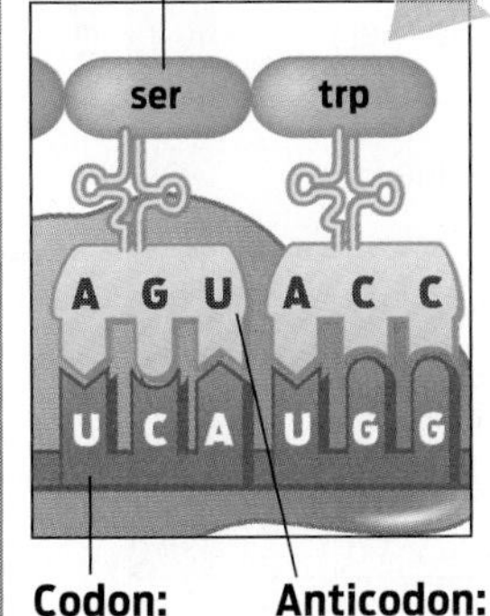

Codon: Triplet code on mRNA

Anticodon: Triplet code on tRNA that recognizes a specific codon

Growing chain

trp thr arg ser

ACC U G A G C A AGU

A U G U C A U G G A C U C G U U C A U G G

3. When the correct tRNA is in place, the specified amino acid is added to the growing chain. The ribosome moves on to the next codon.

Antithrombin protein

AGU UAC

A U G U C A U G G A C U C G U U C A U G G

4. The finished amino acid chain detaches from the ribosome and folds into its three-dimensional shape.

of proteins. The rules by which mRNA codons specify amino acids are known as the **genetic code.**

The genetic code is universal, meaning that it is the same in all living organisms. It is because the code is universal that the mammary cells of a goat carrying the human gene for antithrombin are able to express the gene and produce antithrombin protein in its milk **(Infographic 8.10)**.

GENETIC CODE
The particular amino acids specified by particular mRNA codons.

The Advantages of GMOs and "Pharming"

One of the primary advantages of using transgenic animals to churn out protein drugs is that scientists can produce more complex proteins in a mammal's milk than they can from cell culture—the traditional way that scientists have produced many protein drugs. Because the mammary gland is a natural protein factory, mammalian milk already contains dozens of

INFOGRAPHIC 8.10

The Genetic Code Is Universal

Codons are groups of three-nucleotide sequences within chains of mRNA. Codons specify particular amino acids according to the universal genetic code. Since the genetic code is universal, the same gene will be transcribed and translated into the same protein in all cells and organisms.

Second letter

First letter	U	C	A	G	Third letter
U	UUU, UUC: Phenylalanine (Phe) UUA, UUG: Leucine (Leu)	UCU, UCC, UCA, UCG: Serine (Ser)	UAU, UAC: Tyrosine (Tyr) UAA: Stop codon UAG: Stop codon	UGU, UGC: Cysteine (Cys) UGA: Stop codon UGG: Tryptophan (Trp)	U C A G
C	CUU, CUC, CUA, CUG: Leucine (Leu)	CCU, CCC, CCA, CCG: Proline (Pro)	CAU, CAC: Histidine (His) CAA, CAG: Glutamine (Gln)	CGU, CGC, CGA, CGG: Arginine (Arg)	U C A G
A	AUU, AUC, AUA: Isoleucine (Iso) AUG: Methionine (Met); start codon	ACU, ACC, ACA, ACG: Threonine (Thr)	AAU, AAC: Asparagine (Asn) AAA, AAG: Lysine (Lys)	AGU, AGC: Serine (Ser) AGA, AGG: Arginine (Arg)	U C A G
G	GUU, GUC, GUA, GUG: Valine (Val)	GCU, GCC, GCA, GCG: Alanine (Ala)	GAU, GAC: Aspartic acid (Asp) GAA, GAG: Glutamic acid (Glu)	GGU, GGC, GGA, GGG: Glycine (Gly)	U C A G

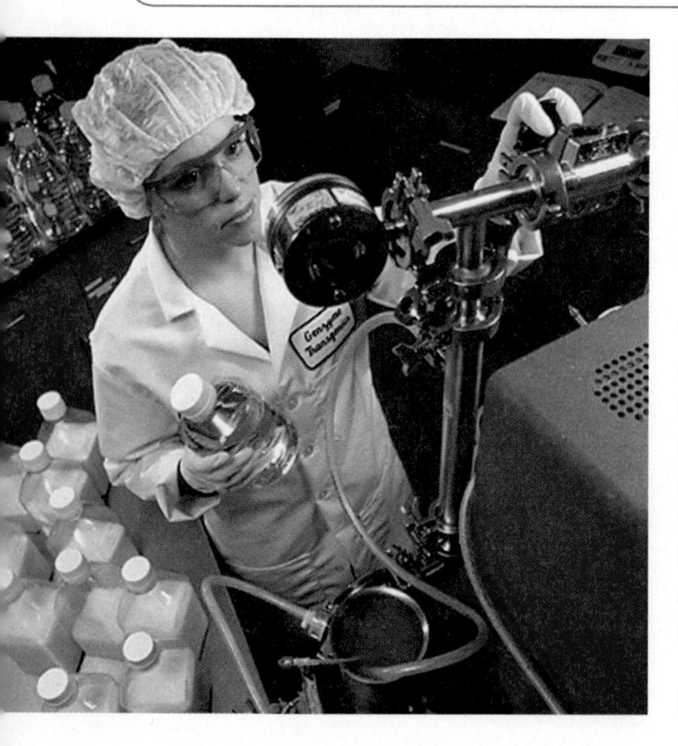

It is easier to scale up medicine derived from milk than to rely on blood donations.

complex proteins that are expressed at high levels.

By contrast, with hamster cells in culture, "You are asking the cell to produce a product in a way and in an environment completely different from what it would naturally produce," according to Thomas Newberry, former vice president for communications at GTC. And when scientists do succeed in getting such cells to produce complex proteins, they are expensive to make in large volumes; consequently the drugs are very expensive.

As an example, Newberry points to another human protein drug, one called factor VIIa. This is a protein that helps blood clot. Some people with the genetic disease hemophilia are born with clotting factor alleles that either encode nonfunctional clotting factor protein or no clotting factor protein at all. Consequently, if a hemophiliac cuts himself, he must be quickly transfused with clotting factor–otherwise he may bleed to death. Doctors usually give patients factor VIIa protein to restore normal coagulation and prevent excessive bleeding.

Companies that sell the protein drug produce it via cell culture. But it is extremely expensive: one milligram of factor VIIa can cost up to $1,000. Fortunately, hemophiliacs don't need much of the clotting factor to stop bleeds. But because of the drug's expense, they are given the drug only in an emergency. Over time, however, such rescue therapy, while minimizing initial cost, can be detrimental to patients' health: joints and other areas where bleeding typically occurs can become stiff and difficult

"I won't be surprised if people start to think that reinjecting blood products into other people is 'barbaric.'" —Thomas Newberry

to move. Transgenic animals can produce human clotting factors in their milk in large volumes for about one tenth of the amount of money it would take to produce the same proteins using cell culture. In fact, GTC Biotherapeutics is working on establishing transgenic animals to produce two other human clotting proteins: factor VIII and factor IX. In the FDA press release announcing the approval of GTC's antithrombin deficiency medication, Dunham was quoted: "I am pleased that this approval makes possible another source of an important human medication."

"We have the potential to build an abundant and controlled supply for any plasma protein," says Newberry, who predicts that protein drugs extracted from human blood may become a thing of the past. "In the future, I won't be surprised if people start to think that reinjecting blood products into other people is 'barbaric.'" ■

Summary

- Genes provide instructions to make proteins. The process of using the information in genes to make proteins is known as gene expression.
- Proteins are folded chains of amino acids that make up cell structures and help cells to function properly.
- Amino acid sequences determine the shape and function of a protein.
- Many drugs act on proteins in the body, or are themselves proteins.
- A change in the DNA sequence of a gene can change the corresponding amino acid sequence, and therefore the function, of a protein.
- Different versions of the same gene, those with different nucleotide sequences, are called alleles.
- Every gene has two parts: a coding sequence and a regulatory sequence. The coding sequence determines the identity of a protein and the regulatory sequence determines where, when, and how much of the protein is produced.
- Gene expression occurs in two stages, transcription and translation, which take place in separate compartments in eukaryotic cells.
- Transcription is the first step of gene expression and copies the information stored in DNA into mRNA. Transcription occurs in the nucleus.
- Translation is the second step of gene expression and uses the information stored in mRNA to assemble a protein. Translation occurs in the cytoplasm.
- Proteins are assembled by ribosomes with the help of tRNA.
- The genetic code is the set of rules by which DNA sequences are translated into protein sequences; the code is shared by all living organisms.
- Through genetic engineering, genes from one species of organism can be inserted into the genome of another species of organism to make a transgenic organism.

Chapter 8 Test Your Knowledge

PROTEIN STRUCTURE AND FUNCTION

Proteins have a unique three-dimensional structure that specifies their function. The structure of a protein is determined by its corresponding gene sequence.

HINT **See Infographics 8.1–8.4.**

KNOW IT

1. What determines a protein's function?

2. The final product of gene expression is
   a. a DNA molecule.
   b. an RNA molecule.
   c. a protein.
   d. a ribosome.
   e. an amino acid.

USE IT

3. Heating can cause a protein to denature, or unfold. What do you think would happen to a protein's function in this case? Explain your answer.

4. Insulin is a protein that is used therapeutically to treat people with diabetes. In your own words, describe the relationship between the insulin gene and the insulin protein.

GENE STRUCTURE

All genes have two key parts: a regulatory sequence and a coding sequence. To review gene structure, refer to Infographics 8.4 and 8.5.

KNOW IT

5. The difference between two alleles of a gene is best ascertained by
   a. examining the amount of protein produced from each allele.
   b. examining the structure of the protein produced from each allele.
   c. examining the amount of mRNA produced from each allele.
   d. examining the nucleotide sequence of each allele.
   e. examining the amount of tRNA produced from each allele.

6. If a functional allele of antithrombin is expressed,
   a. blood clots will be more likely to form in the wrong place.
   b. blood clots will be less likely to form in the wrong place.
   c. functional antithrombin protein will be present in blood.
   d. a and c
   e. b and c

USE IT

7. You are a doctor. Your patient has reduced levels of normal functioning antithrombin. Would you suspect a problem in the regulatory or in the coding sequence of the antithrombin gene? Why?

8. If you wanted to use genetic engineering to increase the amount of antithrombin this patient produces, would you modify the regulatory sequence or the coding sequence? Explain your answer.

MAKING TRANSGENIC ORGANISMS

Transgenic organisms are becomingly increasingly important in agriculture and medicine.

HINT **See Infographics 8.5 and 8.6.**

KNOW IT

9. Melanin is expressed in skin cells and gives skin its color. If you wanted to express a different gene in skin cells, which part of the melanin gene would you use? Why? If you wanted to produce melanin in yeast cells, what part of the melanin gene would you use? Why?

USE IT

10. Explain why scientists used the beta casein regulatory sequence to express human antithrombin in goats' milk.

GENE EXPRESSION

Gene expression is the multistep process of converting the information of DNA into proteins.

HINT **See Infographics 8.7–8.10.**

KNOW IT

11. For each structure or enzyme listed, indicate by N (nucleus) or C (cytoplasm) its active location in eukaryotic cells:
RNA polymerase _____
Ribosome _____
tRNA _____
mRNA _____

12. The sequence of a strand of DNA of a gene is AGATACGAAACA.

a. Write the sequence of the complementary strand of DNA.

b. Write the sequence of the mRNA that is complementary to the original DNA strand.

c. Refer to the genetic code in Infographic 8.10 to translate this short stretch of RNA. How many amino acids does it encode? What are they? (Remember that translation always begins at a start codon.)

USE IT

13. A change in DNA sequence can affect gene expression and protein function. What would be the impact of each of the following changes? How, specifically, would each affect protein or mRNA structure, function, and levels?

a. a change that prevents RNA polymerase from binding to a gene's regulatory region

b. a change in the coding sequence that changes the amino acid sequence of the protein

c. a change in the regulatory region that allows transcription to occur at much higher levels

d. a combination of the changes in b and c.

14. The 18th codon in the coding sequence of a gene's mRNA is CCA.

a. What amino acid is encoded by this codon?

b. What amino acid would be encoded if the codon changed to CCG?

c. What amino acid would be encoded if the codon changed to CUA?

SCIENCE AND ETHICS

15. Some people with diabetes would die without insulin because their bodies can no longer produce this protein. Historically, scientists purified insulin from the pancreas of pigs. Human insulin is now produced by inserting an artificial gene construct into bacteria. What are the ethical pros and cons of each type of insulin?

Milestones in Biology

Sequence Sprint

What You Will Be Learning

The Human Genome Project was a massive undertaking that continues to spur new technology and discoveries in scientific research and medicine.

Paramedic Plants

What You Will Be Learning

Paramedic Plants

Will herbs be the next cancer therapy?

In the late nineties, online chat rooms began buzzing that an herbal supplement called PC-SPES could shrink prostate tumors. This was a "natural" approach that caused fewer side effects than conventional prostate cancer therapy, so it was claimed. PC-SPES was introduced to the U.S. market in 1996, and within a few years, as many as 10,000 men in the United States with prostate cancer were taking the supplement and seeing their tumors shrivel—or so the rumors went.

Although some studies had shown that PC-SPES could kill cancer cells, no one had ever studied the supplement in men with prostate cancer. Concerned that so many cancer patients were taking an unproven therapy, a group of scientists at the National Center for Alternative and Complementary Medicine decided to test the herbal mixture in clinical trials.

To their surprise, initial studies seemed to support the rumors. PC-SPES, which was marketed as a mixture of eight herbs known in China since ancient times, appeared to fight prostate tumors.

But now community chat rooms were buzzing again, this time with suspicions that the supplement was contaminated with toxic metals and perhaps even harmful drugs. These stories made their way to the California Department of Health Services, which decided to investigate. The department's analyses were shocking: in many instances, the quantity of each herb varied considerably from bottle to bottle. Moreover, some bottles were laced with three different prescription drugs, including an estrogen-like drug and a blood-thinner. The FDA—the U.S. Food and Drug Administration—issued a warning to all consumers to stop taking PC-SPES. By late 2002, the company that made PC-SPES had voluntarily taken the product off the market and went bankrupt soon after.

PC-SPES isn't the only herbal supplement that's gotten into trouble with the law. Authorities have found that many supplements are contaminated with dangerous heavy metals or bacteria. While contamination with prescription drugs or other substances appears to

be a rare phenomenon, investigators have found that the amount of active ingredient in many supplements commonly varies from pill to pill and from bottle to bottle. Some melatonin pills, for example, which some people take to ward off jet lag, have been found to contain very little melatonin, and the quantity can vary from pill to pill within the same bottle, according to a 2003 study by researchers at the University of Colorado, Denver.

This lack of consistency makes it difficult for consumers to know if they are getting what they are paying for. In the United States, supplements like echinacea, ginseng, and St. John's wort aren't stringently regulated. Although manufacturers are prohibited from making false statements on their labels, there is no government agency that *certifies* that a supplement actually contains what is listed on its label or does what it claims to do. More significantly, very few herbal supplements have been rigorously studied, and few have been shown to contain consistent levels of active compounds from batch to batch, or proved to prevent or treat illnesses. Authorities typically discover that a product has been falsely advertised or contaminated only after investigating complaints from consumers.

But some of this is changing. In 2004, the FDA put procedures in place that allow companies to apply to the agency to market an herbal supplement almost like a pharmaceutical drug. Not only must a company prove with clinical trial data that a supplement works, it must also prove that it can produce the supplement with consistent quality and quantity from batch to batch. While the new guidelines don't require sellers of supplements to apply to the FDA, they do offer manufacturers a financial incentive to cooperate: a company can sell an FDA-approved supplement by prescription for 5 years without competition from a similar product. In 2006, the FDA approved the first such herbal, a topical green tea extract used to treat genital warts, available by prescription only.

"The payoff will be huge," says Robert Tilton, vice president of science and technology at PhytoCeutica, Inc., a company based in New Haven, Connecticut, that is investigating the use of Chinese herbs in the treatment of cancer. There are so many diseases or afflictions for which available treatments are inadequate, Tilton points out. While researchers are investigating herbs to treat everything from alcohol addiction to heart disease, the next to be offered to patients by prescription may be herbs that help treat cancer. At a time when doctors are actively searching for new cancer therapies, herbal supplements may offer a way to improve existing cancer therapies and make them less toxic.

Cell Division and Cancer

The founders of PhytoCeutica sought to study herbs that might help improve the treatment of **cancer**–a disease of unregulated cell division, for which most existing therapies cause severe side effects. They decided on a mixture of four herbs developed in China more than 1,800 years ago to treat gastrointestinal distress. They dubbed the mixture PHY906.

The notion that an herb can have a medicinal benefit certainly is not new. As the origins of PHY906 attest, traditional cultures have relied on herbs to treat disease for thousands of years. Many modern medicines are also derived from plant sources. Salicylic acid, for example, the primary ingredient in aspirin, was initially extracted from the bark of the willow tree. And the cancer drug paclitaxel was originally extracted from the bark of the Pacific yew tree.

TABLE 9.1

Drugs from Plants

The well-established drugs listed below are among dozens that were developed after scientists began to analyze the chemical constituents of plants used by traditional peoples for medicinal or other purposes.

DRUG	MEDICAL USE	PLANT SOURCE	COMMON PLANT NAME
Aspirin	Reduces pain and inflammation	*Filipendula ulmana*	Meadowsweet
Codeine	Eases pain, suppresses coughing	*Papaver somnifenum*	Opium poppy
Ipecac	Induces vomiting	*Psychotria ipecacuanha*	Ipecacuanha
Pilocarpine	Reduces pressure in the eye	*Pilocarpus jaborandi*	Jaborandi
Pseudoephedrine	Reduces nasal congestion	*Ephedra sinica*	Ephedra, ma huang
Quinine	Combats malaria	*Cinchona pubescens*	Quinine tree
Reserpine	Lowers blood pressure	*Rauvolia serpentina*	Serpentine wood, snakeroot
Scopolamine	Eases motion sickness	*Datura stramonium*	Jimson weed
Theophyline	Opens bronchial passages	*Camellia sinensis*	Tea plant
Paclitaxel	Chemotherapeutic drug	*Taxus brevitolin*	Pacific yew tree

Source: Cox, PA et al. 1994. *Scientific American* 82–87.

Traditional cultures have relied on herbs to treat disease for thousands of years.

Modern drugs tend to contain a single active ingredient that has been highly purified and tested in clinical trials and shown to be safe and effective **(Table 9.1)**.

But herbal supplements aren't nearly as well studied or developed as conventional prescription drugs. Herbal supplements are dried pieces of a plant or fungus that are typically ground up

CANCER
A disease of unregulated cell division: cells divide inappropriately and accumulate, in some instances forming a tumor.

INFOGRAPHIC 9.1

How Conventional Drugs Differ from Herbal Supplements

Conventional Drugs

One specific active ingredient is purified from a plant or fungus or synthesized in the laboratory and concentrated into pill, capsule, or injectable form.

The bark of the Pacific yew, *Taxus brevifolia*

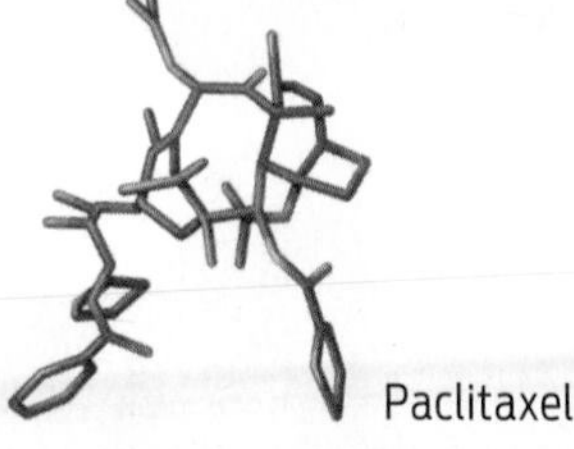

Paclitaxel, commercially known as Taxol, is a single ingredient originally purified from the bark and needles of the Pacific yew tree, *Taxus brevifolia*.

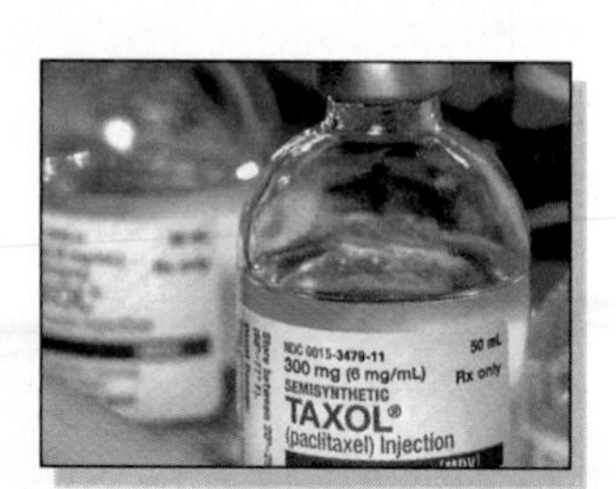

Mandatorily tested in clinical trials and approved by the FDA

Herbal Supplements

Plants, algae, fungi, and combinations of these are used as a tea, an extract, or ground into powder and sold in capsule form. They contain complex mixtures of many different unpurified plant molecules.

Baikal skullcap
Scutellaria baicalensis

PC-SPES Herbal Supplement contains extracts from specific parts of the following eight plants and fungus:

Baikal skullcap (Root)	*Scutellaria baicalensis*
Reishi (Stem)	*Ganoderma lucidum*
Rabdosia (Leaf)	*Rabdosia rubescens*
Dyer's woad (Leaf)	*Isatis indigotica*
Chrysanthemum (Flower)	*Dendrathema morifolium*
Saw palmetto (Berry)	*Serenoa repens*
San-Qi ginseng (Root)	*Panax notoginseng*
Licorice (Root)	*Glycyrrhiza uralensis*

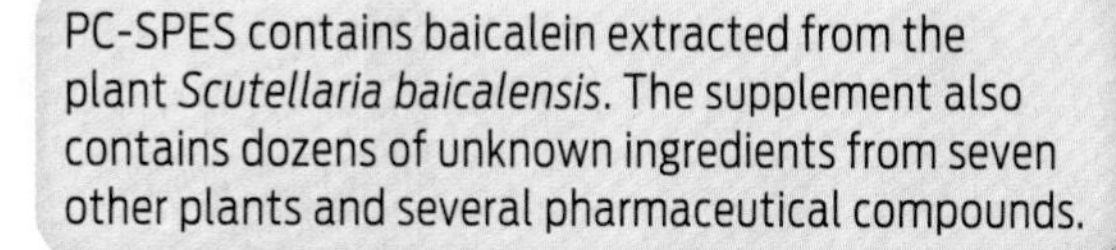

PC-SPES contains baicalein extracted from the plant *Scutellaria baicalensis*. The supplement also contains dozens of unknown ingredients from seven other plants and several pharmaceutical compounds.

Not mandatorily tested in clinical trials or approved by the FDA

> **"Whether botanicals will provide a source of products to treat cancer is hard to say for sure."**
> –K. Simon Yeung

and packaged into pills or capsules, or brewed into tea. Botanical herbs used for supplements often contain a number of different compounds rather than one single active ingredient (**Infographic 9.1**).

Unlike drugs, herbal supplements typically contain several compounds, and they can target diverse biochemical pathways. This is precisely why PhytoCeutica founders believed that PHY906 might prove helpful in treating cancer. Many cancer therapies, especially chemotherapy, cause vomiting and diarrhea as side effects. Because the herbal mixture in PHY906 works on different biochemical pathways, it seemed plausible that the herbs in PHY906 might not only lessen the side effects of chemotherapy but also have other, positive, effects. And indeed, early studies with cancer patients suggest that PHY906 can reduce vomiting and diarrhea and may even make chemotherapy more effective.

"Whether botanicals will provide a source of products to treat cancer is hard to say for sure," says K. Simon Yeung, a research pharmacist

INFOGRAPHIC 9.7

Conventional Cancer Therapy

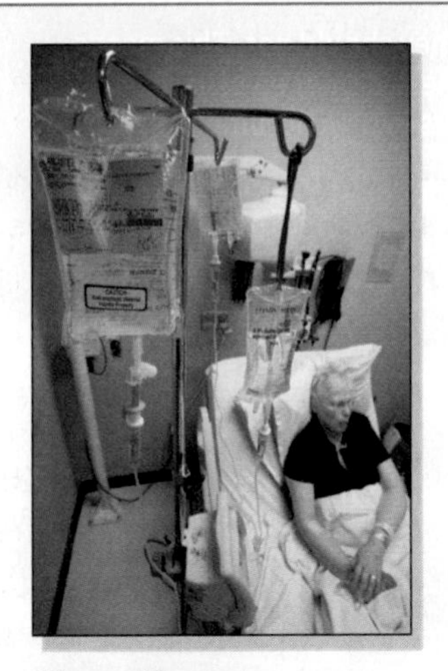

Chemotherapy

Mode of Action:
Chemicals interfere with different parts of cell division. For example, Taxol interferes with the separation of sister chromatids during mitosis; irinotecan interferes with DNA replication.

Side Effects:
Because chemotherapy targets all dividing cells, including healthy ones, the treatment causes side effects. Patients may experience nausea, vomiting, diarrhea, hair loss and a high risk of infection when the treatment interferes with cell division in the digestive tract, hair follicles, and bone marrow.

Radiation

Mode of Action:
High-energy radiation severely damages DNA. Excessive DNA damage will cause cells to die, either by apoptosis or by interrupting DNA replication.

Side Effects:
Because radiation targets all dividing cells in the irradiated area, it causes side effects. If cells in the digestive tract, hair follicles, or bone marrow are irradiated, patients may experience nausea, vomiting, diarrhea, hair loss, and susceptibility to infection, as the radiation interferes with cell division in these locations.

drugs to aid in the fight against this deadly disease, and cancer patients have been surviving longer and longer. Cancer death rates have been slowly declining over the years; the latest research shows that deaths from all cancers dropped 15.8% between 1991 and 2006. Part of the decline in deaths is due to more effective chemotherapeutic drugs.

The downside of both radiation and chemotherapy is that they can cause severe side effects. That's because neither therapy is very specific–both radiation and chemotherapy damage all rapidly dividing cells, including healthy ones. These cancer treatments kill healthy cells lining the intestinal tract, the cells in hair follicles, and cells in bone marrow (which divide rapidly throughout our lives to replace worn-out blood cells), often leading to side effects such as hair loss, vomiting, bruising, and susceptibility to infections. Since healthy cells can repair DNA damage, they aren't as severely affected as cancer cells, which have dysfunctional checkpoints and hobbled repair mechanisms. While scientists are trying to develop cancer therapies that target only cancerous cells, chemotherapy and radiation remain the mainstays of cancer therapy today. Anything that might lessen side effects without hindering a drug's effectiveness would be a boon to patients **(Infographic 9.7)**.

The herbal supplement PHY906 is meant to be taken in addition to chemotherapy. So far, it has been tested in people with colon, liver, and pancreatic cancer who are also being treated

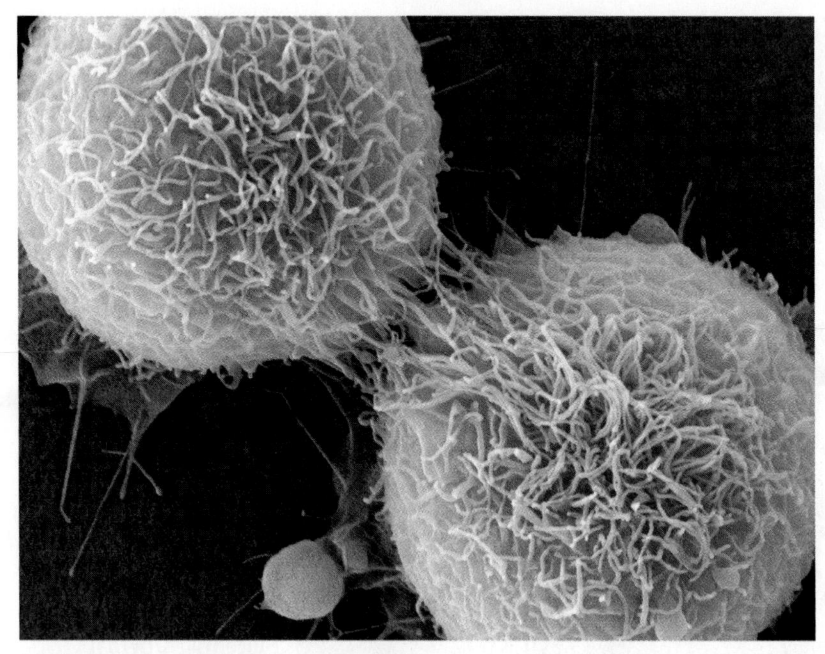

Cervical cancer cells dividing unchecked.

with the chemotherapy drug irinotecan. These initial studies suggest that PHY906 can reduce the side effects of chemotherapy.

Although researchers aren't sure exactly how PHY906 works, they do have some clues. Research in animals suggests that the herb mixture reduces inflammation in the gut. Because chemotherapy kills cells lining the stomach and intestines, gut tissue becomes inflamed. The ability of PHY906 to calm inflammation might account for its ability to reduce nausea, vomiting, and diarrhea during chemotherapy. And while PHY906 does not by itself kill cancer cells, animal studies by the company suggest that the supplement enhances the effect of chemotherapy by making cancer cells more permeable to drugs. (Infographic 9.8). That PHY906 can reduce side

INFOGRAPHIC 9.8

Herbal Supplements May Complement Cancer Therapy

Experiments show that the herbal supplement PHY906 can improve traditional cancer therapy. While PHY906 doesn't appear to kill cancer cells when used alone, it enhanced the ability of a chemotherapeutic agent called irinotecan to shrink colorectal tumors in mice. In other words, the drug and herbal supplement worked synergistically. Since PHY906 does not directly kill cancer, it will likely only complement existing cancer therapy regimens.

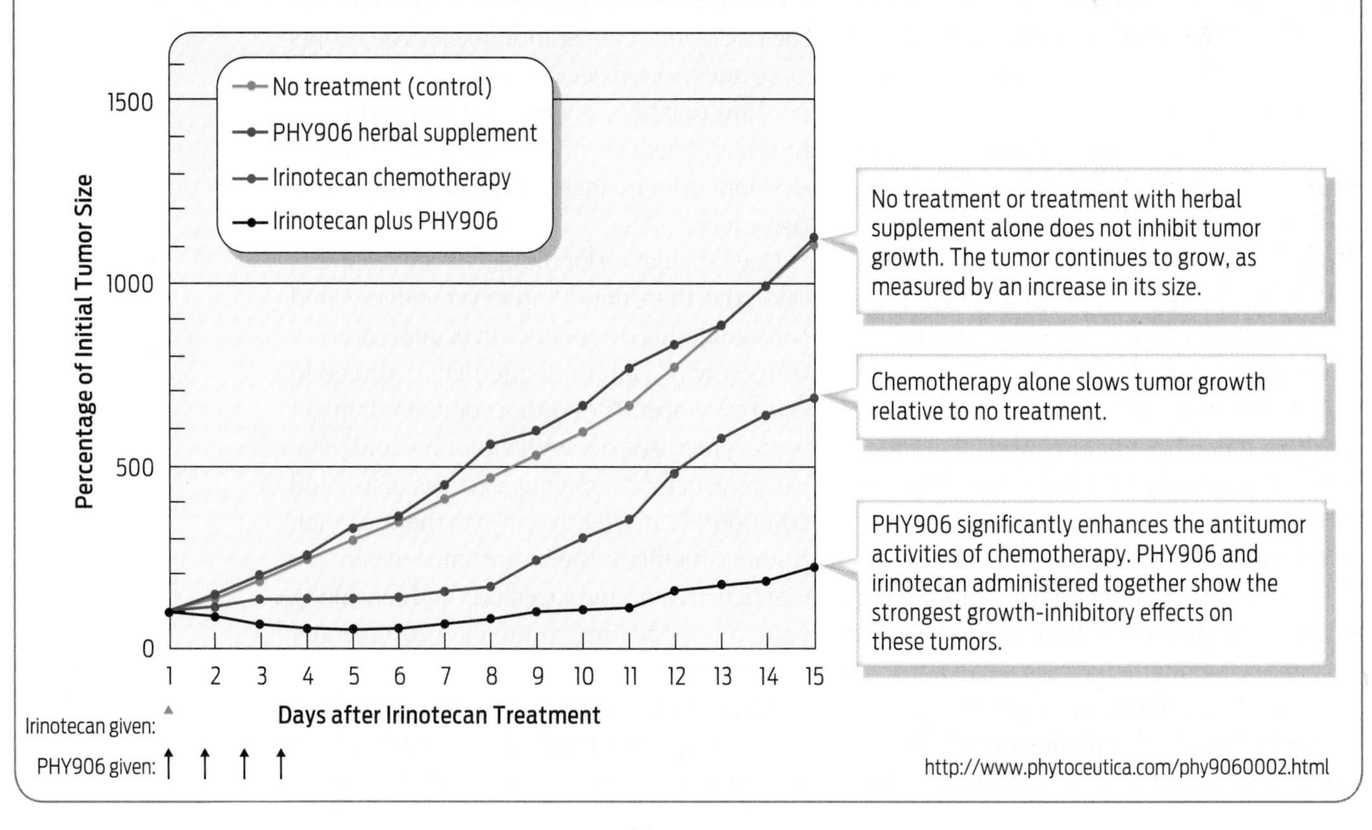
No treatment (control)
PHY906 herbal supplement
Irinotecan chemotherapy
Irinotecan plus PHY906
Percentage of Initial Tumor Size
1500
1000
500
0
1 2 3 4 5 6 7 8 9 10 11 12 13 14 15
Days after Irinotecan Treatment
Irinotecan given:
PHY906 given:
No treatment or treatment with herbal supplement alone does not inhibit tumor growth. The tumor continues to grow, as measured by an increase in its size.
Chemotherapy alone slows tumor growth relative to no treatment.
PHY906 significantly enhances the antitumor activities of chemotherapy. PHY906 and irinotecan administered together show the strongest growth-inhibitory effects on these tumors.
http://www.phytoceutica.com/phy9060002.html

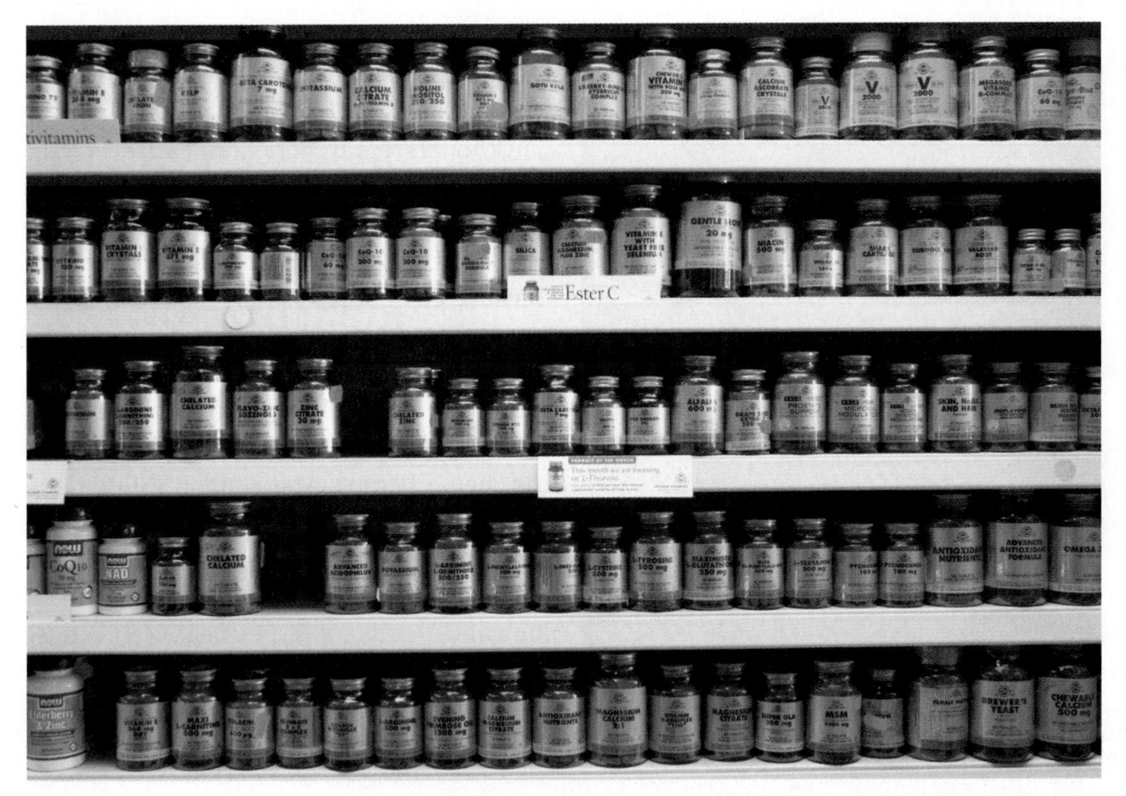

Thanks to recent FDA incentives, herbal supplements may soon join the ranks of some of our most valued prescription drugs.

effects from chemotherapy also raises the possibility that higher doses of chemotherapy could be administered without a corresponding increase in side effects.

PHY906 isn't the only herbal supplement that is showing promise in cancer treatment: some common plant ingredients may also aid the fight against cancer. For instance, scientists at the University of California, Los Angeles, have found in laboratory experiments that green tea extracts, which have been popularly thought to be cancer-fighting agents, can in fact slow down aberrant cell division and increase the likelihood that a damaged cell will go through apoptosis. Studies on the spice turmeric, commonly used in Asian cooking, suggest that it can help fight cancer: in animals, turmeric has been shown to protect the liver, inhibit tumors, and reduce inflammation. A few studies have also shown that the Chinese herb astragalus, combined with another herb extracted from the berries of the glossy privet tree native to Japan and Korea, can boost the immune system and help it fight cancer.

But the news on supplements isn't all good. One of the largest studies to date, which followed almost 30,000 smokers for 8 years, found that beta-carotene supplements actually increased the risk of lung cancer in smokers. The same study found that vitamin E supplements appeared to have no effect in reducing the risk of lung cancer in smokers. The results were unexpected because epidemiologic studies show that people who eat diets rich in vitamin E and beta-carotene have a lower risk of developing lung cancer, suggesting a preventative effect.

One explanation for the discrepancy, experts say, is that there may be other yet undiscovered cancer-fighting compounds in plants that act in concert with vitamin E and beta-carotene to ward off cancer. These other compounds might work synergistically with vitamin E and beta-carotene in foods, making extracts of isolated compounds ineffective. Nevertheless, such findings highlight the importance of eating a diet rich in fruits and vegetables and consulting your doctor before you decide to take any supplement.

While it's too soon to be certain that any herbal supplement will join the ranks of our most valuable prescription drugs, most

In the next few years, herbs might join conventional drugs as therapeutics.

experts are optimistic. After all, many of our most important drugs started their pharmaceutical life as plant extracts, so it's reasonable to think that some of today's herbals will eventually prove useful, too. In the next few years, herbs might join conventional drugs as therapeutics, according to Mary Hardy, associate director of UCLA's Center for Dietary Supplements Research. And with the FDA's new initiative that offers the agency's approval to herbal supplements that have been shown to be effective through clinical trials, we can expect to see greater integration of these supplements into medical treatments in the United States. A similar procedure already is in place in Europe: Germany and the Netherlands require clinical evidence of an herb's efficacy stated on its label, and herbal drugs are prescribed by doctors in those countries.

Some herbs have valid medicinal uses, says Hardy, and studying them will help doctors better advise their patients. "Eventually," she says, "herbs will be integrated into the broader medical paradigm." ■

Summary

- Cell division is a fundamental feature of life, necessary for normal growth, development, and repair of the body.
- The cell cycle is the sequence of steps that a cell undergoes in division. Stages of the cell cycle include interphase, mitosis, and cytokinesis.
- During mitosis, replicated chromosomes segregate to opposite poles of the dividing cell; during cytokinesis, the cell physically divides into two daughter cells.
- Cell cycle checkpoints ensure accurate progression through the cell cycle; repair mechanisms at each checkpoint can fix mistakes that occur, such as DNA damage.
- In the absence of proper checkpoint function, cells can acquire DNA damage during cell division and pass these DNA defects on to daughter cells.
- Mistakes in the course of cell division can lead to cancer, which is unregulated cell division.
- Cancer cells have lost the ability to regulate cell division and reproduce uncontrollably, often eventually forming a tumor.
- Conventional cancer treatments—chemotherapy and radiation—kill all rapidly dividing cells, both cancer cells and healthy cells.
- Many drugs, including some of those used to treat cancer, are extracted from plants.
- Herbal supplements may enhance conventional cancer treatments.

Chapter 9 Test Your Knowledge

BASICS OF CELL DIVISION AND THE CELL CYCLE

Dividing cells progress through a series of stages known as the cell cycle. Checkpoints monitor passage through the cell cycle.

HINT **See Infographics 9.2–9.5.**

KNOW IT

1. Following mitosis and cytokinesis, daughter cells are
- **a.** genetically unique.
- **b.** genetically identical to each other.
- **c.** genetically identical to the parent cell.
- **d.** contain half of the parent cell's chromosomes.
- **e.** b and c

2. During the cell cycle, DNA is replicated during
- **a.** mitosis.
- **b.** G_1.
- **c.** S.
- **d.** G_2.
- **e.** in cytokinesis

3. Explain how embryonic development, wound healing, and replacement of blood cells are related.

USE IT

4. If a cell fails to replicate its DNA completely, what will happen?
- **a.** It will progress through G_2 and mitosis.
- **b.** It will die by apoptosis.
- **c.** It will pause to allow DNA replication to complete.
- **d.** It will stop in S phase and never progress further through the cell cycle.
- **e.** It will stay in interphase indefinitely.

5. Many drugs interfere with cell division. Why shouldn't pregnant women take these drugs?

CANCER AND CANCER THERAPIES

When cells fail to accurately progress through the cell cycle, cancer may arise. Cancer cells may have lost checkpoint function, or may divide even without a signal to do so.

HINT **See Infographics 9.1 and 9.6–9.8.**

KNOW IT

6. A normal cell that sustains irreparable amounts of DNA damage will most likely
- **a.** divide out of control.
- **b.** die by apoptosis.
- **c.** arrest in G_2.
- **d.** immediately go back to S phase.
- **e.** stop in S phase and never progress through the cell cycle.

7. Explain why traditional chemotherapy can cause nausea, diarrhea, and hair loss.

8. Which type of cancer treatment relies on purified chemicals to kill rapidly dividing cells?
- **a.** chemotherapy
- **b.** radiation therapy
- **c.** herbal supplement therapy
- **d.** a combination of chemotherapy and herbal supplement therapy
- **e.** none of the above

USE IT

9. After a bad sunburn, skin usually peels. What process best describes what happens to the burned skin cells?
- **a.** skin cancer
- **b.** metastasis
- **c.** apoptosis
- **d.** checkpoint failure
- **e.** cytokinesis

10. Liver cells and neurons rarely, if ever, divide in normal circumstances. The cells lining the digestive tract are replaced by cell division on a regular basis. Explain why chemotherapy frequently causes digestive symptoms but less frequently causes cognitive symptoms.

11. Your pet mouse has developed colon cancer. Which of the following treatments will likely be most effective?
- **a.** PHY906
- **b.** irinotecan chemotherapy
- **c.** PHY906 plus irinotecan
- **d.** radiation therapy
- **e.** There is no treatment for colon cancer.

12. Look at Infographic 9.8.
- **a.** Does irinotecan actually shrink tumors in the colon? Explain your answer.
- **b.** Does PHY906 plus irinotecan shrink colon tumors? Explain your answer.

13. Why might a beta-carotene supplement not have the same effect on cancer as a diet with lots of food rich in beta-carotene?

A CLOSER LOOK AT MITOSIS AND THE CELL CYCLE

Mitosis is a critical stage of cell division. It ensures that chromosomes accurately separate into daughter cells.

HINT **See Infographic 9.4 and Up Close: The Phases of Mitosis.**

KNOW IT

14. During which stage of the cell cycle do sister chromatids separate from each other?

15. During which stage of the cell cycle are sister chromatids initially produced?

USE IT

16. What would be the result if a cell completed interphase and mitosis but failed to complete cytokinesis? (That is, how many cells would there be, and how many chromosomes relative to the parent cell would those cells have?)

17. Looking at **Up Close: The Phases of Mitosis**, would you say that a drug that stabilizes spindle fibers, preventing them from shortening, would be a valuable chemotherapy drug? Why or why not?

SCIENCE AND ETHICS

18. What are some of the risks of taking an over-the-counter herbal supplement as an alternative to conventional cancer therapy?

19. PHY906 has been tested in mice. What steps would you take to establish its efficacy in humans ethically and safely?

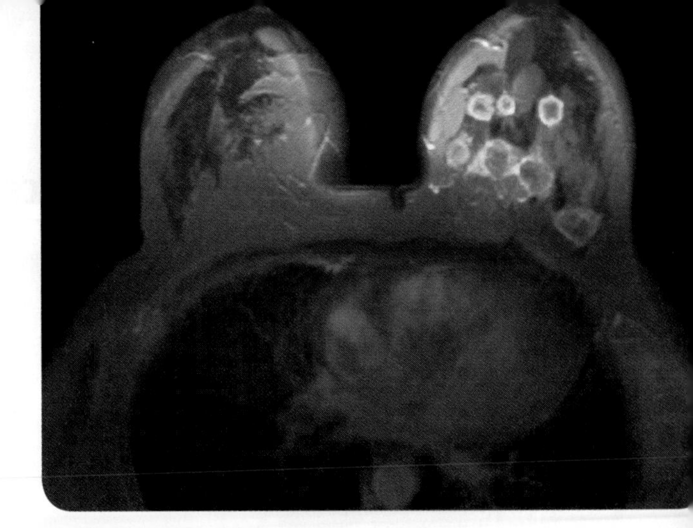

Chapter 10 Mutations and Cancer

Fighting Fate

What You Will Be Learning

10.1 Inheriting One Mutated *BRCA1* Allele Increases the Risk of Cancer

10.2 Mistakes in DNA Replication Can Produce Mutations

10.3 Mutations in DNA Can Alter Protein Function and Cause Cancer

10.4 What Causes Mutations?

10.5 Mutations in Two Types of Cell Cycle Genes Cause Most Types of Cancer

10.6 Tumors Develop in Stages as Mutations Accumulate in a Cell

10.7 *BRCA* Mutation Increases the Risk of Breast Cancer

INFOGRAPHIC 10.3

Mutations in DNA Can Alter Protein Function and Cause Cancer

Mutations alter the nucleotide sequence of DNA. If a mutation changes the coding region of any gene, the corresponding protein may be dysfunctional. When the protein in question helps regulate the cell cycle, cancer may result.

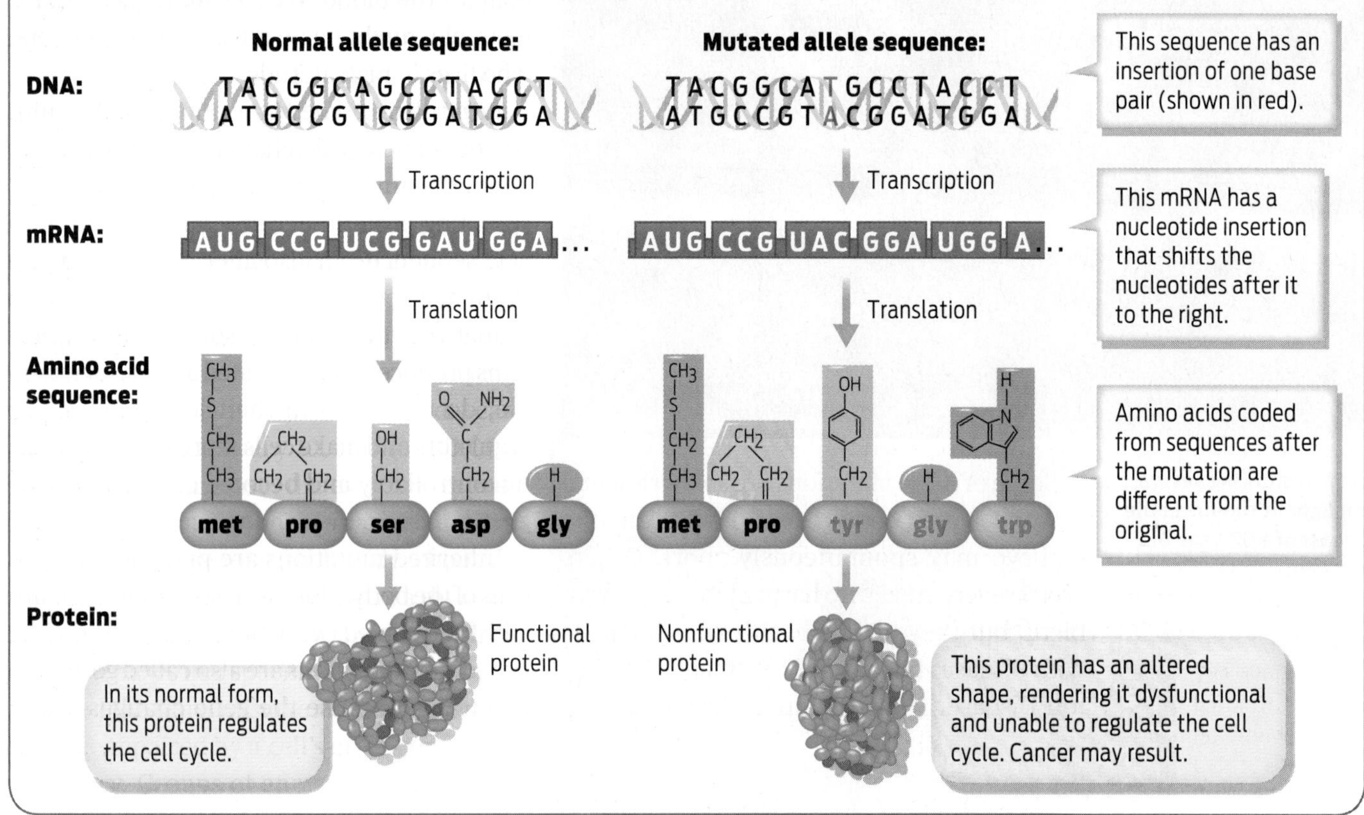

the sperm and egg cells–the germ cells–and can be passed from parent to child each generation.

By contrast, mutations in somatic cells–the cells in the rest of the body–are not passed on to future generations, although they can cause disease. A person who acquires a mutation in a skin cell from too much sun exposure, for example, will not pass this mutation on to his or her children. That's because the mutation did not occur in sperm or egg cells, nor will it affect those cells. This mutation can, however, be passed by mitosis and cell division to daughter cells of the mutated cell and cause disease in the affected person. This is one way *nonhereditary* cancers develop.

Now imagine germ-line mutations accumulating over thousands of years in a population. As long as a mutation does not affect a person's ability to reproduce, it will be passed on to future generations through sexual reproduction. The result is that a single gene such as *BRCA1* can have hundreds of different nucleotide sequences, or alleles, in a population.

Ethnic Groups and Genetic Disease

Ahern descends from a subgroup of Jews called the "Ashkenazi"–the term generally refers to Jews of Eastern European descent. Ahern's father was born in Germany, immigrating to the United States in 1939; her mother was born in the United States; but Ahern's maternal grandfather was born in Russia. But the history of this Jewish subgroup extends much further back than modern Europe.

Recent gene studies support the biblical history of Jews as descended from populations in what is now the Middle East. The Ashkenazi Jews are a subgroup that left the Middle East and began populating parts of Europe more than

TABLE 10.1

Incidence of Hereditary Diseases in Different Populations

HEREDITARY DISEASE	CARRIER RATE IN ASHKENAZI JEWISH POPULATION	CARRIER RATE IN GENERAL POPULATION
Tay-Sachs disease	1 in 25	1 in 250
Canavan disease	1 in 40	Rare/unknown
Niemann-Pick disease, type A	1 in 90	1 in 40,000
Gaucher disease, type 1	1 in 14	1 in 100
Bloom syndrome	1 in 100	Rare/unknown
BCRA mutation	1 in 40	1 in 350–1,000
Familial dysautonomia	1 in 30	Rare/unknown

2,000 years ago. The majority of Ashkenazis, however, migrated into Europe in the 10th century from the region of present-day Israel, settling in the Rhineland, the valley of the Rhine River, in Germany.

A number of historical factors have made the Ashekenazi Jewish population more susceptible to genetic diseases. First, they descend from a small group of people. Second, that population has expanded and contracted over time. Third, and most important, members of the population usually marry within the community. In other words, Ashkenazi Jews have many of the characteristics of an isolated population—new alleles are not frequently introduced by people immigrating into the population.

Consequently, Ashkenazi Jews are an example of an ethnic group that has a more homogeneous genetic background than the general population, and is more likely to pass on certain genetic diseases to future generations. Scientists have discovered more than 1,000 recessive diseases in the general population, but most of them are rare. In Ashkenazi Jews, however, the prevalence of some recessive diseases is increased 100-fold or more. Tay-Sachs disease, Gaucher disease, and Bloom syndrome are genetic diseases that all occur more frequently in this ethnic group than in the general population; approximately 1 in 25 Ashkenazi Jews carry disease alleles for at least one of these disorders (**Table 10.1**).

Ashkenazi Jews are not the only ethnic group to have a higher incidence of certain genetic diseases than occurs in the general population. For example, people from Mediterranean, African, and Asian countries have higher rates of thalassemias, blood disorders that cause anemia. Sickle-cell anemia, another type of hereditary anemia, is more common among people of African descent.

Ashkenazi Jews are also more likely than the general population to carry mutations in *BRCA1* and *BRCA2*. Some studies have found that more than 8% of Ashkenazi women carry a mutated *BRCA1* gene, compared to only 2.2% of other women. These alleles can take the form of changes in one DNA base pair, or in several. In some cases, large DNA segments are rearranged. In mutated *BRCA2* genes, a small number of additional DNA base pairs is inserted into or deleted from the gene. These mutations, or alleles, of these genes arose and became prevalent over thousands of years.

Cancer Genetics

Inheriting a gene that carries a predisposition to a disease such as cancer doesn't mean you will automatically get the disease. Inherited predispositions increase the risk, but they don't definitively determine that the disease will occur. In most cases, there are several other contributing factors. Cancer often occurs only when additional, nonhereditary, mutations in a cell accumulate.

Environmental insults such as chemicals, ultraviolet light, radiation, and other factors can damage our DNA and cause it to mutate. Exposure to ultraviolet light for example, impairs the DNA in our skin cells and can lead to skin cancer. Physical or chemical agents that cause mutations with either positive, negative, or neutral outcomes are known as **mutagens.** Chemicals and other factors such as pesticides and pollutants that can cause cancer are a class of mutagens known as **carcinogens** because

MUTAGEN
Any chemical or physical agent that can damage DNA by changing its nucleotide sequence.

CARCINOGEN
Any chemical agent that causes cancer by damaging DNA. Carcinogens are a type of mutagen.

MUTATIONS AND CANCER

Cancer occurs when cells accumulate several DNA mutations that enable the cells to divide uncontrollably. People with inherited predispositions to cancer develop the disease at an earlier age than others because their cells already have one mutation that hinders their cells' ability to divide normally.

HINT **See Infographics 10.1–10.7.**

KNOW IT

1. What are some differences and some similarities between tumor suppressor genes and oncogenes?

2. What is the role of *BRCA1* in normal cells?

3. In an otherwise normal cell, what happens if one mistake is made during DNA replication?
- **a.** Nothing; mistakes just happen.
- **b.** A cell cycle checkpoint detects the damage and pauses the cell cycle so the error can be corrected.
- **c.** The cell will begin to divide out of control, forming a malignant tumor.
- **d.** A checkpoint will force the cell to carry out apoptosis, a form of cellular suicide.
- **e.** The mutation will be inherited by the individual's offspring.

4. Which of the following can cause cancer to develop and progress?
- **a.** a proto-oncogene
- **b.** an oncogene
- **c.** a tumor suppressor gene
- **d.** a mutated tumor suppressor gene
- **e.** b and d
- **f.** b and c

5. Someone with a *BRCA1* mutation
- **a.** will definitely develop breast cancer.
- **b.** is at increased risk of developing breast cancer.
- **c.** must have inherited it from her mother, because of the link to breast cancer.
- **d.** will also have a mutation in *BRCA2*.
- **e.** b and c

6. Why does wearing sunscreen reduce cancer risk?
- **a.** Sunscreen can repair damaged DNA.
- **b.** Sunscreen can activate checkpoints in skin cells.
- **c.** Sunscreen can reduce the chance of mutations caused by exposure to UV radiation present in sunlight
- **d.** It doesn't; sunscreen causes mutation and actually increases cancer risk.
- **e.** Sunscreen can prevent cells with mutations from being destroyed.

USE IT

7. Lorene Ahern was born with an inherited predisposition to cancer. At the cellular and genetic level, what was she born with? At birth, were cells in her breast genetically identical to cells in her liver? Now that she has breast cancer, are her cancer cells genetically identical to her normal breast cells? Explain your answers.

8. What would you say to a niece if she asked you how she could reduce her risk of getting breast cancer? Assume there is no family history of breast cancer. How might each of your suggestions reduce her risk?

9. If you wanted to change your lifestyle to reduce your risk of developing cancer, which of the following behaviors would be important?
- **a.** limiting alcohol consumption
- **b.** wearing sunscreen
- **c.** avoiding exposure to tobacco
- **d.** by avoiding exposure to pesticides
- **e.** all of the above

10. Who of the following women would be most likely to benefit from genetic testing for breast cancer?
- **a.** a 25-year-old woman whose mother, aunt, and grandmother had breast cancer
- **b.** a healthy 75-year-old woman with no family history of breast cancer
- **c.** a 40-year-old woman who has a cousin with breast cancer
- **d.** a 55-year-old woman whose older sister was just diagnosed with breast cancer
- **e.** All women can benefit from genetic testing for breast cancer.

11. People like Lorene Ahern have inherited a mutated version of *BRCA1*. Why does this mutation pose a problem? Why are these people at high risk of developing breast cancer when they still have a functional *BRCA1* allele? Describe how the protein encoded by normal *BRCA1* compares to that encoded by mutant alleles of *BRCA1*.

SCIENCE AND ETHICS

12. Nellie has a family history similar to Lorene Ahern's. Nellie's mother died at an early age from breast cancer, as did her maternal aunt (her mother's sister). Nellie is not yet 35 but has started having annual mammograms. She has also been tested for *BRCA1* and *BRCA2* mutations. She has a *BRCA2* mutation and is considering prophylactic surgery. Her younger sister, Anne, doesn't want to know the results of Nellie's genetic testing because if Nellie has a *BRCA2* mutation, then there is a chance that Anne could have inherited the same mutation from their mother. Does Nellie or Nellie's doctor have an obligation to tell Anne about the test results? What about Nellie's older brother? Should he be told? There are personal and medical benefits and risks to consider here.

Rock for a Cause

What You Will Be Learning

slight change wreaks havoc on victim's bodies–their lungs, sweat glands, and pancreas no longer function normally **(Infographic 11.1)**.

Today, almost 20 years later, scientists understand the disease better, and this has led to better drugs and therapies to treat symptoms; victims of CF are living longer than ever. But despite scientific advances, there is still much to learn. One aspect of the disease that scientists are studying intensively is that people with identical CF alleles vary in the course of their disease–some have worse symptoms and live shorter lives than others. In recent years, scientists have discovered that there are other genes that contribute to a patient's overall health–so-called modifier genes. That discovery is leading to exciting new therapies that may extend Emily's life and the lives of thousands of other people with CF.

How Is CF Inherited?

When Emily's mother, Debbie, learned that her daughter had CF, she was shocked. She and her husband, Lowell, were both healthy, and they already had two healthy sons. How did their daughter Emily develop a disease that neither Debbie nor her husband had?

The answer is inheritance. Genes, which provide instructions for making proteins, are the units of inheritance, physically transmitted from parents to children. The particular alleles of genes you received from your parents are the reason you resemble your mother and father, and possibly also an uncle or a grandparent. But not every child of a couple receives the exact same parental genes, and so children can and do differ from their parents and from each other.

Consider Emily's parents. Because they are **diploid** organisms, each of their body cells carries two copies of each chromosome–one inherited from mom, the other from dad. Such paired chromosomes are called **homologous chromosomes.** Because chromosomes come in pairs, we have two copies of nearly every gene in our body cells. Genes located on the X and Y chromosome in males do not have a second copy. While the two gene copies have the same general function, the nucleotide sequence of each copy can differ. In other words, a person can carry two different alleles of the same gene, one of which functions differently from the other. In the case of the gene *CFTR,* a person can have one CF-associated allele and remain healthy if his or her other

INFOGRAPHIC 11.1

CF Is Caused by Mutations in the *CFTR* Gene

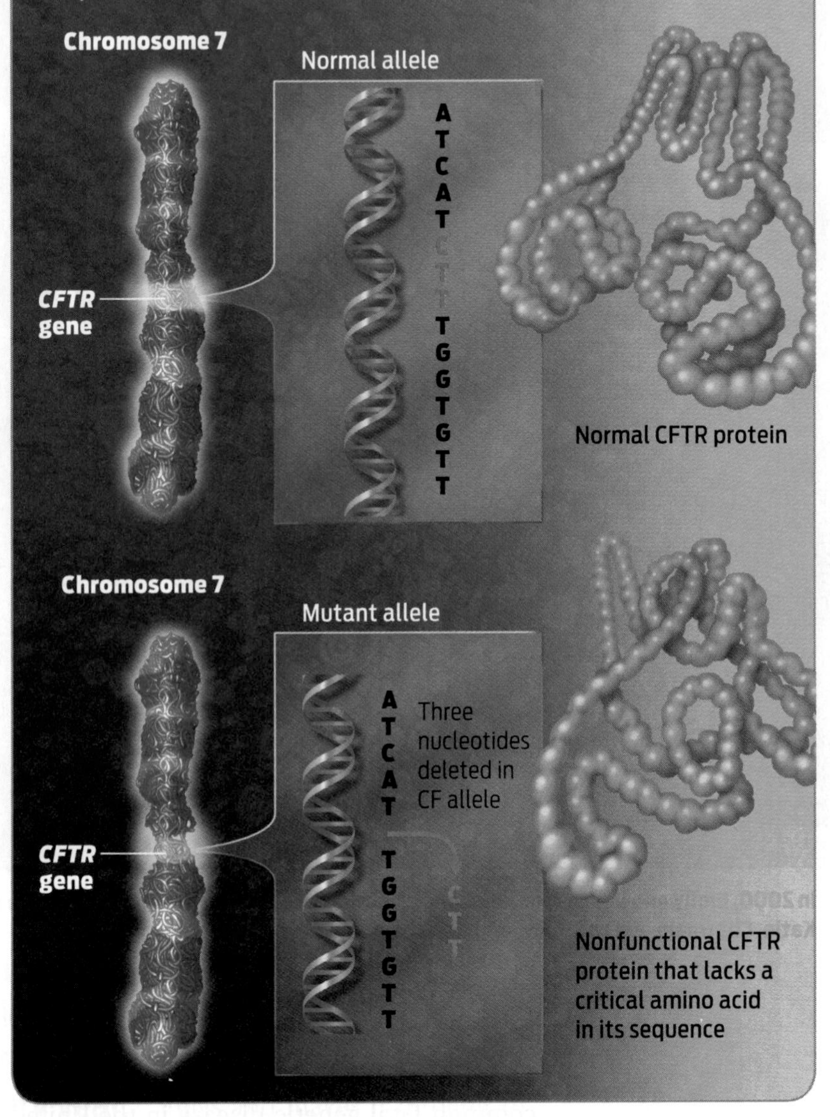
Cystic fibrosis (CF) is caused by a variety of mutations in the cystic fibrosis transmembrane regulator (CFTR) gene that sits on chromosome 7. One such mutation consists of a deletion of three consecutive nucleotides, which creates a mutant allele. Consequently, the mRNA expressed from this gene has a missing codon and the resulting protein lacks an amino acid in a specific location, rendering the protein nonfunctional.
Chromosome 7
Normal allele
A T C A T C T T T G G T G T T
CFTR gene
Normal CFTR protein
Chromosome 7
Mutant allele
A T C A T T G G T G T T
Three nucleotides deleted in CF allele
C T T
CFTR gene
Nonfunctional CFTR protein that lacks a critical amino acid in its sequence

DIPLOID
Having two copies of every chromosome.

HOMOLOGOUS CHROMOSOMES
The two copies of each chromosome in a diploid cell. One chromosome in the pair is inherited from the mother, the other is inherited from the father.

INFOGRAPHIC 11.2

Humans Have Two Copies of Nearly Every Gene

Human cells have 23 pairs of homologous chromosomes. One chromosome of each pair is inherited from mom, one from dad. This makes us diploid, as virtually every cell in the body carries two copies of every gene. Each copy of each gene has two alleles that can either be identical to each other or different. In the case of CF, carrying at least one normal allele is enough to remain healthy.

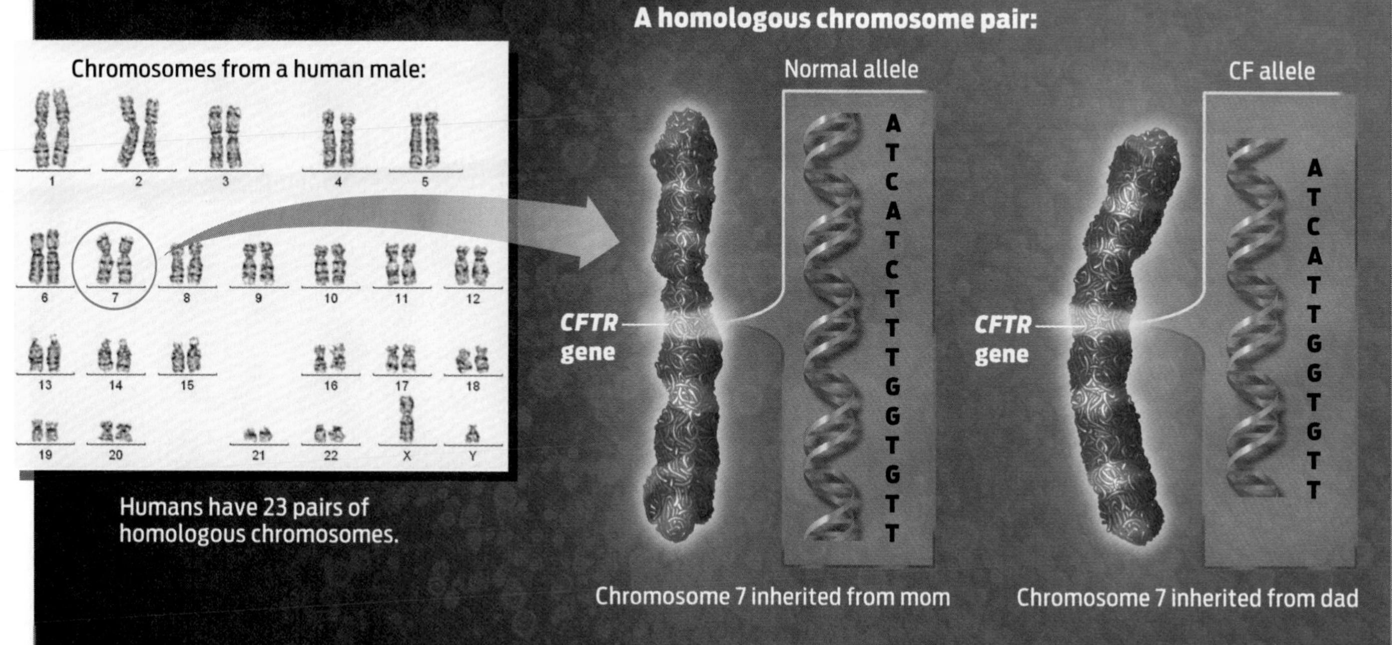

Emily's parents remain healthy because each has one normal allele that makes up for the defective copy.

PHENOTYPE
The visible or measurable features of an individual.

GENOTYPE
The particular genetic makeup of an individual.

GAMETES
Specialized reproductive cells that carry one copy of each chromosome (that is, they are haploid). Sperm are male gametes; eggs are female gametes.

HAPLOID
Having only one copy of every chromosome.

chromosome has a normal allele to make up for the defective copy (**Infographic 11.2**). That's why Emily's parents, even if each of them had a CF-associated gene, could be healthy.

Geneticists make a distinction between a person's observable or measurable traits, or **phenotype,** and his or her genes, or **genotype.** As in the case of Debbie and Lowell, one cannot always determine genotype from phenotype. Both Debbie and Lowell have normal phenotypes, but they both also carry a disease allele as part of their genotype. They each inherited one CF allele from one of their parents and therefore can pass that defective allele along to their children—as they did to Emily.

But not all the Schaller children have the disease—Debbie and Lowell also have two healthy boys. Why didn't these children inherit CF? Sexual reproduction is a bit like shuffling cards. Before parents pass their genes to their offspring, those genes are first mixed up and then the two copies of each gene are separated from each other, so that not every child receives the same combination of alleles. It is the unique combination of maternal and paternal alleles that come together during fertilization that determines a person's genotype and contributes to his or her phenotype.

To reproduce sexually, organisms must first create sex cells called **gametes.** In humans, these are the egg and sperm cells. Unlike the rest of the body's cells, which are diploid, gametes carry only one copy of each chromosome, which makes them **haploid.** To become haploid, the cells that form gametes go through a unique kind of cell division, called

INFOGRAPHIC 11.3

Gametes Pass Genetic Information to the Next Generation

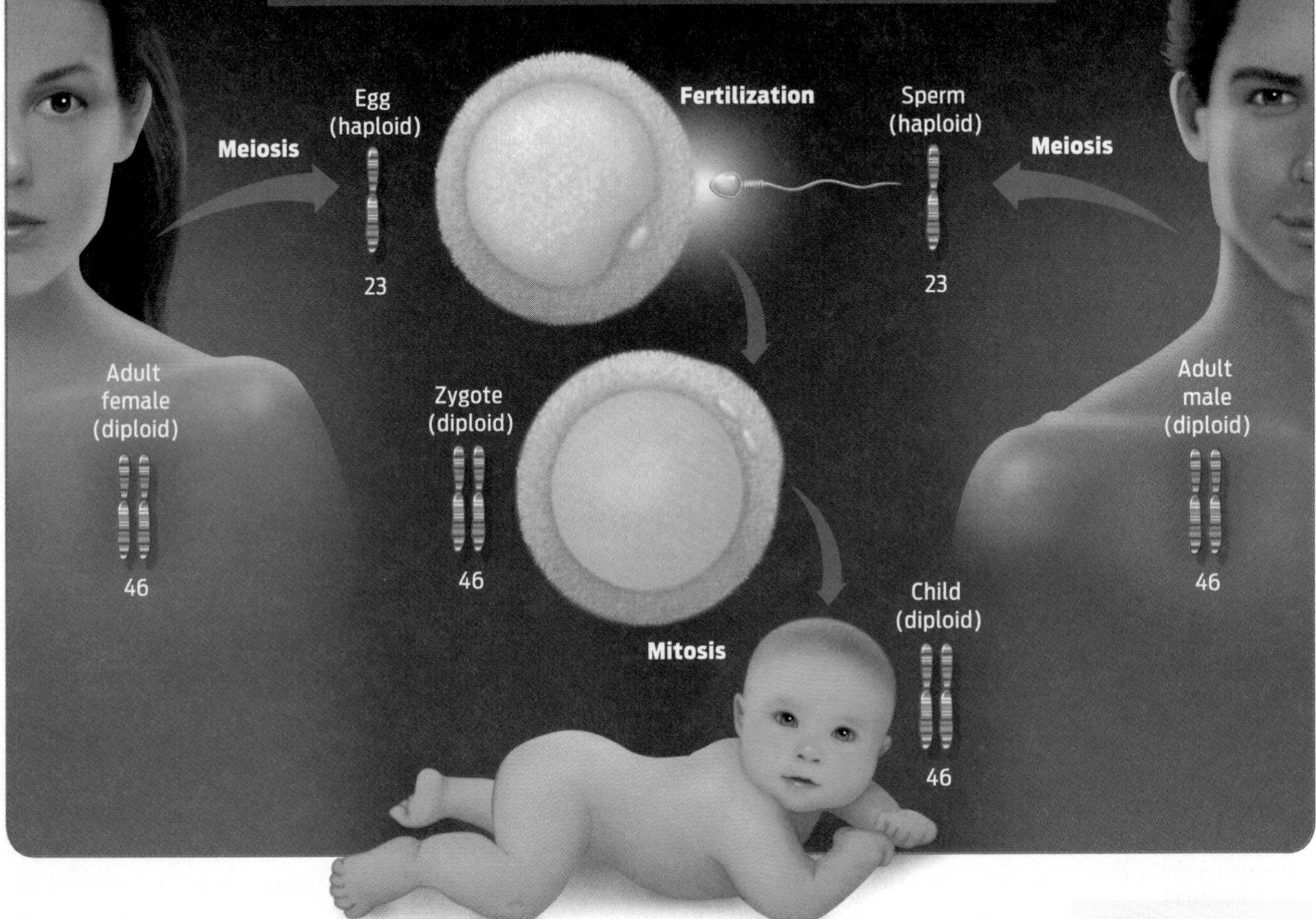

To reproduce sexually, diploid organisms produce specialized sex cells called gametes, which are haploid – they carry only one copy of each chromosome. When a sperm fertilizes an egg the resulting diploid zygote divides by mitotic cell division, eventually generating enough cells to form a baby. The baby is diploid.
The process of cell division that creates gametes is known as meiosis. Men produce sperm, the male gametes, and women produce eggs, the female gametes.
Meiosis
Egg (haploid)
23
Fertilization
Sperm (haploid)
23
Meiosis
Adult female (diploid)
46
Zygote (diploid)
46
Adult male (diploid)
46
Mitosis
Child (diploid)
46

meiosis, which halves the number of chromosomes from 46 to 23. When a haploid sperm fertilizes a haploid egg, the result is a diploid **zygote** that now carries two copies of every gene on 46 chromosomes. In turn, this zygote will divide by mitosis to become an **embryo,** which will eventually grow into a human child **(Infographic 11.3)**.

Meiosis, the cell division that creates sperm and egg, is similar to mitotic cell division (Chapter 9), except that in meiosis there are two separate divisions. The first division separates homologous chromosomes; the second division separates sister chromatids **(Infographic 11.4)**.

Because it unites haploid egg and sperm from two people, sexual reproduction is the primary reason why children don't look and behave exactly like one parent in particular; they inherit alleles from both parents and consequently are genetically a combination of the two.

Besides forming haploid sex cells, meiosis contributes to the genetic diversity of offspring in other ways as well. No two gametes produced by the same parent are identical, and that is because of two major events during meiosis that contribute to the huge variation we see among parents, children, and their siblings. The first is **recombination,** in which homologous

MEIOSIS
A specialized type of cell division that generates genetically unique haploid gametes.

ZYGOTE
A cell that is capable of developing into an adult organism. The zygote is formed when an egg is fertilized by a sperm.

EMBRYO
An early stage of development reached when a zygote undergoes cell division to form a multicellular structure.

INFOGRAPHIC 11.4

Meiosis Produces Haploid Egg and Sperm

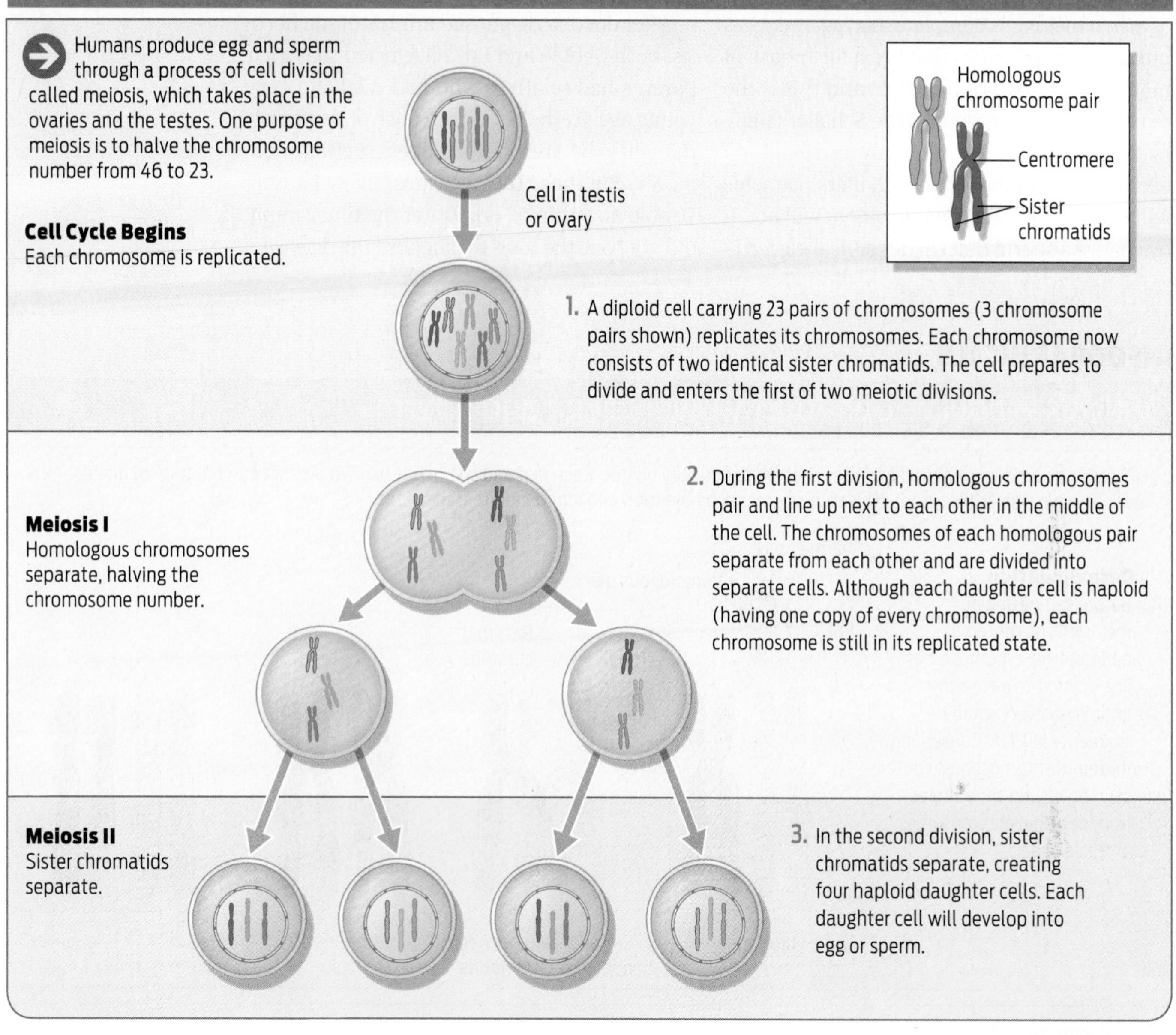

Humans produce egg and sperm through a process of cell division called meiosis, which takes place in the ovaries and the testes. One purpose of meiosis is to halve the chromosome number from 46 to 23.
Homologous chromosome pair
Centromere
Sister chromatids
Cell in testis or ovary
Cell Cycle Begins
Each chromosome is replicated.
1. A diploid cell carrying 23 pairs of chromosomes (3 chromosome pairs shown) replicates its chromosomes. Each chromosome now consists of two identical sister chromatids. The cell prepares to divide and enters the first of two meiotic divisions.
Meiosis I
Homologous chromosomes separate, halving the chromosome number.
2. During the first division, homologous chromosomes pair and line up next to each other in the middle of the cell. The chromosomes of each homologous pair separate from each other and are divided into separate cells. Although each daughter cell is haploid (having one copy of every chromosome), each chromosome is still in its replicated state.
Meiosis II
Sister chromatids separate.
3. In the second division, sister chromatids separate, creating four haploid daughter cells. Each daughter cell will develop into egg or sperm.

RECOMBINATION
The stage of meiosis in which maternal and paternal chromosomes pair and physically exchange DNA segments.

INDEPENDENT ASSORTMENT
The principle that alleles of different genes are distributed independently of one another during meiosis.

maternal and paternal chromosomes pair up and swap genetic information. As a result of recombination, maternal chromosomes actually contain segments (and therefore alleles) from paternal chromosomes and vice versa.

The second vitally important aspect of meiosis is **independent assortment,** which means that alleles of different genes are distributed independently of one another, not as a package. Because the number of possible combinations of alleles is therefore huge, a unique combination of maternal and paternal chromosomes is distributed into each sperm and each egg cell. This distribution occurs at the first division of meiosis (known as meiosis I), when maternal and paternal chromosomes line up along the midline of the cell and segregate into newly forming cells. Because maternal and paternal chromosomes line up randomly (sometimes on the "left," sometimes on the "right,"), the exact combination of maternal and paternal chromosomes that each sperm or egg inherits differs every time meiosis occurs (**Infographic 11.5**).

When meiosis is complete, each gamete has only 23 chromosomes that are a mishmash of maternal and paternal alleles–and this is the reason that not everyone in the Schaller family has CF. Because alleles randomly distribute into each gamete, some of the Schallers' gametes will carry the CF allele while others will not. If by chance a sperm that carries a CF allele fertilizes an egg that also carries a CF allele, the resulting child will have CF.

After doctors diagnosed Emily's cystic fibrosis, both Debbie and Lowell learned that their parents had relatives who had died at a very young age. At the time, the cause of death was thought to be a respiratory illness such as pneumonia. But these relatives most likely had CF, Debbie now thinks; doctors at the time simply did not have the tools to diagnose the disease.

INFOGRAPHIC 11.5

Meiosis Produces Genetically Diverse Egg and Sperm

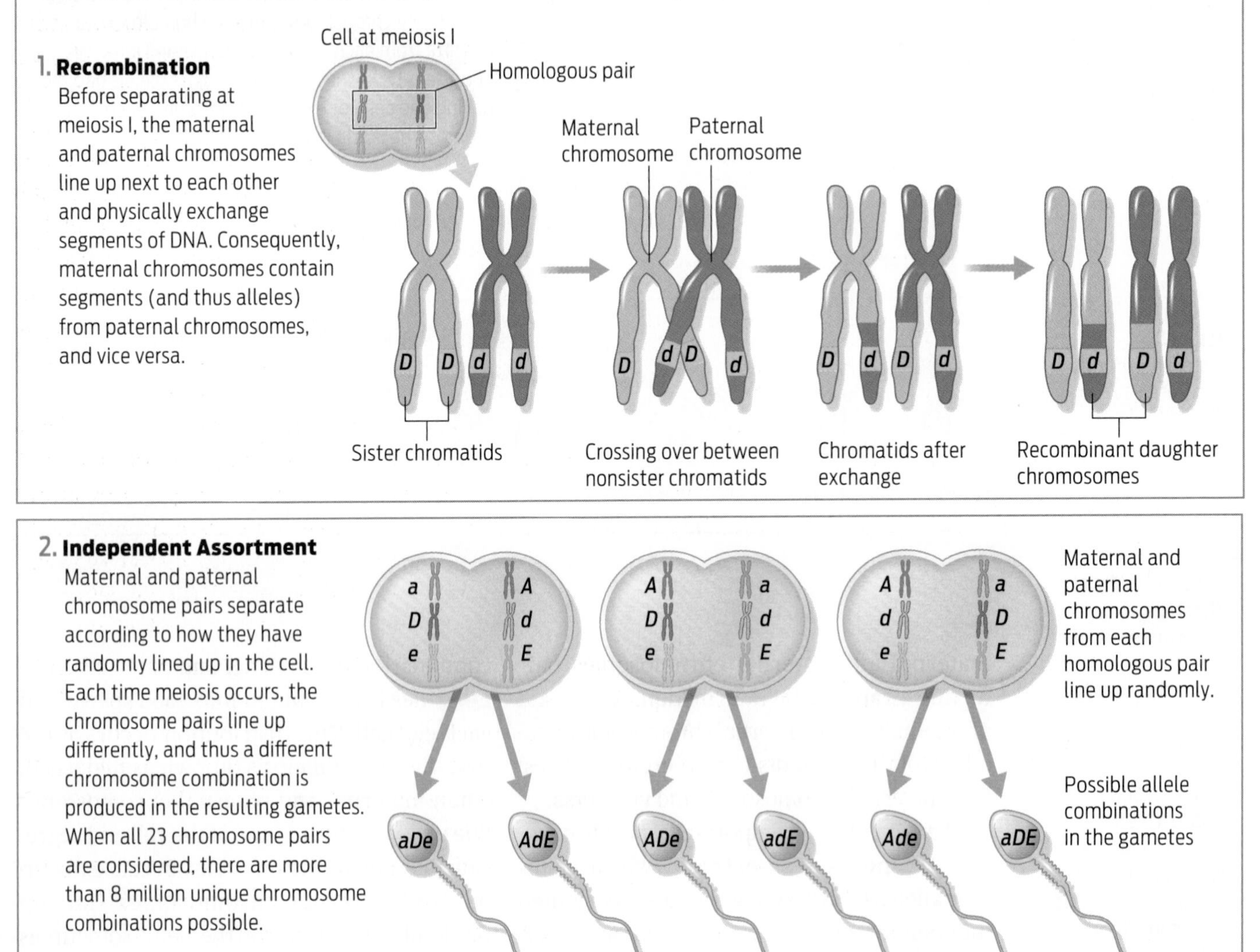

Meiosis produces haploid gametes that are genetically unique. Each egg and sperm has its own distinct combination of alleles. The two events that create this diversity are recombination and independent assortment.
Cell at meiosis I
1. Recombination
Before separating at meiosis I, the maternal and paternal chromosomes line up next to each other and physically exchange segments of DNA. Consequently, maternal chromosomes contain segments (and thus alleles) from paternal chromosomes, and vice versa.
Homologous pair
Maternal chromosome
Paternal chromosome
D
D
d
d
D
d
D
d
D
d
D
d
D
d
D
d
Sister chromatids
Crossing over between nonsister chromatids
Chromatids after exchange
Recombinant daughter chromosomes
2. Independent Assortment
Maternal and paternal chromosome pairs separate according to how they have randomly lined up in the cell. Each time meiosis occurs, the chromosome pairs line up differently, and thus a different chromosome combination is produced in the resulting gametes. When all 23 chromosome pairs are considered, there are more than 8 million unique chromosome combinations possible.
a
D
e
A
d
E
A
D
e
a
d
E
A
d
e
a
D
E
Maternal and paternal chromosomes from each homologous pair line up randomly.
aDe
AdE
ADe
adE
Ade
aDE
Possible allele combinations in the gametes

Cystic fibrosis patients like Emily wear vibrating vests to loosen the mucus in their lungs while inhaling a saltwater solution to thin out the mucus.

The Schallers now understood that the disease ran in both their families. But they could still not help Emily. "They told us she would only live to be about 12 years old," Debbie recalls, adding, "We just put ourselves in the hands of medical professionals."

Living with the Disease

Growing up, Emily was scarcely aware of her own disability. The visits to doctors and periodic stays in the hospital were just a part of life. All her teachers and friends knew that she had CF. "My family and friends were all so supportive," she says. In high school she played volleyball, basketball, and soccer, and participated in many walkathons to raise money for CF research. Thanks to medical progress, Emily has outlived doctors' original expectations by more than a decade.

"They told us she would only live to be about 12 years old. We just put ourselves in the hands of medical professionals." –Debbie Schaller

But she deals daily with the legacy of her genetic inheritance. In healthy people, the CFTR protein acts as a channel within a cell's membrane that allows certain ions in and out of the cell, keeping the cell's chemistry in balance. But in people with CF, the channel is distorted or dysfunctional and the mechanism goes awry. The result is that mucus–a slippery substance that lubricates and protects the linings of the airways, digestive system, reproductive system, and other tissues–becomes abnormally thick **(Infographic 11.6)**.

This abnormal mucus blocks ducts throughout the body. The most problematic symptom, however, is that thick mucus builds up in the lungs. Patients have trouble breathing, and the mucus provides fertile ground for bacteria and other organisms. Over time, repeated infections permanently damage the lungs. Suffocation often kills CF victims as they slowly lose their ability to breathe.

To avoid lung damage, every morning Emily dons an inflatable vest that vibrates to loosen mucus in her lungs. During this 30-minute therapy she inhales a saltwater solution and another medication to thin her mucus, which she then coughs out periodically. To that regime she adds two other medications three times a week to keep her lungs from becoming inflamed and to kill off infections. But despite her best efforts, Emily has been hospitalized more frequently in recent years because of serious lung infections that hinder her ability to breathe.

Emily remains undaunted. "I just live each day at a time," she says. She works about 30 hours a week at a retail shop in downtown Detroit and spends her evenings practicing with her band, performing at concerts, playing guitar, or hanging out with friends. While she doesn't plan too far ahead into the future, she hopes her band's fame and success will grow. If the band's following expands beyond Detroit, she hopes to tour Europe. Emily hasn't ruled out having a family of her own one day. Even though Emily has CF, her children will not necessarily have the disease.

Why not? Remember that since Emily has CF, her parents, Lowell and Debbie, both must carry disease alleles. But as neither of them has the disease, the CF alleles must be "hidden." When one allele masks the effect of

INFOGRAPHIC 11.6

The CFTR Protein and Cystic Fibrosis

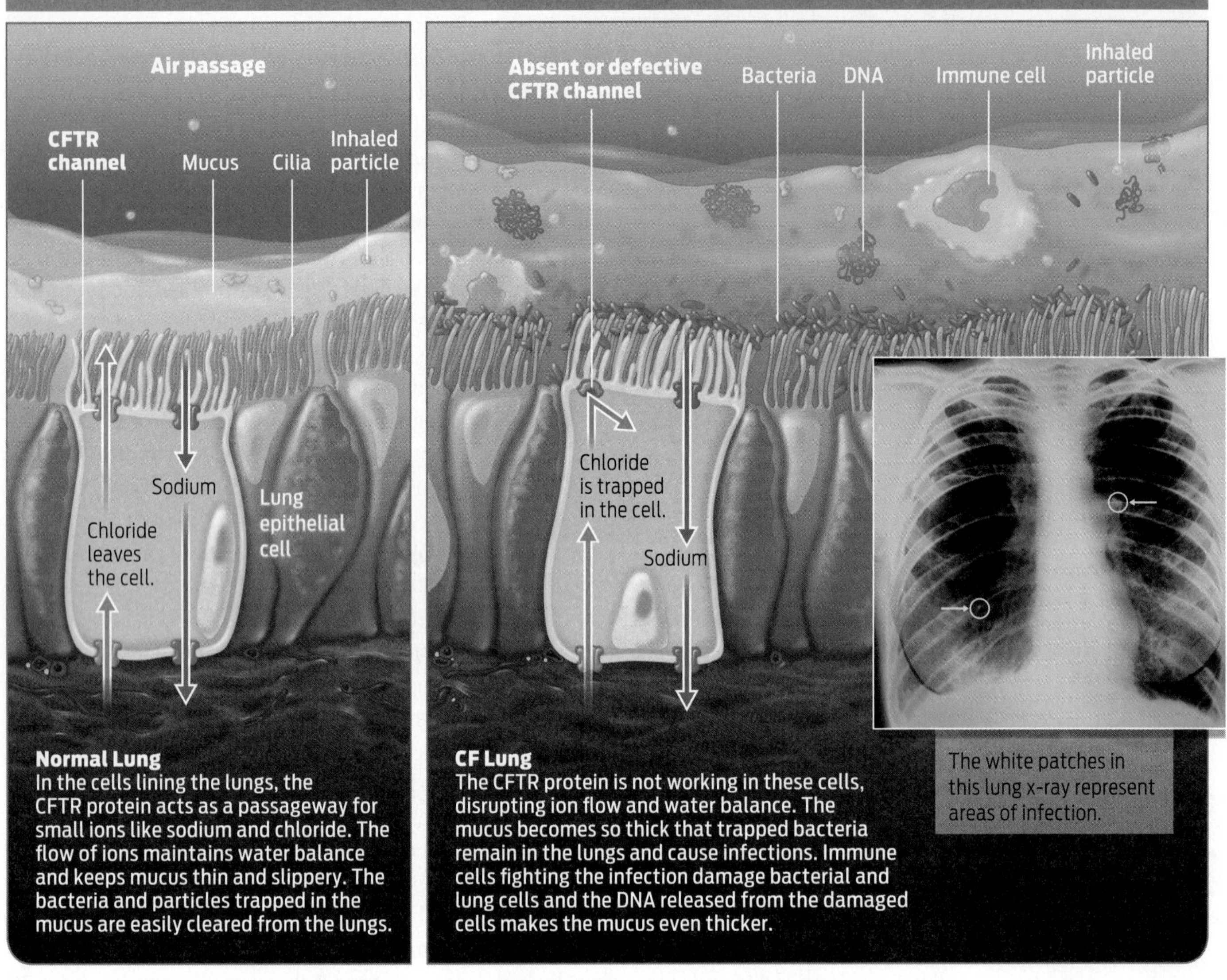

Air passage
CFTR channel
Mucus
Cilia
Inhaled particle
Sodium
Chloride leaves the cell.
Lung epithelial cell
Absent or defective CFTR channel
Bacteria
DNA
Immune cell
Inhaled particle
Chloride is trapped in the cell.
Sodium
Normal Lung
In the cells lining the lungs, the CFTR protein acts as a passageway for small ions like sodium and chloride. The flow of ions maintains water balance and keeps mucus thin and slippery. The bacteria and particles trapped in the mucus are easily cleared from the lungs.
CF Lung
The CFTR protein is not working in these cells, disrupting ion flow and water balance. The mucus becomes so thick that trapped bacteria remain in the lungs and cause infections. Immune cells fighting the infection damage bacterial and lung cells and the DNA released from the damaged cells makes the mucus even thicker.
The white patches in this lung x-ray represent areas of infection.

another, the hidden allele is described as **recessive** (designated by a lower-case letter, e.g., *a*). The normal allele, which conceals the effect of the recessive allele, is known as the **dominant allele** (designated by a capital letter, e.g., *A*). Debbie and Lowell are healthy because they each have a dominant normal allele that compensates for their defective recessive CF allele. Geneticists call their genotype **heterozygous.** Their two healthy sons either are heterozygous like their parents, or have two normal alleles–that is, their genotype is **homozygous.** A genotype made up of two dominant alleles is known as homozygous dominant. Emily's genotype, however, is homozygous recessive: she inherited one recessive CF allele from each parent, which is why she has the disease.

What were the chances that Debbie and Lowell would have a child with CF? To figure out the likelihood that parents will have a child with a particular trait, we can plot the possibilities on a **Punnett square,** a tool named for the British geneticist Reginald C. Punnett, who devised it. A Punnett square matches up the possible parental gametes and shows the likelihood that particular parental alleles will combine. As heterozygous individuals, Debbie and Lowell each have a 50% chance of passing on their CF allele to a child, which means they have a

RECESSIVE ALLELE
An allele that reveals itself in the phenotype only if the organism has two copies of that allele.

DOMINANT ALLELE
An allele that can mask the presence of a recessive allele.

HETEROZYGOUS
Having two different alleles.

HOMOZYGOUS
Having two identical alleles.

PUNNETT SQUARE
A diagram used to determine probabilities of offspring having particular genotypes, given the genotypes of the parents.

CARRIER
An individual who is heterozygous for a particular gene of interest, and therefore can pass on the recessive allele without showing any of its effects.

25% chance of having a child with CF and a 75% chance of having a healthy child. The chance that a child will be a heterozygous **carrier**–that is, that the child will carry the recessive allele for CF but will not have the disease because the allele's effect is masked by the dominant allele–is 50% **(Infographic 11.7)**.

Just as Emily's genotype is different from her parents' genotype, Emily's children will have different genotypes from her own. Whether or not her children develop CF depends on the father's genotype. Since Emily is homozygous, she can contribute only recessive CF alleles to her children. If Emily were to have children with a man who had two normal alleles, for example, none of her children would have the disease–they would all have a heterozygous genotype but a normal phenotype. But as carriers they could pass on the disease to their children. If she had children with a man who was heterozygous for the CF gene, then her children would have a 1 in 2, or 50%, chance of having CF.

Not all recessive alleles cause disease. Physical traits such a blue eyes, for example, result from the inheritance of two recessive alleles of the same gene. And not all genetic diseases are caused by recessive alleles; some, such as the neurodegenerative disorder called Huntington disease, are determined by dominant alleles. Diseases caused by dominant alleles, however,

INFOGRAPHIC 11.7

How Recessive Traits Are Inherited

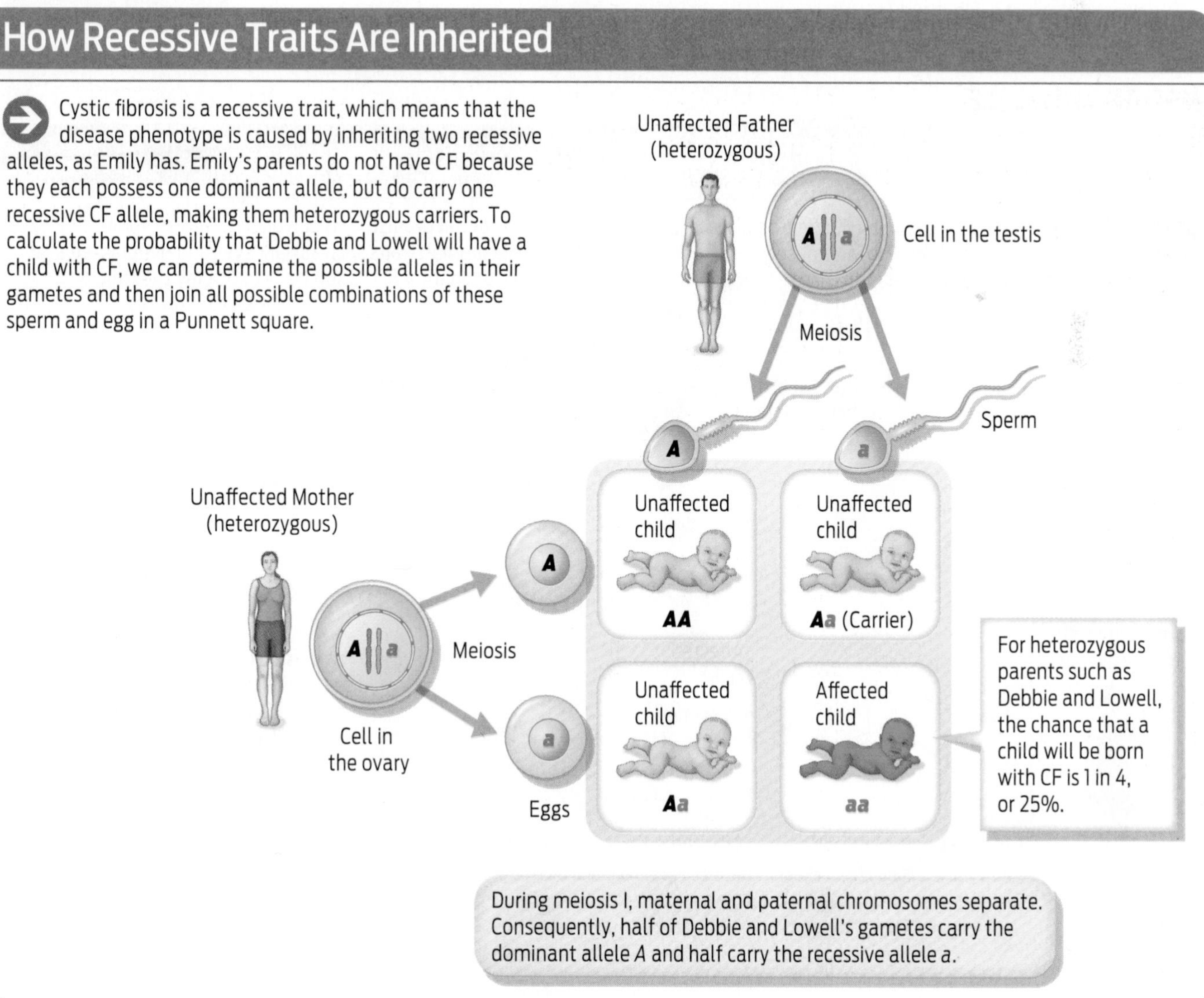
Cystic fibrosis is a recessive trait, which means that the disease phenotype is caused by inheriting two recessive alleles, as Emily has. Emily's parents do not have CF because they each possess one dominant allele, but do carry one recessive CF allele, making them heterozygous carriers. To calculate the probability that Debbie and Lowell will have a child with CF, we can determine the possible alleles in their gametes and then join all possible combinations of these sperm and egg in a Punnett square.
Unaffected Father (heterozygous)
A a
Cell in the testis
Meiosis
Sperm
A
a
Unaffected Mother (heterozygous)
A a
Cell in the ovary
Meiosis
A
a
Eggs
Unaffected child
AA
Unaffected child
Aa (Carrier)
Unaffected child
Aa
Affected child
aa
For heterozygous parents such as Debbie and Lowell, the chance that a child will be born with CF is 1 in 4, or 25%.
During meiosis I, maternal and paternal chromosomes separate. Consequently, half of Debbie and Lowell's gametes carry the dominant allele A and half carry the recessive allele a.

have a high probability of being passed to the next generation **(Infographic 11.8)**.

In all cases, anyone with a genetic disease is at risk for passing it on to his or her children. The risk merely varies, depending on whether the alleles are dominant or recessive and on the genotype of the partner **(Table 11.1)**.

Couples who carry disease genes needn't feel that having children is a roll of the dice.

Couples who carry disease genes needn't feel that having children is a roll of the dice. There are ways to ensure that their children won't develop the diseases they could otherwise inherit. Many couples in this situation use a technology called pre-implantation genetic diagnosis to detect and select embryos that do not carry defective alleles. Through in vitro fertilization, a man's sperm can fertilize a woman's eggs outside the body. The genes of each resulting embryo are then examined for specific alleles, and then only embryos that don't contain defective alleles are implanted into the mother. Hundreds of thousands of babies have already been born by this technique.

New Research in the Pipeline

Some couples, however, may choose not to undergo assisted reproduction because of religious or other reasons. In the case of CF, there are new treatments in the pipeline that could help Emily, and her children and grandchildren, too.

Furthest along are a class of medications that, when inhaled, can restore the balance of ions inside affected cells. Scientists are presently

INFOGRAPHIC 11.8

How Dominant Traits Are Inherited

Some genetic conditions, such as Huntington disease, a degenerative neurological disease, and polydactyly, having more than five fingers or toes per limb, are caused by dominant alleles. Plenty of common traits such as dark eyes and dimples are also determined by dominant alleles. In these cases, inheriting one copy of the dominant allele is sufficient to display the trait.

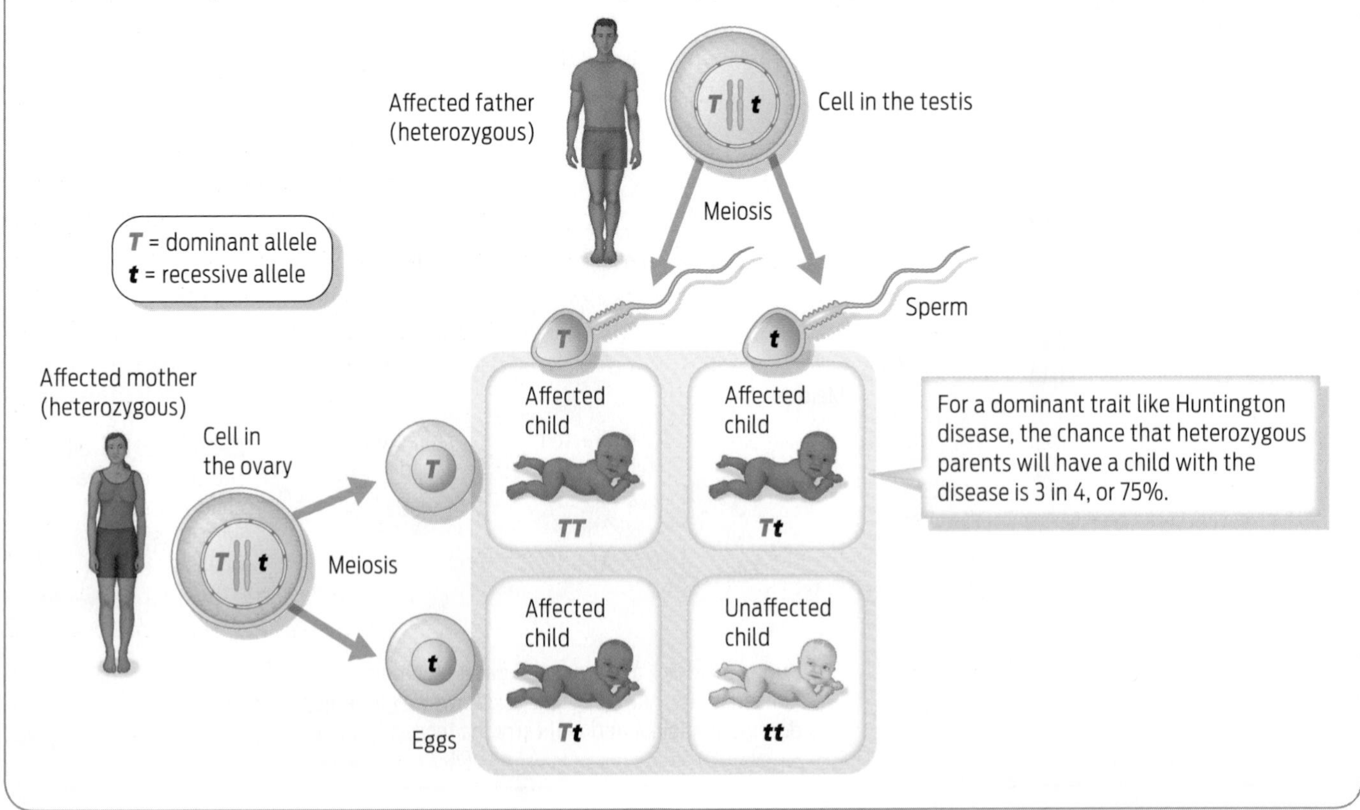

TABLE 11.1

Inherited Genetic Conditions in Humans

RECESSIVE TRAITS	PHENOTYPE
Albinism	Lack of pigment in skin, hair, and eyes
Cystic fibrosis	Excess mucus in lungs, digestive tract, and liver; increased susceptibility to infections
Sickle-cell disease	Sickled red blood cells; damage to tissues
Tongue rolling	Ability to curl tongue into a U-shape
Tay-Sachs disease	Lipid accumulation in brain cells; mental deficiency, blindness, and death in childhood
DOMINANT TRAITS	**PHENOTYPE**
Huntington disease	Mental deterioration and uncontrollable movements; onset at middle age
Freckles	Pigmented spots on skin, particularly on face and arms
Polydactyly	More than five digits on hands or feet
Dimples	Indentation in the skin of the cheeks
Chin cleft	Indentation in chin

testing at least six different experimental drugs in humans.

Through basic research, scientists continue to learn more about the disease. Over the past 20 years, scientists have discovered more than 1,000 different alleles of the *CFTR* gene. The most common is *ΔF508,* which accounts for about 70% of all CF alleles. This particular CF allele is associated with more severe disease. But researchers have long puzzled over why the disease varies in two people with identical CF alleles—even two people homozygous for the *ΔF508* allele will vary in how their disease progresses. Researchers long thought that perhaps environmental factors such as diet, social relationships, and exercise might be responsible.

Over the past 20 years, scientists have discovered more than 1,000 different alleles of the *CFTR* gene.

But in recent years, scientists have learned that there is more to the story. Researchers have discovered other genes on different chromosomes that contribute to the severity of CF symptoms. The genes so far discovered predominantly influence a person's immune system, which helps the body fight off infections.

For example, scientists have found that one allele of a gene called *TGFB1,* located on chromosome 19, is associated with more severe lung disease in CF patients. This gene influences a person's immune response to infection. Scientists suspect that CF patients with certain *TGFB1* alleles mount a more vigorous response to infections than those with other alleles. Such a heightened immune response can cause lung tissue to scar. So if a CF patient also inherited this specific allele of *TGFB1,* his or her lungs are more likely to scar in response to infections. The impact of such modifier genes on the CF phenotype makes it more complicated to assess how disabling any particular person's CF disease will be—but it is not impossible.

Parents who are heterozygous carriers of CF, for example, have a 1 in 4, or 25%, chance of having a child who has CF. If these two parents are also heterozygous for *TGFB1,* then the probability that their child will be homozygous recessive for *TGFB1* is also 1 in 4 (25%). The chance of two independent events occurring together is calculated by multiplying the two independent chances together. So the probability of being homozygous recessive for both *CFTR* and *TGFB1* is $\frac{1}{4} \times \frac{1}{4}$, or 1 in 16. This probability can also be calculated using a Punnett square **(Infographic 11.9)**.

Understanding how these modifier genes contribute to the disease may point the way to even more therapies. In some cases, existing drugs may prove useful. Drugs that reduce inflammation by targeting the TGFβ1 protein, for example, may help reduce scarring in the lungs.

Emily recently had her genotype tested. While she doesn't carry *ΔF508,* she does carry another allele associated with severe disease. Her second CF allele, however, is much rarer, and scientists can't assess how these two particular alleles along with alleles of Emily's other genes will interact as she gets older. So they can't predict how quickly her disease will progress.

Keenly aware of how medical progress has extended her life, Emily conducts her own share of fund raising and education. After that fateful New Year's Eve when she and her friends opened for her brother's band, Self Normal, Hellen's following grew. The band wrote more songs and refined their sound. "I favor the old stuff, AC-DC, Led Zeppelin, the Ramones," Emily remarks. And fans just

INFOGRAPHIC 11.9

Tracking the Inheritance of Two Genes

People with CF differ in the severity of their disease. Some of this variability is influenced by alleles of other genes that sit on other chromosomes. One such gene, called *TGFB1*, is located on chromosome 19, shown here with symbol *D*. We can also use a Punnett square to follow the inheritance of two genes, as in the example below.

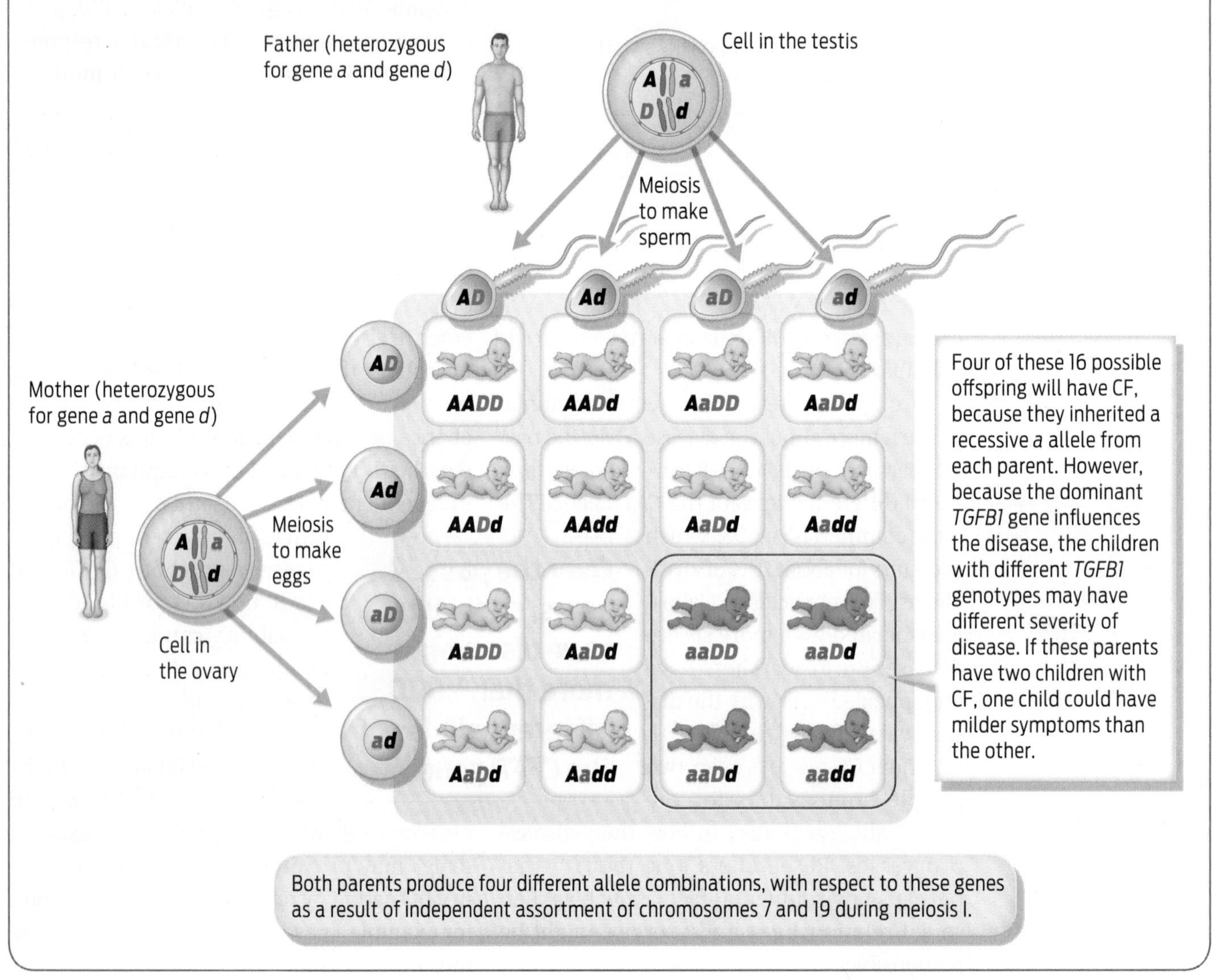

Father (heterozygous for gene *a* and gene *d*)
Cell in the testis
A a
D d
Meiosis to make sperm
AD
Ad
aD
ad
Mother (heterozygous for gene *a* and gene *d*)
Meiosis to make eggs
Cell in the ovary
AADD
AADd
AaDD
AaDd
AADd
AAdd
AaDd
Aadd
AaDD
AaDd
aaDD
aaDd
AaDd
Aadd
aaDd
aadd
Four of these 16 possible offspring will have CF, because they inherited a recessive *a* allele from each parent. However, because the dominant *TGFB1* gene influences the disease, the children with different *TGFB1* genotypes may have different severity of disease. If these parents have two children with CF, one child could have milder symptoms than the other.
Both parents produce four different allele combinations, with respect to these genes as a result of independent assortment of chromosomes 7 and 19 during meiosis I.

Every year Emily plays a concert to benefit CF.

couldn't seem to get enough of the girls' music. Each year, the number of fans has grown.

Sparked by a Self Normal hit single called "Just Breathe," Emily had the idea to organize a concert to benefit CF research. The song has nothing to do with CF, but Emily thought that it might be a good theme song for a concert. Besides, she says, "We were tired of walk-athons and black tie events with tickets that cost $300 each."

Emily and her brother organized the first benefit concert in 2004. Called "Just Let Me Breathe," it featured four Detroit bands. The concert sold out and raised about $9,000. With all of her fund-raising activities, Emily has raised more than $150,000 so far. But she doesn't plan on stopping there. The concert was so successful that Emily and her brother have planned others at bigger venues in coming years. She hopes to draw chart topping Detroit bands, raise even more money, and "rock CF." ■

Summary

- An organism's physical traits constitute its phenotype, while its genes constitute its genotype. A person's genotype can't always be determined from his or her phenotype.
- Genes, which code for proteins, are the units of inheritance, physically passed down from parents to offspring.
- Different versions of the same gene are known as alleles. Alleles arise from mutations that change the nucleotide sequence of a gene.
- Alleles may be dominant or recessive. Dominant alleles can mask the effects of recessive alleles, which can be hidden.
- Many traits result from carrying two recessive alleles, while others result from carrying one dominant allele.
- Meiosis is a type of cell division that produces genetically distinct sperm and egg.
- Homologous chromosomes recombine and assort independently during meiosis to generate genetically diverse sperm and eggs.
- Haploid gametes fuse randomly during fertilization, generating genetically unique zygotes.
- Cystic fibrosis (CF) is a recessively inherited genetic disease. Alterations in the gene *CFTR* cause disease by interfering with ion and water balance.
- A Punnett square can help predict a child's genotype and phenotype when the pattern of inheritance, dominant or recessive, is known.

Chapter 11 Test Your Knowledge

GENES, CHROMOSOMES, ALLELES

Humans have two copies of nearly every gene, located on pairs of chromosomes.

HINT See Infographics 11.1 to 11.3.

KNOW IT

1. How many chromosomes are present in one of your liver cells?

2. How many chromosomes are present in one of your gametes?

USE IT

3. How many copies of the CF-associated allele does a person with CF have in one of his or her lung cells? How does this compare to someone who is a carrier for CF? How does it compare to someone who is homozygous dominant for the gene *CFTR?*

4. Strictly on the basis of the following *CFTR* genotypes, what do you predict the phenotype of each to be?

a. heterozygous
b. homozygous dominant
c. homozygous recessive

5. From the discussion in this chapter, why might a person with a homozygous recessive *CFTR* genotype have a somewhat different phenotype from someone with a homozygous recessive *CFTR* genotype?

MEIOSIS

Meiotic cell division is critical for making gametes. Two separate events, recombination and independent assortment, occur during meiosis to produce genetic diversity.

HINT See Infographics 11.4 and 11.5.

6. A human female has _____ chromosomes in each skin cell and ______ chromosomes in each egg.

a. 46; 46
b. 23; 46
c. 46; 23
d. 23; 23
e. 92; 46

7. A woman is heterozygous for the CF-associated gene (the alleles are represented here by the letters *A* and *a*). Assuming that meiosis occurs normally, which of the following represent eggs that she can produce?

a. *A*
b. *a*
c. *Aa*
d. *AA*
e. *aa*
f. either *A* or *a*
g. any of *A*, *a*, or *Aa*

8. Draw a maternal version of chromosome 7 in one color and a paternal version of chromosome 7 in another color. Maintaining this color distinction, now draw a possible version of chromosome 7 that could end up in a gamete following meiotic division.

USE IT

9. An alien has 82 total chromosomes in each of its body cells. The chromosomes are paired, making 41 pairs. If the alien's gametes undergo meiosis, what are the number and arrangement (paired or not) of chromosomes in one of its gametes? Give the reason for your answer.

10. Describe at least two major differences between mitosis (discussed in Chapter 9) and meiosis.

11. If meiosis were to fail and a cell skipped meiosis I, so that meiosis II was the only meiotic division, how would you describe the resulting gametes?

PREDICTING PATTERNS OF INHERITANCE

Cystic fibrosis is a genetic disease with one pattern of inheritance; other genetic diseases have different inheritance patterns.

HINT See Infographics 11.6 to 11.9.

KNOW IT

12. What is the genotype of a person with CF?

a. homozygous dominant
b. homozygous recessive
c. heterozygous
d. any of the above
e. none of the above

13. A person has a heterozygous genotype for a disease gene and no disease phenotype. Does this disease have a dominant or a recessive inheritance pattern?

14. Women can inherit alleles of a gene called *BRCA1* that makes them susceptible to breast cancer. The alleles associated with elevated cancer risk are dominant. Of the genotypes listed below, which has the lowest genetic risk of developing breast cancer?

- **a.** *BB*
- **b.** *Bb*
- **c.** *bb*
- **d.** *BB* and *Bb* have less risk than *bb*.
- **e.** All have equal risk.

USE IT

15. Assume that Emily (who has CF) decides to have children with a man who does not have CF and who has no family history of CF.

- **a.** What combination of gametes can each of them produce?
- **b.** Place these gametes on a Punnett square and fill in the results of the cross.
- **c.** On the basis of the Punnett square results, what is the probability that they will have a child with CF?
- **d.** On the basis of the Punnett square results, what is the probability that they will have a child who is a carrier for CF?

16. Your friend's mother has Huntington disease and her mother's mother does not have Huntington disease. If your friend's father does not have Huntington disease, what is the probability that your friend will develop Huntington disease? (Hint: draw a Punnett square.)

SCIENCE AND ETHICS

17. Emily took a genetic test to determine which CF alleles she inherited. The results revealed she has one allele about which very little is known. Although genetic testing can predict whether a person will develop CF and drugs can prolong life, for some other genetic diseases, such as Huntington disease, treatment is limited and there is no cure. If you were faced with the decision to take a genetic test, especially for a disease for which there is no cure, would you take the test? Why or why not?

Milestones in Biology

Mendel's Garden

What You Will Be Learning

By studying the inheritance of traits in pea plants, Mendel unknowingly discovered genes and the chromosomal basis of inheritance.

Mendel's Law of Independent Assortment

Mendel went on to study how multiple traits are inherited. For example, he studied plants that had different seed color and seed texture and how those traits passed to the next generation. Tracing two traits at a time helped him form the Law of Independent Assortment. This law posits that two alleles for any given trait will segregate independently from any other alleles when passed on to gametes. Consequently, each gamete may acquire all possible allele combinations and traits.

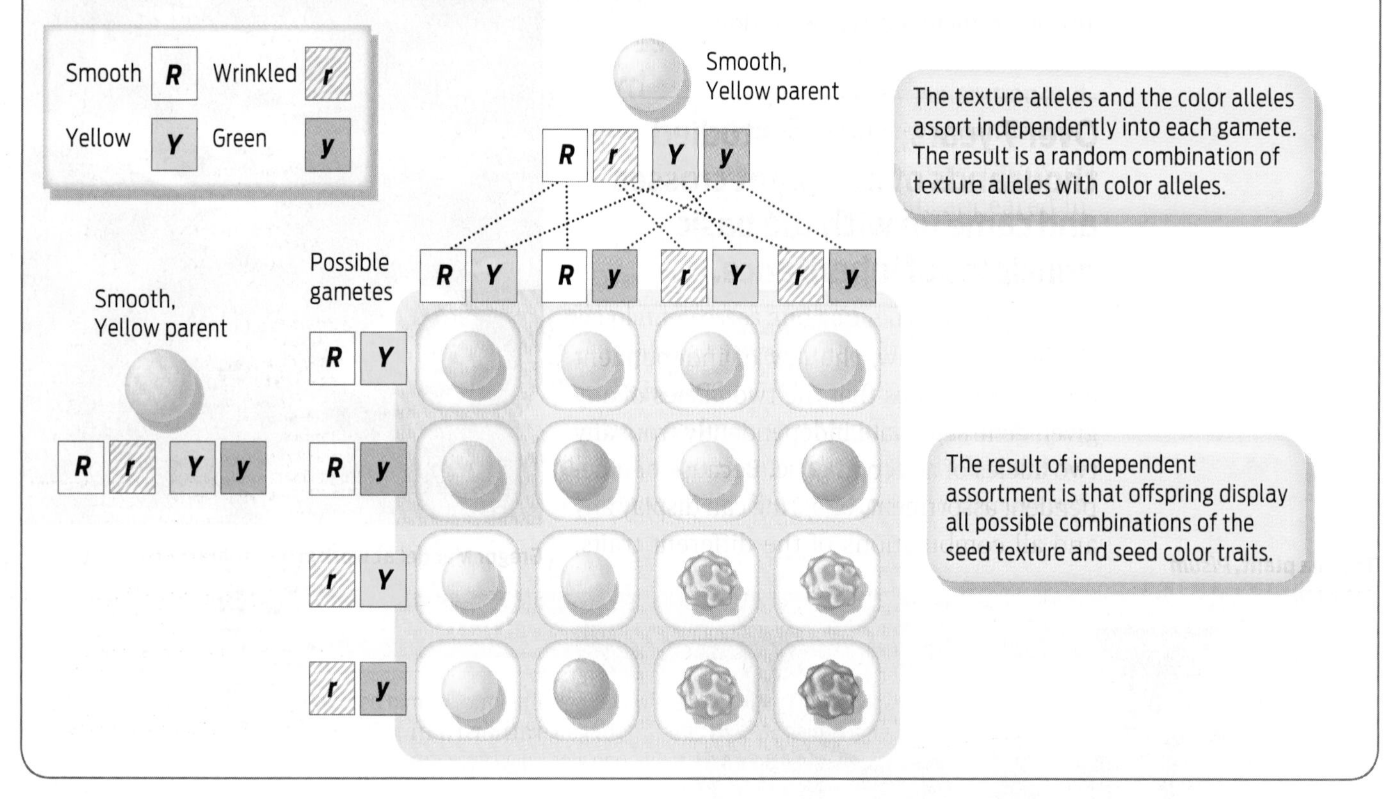

Smooth R
Wrinkled r
Yellow Y
Green y
Smooth, Yellow parent
R r Y y
The texture alleles and the color alleles assort independently into each gamete. The result is a random combination of texture alleles with color alleles.
Possible gametes
R Y
R y
r Y
r y
Smooth, Yellow parent
R r Y y
The result of independent assortment is that offspring display all possible combinations of the seed texture and seed color traits.

Q & A: Genetics

What You Will Be Learning

appear male. Often intersex babies are born with ambiguous genitalia.

Debate over the case of David Reimer, and similar cases, as well as research showing how strongly biology influences sexual identity, has changed the care of intersex babies. Today, such babies are often assigned a sex by parents and doctors only after a period of observation to assess behavior patterns. Surgeons then perform surgery to create either male or female genitalia. Or, parents may forgo surgery, preferring that their child remain as is.

Genetically speaking, disorders of sex development can arise from a number of genetic mutations. For example, if the Y chromosome has a mutation in a gene called *SRY,* the embryo is likely to have undeveloped gonads with external female genitalia, even though it carries an XY chromosome pair.

There are also cases of people with XY sex chromosomes who are missing genes that code for androgen (a male hormone) receptors. So even though they carry a functional *SRY* gene and have internal testes, a mutation in a gene carried on the X chromosome causes a failure of their cells to respond to male hormones like testosterone. As a result, complete male external genitalia do not develop, and these people appear to be female.

Similarly, there are people with XX sex chromosomes who have male genitalia. In some cases, this is caused by a condition called congenital adrenal hyperplasia. These individuals have one or more mutations in genes on autosomal chromosomes. One result is excessive production of male hormones. These people have ovaries but may have genitals that appear more male than female.

INFOGRAPHIC 12.1

X and Y Chromosomes Determine Human Sex

Males and females differ by virtue of a pair of sex chromosomes. Females have two X chromosomes and males have a single X and a single Y chromosome. Every person must have at least one X chromosome, but it's the presence of a gene on the Y chromosome that initiates male development.

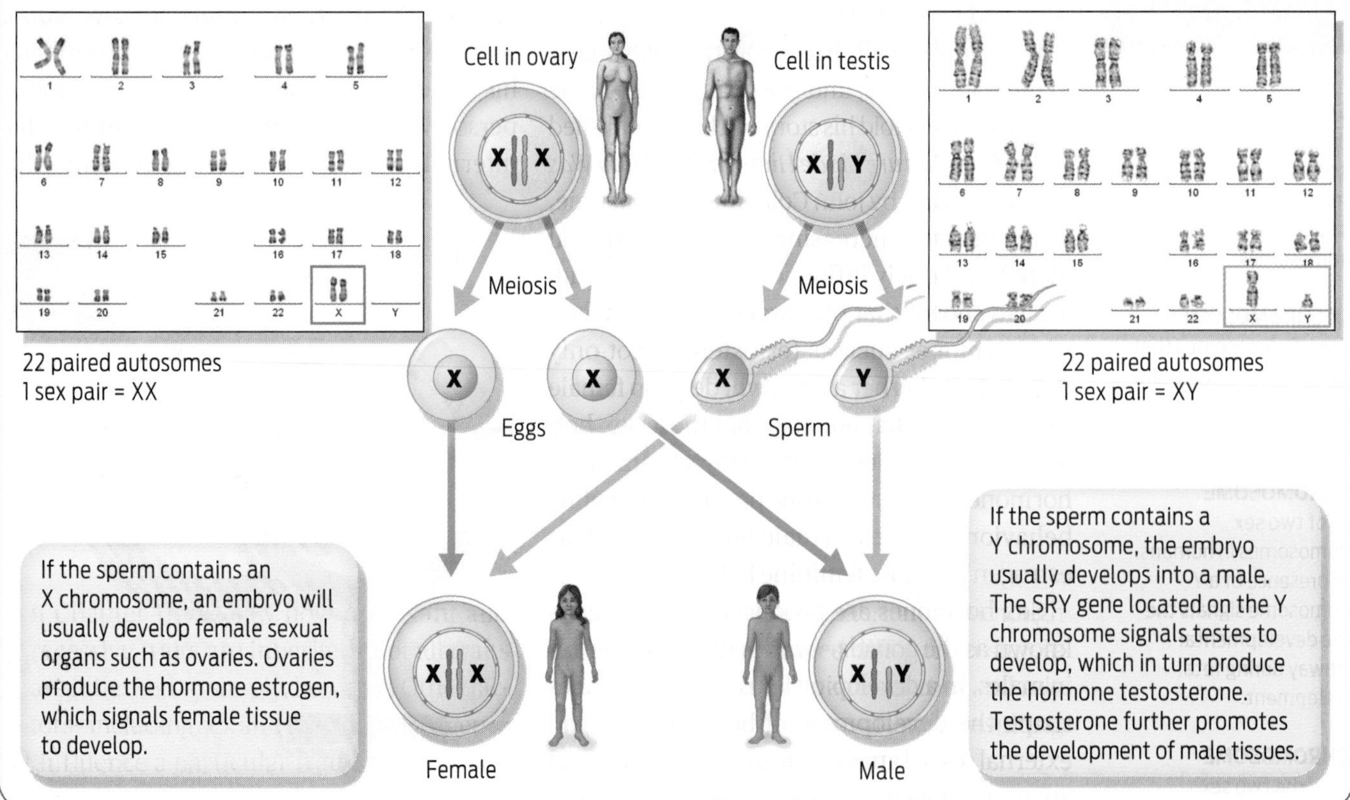

Cell in ovary
Cell in testis
Meiosis
Meiosis
22 paired autosomes
1 sex pair = XX
22 paired autosomes
1 sex pair = XY
Eggs
Sperm
If the sperm contains an X chromosome, an embryo will usually develop female sexual organs such as ovaries. Ovaries produce the hormone estrogen, which signals female tissue to develop.
If the sperm contains a Y chromosome, the embryo usually develops into a male. The SRY gene located on the Y chromosome signals testes to develop, which in turn produce the hormone testosterone. Testosterone further promotes the development of male tissues.
Female
Male

TABLE 12.1

How Many Sexes Are There?

Each of the following individuals has 22 pairs of autosomal chromosomes. Discrepancies in the sexual phenotype may result from environmental factors, hormone imbalance, or having too many or too few sex chromosomes.

SEX CATEGORY	CHROMOSOMES	GONADS	GENITALIA	OTHER CHARACTERISTICS
Female	XX	Ovaries	Female	
Male	XY	Testes	Male	
Female pseudo-hermaphroditism	XX	Ovaries	Male	Infertile
Male pseudo-hermaphroditism	XY	Testes	Female or Ambiguous	Infertile
True gonadal intersex	XX and/or XY	Ovaries and testes	Male or Female or ambiguous	Infertile; historically called true hermaphrodites
Triple X syndrome	XXX	Ovaries	Female	Fertile, taller than average, learning disabilities
Klinefelter syndrome	XXY	Testes	Male	Infertile, enlarged breast tissue
47, XYY syndrome	XYY	Testes	Male	Fertile, taller, learning and emotional disabilities
Turner syndrome	X	Ovaries	Female	Infertile, broad chest, webbed neck

Some people have only one sex chromosome, and others may have three sex chromosomes. Every person must have at least one X chromosome (the only sex chromosome the mother can contribute). Because of errors in chromosome segregation during meiosis, a variety of other X and Y combinations are possible: XXY men, women with only a single X chromosome, XXX females, and XYY males. In many of these cases, a person's physical traits and genitalia reveal that they do not have the usual makeup of sex chromosomes, but not always. These are just a few examples of the many chromosomal possibilities that determine sex. Environmental factors, like exposure to chemicals or abnormal levels of hormones during sexual development, can also play a role in defining a person's sex in terms of external genitalia **(Table 12.1)**.

The question of sex goes beyond a person's genitalia or genotype, however. Further, defining what counts as "masculine" and "feminine" can be even more complicated. For example, some men have characteristics that we typically identify as female, such as a high voice and sparse body hair, yet they are genetically and anatomically male. And many women have what are considered to be more masculine features, such as angular faces and more muscle as compared to body fat. Yet, they, correspondingly, are genetically and anatomically female. In other words, there is no set of physical or mental characteristics that is entirely male or entirely female. In addition, some people may mentally identify with one sex even though their genitalia and chromosomal makeup classify them as the other. Consequently, "sexual identity" is more complicated than simply having an X or a Y chromosome, or male or female genitalia.

SEX-LINKED INHERITANCE

Q Why do some genetic conditions affect sons more often than daughters?

A Some 10 million American men–about 7% of the male population–either cannot distinguish red from green, or see these two colors as different hues from the ones most other people perceive. But such red-green color blindness affects only 0.4% of women. Similarly, 1 in 5,000 boys worldwide is born with hemophilia, a blood-clotting disorder; yet hemophilia rarely afflicts girls.

Why this disparity? Some genetic conditions are more common in boys than in girls. These conditions are caused by genes found on the X chromosome. When a gene is located on either of the sex chromosomes, daughters and sons don't share the same probability of inheriting it.

Take the neuromuscular condition Duchenne muscular dystrophy (DMD), for example. DMD is a disease in which muscles slowly degenerate, leading to paralysis. About 1 in 2,400 boys worldwide is born with the condition each year. Most affected boys are in wheelchairs by the time they become teenagers, and they rarely live longer than 30 years. Why does DMD primarily affect men? Recall that a female has two X chromosomes. For a recessive trait like DMD, a normal copy on one X chromosome masks the recessive disease allele on the other X chromosome. A male, on the other hand, has a single X chromosome, and so will show the effects of any recessive alleles located on his X chromosome. Because females can carry the disease allele without showing it, they may not even know they are at risk of passing it on to their sons.

So, for example, if you are female, and DMD runs in your family–say, a male cousin has the disease–you could in theory pass the disease allele on to your children. Whether or not your children will have the disease depends on which X chromosome they inherit from you. Note that a woman always passes one of her two X chromosomes to each of her children. A man, on the other hand, passes his single X chromosome to his daughters and his Y chromosome to his sons. So if a male carries a disease allele on his X chromosome, he can't pass it to his sons. He can, however, pass the disease allele to his daughters. Diseases and other traits such as DMD that are inherited on X chromosomes are called **X-linked traits.** By contrast, boys and girls share the same probability of inheriting diseases that, like cystic fibrosis, are carried on autosomes (Infographic 12.2).

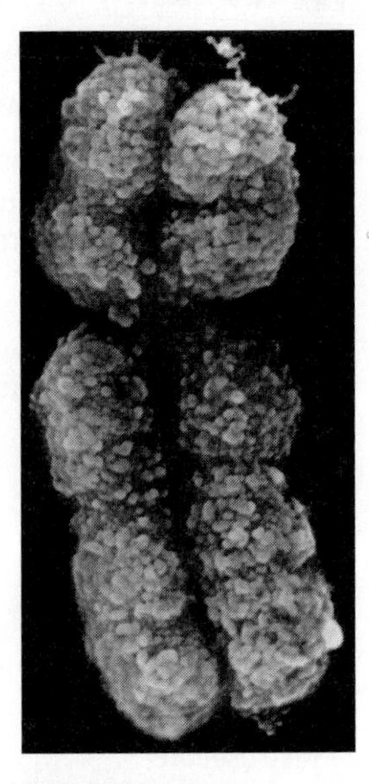

X-linked traits are passed from mothers to children on their X chromosome.

If you are female, and your male cousin has DMD, your aunt was likely a DMD carrier–a person who has a recessive gene for a particular disease, in this case on one of her X chromosomes. Her son, your cousin, inherited her DMD-carrying X chromosome. Because males have only one X chromosome, your cousin doesn't have another allele to mask his defective one.

Your children's risk depends on whether or not you carry a defective DMD gene, which you might have inherited from your mother if she, too, carries a defective DMD gene. If you are a carrier and your husband is healthy, a son who inherits a diseased DMD gene from you would have the disease. This pattern is typical of X-linked traits, which pass down through generations to boys via their mothers (Infographic 12.3).

CHROMOSOME ANALYSIS

Q Did Thomas Jefferson father children with a slave?

A Thomas Jefferson was the third president of the United States, the principal architect of the Declaration of Independence, and founder of the University of Virginia. He was

X-LINKED TRAIT
A phenotype determined by an allele on an X chromosome.

INFOGRAPHIC 12.2

X-linked Traits Are Inherited on X Chromosomes

Duchenne Muscular Dystrophy (DMD) is an example of an X-linked trait. Recessive mutations of the dystrophin gene on the X chromosome cause the disease. DMD primarily affects males because they inherit only one copy of the X chromosome (from their mothers). Therefore, the single DMD allele they inherit determines their phenotype. Since females have two X chromosomes, they may carry the DMD allele, but have a healthy phenotype.

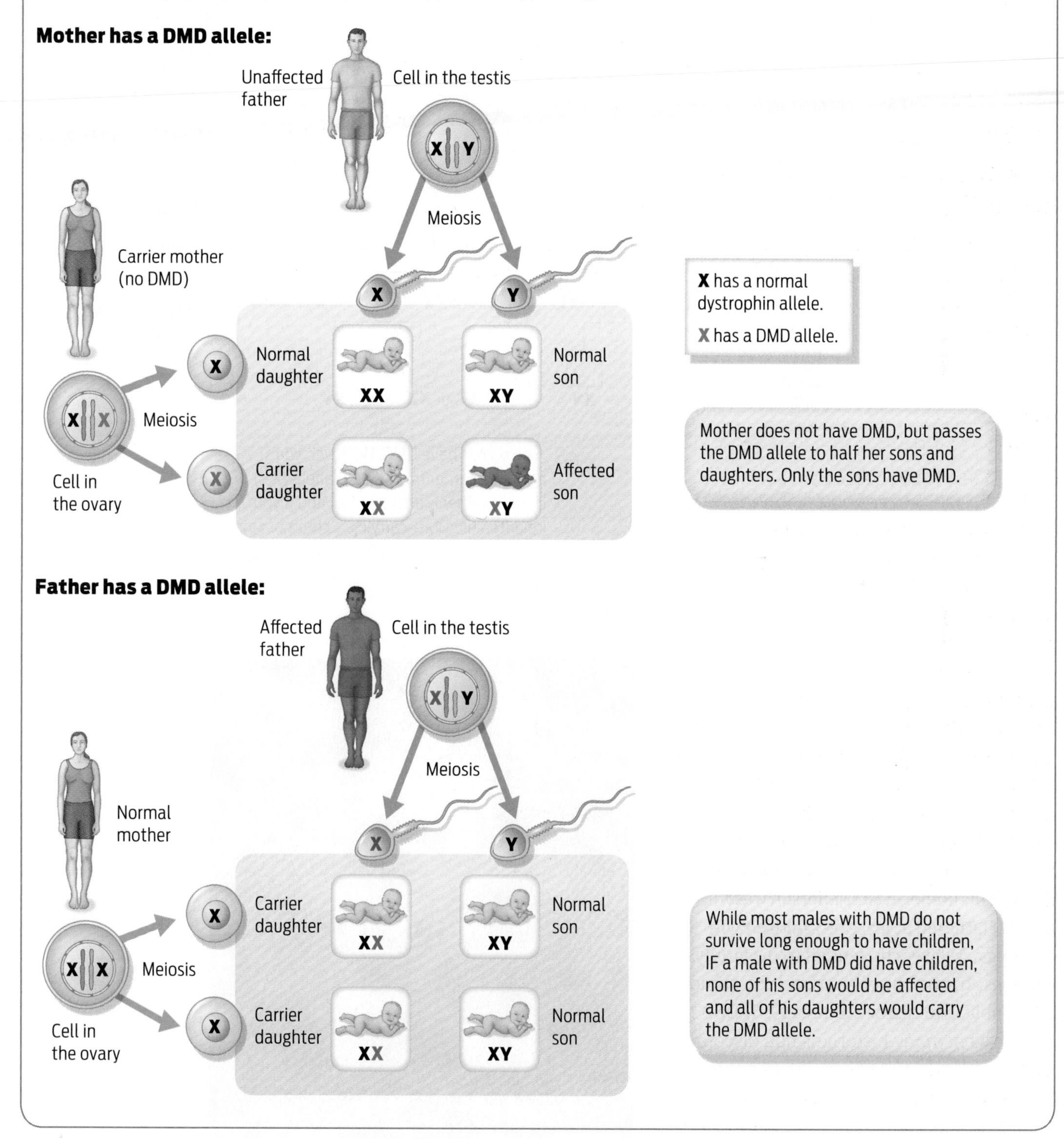

Mother has a DMD allele:
Unaffected father
Cell in the testis
X
Y
Meiosis
Carrier mother (no DMD)
X has a normal dystrophin allele.
X has a DMD allele.
Normal daughter
XX
XY
Normal son
Meiosis
Carrier daughter
XX
XY
Affected son
Cell in the ovary
Mother does not have DMD, but passes the DMD allele to half her sons and daughters. Only the sons have DMD.
Father has a DMD allele:
Affected father
Cell in the testis
Meiosis
Normal mother
Carrier daughter
XX
XY
Normal son
Meiosis
Carrier daughter
XX
XY
Normal son
Cell in the ovary
While most males with DMD do not survive long enough to have children, IF a male with DMD did have children, none of his sons would be affected and all of his daughters would carry the DMD allele.

INFOGRAPHIC 12.3

Female Carriers Can Pass Disease Alleles to Their Children

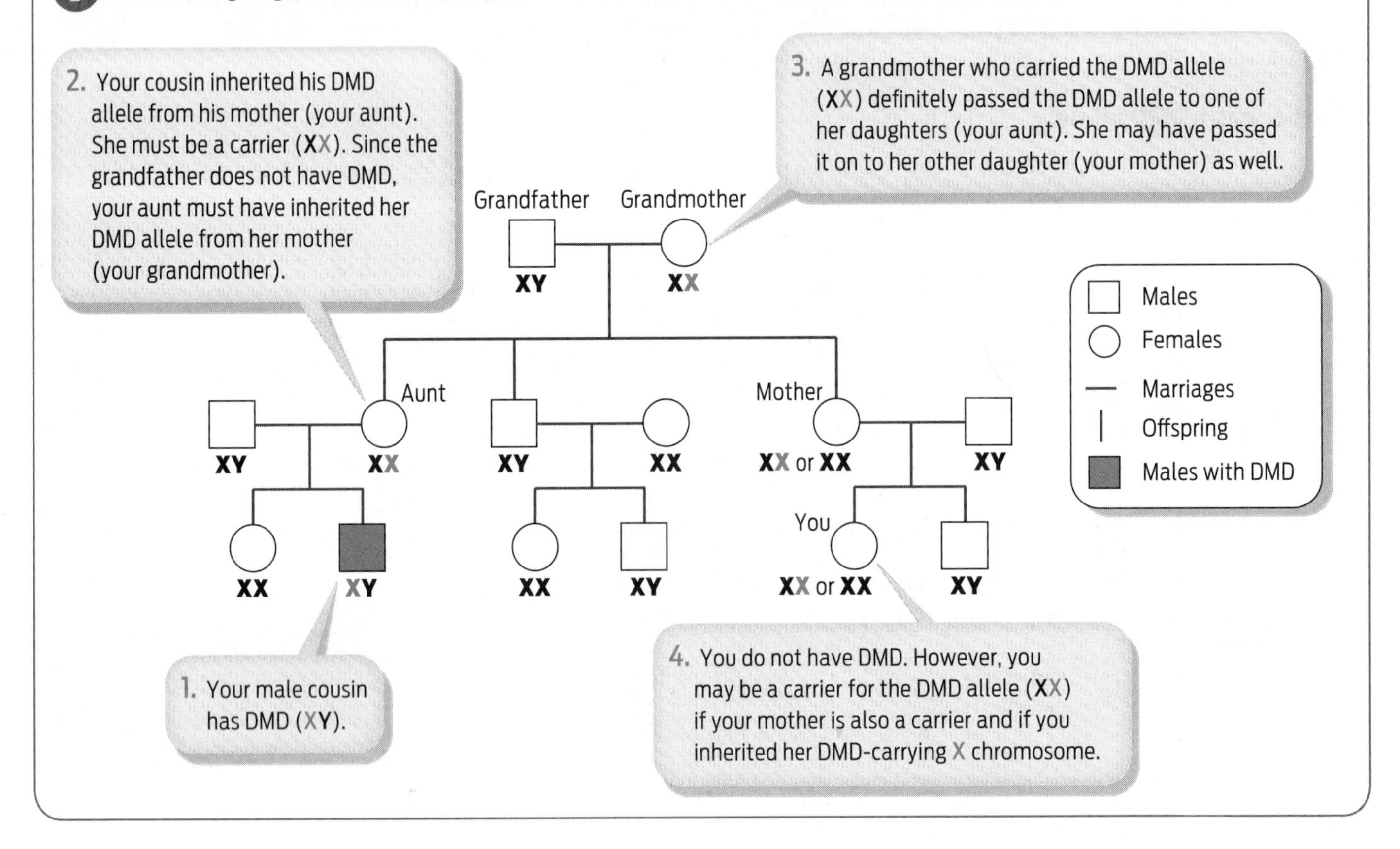

The following diagram, known as a pedigree, shows how an X-linked trait passes through generations.
2. Your cousin inherited his DMD allele from his mother (your aunt). She must be a carrier (XX). Since the grandfather does not have DMD, your aunt must have inherited her DMD allele from her mother (your grandmother).
3. A grandmother who carried the DMD allele (XX) definitely passed the DMD allele to one of her daughters (your aunt). She may have passed it on to her other daughter (your mother) as well.
Grandfather
Grandmother
XY
XX
Aunt
Mother
XY
XX
XY
XX
XX or XX
XY
You
XX
XY
XX
XY
XX or XX
XY
Males
Females
Marriages
Offspring
Males with DMD
1. Your male cousin has DMD (XY).
4. You do not have DMD. However, you may be a carrier for the DMD allele (XX) if your mother is also a carrier and if you inherited her DMD-carrying X chromosome.

also a slave holder. Historians have long debated the meaning of these and other seeming contradictions in the founding father's life and politics. For example, although Jefferson's writings clearly show that he did not believe in the institution of slavery, he owned at least 200 slaves. He made disparaging comments about slaves, yet maintained close relationships with those living in his home. In fact, Jefferson was rumored to have fathered at least six children with Sally Hemings, a slave who tended to his family. For decades, historians discredited the rumor as unreliable oral history. But in 1997, DNA supported what many historians had discounted. Scientists tested the DNA of both Hemings's and Jefferson's descendants using a technique called Y-chromosome analysis. The results: Jefferson could have fathered at least one of Hemings's children.

Third U.S. president Thomas Jefferson.

Are these people descendants of Thomas Jefferson and Sally Hemings?

Y-chromosome analysis is commonly used to study ancestry and to identify paternity. It is just one of several ways that science can complement history. Scientists can use it to verify, discredit, or fill in missing pieces of historical information.

For example, scientists have used Y-chromosome analysis to show that 90% of the Cohanim—members of the Jewish priesthood—are related, supporting oral and written histories claiming the Cohanim are all descended from Aron, brother of Moses. They've also used Y-chromosome analysis to support the oral history of the Lemba, an African tribe, who claim they are descended from Jews. And Y-chromosome analysis suggests that about 8% of Eastern European and Asian men are descended from Genghis Khan or his family.

How does Y-chromosome analysis work? Sons inherit their Y chromosome from their biological fathers. These Y chromosomes are passed through generations largely intact. That's because Y chromosomes have no homologous partner chromosome with which to pair and exchange DNA during meiosis. In other words, the Y chromosome rarely undergoes genetic recombination. Consequently, the Y chromosome that a son inherits from his father is almost identical to the Y chromosome that his father inherited from his father. In this way, Y chromosomes are transmitted essentially unchanged from fathers to sons.

Comparing DNA sequences on Y chromosomes can reliably establish paternity. In 1998, a team led by Eugene A. Foster, a pathologist, compared the Y chromosomes of four groups of men: descendants of Thomas Jefferson's grandfather Field Jefferson; descendants of Thomas Woodson, a man who claimed he was Sally Hemings and Thomas Jefferson's

INFOGRAPHIC 12.4

Y Chromosomes Pass Largely Unchanged from Fathers to Sons

Paternity testing (Y-chromosome analysis) relies upon the fact that the Y chromosome does not undergo recombination during meiosis and so passes unchanged from the father to his sons.

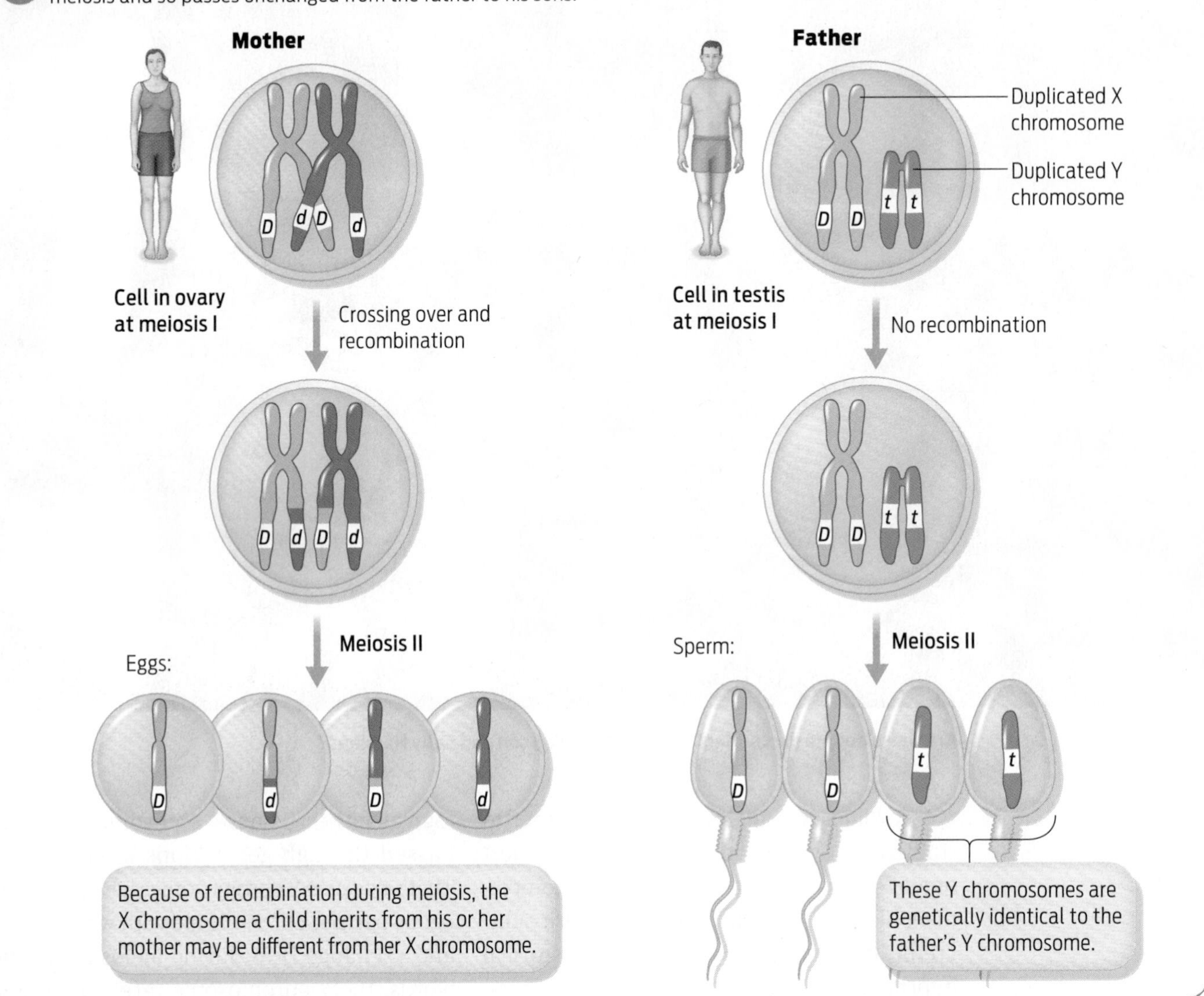

first child; descendants of Eston Hemings, Sally Hemings's son; and descendants of John Carr, Jefferson's sister's son. Since Jefferson's only surviving child from his wife was a daughter, he did not have any direct male descendants, which is why scientists tested descendants of Jefferson's grandfather **(Infographic 12.4)**.

The study analyzed 11 short tandem repeats (STRs) on the Y chromosome. (Recall from Chapter 7 that STRs are regions of noncoding DNA that show differences in the number of times a short DNA sequence is repeated among different people.) The results showed that the man descended from Eston Hemings has the same Y chromosome as the descendants of Field Jefferson. Consequently, Thomas Jefferson could have fathered Eston Hemings. However, any male Jefferson could have fathered Hemings's son Eston **(Infographic 12.5)**.

In fact, some historians have argued that Thomas's brother Randolph Jefferson fathered Eston. But other experts have argued that historical evidence places Thomas himself rather than Randolph under the same roof as Sally at the time of her conceptions.

INFOGRAPHIC 12.5

DNA Links Sally Hemings's Son to Jefferson

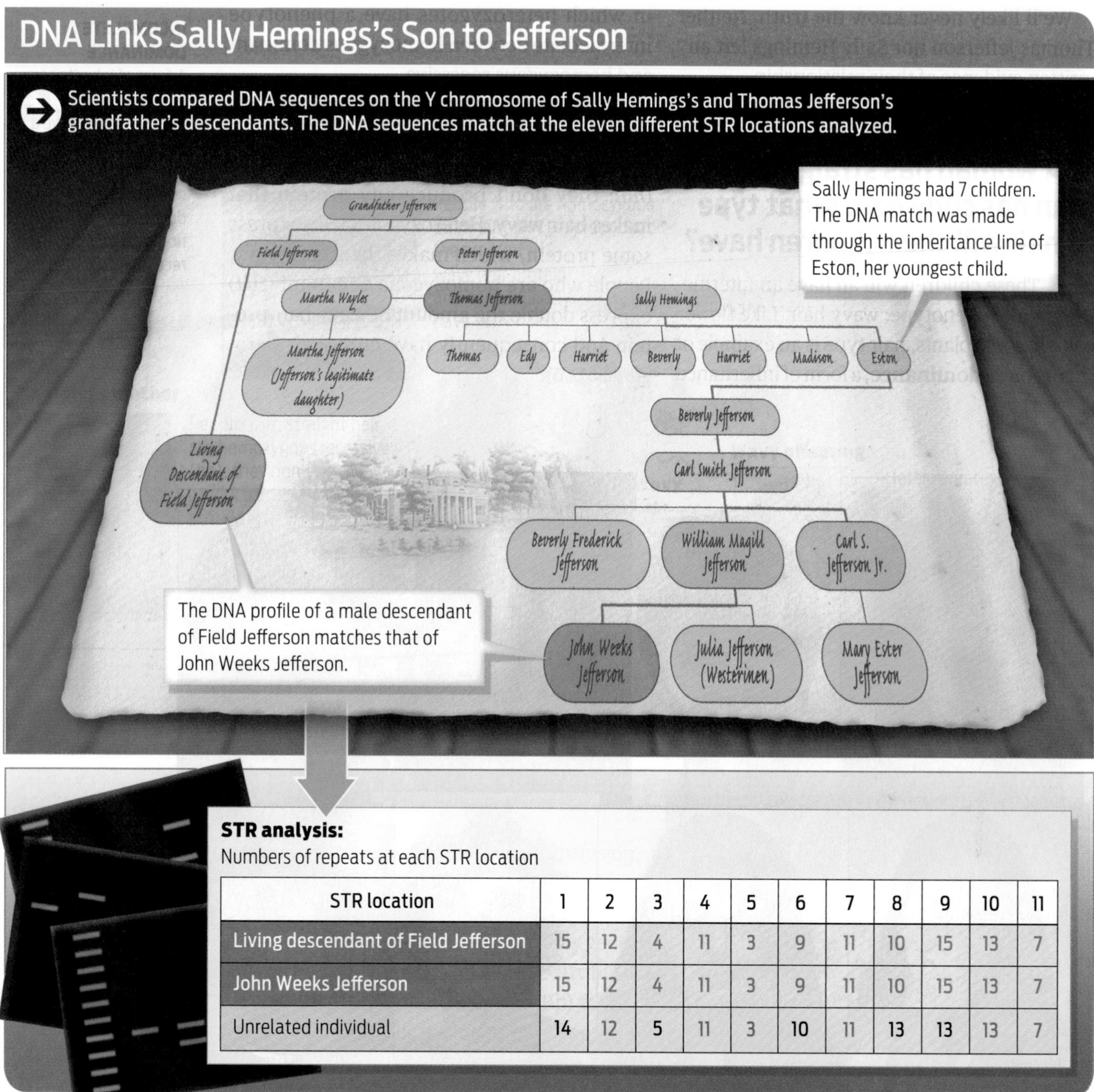

STR analysis:
Numbers of repeats at each STR location

STR location	1	2	3	4	5	6	7	8	9	10	11
Living descendant of Field Jefferson	15	12	4	11	3	9	11	10	15	13	7
John Weeks Jefferson	15	12	4	11	3	9	11	10	15	13	7
Unrelated individual	**14**	12	**5**	11	3	**10**	11	**13**	**13**	13	7

For the descendants of Eston Hemings, the DNA study was powerful vindication. They had long argued that they were descended from Thomas Jefferson, but without hard evidence, most historians disregarded their claims. "I feel wonderful about it," Julia Jefferson Westerinen, a Staten Island artist and Eston's great-great-granddaughter told the *New York Times* when the study results were published. "I feel honored."

As for the relationship between Jefferson and Sally Hemings, historians continue to debate whether it was consensual or forced. "I was a history major," said Jefferson Westerinen, "And we learned not to say, 'I feel this, I think that,' without knowing the facts. They had a relationship of 38 years. I would like to think they were in love, but how would I know?"

INFOGRAPHIC 12.9

Human Height Is Both Polygenic and Multifactorial

Multiple genes as well as environmental factors such as diet, nutrition, and overall health act together to determine how tall we become.

Polygenic (blue bars on graph):
Many genes contribute to determine one's height. The combination of alleles a person inherits (*aabbcc*, *AabbCc*, etc.) predicts a distinct height phenotype.

Multifactorial (red line on graph):
Human populations show a continuous range of heights, however, rather than the genetically predicted number of distinct phenotypes. This is due to environmental influences interacting with one's genetically determined potential for height.

A mating between two people with medium height (where three genes control height): *AaBbCc* × *AaBbCc* produces seven distinct phenotypes determined by the number of dominant genes inherited.

Female gametes \ Male gametes	ABC	ABc	AbC	aBC	Abc	aBc	abC	abc
ABC	AABBCC	AABBCc	AABbCC	AaBBCC	AABbCc	AaBBCc	AaBbCC	AaBbCc
ABc	AABBCc	AABBcc	AABbCc	AaBBcC	AABbcc	AaBBcc	AaBbCc	AaBbcc
AbC	AAbBCC	AAbBCc	AAbbCC	AabBCC	AAbbCc	AabBCc	AabbCC	AabbCc
aBC	aABBCC	AaBBCc	aABbCC	aaBBCC	aABbCc	aaBBCc	aaBbCC	aaBbCc
Abc	AABBcC	AAbBcc	AAbbcC	AabBcC	AAbbcc	AabBcc	AabbcC	Aabbcc
aBc	aABBcC	aABBcc	aABbcC	aaBBcC	aABbcc	aaBBcc	aaBbcC	aaBbcc
abC	aAbBCC	aAbBCc	aAbbCC	aabBCC	aAbbCc	aabBCc	aabbCC	aabbCc
abc	aAbBcC	aAbBcc	aAbbcC	aabBcC	aAbbcc	aabBcc	aabbcC	aabbcc

An individual inheriting the genotype *aaBbCc* falls in the 5.0 ft phenotype.

That same individual may actually grow to a height of 5.3 ft depending upon environmental influences.

Number of individuals: 0, 200, 400, 600, 800

Human height (feet): 4.0, 4.5, 5.0, 5.5, 6.0, 6.5, 7.0

children, and so on. In developing countries, where many people are still malnourished, environment plays a larger role. Another way of looking at this is that more people in developed countries have reached their genetic potential than people in developing countries because most of us in the developed world have access to adequate nutrition. In developing countries, access to nutrition varies much more and this variation is reflected in larger variations in height between a person and his or her direct relatives. In fact, the average height of the U.S. population has almost leveled off in the past decade, suggesting that the environment has almost maximized the genetic potential of height in this country.

When both genes and environment work together to influence a given trait, the trait is described as multifactorial. So height is both polygenic and multifactorial **(Infographic 12.9)**.

Multifactorial inheritance is a common pattern of inheritance. Diseases such as asthma, diabetes, and heart disease are all caused by a combination of several genes and their interaction with the environment. For example, some studies have found that cigarette smoke, air pollution, and ozone can exacerbate asthma. Other studies have shown that people who carry the *E4* allele of a gene called *APO* have an increased chance of developing heart disease if they smoke and don't exercise, compared to people with other *APO E* alleles who smoke and don't exercise. Even for traits that are largely genetically determined, the environment plays a very important role in influencing our phenotypes.

MULTIFACTORIAL INHERITANCE
An interaction between genes and the environment that contributes to a phenotype or trait.

Depression cannot be explained by genetic or environmental factors alone, but by an interaction between the two.

MULTIFACTORIAL INHERITANCE

Q Can people be genetically predisposed to depression?

A In the early 1990s, Stephen Suomi and Dee Higley, researchers at the National Institute of Child Health and Human Development, were studying how stress affected the mental development of infant monkeys. More specifically, they were looking at whether certain alleles of a specific gene, called serotonin transporter, made infant monkeys more vulnerable to stress early in life.

The serotonin transporter gene is located on chromosome 17; it exists as two alleles, a short version and a long version. The long version contains about 44 extra base pairs. Previous research had suggested that people who had at least one copy of a short version of this gene were much more likely to have an anxiety disorder.

Higley and Suomi showed that infant monkeys exposed to stress, such as being deprived of their mothers, and who carried short versions of this allele, behaved differently from their counterparts: they were more anxious, aggressive, and some even became alcoholics as adults.

Despite this finding in monkeys, it quickly became clear that having short versus long alleles could not explain why some people become severely depressed while others are more resilient. Researchers could not find a clear association between any particular allele and depression in people.

Taking their cue from Higley and Suomi, in 2003 Terrie Moffitt and Avshalom Caspi, a husband-and-wife team of psychologists at King's College London, decided to test whether environmental influences might also contribute to depression in people. Moffitt and Caspi turned to a long-term study of almost 900 New Zealanders, identified these subjects' serotonin transporter alleles, and interviewed them about traumatic experiences in early adulthood—experiences such as a major breakup, a death in the family, or serious injury—to see if these difficulties brought out an underlying genetic tendency toward depression.

The results were striking: clinical depression was diagnosed in 43% of subjects who had two copies of the short allele and who had experienced four or more tumultuous events. By contrast, only 17% of subjects who had two copies of the long allele and who had endured four or more stressful events had become depressed—this was no more than the rate of depression in the general population. Subjects with the short allele who experienced no stressful events fared pretty well, too—they also became depressed at the average rate. Clearly, it was the combination of hard knocks

INFOGRAPHIC 12.10

Serotonin Transporter Function Is Linked to Depression

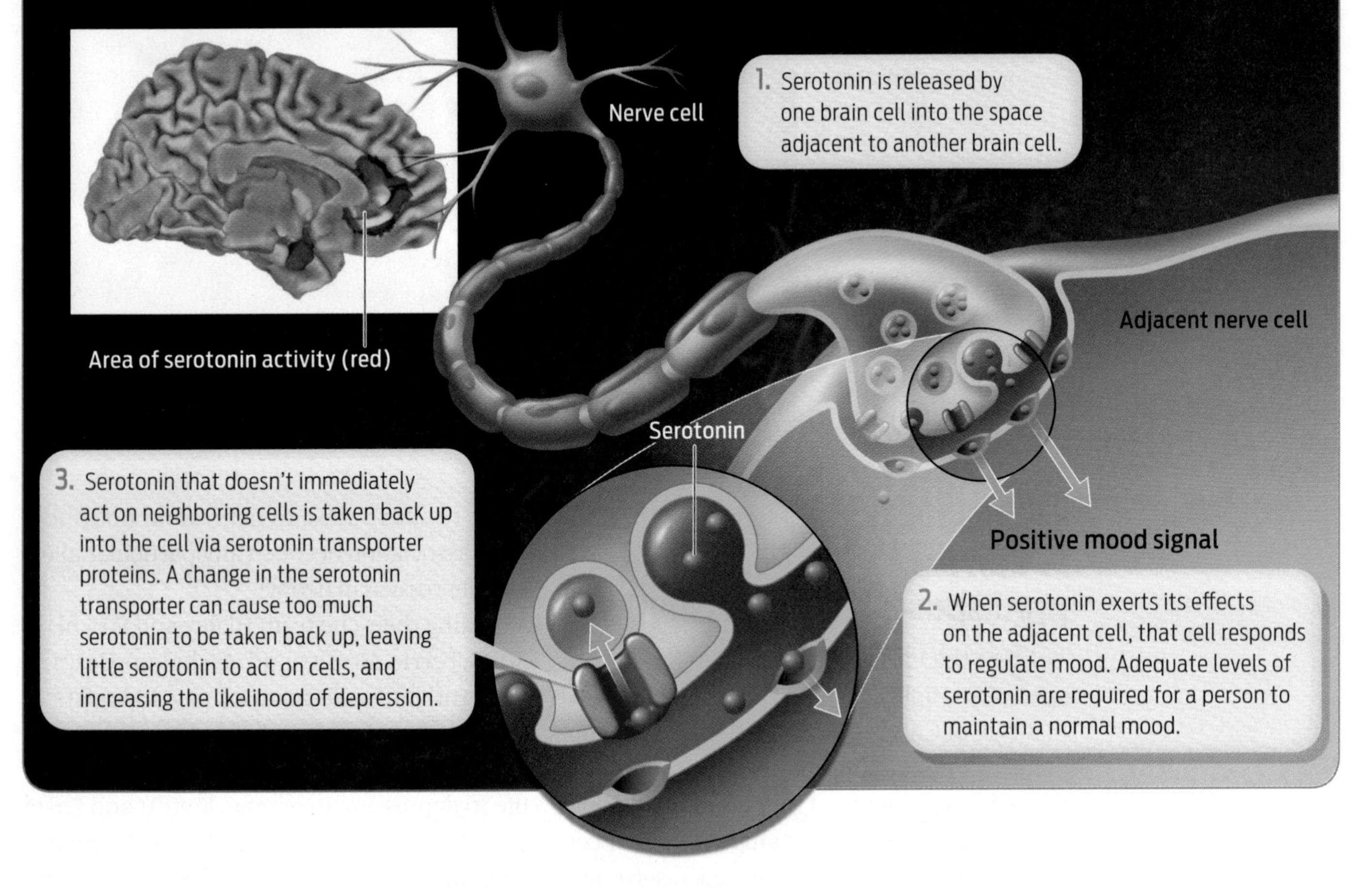

Serotonin is an important signaling molecule in the brain. The serotonin transporter influences the levels of serotonin available for signaling. People with depression have lowered levels of serotonin in the spaces between cells.
Nerve cell
1. Serotonin is released by one brain cell into the space adjacent to another brain cell.
Adjacent nerve cell
Area of serotonin activity (red)
Serotonin
3. Serotonin that doesn't immediately act on neighboring cells is taken back up into the cell via serotonin transporter proteins. A change in the serotonin transporter can cause too much serotonin to be taken back up, leaving little serotonin to act on cells, and increasing the likelihood of depression.
Positive mood signal
2. When serotonin exerts its effects on the adjacent cell, that cell responds to regulate mood. Adequate levels of serotonin are required for a person to maintain a normal mood.

and short alleles that more than doubled the risk of depression.

Since the early 1990s, researchers have shown that the serotonin transporter gene influences the levels of serotonin present in the spaces between brain cells in humans and other animals, and that low levels of serotonin in these spaces is one biological hallmark of depression in people (**Infographic 12.10**). But there are likely other factors that contribute to a person's risk of depression—there are many people who carry long alleles and who suffer from depression, as well as people who carry short alleles who do not, despite having gone through distressing experiences.

Nevertheless, Caspi and Moffitt's study was one of the first to examine the combined effects of genetic predisposition and experience on a specific trait—psychiatrists were delighted. While scientists had been trying to tease apart environmental from genetic influences on physical diseases like cancer, this was the first study to investigate this relationship in a mental disorder. Moreover, the findings reinforced the emerging view that the majority of mental illnesses and other complex diseases cannot be explained by genetic or environmental factors alone, but often arise from an interaction between the two. That is, mental illnesses exhibit multifactorial inheritance (**Infographic 12.11**).

NONDISJUNCTION

Q Why does the risk of having a baby with Down syndrome go up as a woman ages?

A At age 25, a woman's risk of having a baby with Down syndrome is 1 in 1,250 births. At age 40 her risk skyrockets to 1 in 100 births.

INFOGRAPHIC 12.11

Depression Is a Multifactorial Trait

In 2003, Terrie Moffitt and Avshalom Caspi showed that a specific allele of the serotonin transporter gene—a gene that influences levels of the signaling molecule serotonin in the brain—in combination with stressful life events can cause depression. The gene comes in long and short versions.

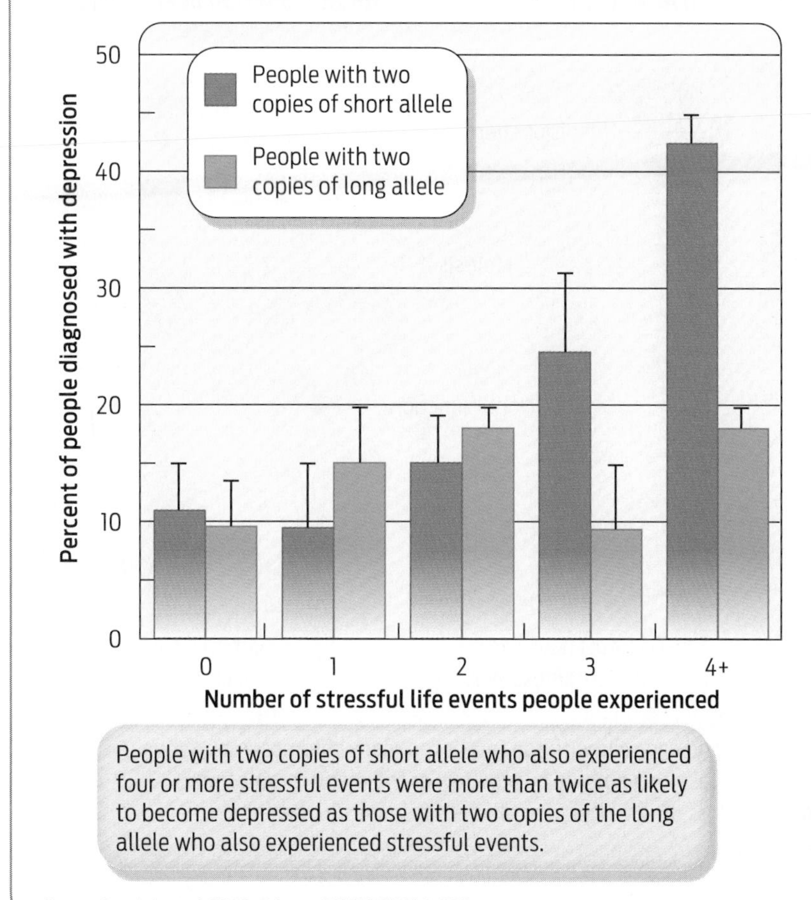

People with two copies of short allele who also experienced four or more stressful events were more than twice as likely to become depressed as those with two copies of the long allele who also experienced stressful events.

Source: Caspi, A. et al. 2003. *Science.* 301(5631):386–389.

an error in chromosome segregation, leading to a chromosomal abnormality.

A chromosomal abnormality means that a developing fetus carries a chromosome number that differs from the usual 46. The most common abnormalities in humans are called **aneuploidies,** deviations from the normal number of chromosomes because single chromosomes are either duplicated or deleted. Most aneuploidies arise during meiosis in the parents' sex cells. If a gamete makes a mistake when chromosomes segregate, an occurrence called **nondisjunction,** it will either lack a chromosome or carry an extra copy. When that egg is fertilized by a normal gamete, the resulting zygote can have an abnormal number of chromosomes. In most cases, the abnormality is so severe the zygote spontaneously aborts.

There are, however, cases in which the abnormality is not life threatening but does cause severe disability—the most common is **trisomy 21,** also called Down syndrome. Trisomy 21 results when an embryo inherits an extra copy of chromosome 21. Anyone can conceive a child with the abnormality, but older women are at exceptionally high risk (Infographic 12.12).

Most Down syndrome children have learning disabilities that range from mild to moderate, but some have profound mental disability. They

ANEUPLOIDY
An abnormal number of one or more chromosomes (either extra or missing copies).

NONDISJUNCTION
Failure of chromosomes to separate accurately during cell division; nondisjunction in meiosis leads to aneuploid gametes.

TRISOMY 21
Carrying an extra copy of chromosome 21; also known as Down syndrome.

In fact, as women age, the risk of giving birth to a baby with any chromosomal abnormality increases. That's because as a woman ages, so do her eggs. All the eggs that a woman will ever have were formed before she was born, and they have been aging like the rest of the cells in her body. Until puberty, a woman's eggs are "paused" in the middle of meiosis (at meiosis I); they haven't yet completed their cell division. During a menstrual cycle, one egg resumes meiosis and is ovulated. In older women, when these eggs complete meiosis and are ovulated, they are more likely to have

INFOGRAPHIC 12.12

Chromosomal Abnormalities: Aneuploidy

Birth defects can arise when chromosomes fail to separate normally during meiosis, a phenomenon called nondisjunction. The resulting gametes carry an abnormal number of chromosomes, a condition called aneuploidy.

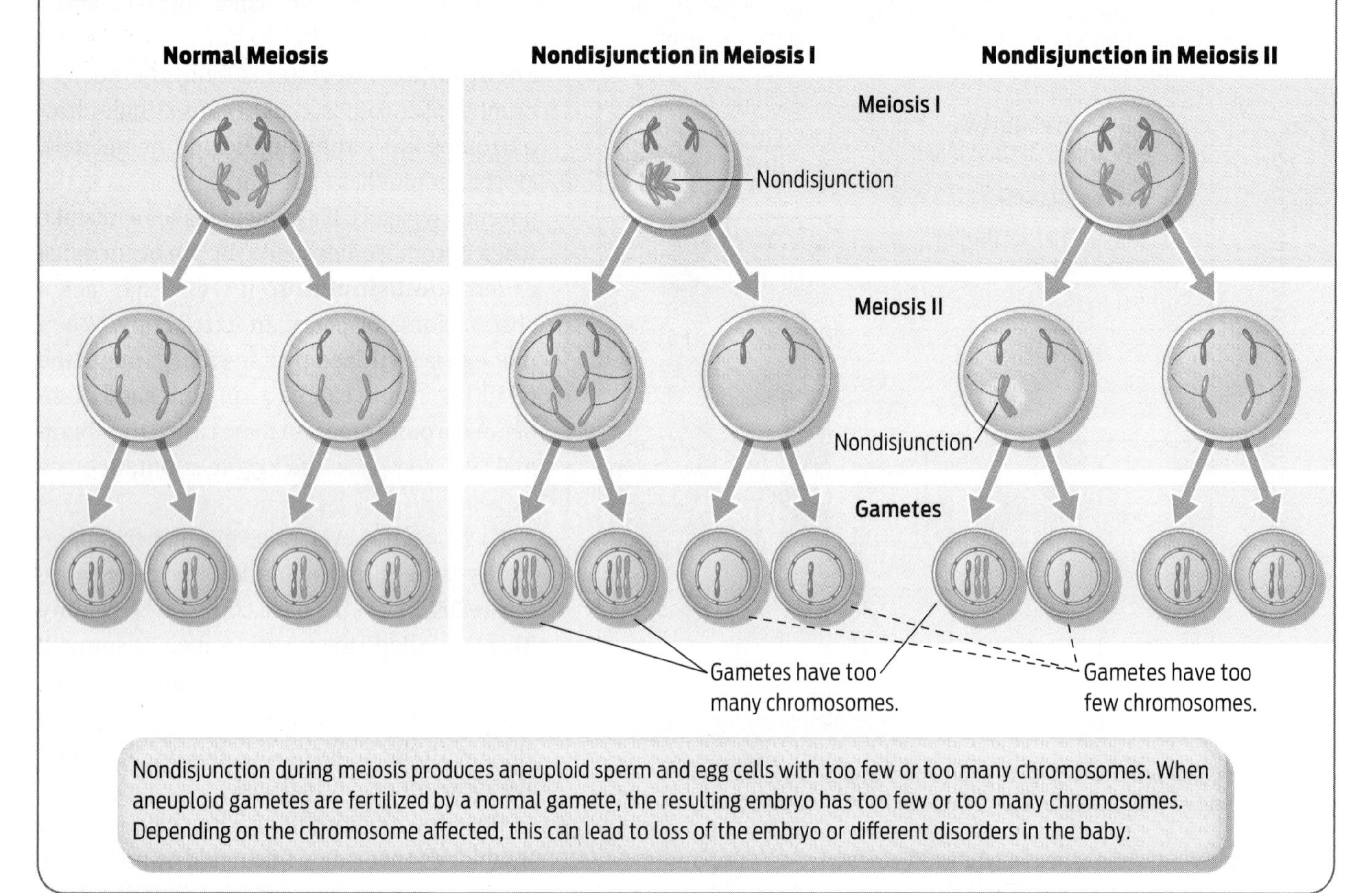

Nondisjunction during meiosis produces aneuploid sperm and egg cells with too few or too many chromosomes. When aneuploid gametes are fertilized by a normal gamete, the resulting embryo has too few or too many chromosomes. Depending on the chromosome affected, this can lead to loss of the embryo or different disorders in the baby.

are also at higher risk for other diseases and typically don't live beyond 50 years of age.

Down syndrome, as well as other chromosomal abnormalities, can be diagnosed by **amniocentesis.** This procedure is usually performed between 14 and 20 weeks of pregnancy, although some medical centers may perform it as early as 11 weeks. The procedure is quick. A long, thin, hollow needle is inserted through a woman's abdominal wall and into her uterus. Through the needle, the equivalent of 2 to 4 teaspoons of amniotic fluid, which surrounds the growing fetus, is removed. This fluid contains fetal cells that contain the fetus's DNA. From that fluid, technicians analyze the fetal **karyotype**–that is, the chromosomal makeup in its cells (Infographic 12.13).

The reasons to undergo amniocentesis vary from couple to couple. But if a test comes back positive, couples have options: they can begin to plan for a disabled child, or make the decision not to carry the child to term.

Although scientists have linked some of the most obvious birth defects to the age of a woman's eggs, recent research also shows that a man's age affects his sperm quality. Men who father children after age 45 are more likely to have children with cognitive disorders such as autism, for example. Male fertility declines over time, too, although much more gradually than does female fertility. Research shows that the older the man, the more likely he is to produce sperm with genetic defects. ■

AMNIOCENTESIS
A procedure that removes fluid surrounding the fetus to obtain and analyze fetal cells to diagnose genetic disorders.

KARYOTYPE
The chromosomal makeup of cells. Karyotype analysis can be used to detect trisomy 21 prenatally.

INFOGRAPHIC 12.13

Amniocentesis Provides a Fetal Karyotype

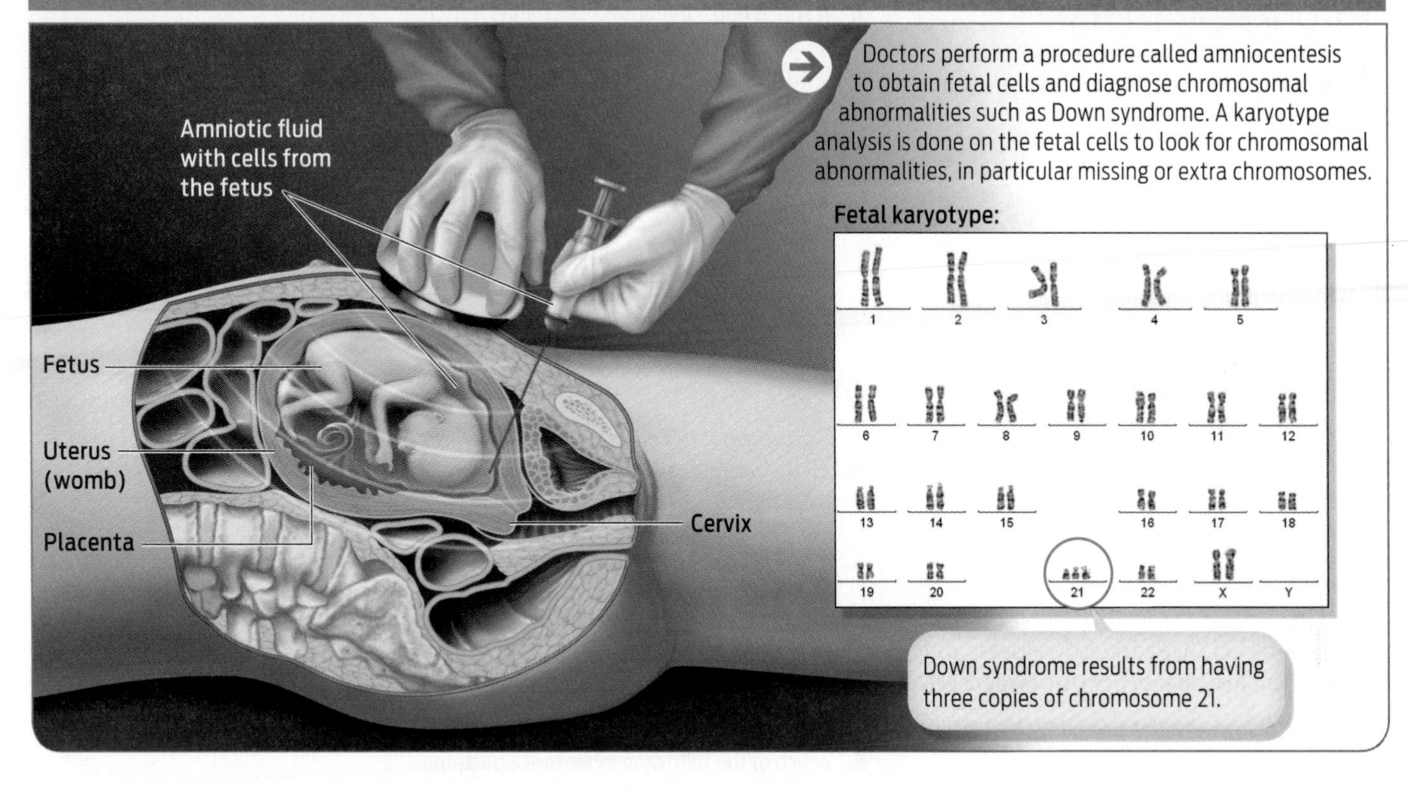

Doctors perform a procedure called amniocentesis to obtain fetal cells and diagnose chromosomal abnormalities such as Down syndrome. A karyotype analysis is done on the fetal cells to look for chromosomal abnormalities, in particular missing or extra chromosomes.
Amniotic fluid with cells from the fetus
Fetus
Uterus (womb)
Placenta
Cervix
Fetal karyotype:
1
2
3
4
5
6
7
8
9
10
11
12
13
14
15
16
17
18
19
20
21
22
X
Y
Down syndrome results from having three copies of chromosome 21.

Summary

- Humans have 23 pairs of chromosomes. One of these pairs is the sex chromosomes: XX in females and XY in males. It is the presence of the Y chromosome that determines maleness.
- Because the Y chromosome in a male does not have a homologous partner, it does not experience recombination during meiosis. The Y chromosome a son inherits from his father is essentially identical to the Y chromosome his father inherited from his father (the grandfather), a fact that can be used to establish paternity.
- Some genes are located on the X chromosome; these are known as X-linked genes. Disorders inherited on X chromosomes are called X-linked disorders, and are more common in males than in females.
- Hair type is an example of incomplete dominance, a form of inheritance in which heterozygotes have a phenotype intermediate between homozygous dominant and homozygous recessive.
- ABO blood type is an example of a codominant trait—both maternal and paternal alleles contribute equally and separately to the phenotype.
- Many traits are polygenic—that is, they are influenced by the additive effects of multiple genes. Polygenic traits show a normal distribution in the population.
- In many cases, a person's phenotype is determined by both his or her genotype at a number of different genes as well as by environmental influences; this type of inheritance is described as multifactorial. Human height, cardiovascular disease, and depression are examples of multifactorial inheritance.
- Some genetic disorders result from having a chromosome number that differs from the usual 46. Down syndrome, or trisomy 21, is caused by having an extra copy of chromosome 21.

Chapter 12 Test Your Knowledge

SEX-LINKED INHERITANCE

The two human sex chromosomes are X and Y. Genes located on the sex chromosomes are said to be sex linked.

HINT See Infographics 12.1–12.5.

KNOW IT

1. Which of the following most influences the development of a female fetus?

- **a.** the presence of any two sex chromosomes
- **b.** the presence of two X chromosomes
- **c.** the absence of a Y chromosome
- **d.** the presence of a Y chromosome
- **e.** either b or c

2. Why are more males than females affected by X-linked recessive genetic diseases?

3. If a man has an X-linked recessive disease, can his sons inherit that disease from him? Why or why not?

USE IT

4. Which of the following couples could have a boy with Duchenne muscular dystrophy?

- **a.** a male with Duchenne muscular dystrophy and a homozygous dominant female
- **b.** a male without Duchenne muscular dystrophy and a homozygous dominant female
- **c.** a male without Duchenne muscular dystrophy and a carrier female
- **d.** a and c
- **e.** none of the above

5. Predict the sex of a baby with each of the following pairs of sex chromosomes. (You may want to use this question to go back and check your answer to Question 1.)

- **a.** XX
- **b.** XXY
- **c.** XY
- **d.** X

6. Consider your brother and your son.

- **a.** If you are female, will your brother and your son have essentially identical Y chromosomes? Explain your answer.
- **b.** If you are male, will your brother and your son have essentially identical Y chromosomes? Explain your answer.

7. A wife is heterozygous for Duchenne muscular dystrophy alleles and her husband has a dominant allele on his X chromosome. What percentage of their sons, and what percentage of their daughters, will have:

- **a.** Duchenne muscular dystrophy (which is determined by a recessive allele on the X chromosome)
- **b.** an X-linked dominant form of rickets (a bone disease)

OTHER PATTERNS OF INHERITANCE

Not all traits are inherited in simple dominant and recessive patterns.

HINT See Infographics 12.6–12.11.

KNOW IT

8. What aspects of height make it a polygenic trait?

9. Which of the following inheritance patterns includes an environmental contribution?

- **a.** polygenic
- **b.** X-linked recessive
- **c.** X-linked dominant
- **d.** multifactorial
- **e.** none of the above

10. What is the difference between polygenic inheritance and multifactorial inheritance?

11. How does incomplete dominance differ from codominance?

12. If you are blood type A-positive, to whom can you safely donate blood? Who can safely donate blood to you? List all possible recipients and donors and explain your answer.

USE IT

13. If two women have identical alleles of the suspected 20 height-associated genes, why might one of those women be 5 feet 5 inches tall and the other 5 feet 8 inches tall?

14. Look at Infographic 12.11. How do the data given support the hypothesis that both genes and the environment influence at least some cases of clinical depression?

15. Look at Infographic 12.11. At approximately how many stressful experiences does the homozygous short genotype begin to influence the depression phenotype?

16. From information in this chapter, how can you account for two people with the same genotype for a predisposing disease allele having different phenotypes?

CHROMOSOMAL ABNORMALITIES

Improper chromosome segregation during cell division can lead to birth defects.

HINT **See Infographics 12.12 and 12.13.**

KNOW IT

17. What is the normal chromosome number for each of the following:

- **a.** a human egg
- **b.** a human sperm
- **c.** a human zygote

18. When looking at a karyotype, for example to diagnose trisomy 21 in a fetus, is it possible to use that analysis also to tell if the fetus has inherited a cystic fibrosis allele from a carrier mother?

USE IT

19. Which of the following can result in trisomy 21?

- **a.** an egg with 23 chromosomes fertilized by a sperm with 23 chromosomes
- **b.** an egg with 22 chromosomes fertilized by a sperm with 23 chromosomes
- **c.** an egg with 24 chromosomes, two of which are chromosome 21, fertilized by a sperm with 23 chromosomes
- **d.** an egg with 23 chromosomes fertilized by a sperm with 24 chromosomes, two of which are chromosome 21

20. From information in this chapter, which of the possibilities in Question 19 is most likely? Explain your answer.

SCIENCE AND ETHICS

21. What factors would lead you to consider prenatal genetic testing? In your opinion, what is the value of having this information?

Grow Your Own

What You Will Be Learning

Grow Your Own

Stem cells could be the key to engineering organs

In 1995, Charles Vacanti, an anesthesiologist, and Linda Griffith-Cima, then an assistant professor of chemical engineering at the Massachusetts Institute of Technology (MIT), amazed the world with an unusual and important experiment. Under the skin of a laboratory mouse, they injected cow cartilage cells into an implanted and biodegradable mold shaped like a human ear. The result? A structure of cartilage, shaped like a human ear, grew on the mouse's back. The sensational image splashed across tabloids, and many hailed the feat as a great scientific accomplishment–but some likened it to creating Frankenstein's monster. The experiment had a serious purpose, however: the mouse's body nurtured the ear as it grew, and once the ear was large enough, a surgeon could remove it and attach it to someone whose ear was missing. Doctors never actually transplanted the ear; the scientists merely intended to demonstrate the possibilities of tissue engineering. Lose an ear because of an accident? Doctors can grow you a new one.

Today, more than a decade later, we know that the mouse experiment helped pave the way for an entirely new kind of transplanted organ, one grown from a patient's own cells. In 2006, Anthony Atala, director of the Wake Forest University Institute for Regenerative Medicine,

Cartilage cells grow on a mouse back into the shape of a human ear.

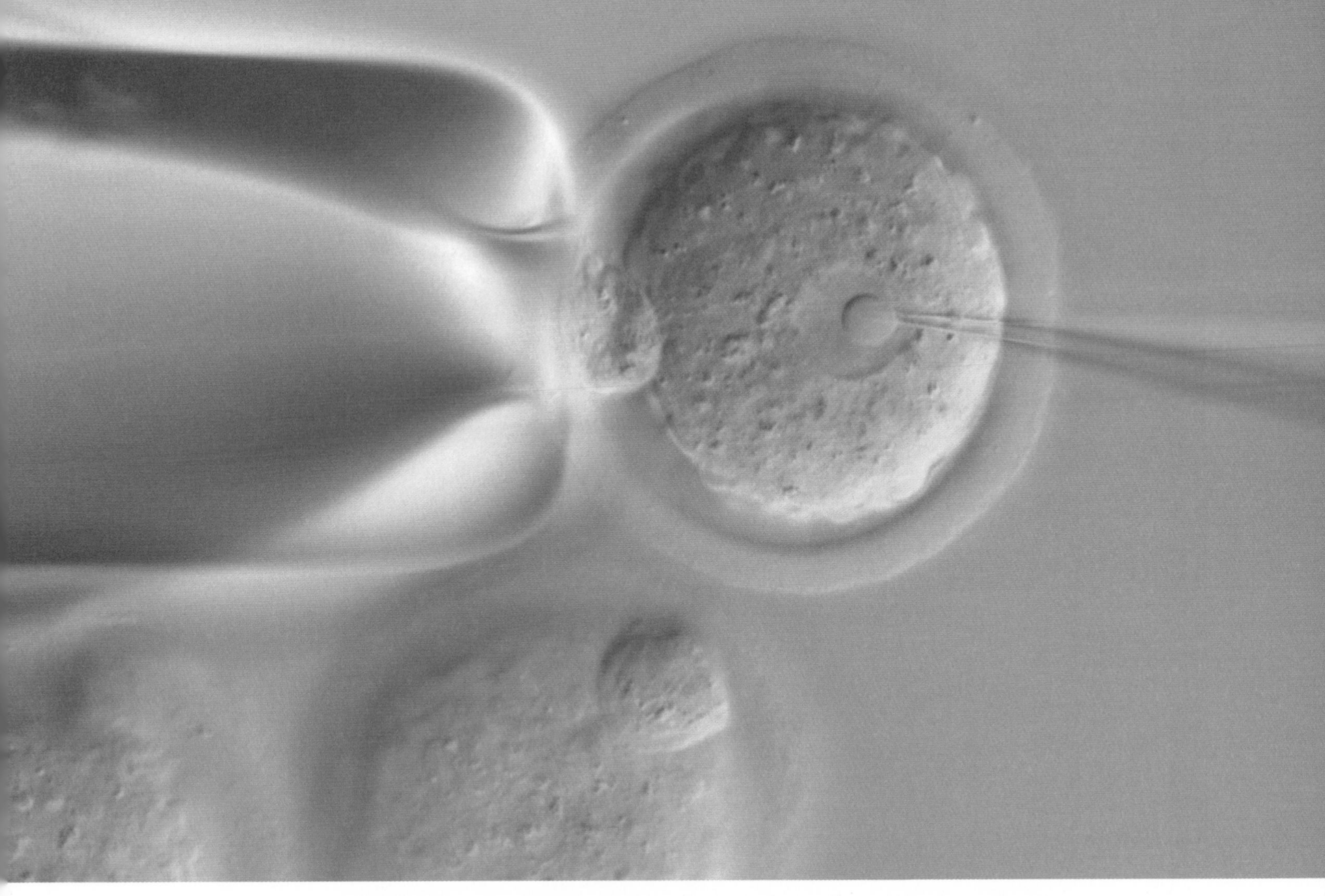

Scientists can manipulate the nucleus of a single cell to unlock its therapeutic potential.

The mouse experiment helped pave the way for an entirely new kind of transplanted organ.

announced that he and his colleagues had successfully transplanted engineered human bladders into several children and teenagers. Although scientists had been growing human tissue outside the body for years, the bladder transplants were the first time that a damaged organ was repaired with a person's own cells.

"It was very significant work," says William Wagner, deputy director of the McGowan Institute for Regenerative Medicine at the University of Pittsburgh. "He's overcome a huge number of challenges."

The potential applications of the technique are enormous. For years transplant surgeons have worked to help patients suffering from organ failure who are in need of an organ donor. But each year, the demand for organs such as hearts, livers, and kidneys vastly exceeds supply. Last year, for example, surgeons transplanted about 30,000 organs, according to the Organ Procurement and Transplantation Network. Meanwhile, there are about 100,000 people waiting for an organ transplant. And even when an organ does become available, the recipient's body may reject the organ because the donor and recipient immune systems are not compatible–leaving the patient sicker than before the transplant.

Growing organs from a person's own cells would not only sidestep organ rejection, it would also eliminate the need for donors. A decade ago, most scientists considered such a

INFOGRAPHIC 13.1

Cells Are Organized into Tissues, Organs, and Systems

Tissues are integrated groups of specialized cells working together. Multiple tissues combine to form organs, which in turn cooperate as part of a single functioning organ system.

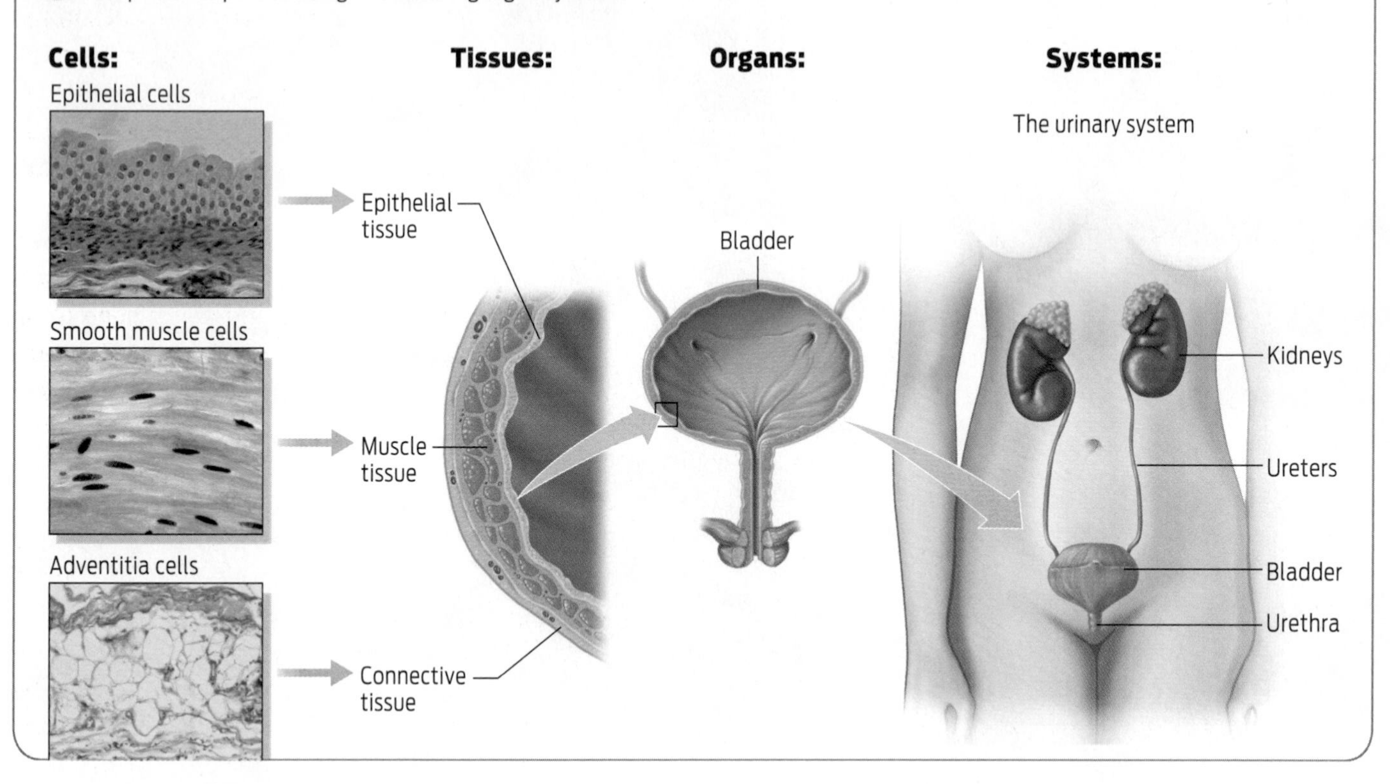

feat a pipe dream. They knew that human **tissues** consisted of integrated groups of different and specialized cells that together perform a specific function (Infographic 13.1). But controlling tissues to repair organs seemed impossible.

More recently, researchers discovered that most body tissues contain pools of **stem cells**—immature cells that can spontaneously divide repeatedly and give rise to more specialized cell types in the body. For example, stem cells found in bone, heart, and brain tissues help regenerate those tissues and organs (Infographic 13.2).

Lose an ear because of an accident? Doctors can grow you a new one.

The discovery has fueled a search for ways to harness the regenerative potential of stem cells to treat ailing patients. In addition to using stem cells to create organs for transplant, scientists also hope one day to tap into the body's natural healing processes and coax stem cells into healing damaged tissue.

From Ears to Bladders

The effort to engineer human tissue for transplants dates back almost 40 years and is based on the knowledge that the majority of our cells continuously die and are replaced by new cells. Without such cell division, an organism would neither grow nor heal. During early development, for example, a single fertilized egg cell divides to begin to form an embryo, and these cells divide again and again to form millions of cells by the time the embryo becomes a fetus. As we age, the body discards old cells and generates new ones in their place. And when we cut or injure ourselves, cells in the area undergo cell division to heal the injury. Transplant science tries to

TISSUE
An organized group of different cell types that work together to carry out a particular function.

STEM CELLS
Immature cells that can divide and differentiate into specialized cell types.

INFOGRAPHIC 13.2

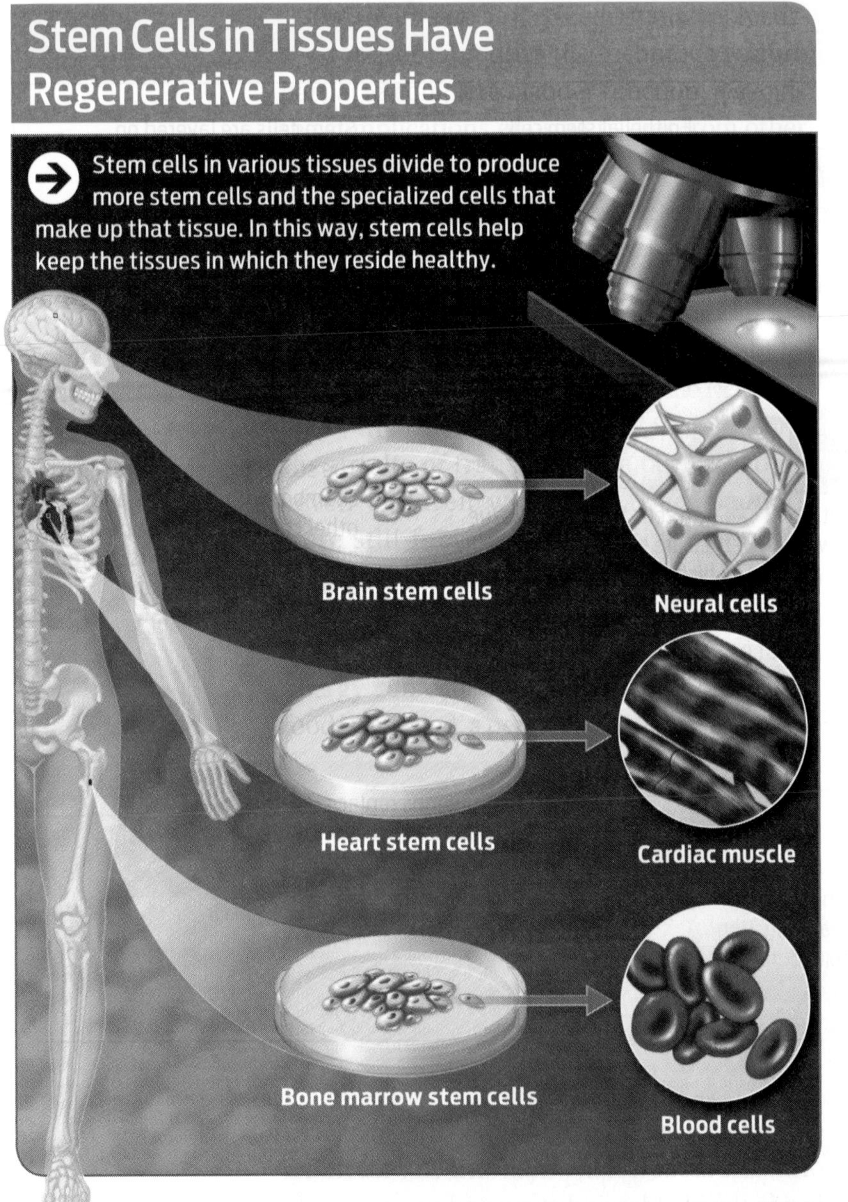
Stem Cells in Tissues Have Regenerative Properties
Stem cells in various tissues divide to produce more stem cells and the specialized cells that make up that tissue. In this way, stem cells help keep the tissues in which they reside healthy.
Brain stem cells
Neural cells
Heart stem cells
Cardiac muscle
Bone marrow stem cells
Blood cells

harness this natural ability of the body to grow and heal.

So far, progress has been incremental but steady. In the early 1970s, W. T. Green, an orthopedist, tried to grow cartilage tissue outside the body. He placed cartilage cells onto scaffolds made out of bone to try to get the cells to grow in specific formations. Although he was unsuccessful, his experiments set the stage for growing cells on scaffolds. Several years later, John Burke of Harvard Medical School and Ioannis Yannas of MIT developed a method using scaffolds transplanted into wound areas that stimulated both the dermis and the epidermis of the skin to grow to cover the wound. Their method is still used today to treat burn victims, patients who undergo plastic surgery, and patients with recurrent skin wounds.

But the field made its largest strides in the late 1980s, when Joseph Vacanti of Boston's Children's Hospital teamed up with Robert Langer at MIT to engineer tissues. The pair wanted to design synthetic biodegradable scaffolds that could be molded into particular shapes–a human ear, for example–and then coat the mold with cells that would grow into a tissue. The scaffold would never need to be removed–it would in time dissolve. Vacanti's brother, Charles, and Linda Griffith-Cima used this technique to grow a "human" ear on a mouse's back.

About the time of Joseph Vacanti and Langer's achievement, Anthony Atala, who had collaborated with Joseph Vacanti and Langer, applied some of this research on biodegradable scaffolds to his own work on engineered bladders. Atala, a urologist, sought to help his patients whose bladders were not functioning normally because of cancer, injury, or an inborn defect. For about a century, doctors have treated such patients by using pieces of their stomach or bowel to reconstruct their bladders. But because the procedure requires surgically removing pieces of healthy tissue, it is not an ideal treatment. A better option would be to grow a piece of new bladder tissue to repair the organ.

Langer was one of the first to devise a way to grow tissue on biodegradable scaffold.

Chapter 13 Test Your Knowledge

TISSUES AND CELL DIFFERENTIATION

Tissues are made up of a variety of specialized types of cells. Each cell type differentiates from an unspecialized stem cell.

HINT See Infographics 13.1–13.4 and Table 13.1.

KNOW IT

1. Does a 5-year-old child have adult stem cells in his or her tissues? Explain your answer

2. Relative to one of your liver cells, one of your skin cells

a. has the same genome (that is, the same genetic material).
b. has the same function.
c. has a different pattern of gene expression.
d. a and c
e. b and c

3. You shed dead skin cells every day. How are those cells replaced?

a. by mitotic division and specialization of embryonic stem cells
b. by differentiation of neighboring neurons into skin cells
c. by differentiation of red blood cells that leave the circulation and migrate into deeper layers of the skin
d. by mitotic division and differentiation of skin stem cells

4. The brain and spinal cord are made up of nervous tissue. This tissue includes neurons, cells that fire electrical impulses and communicate information in the brain. Nervous tissue also includes glial cells, cells that support neurons. Some glial cells enable the electrical impulse to travel faster. What characteristics of glial cells and neurons tell you that they both make up nervous tissue?

USE IT

5. From information in Question 4, would it be sufficient to replace the damaged neurons in someone who had suffered nervous-tissue damage? Why or why not?

6. Different cells have different functions: muscles contract because of the sliding action of actin and myosin proteins in muscle cell fibers; a protein known as retinal makes up the light detecting photoreceptor cells in the retina of the eye; helper T cells of the immune system have a protein on their surface known as CD4 which helps in mounting an immune response. From this information, complete the following table.

	Photo-receptor cells of the retina	**Heart muscle fibers**	**Helper T cells**
Myosin gene present?			
Myosin mRNA present?			
Myosin protein present?			
Retinal gene present?			
Retinal mRNA present?			
Retinal protein present?			
CD4 gene present?			
CD4 mRNA present?			
CD4 protein present			

7. Do stem cells have a larger genome than specialized cells?

a. yes, because they need the genes found in every cell type, whereas specialized cells need only a subset of all the genes
b. yes, because they express more genes than do specialized cells
c. no, because all cells in a person have the identical set of genes in their genome
d. no, they have a smaller genome, because stem cells are equivalent to gametes (which are haploid) in that they can potentially create an entire individual
e. no, they have a smaller genome because stem cells only express a subset of genes

STEM CELLS AND REGENERATIVE MEDICINE

Stem cells can potentially repair damaged tissue. The challenges are how to stimulate existing stem cells to divide, or how to transfer stem cells to the area of damage to promote repair.

HINT See Infographics 13.2 and 13.5–13.8.

KNOW IT

8. List and then describe several advantages of using one's own cells to regenerate an organ over receiving a transplant from an organ donor.

9. Describe at least two differences between embryonic stem cells and somatic (that is, adult) stem cells.

10. Why does the recipient of a liver transplant have a high risk of bacterial infections?

- **a.** because the liver plays a critical role in the immune response
- **b.** because donor livers are often contaminated with disease-causing bacteria
- **c.** because transplant recipients have to take drugs that suppress their immune systems
- **d.** because the surgery poses a high risk for introducing bacteria into the recipient
- **e.** because the immune system may reject the liver

USE IT

11. Why is engineering a bladder more challenging than engineering skin?

12. Which of the following populations of adult stem cells (if any) could be stimulated to divide to treat each of the following conditions?

- **a.** heart attack (damage to the heart muscle)
- **b.** cancer (unregulated cell growth)
- **c.** type I diabetes (destruction of insulin-producing cells in the pancreas)
- **d.** Parkinson disease (loss of specific neurons in the brain)

13. List and then describe some of the successes and challenges associated with using adult stem cells in comparison with embryonic stem cells for stem cell therapy.

14. If all specialized cells have the same genes in their genomes (including stem cell genes), why did James Thompson have to add genes into a mature cell in order to get it turn into a stem cell?

SCIENCE AND ETHICS

15. If you were head of the National Institutes of Health and responsible for allocating research funds to different avenues of research, which line(s) of stem cell research would you fund? Why?

16. Most people make a distinction between therapeutic cloning and reproductive cloning, at least for humans. There have been several reports of cloned pets and other animals. How does the science differ between reproductive cloning and therapeutic cloning? Do you think reproductive cloning should be legal or illegal?

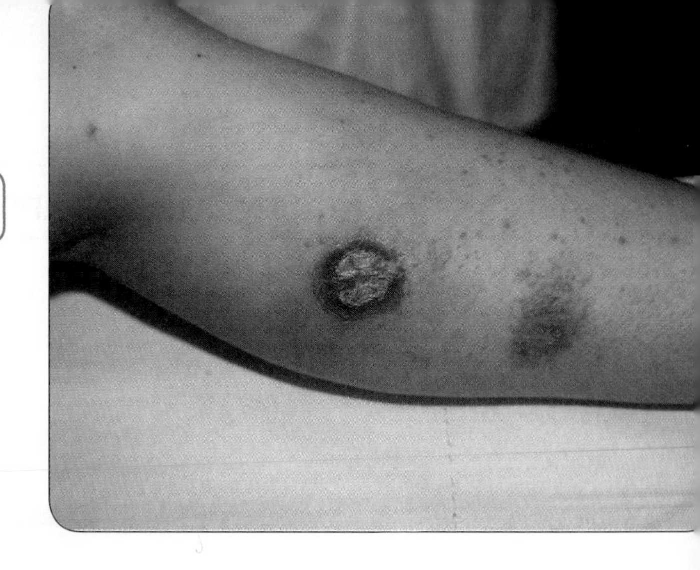

Bugs That Resist Drugs

What You Will Being Learning

Bugs That Resist Drugs

Drug-resistant bacteria are on the rise. Can we stop them?

In January 2008, sixth-grader Carlos Don, an active footballer and skateboarder, boarded a bus headed for a class trip, happy and healthy. A month later Carlos was dead.

In April 2006, 17-year-old Rebecca Lohsen was a model student at her high school; she was on the honor roll and was a member of the swim team. Four months later Rebecca was dead.

In December 2003, Ricky Lannetti was a college senior, a star football player and all-around athlete. A few weeks later Ricky was dead.

The list of surprising deaths like these goes on and on. But contrary to what you might think, these young people weren't killed in accidents, nor by violence; they were all killed by methicillin-resistant *Staphylococcus aureus* (MRSA)—an infectious bacterium that has become widespread in recent years and that is difficult to treat with most existing antimicrobial drugs.

MRSA sickens some 94,000 people in the United States each year and kills almost 19,000.

MRSA sickens some 94,000 people in the United States each year and kills almost 19,000, according to a 2007 study by Monina Klevens and her colleagues at the Centers for Disease Control and Prevention (CDC). Formerly, outbreaks of MRSA were confined mainly to hospitals. But since the late 1990s, growing numbers of healthy people are becoming infected outside hospitals. In addition, there are new high-risk groups that never had high rates of infection before: day care attendees, the prison population, men who have sex with men, and certain ethnic groups now are showing MRSA infections at a higher rate than

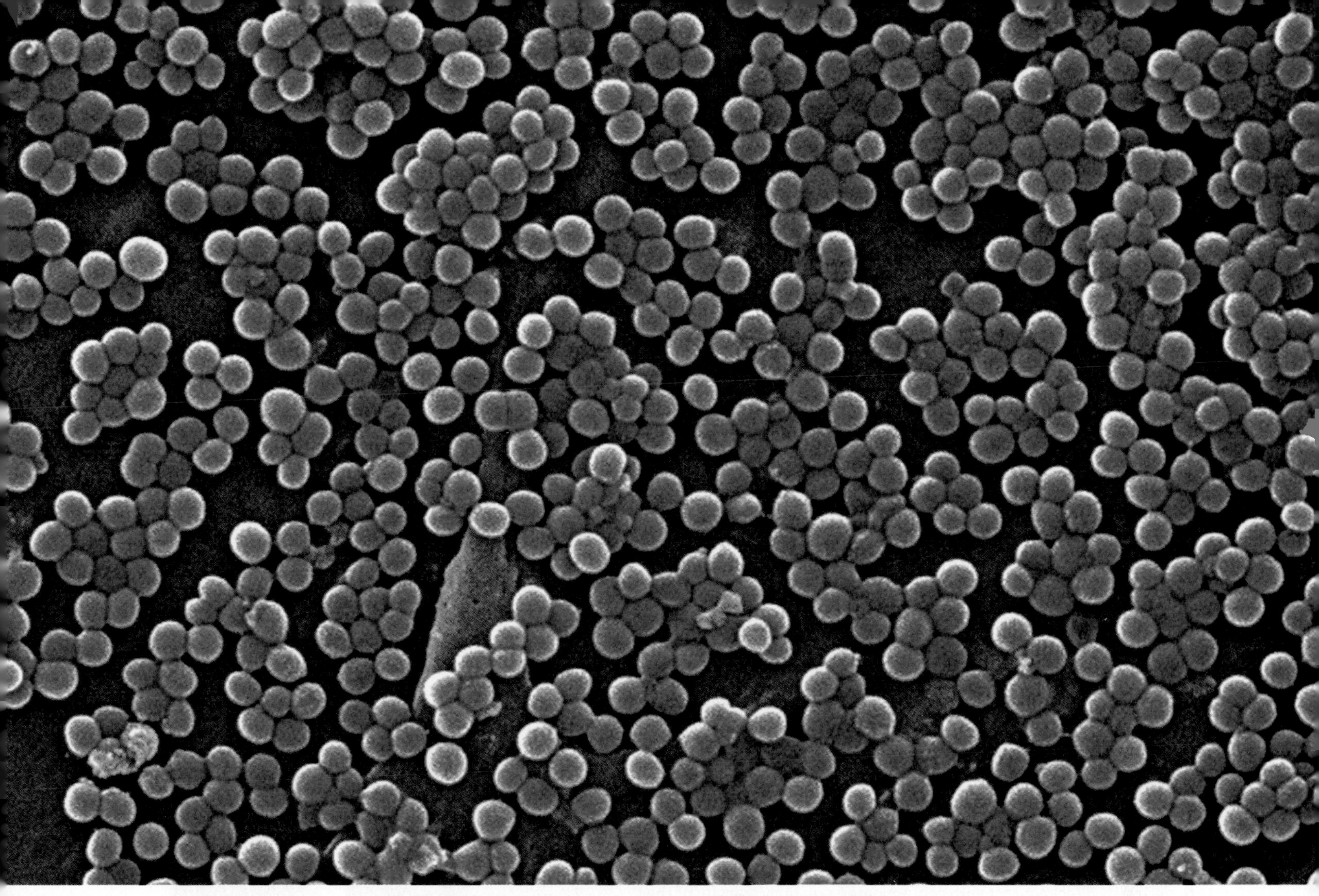

A resistant strain of *Staphylococcus aureus* at 4780x magnification.

the general population. Schools nationwide have been reporting outbreaks and young, healthy people are getting sick.

"This is a major public health imperative," says Robert Daum, professor of microbiology at the University of Chicago and a member of the Infectious Diseases Society of America. "We need a plan of attack now."

Staph the Microbe

MRSA infection is caused by the *Staphylococcus aureus* bacterium—often simply called "staph." Although several species of staph bacteria can cause human disease, the medical community is especially concerned about those, such as *S. aureus,* that have developed resistance to antibiotic drugs that once effectively killed them. "MRSA" is actually a misnomer because methicillin is no longer used to treat staph infections. In fact, drug-resistant strains of staph are usually resistant to several different types of antibiotics. Some people use the terms "MRSA" and "drug-resistant staph" interchangeably to refer to staph strains that are resistant to the common classes of antibiotics—penicillins and cephalosporins—that are used to treat staph infections.

Staph bacteria are harmless to most people who carry them. Between 30% and 40% of the population carries staph on their skin or in their noses, and about 1% of the population carries drug-resistant strains, according to the Centers for Disease Control and Prevention. If you carry staph of any strain but aren't sick, you are "colonized" but not infected. Healthy people can be colonized with any staph strain, including

MRSA, and not become ill. However, they can pass the staph to others via skin-to-skin contact or through shared, contaminated items such as towels and bars of soap. Infections typically occur when the bacteria come into contact with a wound. Athletes who have cuts and scrapes may acquire a staph infection in locker rooms or during contact sports **(Infographic 14.1)**.

"Every one of us has probably had a staph infection at some point," explains Daum. "Staph ranges from the commonest cause of infected fingernails all the way to a severe syndrome with rapid death, and everything in between. Most staph infections don't even result in a medical encounter."

In otherwise healthy people, staph infection usually causes only minor skin eruptions such as boils or pustules that can resemble spider bites. A healthy person can, however, become more severely ill if he or she undergoes a medical procedure that either weakens the immune system or creates a break in the skin that becomes infected with staph. The elderly, who may have weakened immune systems, and children, whose immune systems are still immature, are at especially high risk of developing severe diseases such as pneumonia, infections of the bloodstream, or infections of surgical wounds caused by staph. When bacteria such as staph do cause illness, they do so by multiplying on or in human tissues. They can also secrete toxic substances that harm human cells or interfere with essential cellular processes.

Staph bacteria can cause such a range of disease because the bacteria exist as many different strains. Each strain differs from all others in

INFOGRAPHIC 14.1

The Bacterium *Staphylococcus aureus*

Staphylococcus aureus is a spherical bacterium that can cause pimples, boils, and wound infections in healthy people. *S. aureus* can be passed from person to person by direct contact with contaminated skin or by transfer of the bacteria via contaminated objects or surfaces.

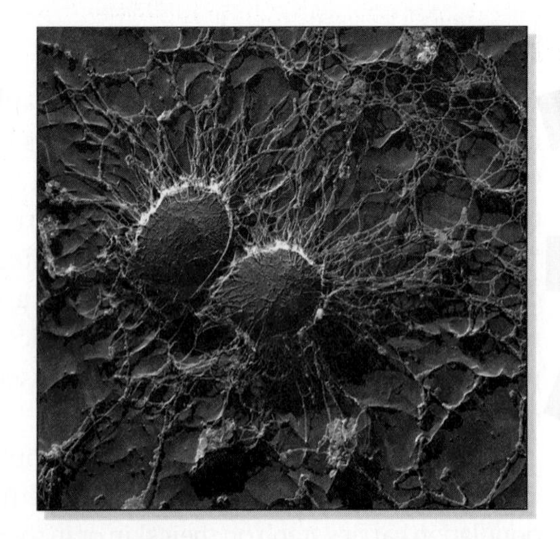

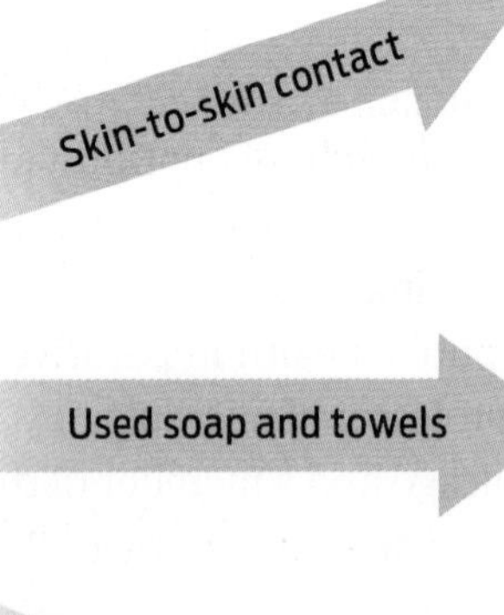

Experts estimate that about one-third of the U.S. population (nearly 90 million people) is colonized by *S. aureus*. Nearly 1% of the U.S. population (just over 2 million people) is colonized with a strain called methicillin-resistant *Staphylococcus aureus* (MRSA), which can cause more severe illness and that is difficult to treat with existing medications.

its genetic makeup. MRSA, for example, is composed of a number of unique strains of staph bacteria, and some cause more serious disease than others. In recent years there have been several cases of healthy people becoming severely ill from MRSA infection, most likely because they were infected by an especially deadly strain of drug-resistant staph.

Ricky Lannetti, for example, was a perfectly healthy 21-year-old football player at Lycoming College in Williamsport, Pennsylvania. "He was strong as an ox and he ran like a deer," says his mother, Theresa Drew. A few days before Ricky died, he had come down with a bout of flu. Ricky wasn't recovering, however, and on the morning of December 6, 2003, Drew drove her son to Williamsport Hospital. By the time he was admitted, his blood pressure had dropped dangerously low and his body temperature was erratic. As each hour passed, his condition worsened. His lungs began to fail. Doctors tried five different antibiotics, in vain. When his heart began to weaken, his doctors prepared him to be flown to the cardiac center at a bigger hospital in Philadelphia. But it was too late. Ricky died that night.

It was only after an autopsy was performed that it was known what had killed him: MRSA that had infected Ricky's bloodstream. Although doctors couldn't be sure how Ricky contracted MRSA, they reasoned that he had inhaled it–the fact that his lungs were so damaged suggested the lungs as the first place of infection. Since MRSA can colonize people's noses, it can travel to their respiratory systems, where it can cause severe damage.

"Doctors tried every antibiotic imaginable, including vancomycin," says his father, Rick Lannetti. But the treatment was too late. Ricky's immune system was already weak because of the flu. When he contracted MRSA at the same time, his body was unable to fight back as well as it otherwise would have. "In the end," his father says, "MRSA had broken every one of his organs beyond repair."

ANTIBIOTICS
Chemicals that either kill bacteria or slow their growth by interfering with the function of essential bacterial cell structures.

The Antibiotic Revolution

Bacterial infections were a common cause of death before the 1940s, when scientists developed the first antibiotics. **Antibiotics** are chemicals that either kill bacteria or slow their growth by interfering with the function of essential bacterial cell structures. Research in the early twentieth century had revealed that certain microorganisms, such as the fungus *Penicillium*, produce compounds that can kill bacteria. In 1928, the Scottish biologist Alexander Fleming isolated the antibiotic penicillin, although it took more than a decade of research by Fleming and others to develop it into a usable drug. The 1940s saw a major search for other drugs to treat infections, and within a few years scientists successfully purified more antibiotics. Though the original antibiotics were derived from microorganisms, many are now synthesized in the laboratory.

Over the decades, antibiotics have been effective in treating most common bacterial infections, including staph, and have saved thousands of lives. But soon after antibiotics were in general use, microorganisms that could survive antibiotics–drug-resistant "bugs"–began to emerge. Within the last decade drug-resistant bacterial strains have become much more common. Although people infected with drug-resistant bacterial strains are treatable, they have fewer treatment options. And sometimes–as in Ricky Lannetti's case–existing drugs are completely ineffective.

Drug-resistant strains of staph, for example, are typically resistant to an entire class of antibiotic drugs called the beta-lactams. Beta-lactams include penicillin and the cephalosporin antibiotics, such as methicillin and cephalexin. Beta-lactams are the most commonly prescribed class of antibiotics. They work by interfering with a bacterium's ability to synthesize cell walls. A variety of non-beta-lactam classes of antibiotics can treat MRSA infections, and vancomycin, an intravenous non-beta-lactam drug, is the antibiotic of choice when a serious or severe MRSA infection is confirmed. But even vancomycin isn't always effective; there are now staph strains resistant to vancomycin, too **(Infographic 14.2)**.

Ricky Lannetti did not respond to vancomycin. Nor did Rebecca Lohsen, the 17-year-old

Chapter 14 Test Your Knowledge

BACTERIA AND DISEASE

Bacteria live in and on humans and may or may not cause disease.

HINT See Infographics 14.1 and 14.2.

KNOW IT

1. The term "MRSA" as it is used today refers to
a. *Staphylococcus aureus* bacteria that are resistant to many antibiotics.
b. a collection of skin and other infections, caused by a type of bacteria.
c. *Staphylococcus aureus* bacteria that are found only in humans with certain types of skin infections.
d. *Staphylococcus aureus* bacteria that are normal residents of human skin in the vast majority of the human population.
e. all bacteria that are resistant to antibiotics.

2. What is the difference between a *Staphylococcus aureus* colonization and a *Staphylococcus aureus* infection?

3. Where is MRSA most likely to be a problem?
a. on the surface of the skin
b. in nasal passages
c. in the bloodstream
d. on your fingernails
e. The presence of MRSA in any of those locations indicates a serious infection.

USE IT

4. A young athlete has a nasty skin infection caused by MRSA. How might he have contracted this infection?

5. For the patient in Question 4, which antibiotic would you choose to treat the infection? What other measures would you recommend to prevent spread of MRSA to the athlete's teammates and family? Explain your answer.

6. Why do the beta-lactam antibiotics affect sensitive bacterial cells but not eukaryotic cells? (You may need to review cell structure to answer this question.)

EVOLUTION BY NATURAL SELECTION

Evolution is a change in allele frequencies in a population over time. When genetically diverse individuals differ in their ability to survive and reproduce, evolution by natural selection occurs.

HINT See Infographics 14.3–14.8.

KNOW IT

7. What are the two major mechanisms by which bacterial populations generate genetic diversity?
a. mutation and meiosis
b. binary fission and evolution by natural selection
c. gene exchange and mutation
d. mutation and binary fission
e. gene exchange and replication

8. What is the environmental pressure in the case of antibiotic resistance?
a. the growth rate of the bacteria
b. how strong or weak the bacterial cell walls are
c. the relative fitness of different bacteria
d. the presence or absence of antibiotics in the environment
e. the temperature of the environment

9. What is the evolutionary meaning of the term "fitness"?

10. The evolution of antibiotic resistance is an example of
a. directional selection.
b. diversifying selection.
c. stabilizing selection.
d. random selection.
e. steady selection

11. In humans, very-large-birth-weight babies and very tiny babies do not survive as well as midrange babies. What kind of selection is acting on human birth weight?
a. directional selection
b. diversifying selection
c. stabilizing selection
d. random selection

USE IT

12. Binary fission is asexual. What does this mean? How could two daughter cells end up with different genomes at the end of one round of binary fission?

13. In what sense do bacteria "evolve faster" than other species?

14. If we take the most fit bacterium from one environment—one in which the antibiotic amoxicillin is abundant, for example—and place it in an environment in which a different antibiotic is abundant, will it retain its high degree of fitness?

a. yes, fitness is fitness, regardless of the environment

b. yes, once a bacterium is resistant to one antibiotic it is resistant to all antibiotics

c. not necessarily; fitness depends on the ability of an organism to survive and reproduce, and it may not do this as well in a different environment

d. no, what is fit in one environment will never be fit in another environment

15. If a single bacterial cell that is sensitive to an antibiotic—for example, vancomycin—is placed in a growth medium that contains vancomycin, it will die. Now consider another single bacterial cell, also sensitive to vancomycin, that is allowed to divide for many generations to become a larger population. If this population is placed into vancomycin-containing growth medium, some bacteria will grow. Why do you see growth in this case, but not with the transferred single cell?

16. If evolution by natural selection is a change in allele frequencies in a population, then why do we detect the process of evolution by natural selection as a change in *phenotype* frequencies in the population?

17. Imagine that a genetically diverse population of garden snails occupies your backyard, in which the vegetation is a variety of shades of green with some brown patches of dry grass.

a. If birds like to eat snails, but they can see only the snails that stand out from their background and don't blend in, what do you think the population of snails in your backyard will look like? Explain your answer.

b. Suppose you move the population of snails to a new environment, one with patches of dark brown pebbles and patches of yellow ground cover. Will individual snails mutate to change their color immediately? As the population evolves and adapts to the new environment, what do you predict will happen to the phenotypes in your population of snails after several generations in this new environment? How did this occur? Include the terms *gametes, mutation, fitness, phenotype,* and *environmental selective pressure* in your answer.

SCIENCE AND ETHICS

18. Your friend has had a virus-caused cold for 3 days and is still so stuffy and hoarse that he is hard to understand. He seems to be telling you that his doctor called in a prescription for an antibiotic for him to pickup at his pharmacy. You hope that you misunderstood him, but you realize that you heard him perfectly well.

a. Why are you dismayed to hear his story?

b. Will the antibiotic help your friend's cold?

c. What are the risks to your friend if he takes the antibiotic? (Think about what might happen if he should develops a wound infection.) What are the risks to you, as his friend?

19. Your roommate has been prescribed an antibiotic for bacterial pneumonia. She is feeling better and stops taking her antibiotic before finishing the prescribed dose, telling you that she will save them to take the next time she becomes sick. What can you tell your roommate to convince her that this is not a good plan?

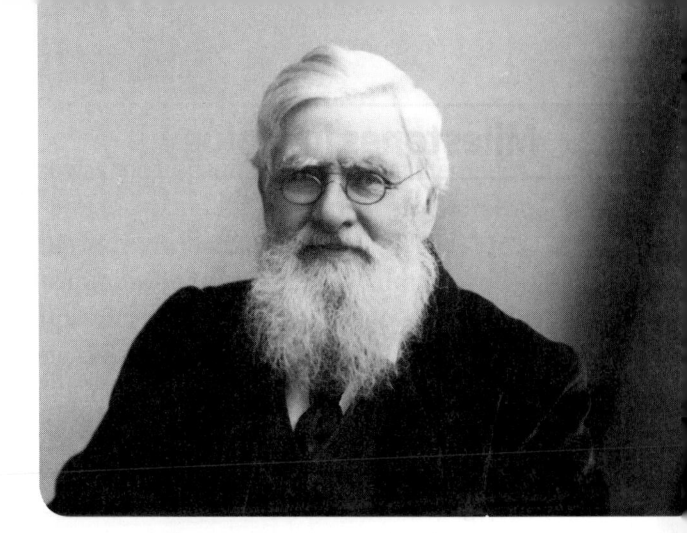

Milestones in Biology

Adventures in Evolution

What You Will Be Learning

Informed by extensive reading, travel, and observations of the natural world, Darwin and Wallace independently proposed the mechanism for evolution.

Journey to an Idea

Though Darwin is the most famous figure associated with the theory of evolution, he did not invent the idea. Nor was he alone among his contemporaries in studying it. In fact, the notion that species change gradually over time had been around for generations. To be sure, most people in the 1830s (Darwin included) still assumed that species were fixed and unchanging, created perfectly by God. But evidence to the contrary had been accumulating for some time. Explorers and naturalists were traveling to faraway lands, discovering a host of never-before-seen plants and animals. Fossils were being unearthed, providing evidence that some species no longer seen on earth had lived in the past. And anatomists were noting uncanny physical resemblances between different species, including chimpanzees and humans. Evolution was in the air when Darwin began thinking about it. His own grandfather, Erasmus Darwin, had even written a book about evolution in the 1790s.

However, the ideas that people in Darwin's time had proposed to explain *how* species changed were flawed. One common misconception was Lamarckianism, named after the French naturalist Jean-Baptiste Lamarck, who suggested that species could change through the inheritance of acquired characteristics. In the Lamarckian view, giraffes, for example, developed their long necks by continually stretching them to feed on tall trees. Once it acquired its long neck, a giraffe could then pass that advantageous trait on to its offspring. This idea of the inheritance of acquired character-

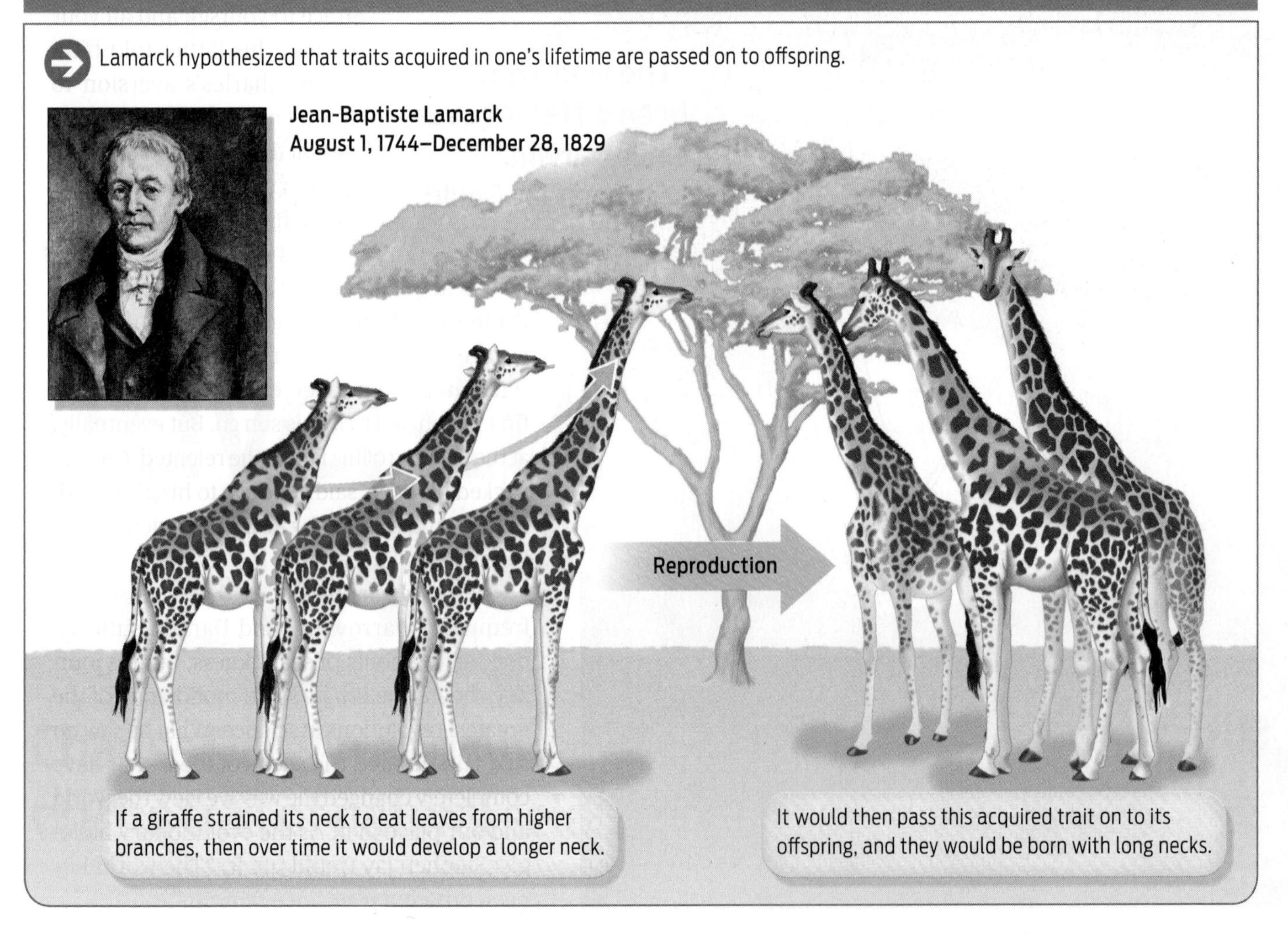

Lamarckianism: An Early (and Incorrect) Idea about Evolution
Lamarck hypothesized that traits acquired in one's lifetime are passed on to offspring.
Jean-Baptiste Lamarck
August 1, 1744–December 28, 1829
Reproduction
If a giraffe strained its neck to eat leaves from higher branches, then over time it would develop a longer neck.
It would then pass this acquired trait on to its offspring, and they would be born with long necks.

The Evolution of Darwin's Thought

Darwin was influenced by the work of others, which informed the way he interpreted his own research and collections.

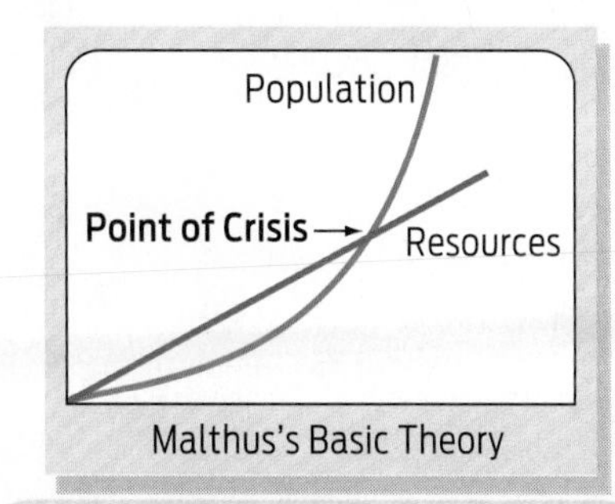

Malthus's work (1798):
- Populations are limited by a number of factors, including food, water, and disease.
- Some individuals die and some survive.

Charles Darwin (February 12, 1809–April 19, 1882)

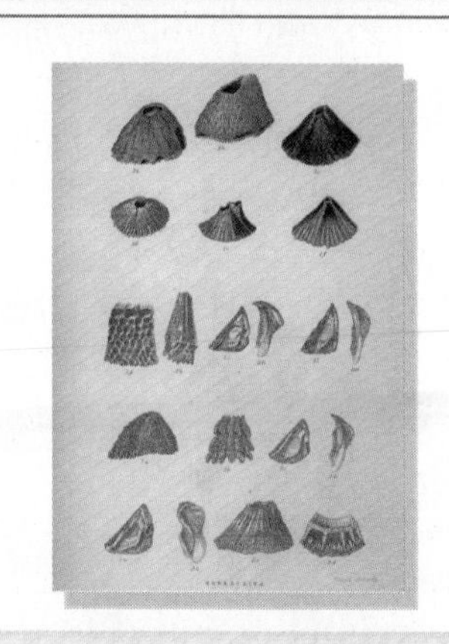

Barnacle research (1846–1854):
- Darwin characterized differences between different groups of barnacles, and developed ideas about how these differences arose.

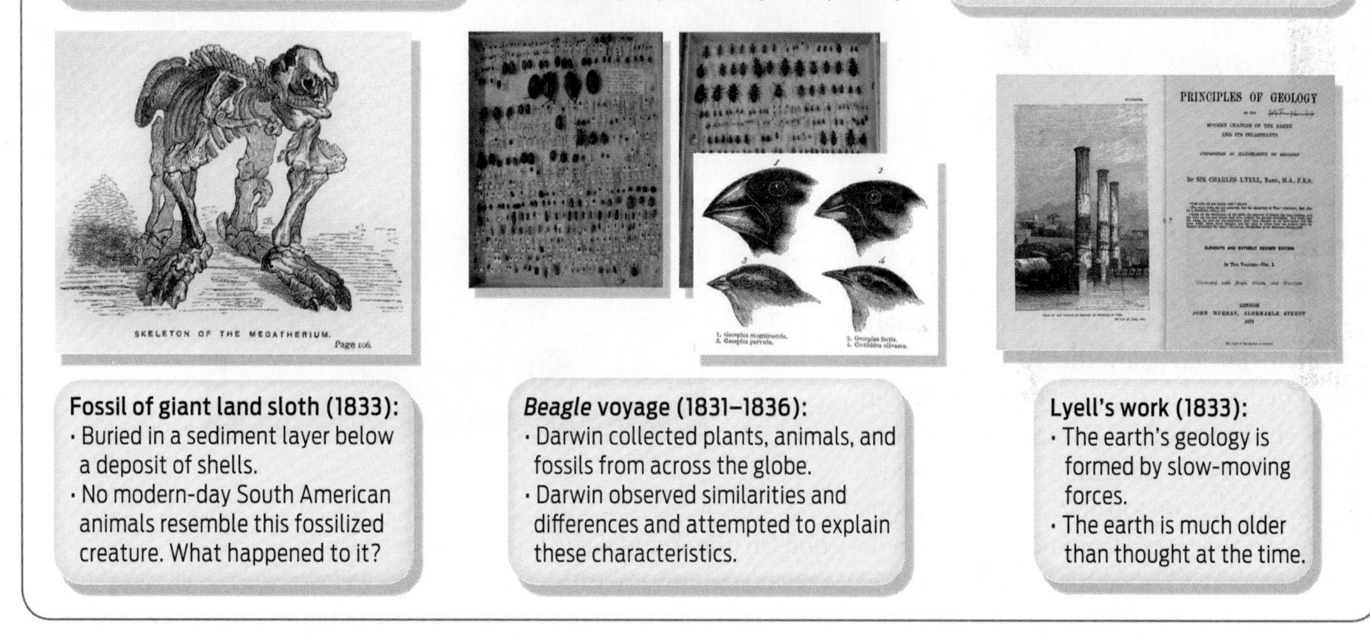

Fossil of giant land sloth (1833):
- Buried in a sediment layer below a deposit of shells.
- No modern-day South American animals resemble this fossilized creature. What happened to it?

***Beagle* voyage (1831–1836):**
- Darwin collected plants, animals, and fossils from across the globe.
- Darwin observed similarities and differences and attempted to explain these characteristics.

Lyell's work (1833):
- The earth's geology is formed by slow-moving forces.
- The earth is much older than thought at the time.

istics, while incorrect, was a popular one in Darwin's time—one that even Darwin himself found it hard to fully shake off in his writings.

While at sea, Darwin had plenty of time to read and think about the ideas then being discussed in scientific circles. He read, for instance, the work of the geologist Charles Lyell. Lyell's *Principles of Geology* (1833) argued that the earth was much older than the 6,000 years popularly accepted at the time (a figure based on a literal reading of the Bible), and that its geology had been shaped entirely by incremental forces operating over a vast expanse of time. "No vestige of a beginning, no prospect of an end," was how James Hutton, Lyell's mentor, had put it. With such thoughts percolating through his mind, Darwin studied the plants, animals, and geology at each stop on his trip, collecting fossils and specimens of local flora and fauna wherever he went.

While exploring the shore of Argentina in August 1833, Darwin unearthed a particularly prized find: the complete fossil of a giant sloth embedded in a cliff, below a layer of

Darwin's Voyage on the *Beagle*

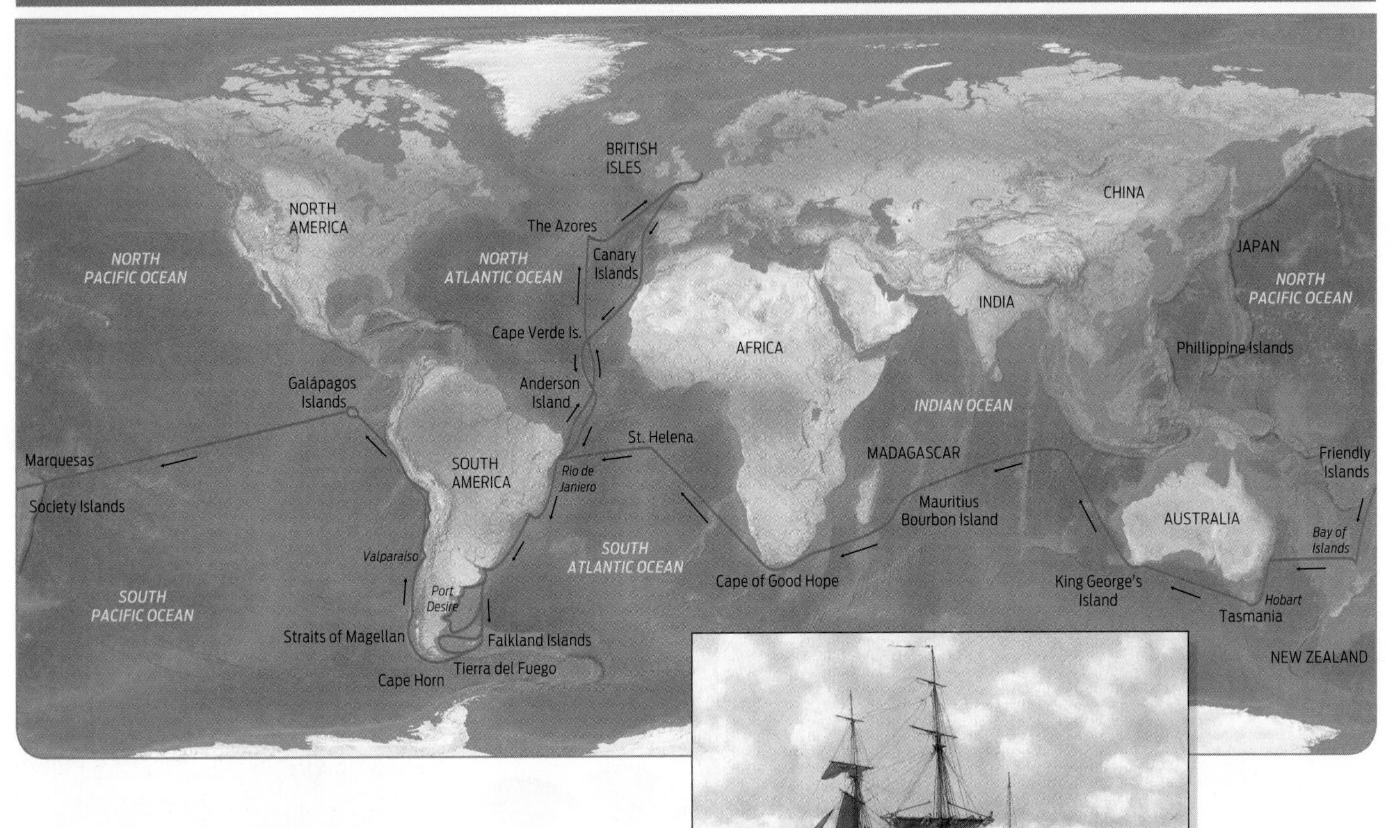

seashells. Darwin realized that the presence of shells at a distinct point, and the dead animal's position in the cliff below them, meant that the animal had lived in the area before it was an ocean environment. The animal had also clearly lived a very long time ago, since many layers of earth sat on top of it. More mysterious was the fact that there were no living animals in the present-day region that looked remotely like the creature he found. Where had they gone? Perhaps, Darwin reasoned, when the landscape changed the animals had been unable to adapt and had become extinct. Darwin also noticed that some South American species resembled European ones, despite living in an entirely different environment. Taken together, these pieces of evidence seemed to conflict with the notion that God had created each species perfectly adapted to its environment. Darwin was beginning to question the common wisdom of his day.

In 1835, the young naturalist stepped ashore on the Galápagos Islands, off the coast of Ecuador. On this archipelago, Darwin observed and collected many creatures, among them a variety of small birds. Months later, while studying the specimens when he was back in England, he realized they were all closely related species of finch. Each species was distinguishable by a different size and shape of beak. He later wrote in the second edition of *The Voyage of the Beagle* (1845), "One might really fancy that, from an original paucity of birds in this archipelago, one species had been taken and modified for different ends." His ideas were taking shape.

After returning home from his eye-opening voyage, Darwin made detailed notes of the evidence supporting his speculations. A key insight came to him in September 1838 while reading the work of the political economist Thomas Malthus, whose pessimistic book, *An Essay on the Principle of Population* (1798), described how hunger, starvation, and disease would ultimately limit human population growth. Darwin realized that, for animals, such limitations would lead to competition for resources that would put weaker individuals at a disadvantage. In these circumstances, Darwin wrote in his notebook, "favourable variations would tend to be preserved, and unfavourable ones to be destroyed." Competition for survival and reproduction among members of a species, he realized, would lead gradually to the species becoming more adapted to its surroundings. In effect, the environment was "selecting" for favorable traits, much as plant and animal breeders selected and perpetuated desirable varietals—a plant with especially large fruit, for instance. This idea of "natural selection" was Darwin's original contribution to the theory of evolution—what he called "descent with modification." (Darwin avoided using the term "evolution," which he thought gave a mistaken idea of progressive development, preferring instead this more descriptive phrase.) Others had speculated at length about species change—most notably Robert Chambers in *Vestiges of the Natural History of Creation* (1844)—but Darwin was the first to provide a clear *mechanism* of evolution. The philosopher of science Daniel Dennett has called the theory of natural selection "the single best idea anyone has ever had."

"One might really fancy that . . . one species had been taken and modified for different ends."
—Charles Darwin

By 1844, Darwin had developed his ideas into a 200-page manuscript that he hoped would be the definitive word on the subject. He did not rush his ideas about natural selection into print, however. He knew that his ideas would be controversial, contradicting as they did strongly held beliefs about God and the special creation of all animals, including humans. Other scientists with evolutionary ideas were causing quite a stir in England and being openly ridiculed (for this reason Robert Chambers had published his book anonymously). Even sharing his theory of evolution by natural selection with trusted colleagues, Darwin said, was "like confessing a murder." To withstand challenges, he knew he would need more detailed evidence.

And so, at age 37, Darwin began to investigate closely one group of animals: barnacles, the small invertebrates that cling to ships or marine life. Darwin spent 8 years, from 1846 to 1854, carefully analyzing the barnacles' tiny adaptations. It was tedious work, leading Darwin to write, "I hate a Barnacle as no man ever did before." Yet the work proved valuable, putting detailed meat on the bones of his skeletal idea. "What he found in barnacles," wrote Janet Browne, a professor of the history of science at Harvard and the author of *Darwin's Origin of Species: A Biography* (2007), "brought important shifts in his biological understanding, strengthened his belief in evolution and provided an essential backdrop to *Origin of Species*."

Darwin was hard at work fleshing out his idea in painstaking detail when he received a letter from a young naturalist with whom he had a casual acquaintance, a collector named Alfred Wallace who made a living selling rare butterflies and birds to other collectors and museums. The envelope was postmarked from an island in Indonesia. Inside was a 20-page manuscript describing the author's bold new idea about how species change over time, which he wanted Darwin to read and have published. Darwin, it seemed, had been scooped.

In Darwin's Shadow

Although we often credit Charles Darwin with the discovery of evolution by natural selection, he was not alone in charting this intellectual territory. Another British naturalist was also hot on the trail. Like Darwin, Wallace was fascinated by natural history and had a thirst for adventure. In other ways, though, the two men couldn't have been more different. Darwin came from a wealthy, upper-class family and had received a prestigious Cambridge education. He was greeted as a minor celebrity when he returned from his trip around the world. Wallace, on the other hand, was a man of much more humble origins, for whom nothing in life had come easily.

The eighth of nine children, Wallace could not afford a university education, and he was plagued by financial difficulties his whole life. He attended night school and supported himself as a builder and railroad surveyor. His budding fascination with natural history, though, led him to read widely. Like Darwin, he read Lyell's work on geology, Malthus's work on human population, and Chambers's *Vestiges*. And, of course, he devoured Darwin's travel account, *The Voyage of the Beagle* (1839), which kindled his sense of adventure.

In 1848, having scrimped and saved, the 25-year-old Wallace set sail for Brazil, to the

The Evolution of Wallace's Thought

Like Darwin, Wallace was influenced by the writings and work of others, which shaped his interpretations of his own observations and research.

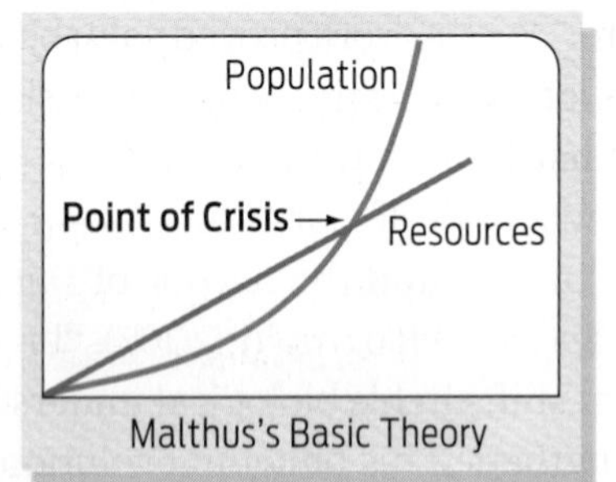

Malthus's work (1798):
- Populations are limited by a number of factors, including food, water, and disease.
- Some individuals die and some survive.

Alfred Russel Wallace (January 8, 1823–November 7, 1913)

Wallace–Darwin correspondence (185?–1858):
- Wallace began corresponding with Darwin several years before sending Darwin his completed manuscript.
- The two men were clearly developing very similar ideas about the nature of evolutionary change.

Amazon trip (1848–1852):
- Wallace observed that related (or "closely allied") species occupied neighboring geographic areas.
- He noted the role of physical barriers (such as the Amazon River) in separating related species from one another.

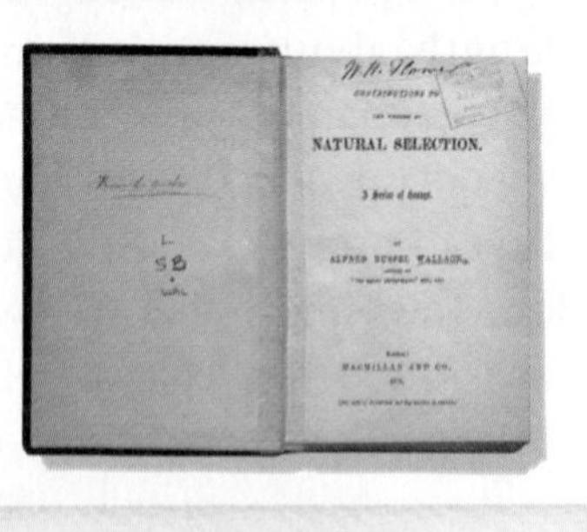

First publication (1855):
- Wallace's publication dealt with the physical distribution of species.
- He introduced the idea that new species are temporally and spatially connected to a related species.

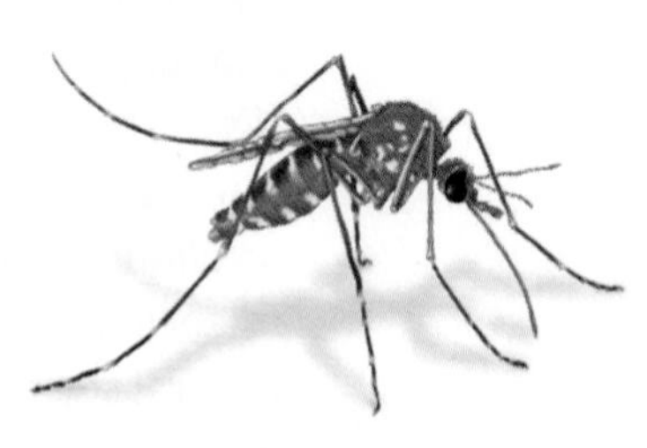

Disease and famine (1858):
- While suffering from malaria, Wallace pondered the role of disease and famine in keeping human populations in check.
- He wondered how these factors could apply to the evolution of animal species.

mouth of the Amazon River. There he hoped to earn his reputation as a respectable scientist. Wallace was an unusually keen observer of the natural world. Exploring the rain forest of the Amazon, he was struck by the distribution of distinct yet similar-looking ("closely allied") species, which were often separated by a geographic barrier, such as a canyon or river. For example, he noted that different species of sloth monkey were found on different banks of the Amazon River. Over the course of his 4-year trip, Wallace scoured the Amazon and collected thousands of specimens.

Wallace was on his way home to London with his specimens in 1852 when disaster struck: his ship caught fire and sank. Wallace survived, but he lost everything–his notes, sketches, journals, and all his specimens. In spite of this catastrophe, Wallace was undeterred. He was, as his biographer Michael Shermer noted in his book *In Darwin's Shadow* (2002), "a veritable scientific and literary engine," a man who was singularly devoted to his research. Less than 2 years later, he was off on another collecting expedition, this time to the islands of Southeast Asia.

"Every species has come into existence coincident both in space and time with a pre-existing closely allied species." –Alfred Russel Wallace

Wallace's first paper, "On the Law Which Has Regulated the Introduction of New Species," was published in September 1855. Based on his island work, it focused on the similar geographical distribution of closely allied species. For example, he wrote, "the Galápagos Islands . . . contain little groups of plants and animals peculiar to themselves, but most nearly allied to those of South America." From these observations, Wallace deduced this law, as he called it: "Every species has come into existence coincident both in space and time with a pre-existing closely allied species."

Wallace's article was groundbreaking, foreshadowing Darwin in a number of ways, but it lacked an explanation–a mechanism–of exactly how one species might have evolved from another.

Wallace continued his research, but in early 1858, disaster struck again: while traveling in what is now Malaysia, he contracted malaria. Not one to waste a perfectly good research opportunity, Wallace turned his convalescence into a sabbatical. As he later recalled, "I had nothing to do but to think over any subjects then particularly interesting me." Acutely aware of his own illness, he thought about what Malthus had written about disease and how it kept human populations in check. He also ruminated on Lyell's recent discoveries, which suggested that the earth was much older than previously thought. How might these forces of disease and death, multiplied over time, influence the composition of different populations, he wondered? As his fever waned, inspiration struck–like "friction upon the specially-prepared match," he later said: in every generation, weaker individuals will die and those with the fittest variations will remain and reproduce; as a result the species will become better adapted to its environment. Wallace had worked out the mechanism for evolution that was missing from his earlier work. He quickly wrote out his idea and sent it to the one naturalist he thought might be able to appreciate it. Wallace's paper arrived on Darwin's doorstep on June 18, 1858.

Darwin was stunned. For 20 years he had been working diligently on the same idea and now it seemed someone else might get credit for it. "All my originality will be smashed," he wailed to Lyell, asking him what he thought he should do. Recognizing the delicacy of Darwin's situation, Lyell and other colleagues devised a plan that would clearly establish Darwin's intellectual precedence: they would arrange to have papers by both men presented at a meeting of the Linnaean Society in London. The meeting took place on July 1, 1858. The papers were dutifully read, but there was no discussion or fanfare. In fact, neither of the authors was even present: Wallace was still traveling in Malaysia and Darwin was mourning the recent death of his young son and too distraught to attend.

The scientific meeting secured Darwin's reputation, but still he was unsettled. Wallace's communication had lit a fire under his feet. He needed to finish his book. That work, *On the Origin of Species by Means of Natural Selection,* was published in November 1859. It would become one of the most famous books of all time, going through six editions by 1872. Just as every species is a product of its predecessors, so are ideas: in his book, Darwin credited Malthus and Lamarck, as well as Wallace.

Although it may seem that Wallace was cheated of his rightful recognition as a discoverer of evolution by natural selection, he was never bitter. On the contrary, he was delighted when he heard about his copublication with Darwin. He fully accepted that Darwin had formulated a more complete theory of natural selection before he did, and there is no trace of resentment in his later writings. In fact, Wallace titled his major work *Darwinism,* in recognition of the other man's intellectual influence.

After the presentation of 1858, Wallace stayed in the Malay archipelago for 4 more years, systematically recording its fauna and flora, and securing his reputation as both the greatest living authority on the region and an expert on speciation. In fact, Wallace is responsible for our modern-day definition of "species." In work on butterflies, he defined "species" as groups of individuals capable of interbreeding with other members of the group but not with individuals from outside the group. This idea–known today as the biological species concept–remains one of the most important in evolutionary theory. ■

Chapter 15 Nonadaptive Evolution and Speciation

Evolution in the Fast Lane

What You Will Be Learning

UP CLOSE Calculating Hardy-Weinberg Equilibrium

How do we know if a population is evolving? To find out, we can use a mathematical formula called the **Hardy-Weinberg equation**, which calculates the frequency of genotypes you would expect to find in a nonevolving population. For a gene with one dominant and one recessive allele, *p* and *q*, this formula can be written as:

p^2	+	$2pq$	+	q^2	$= 1$
Frequency of homozygous dominants		Frequency of heterozygotes		Frequency of homozygous recessives	

By definition, a population is not evolving (and is therefore in Hardy-Weinberg equilibrium) when it has stable allele frequencies and, therefore, stable genotype frequencies from generation to generation. This can only be achieved when *all five* of the following conditions are met:

1. No mutation introducing new alleles into the population
2. No natural selection favoring some alleles over others
3. An infinitely large population size (and therefore no genetic drift)
4. No influx of alleles from neighboring populations (i.e., no gene flow)
5. Random mating of individuals

In nature, no population can ever be in strict Hardy-Weinberg equilibrium, since it will never meet all five conditions. In particular, because no real population is infinitely large, genetic drift will always occur. In other words, all natural populations are evolving. Nevertheless, by describing the pattern of genotypes in a nonevolving population, Hardy-Weinberg equilibrium provides a baseline from which to measure evolution.

To see how the Hardy-Weinberg equation can be used to detect evolutionary change, consider the following example. Say you have a population of manatees with two possible phenotypes for hide color, gray and white. The allele for gray hide color, *G*, is dominant; the allele for white hide color, *g*, is recessive. As every individual in the population has two alleles for the hide-color gene (one maternal and one paternal), there are twice as many alleles as there are members of the population. So a population of 500 manatees has 1,000 alleles of the gene for hide color.

In this population, assume there are 800 *G* alleles, and 200 *g* alleles. We would then say that the frequency of the dominant allele is 0.8 (800/1,000) and the frequency of the recessive allele is 0.2 (200/1,000). Since there are only two alleles in the population, their combined frequencies must add up to 1. If we use *p* to denote the frequency of the dominant allele and *q* to denote the frequency of the recessive allele, then we can say that $p + q = 1$.

Suppose we want to use those allele frequencies to calculate the expected frequency of white-hided (*gg*) individuals in the population. If the frequency of *g* in the population is *q*, then we know from the Hardy-Weinberg equation that the frequency of *gg* is $q^2 = (.2)(.2) = .04$. Thus, in our population of manatees, 4%, or 20 manatees, will have white hides. If we find out that the *actual* number of white manatees in the population is appreciably more or less than this number, then we know that our population is evolving, and we can begin to investigate why.

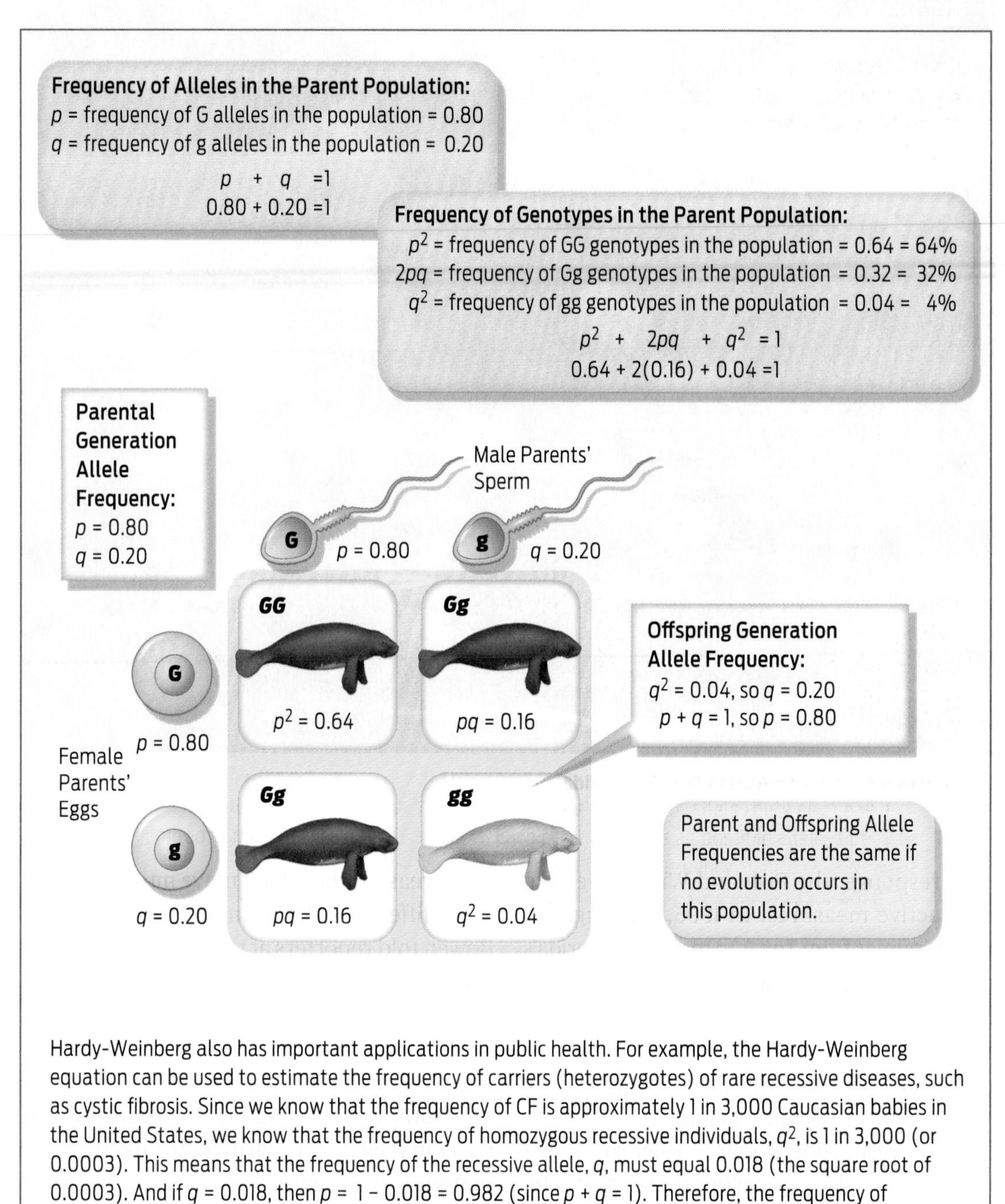

Frequency of Alleles in the Parent Population:
p = frequency of G alleles in the population = 0.80
q = frequency of g alleles in the population = 0.20
p + q = 1
0.80 + 0.20 = 1
Frequency of Genotypes in the Parent Population:
p^2 = frequency of GG genotypes in the population = 0.64 = 64%
2pq = frequency of Gg genotypes in the population = 0.32 = 32%
q^2 = frequency of gg genotypes in the population = 0.04 = 4%
p^2 + 2pq + q^2 = 1
0.64 + 2(0.16) + 0.04 = 1
Parental Generation Allele Frequency:
p = 0.80
q = 0.20
Male Parents' Sperm
G
p = 0.80
g
q = 0.20
GG
p^2 = 0.64
Gg
pq = 0.16
Offspring Generation Allele Frequency:
q^2 = 0.04, so q = 0.20
p + q = 1, so p = 0.80
G
p = 0.80
Female Parents' Eggs
g
q = 0.20
Gg
pq = 0.16
gg
q^2 = 0.04
Parent and Offspring Allele Frequencies are the same if no evolution occurs in this population.

Hardy-Weinberg also has important applications in public health. For example, the Hardy-Weinberg equation can be used to estimate the frequency of carriers (heterozygotes) of rare recessive diseases, such as cystic fibrosis. Since we know that the frequency of CF is approximately 1 in 3,000 Caucasian babies in the United States, we know that the frequency of homozygous recessive individuals, q^2, is 1 in 3,000 (or 0.0003). This means that the frequency of the recessive allele, q, must equal 0.018 (the square root of 0.0003). And if $q = 0.018$, then $p = 1 - 0.018 = 0.982$ (since $p + q = 1$). Therefore, the frequency of heterozygotes, $2pq$, is 0.035, or 3.5%. Knowing the frequency of the disease and carriers in the population helps health workers offer genetic counseling and plan for interventions.

Chapter 15 Test Your Knowledge

POPULATIONS AND GENE POOLS

The genetic diversity of a population is reflected in its collective bank of alleles, or gene pool. The amount of genetic diversity in a population has implications for its evolution.

HINT See Infographic 15.2 and Up Close: Calculating Hardy-Weinberg Equilibrium.

KNOW IT

1. Genetic diversity is measured in terms of allele frequencies (the relative proportion of different alleles in a gene pool). A population of 3,200 manatees has 4,200 dominant *G* alleles and 2,200 recessive *g* alleles. What is the frequency of *g* alleles in the gene pool?

2. Of the three populations described below, each of which has 1,000 members, which population has the highest genetic diversity? Note that only one gene is being presented, and that this gene has three possible alleles: *A1*, *A2*, and *a*.

Population A: 70% have an *A1/A1* genotype, 25% have an *A1/A2* genotype, and 5% have an *A1/a* genotype.
Population B: 50% have an *A1/A1* genotype, 20% have an *A2/A2* genotype, 10% have an *A1/A2* genotype, 10% have an *A2/a* genotype, and 10% have an *a/a* genotype.
Population C: 80% have an *A1/A1* genotype and 20% have an *A1/a* genotype.

3. A starting population of bacteria has two alleles of the *TUB* gene: *T* and *t*. The frequency of *T* is 0.8 and the frequency of *t* is 0.2. The local environment undergoes an elevated temperature for many generations of bacterial reproduction. After 50 generations of reproduction at the elevated temperature, the frequency of *T* is 0.4 and the frequency of *t* is 0.6. Has evolution occurred? Explain your answer.

USE IT

4. Question 2 looked at the allele frequencies of populations A, B, and C. From your answer to that question, which population would you predict to have the greatest chance of surviving an environmental change? Explain your answer.

5. Which of the four populations in the table below would you be concerned about from a conservation perspective? Why would you be concerned?

Population	Number of individuals	Number of alleles, gene 1	Number of alleles, gene 2	Number of alleles, gene 3
1	50	1	7	5
2	1,000	1	5	7
3	50	3	2	2
4	1,000	1	1	2

6. Phenylketonuria (PKU) is a rare, recessive genetic condition that affects approximately 1 in 15,000 babies born in the United States. (You may have noticed on products that contain aspartame the statement "Phenylketonurics: contains phenylalanine," a warning for people with PKU that they should avoid consuming that product.) Calculate the expected frequency of carriers (that is, of heterozygotes) in the U.S. population, based on the information provided about rates of PKU among U.S. births. (Remember the Hardy-Weinberg equation.)

7. Assume a population of 100 individuals. Five are homozygous dominant (*AA*), 80 are heterozygous (*Aa*), and 15 are homozygous recessive (*aa*) for the *A* gene. Determine *p* and *q* for this population. Now use those values for *p* and *q* and plug them into the Hardy-Weinberg equation. Is this population in Hardy-Weinberg equilibrium? Why or why not?

GENETIC DRIFT

Genetic drift can alter the allele frequency of a population. Genetic drift tends to lower the genetic diversity of a population.

HINT See Infographics 15.2 and 15.3.

KNOW IT

8. Which of the following are examples of genetic drift?

a. founder effect
b. bottleneck effect
c. inbreeding
d. a and b
e. a, b, and c

9. A bottleneck is best described as
a. an expansion of a population from a small group of founders.
b. a small number of individuals leaving a population.
c. a reduction in the size of an original population followed by an expansion in its size as the surviving members reproduce.
d. the mixing and mingling of alleles by mating between members of different populations.
e. an example of natural selection.

10. A population of manatees has 12 different alleles, *A* through *L*, of a particular gene. A drunk motorboat driver recklessly tears through the water where the manatees live, killing 90% of them. The surviving manatees are all homozygous for allele *B*.
a. What is the impact of this event on the frequency of alleles *A* through *L*?
b. What type of event is this?

USE IT

11. Why is genetic drift considered to be a form of evolution? How does it differ from evolution by natural selection?

12. In humans, founder effects may occur when a small group of founders immigrates to a new country, for example to establish a religious community. In this situation, why might the allele frequencies in succeeding generations remain similar to those of the founding population rather than gradually becoming more similar to the allele frequencies of the population of the country to which they immigrated?

GENE FLOW AND SPECIATION

Gene flow can alter the genetic diversity of a population as individuals from neighboring populations mix and mate with the original population. Barriers to gene flow contribute to speciation.

HINT See Infographics 15.5, 15.7, and 15.8.

13. The biological species concept defines a species
a. on the basis of similar physical appearance.
b. on the basis of close genetic relationships.
c. on the basis of similar levels of genetic diversity.
d. on the basis of the ability to mate and produce fertile offspring.
e. on the basis of recognizing one another's mating behaviors.

14. How does geographic isolation contribute to speciation events?

USE IT

15. Two populations of rodents have been physically separated by a large lake for many generations. The shore on one side of the lake is drier and has very different vegetation from that on the other side. The lake is drained by humans to irrigate crops, and now the rodent populations are reunited. How could you assess if they are still members of the same species?

16. Why is inbreeding detrimental to a population?

17. If geographically dispersed groups all converge at a common location during breeding season, then return to their home sites to bear and rear their young, what might happen to the gene pools of the different groups over time?

18. A small population of 25 individuals has five alleles, *A* through *E*, for a particular gene. The *E* allele is only represented in one homozygous individual:

5 individuals are *D/A* heterozygotes
5 individuals are *A/A* homozygotes
5 individuals are *A/B* heterozygotes
5 individuals are *C/D* heterozygotes
4 individuals are *C/C* homozygotes
1 individual is an *E/E* homozygote.

If five *A/E* heterozygotes migrate into the population, what will be the impact on the allele frequencies of each of the five alleles?

SCIENCE AND ETHICS

19. Consider the situation of Florida manatees.
a. What is the difference between an endangered and a threatened species, according to the classification established by the U.S. Endangered Species Act? At the present time, what is the status of the Florida manatee?
b. For a species like the Florida panther, why is a habitat conservation approach not sufficient to ensure a healthy recovery?
c. What approach could be taken to try to restore genetic diversity to a species such as the cheetah, given that all cheetahs are survivors of a bottleneck and are essentially genetically identical?

A Fish with Fingers?

What You Will Be Learning

A Fish with Fingers?

A transitional fossil sheds light on how evolution works

For 5 years, biologists Neil Shubin and Ted Daeschler spent their summers trekking through one of the most desolate regions on earth. They were fossil hunting on the remote island of Ellesmere, in the Canadian Arctic, about 600 miles from the north pole. Even in summer, Ellesmere is a forbidding place: a windswept, frozen desert where sparse vegetation grows no more than a few inches tall, where sleet and snow fall in the middle of July, and where the sun never sets. Only a handful of wild animals survive here, but those that do make for dangerous working conditions. Hungry polar bears and charging herds of muskoxen are hazards of working in the Arctic, says Daeschler. The team carried shotguns for protection.

Braving these conditions, the researchers drilled, chiseled, and hammered their way through countless tons of rock looking for fossils. Not just any rocks and fossils, but ones dating from 375 million years ago, when animals were taking their first tentative steps on land. For three summers, they scoured the site of what was once an active streambed but found little of interest, mostly pieces of ancient fish.

Tiktaalik "splits the difference between something we think of as a fish and something we think of as a limbed animal." –Ted Daeschler

Then, in 2004, the team made a tantalizing discovery: the snout of a curious-looking creature protruding from a slab of pink rock. Further excavation revealed the well-preserved remains of several flat-headed animals between 4 and 9 feet long. In some ways, the creatures resembled giant fish—they had fins and scales. But they also had traits that resembled those of

VERTEBRATE
An animal with a bony or cartilaginous backbone.

A model of *Tiktaalik roseae*, the fossil discovery that represents a critical phase in the evolution of four-legged, land-dwelling animals.

The *Tiktaalik roseae* fossil.

land-dwelling amphibians–notably, a neck, wrists, and fingerlike bones. They named the new species *Tiktaalik roseae*; *tiktaalik* (pronounced tic-TAH-lick) is a native word meaning "large freshwater fish." This hybrid animal no longer exists today, but it represents a critical phase in the evolution of four-legged, land-dwelling **vertebrates**–including humans.

Tiktaalik "splits the difference between something we think of as a fish and something we think of as a limbed animal," says Daeschler, a curator of vertebrate paleontology at the Academy of Natural Sciences in Philadelphia. "In that sense, it is a wonderful transitional fossil between two major groups of vertebrates."

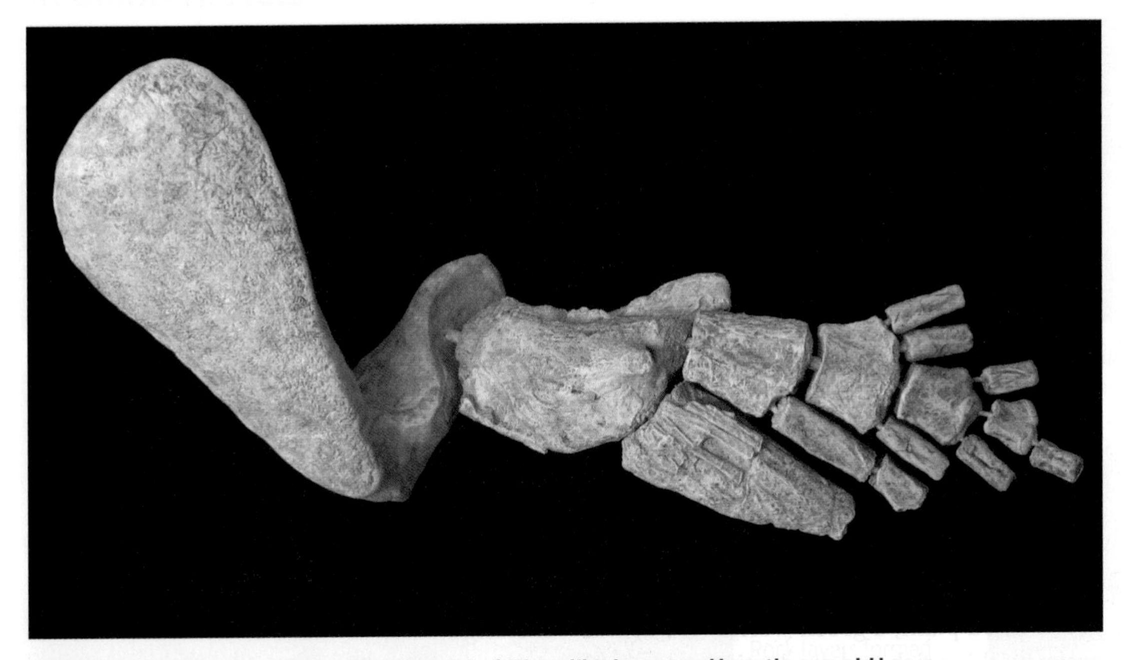

Shubin and Daeschler created a digital model of Tiktaalik's bones and how they would have moved relative to one another.

able to create a model of how the bones would have moved relative to one another, and they are now modeling these movements digitally. The models show that the bones and joints were strong enough to support the body and worked like those of the earliest known tetrapods–the early amphibians. "This animal was able to hold its fin below its body, bend the fin out toward what we think of as a wrist, and bend the elbow," explains Daeschler. In other words, it was a fish that could do a push-up.

With this hybrid anatomy, *Tiktaalik* was not galloping on land, of course. It probably lived most of the time in water, but Shubin and Daeschler suspect that *Tiktaalik* may have used its supportive fins to pull itself out of the water for brief periods. "This is a fish that can live in the shallows or even make short excursions onto land," says Shubin. The ability to crawl onto land would certainly have been a useful trait in the Devonian, when open water was a brutal fish-eat-fish world, whereas land was a predator-free paradise, full of nourishing bugs.

There was, of course, no forethought involved in this process. Fish did not develop limbs for the *purpose* of walking on land. Rather, limbs first evolved in shallow water, where they proved adaptive and were thus retained in the descendants of the organisms who first developed them. Then, when there was an opportunity to take advantage of a tantalizing new habitat–the land–the amphibious creatures already had the skeletal "toolkit."

"This is a fish that can live in the shallows or even make short excursions onto land." –Neil Shubin

For all its amphibian-like adaptations, *Tiktaalik* is still a fish because its limbs lack the true jointed fingers and toes that define tetrapods. But it's by far the most tetrapod-like of all the fishes so far discovered. Scientists have jokingly referred to it as a "fishapod" **(Infographic 16.4)**.

And that's what makes *Tiktaalik* such an important find: it embodies a previously unknown midpoint between fish and tetrapods. Such intermediate, or transitional, fossils document important steps in the evolution of life on earth. They help biologists understand how groups of organisms evolved, through natural selection, from one form into

HOMOLOGY
Anatomical, genetic, or developmental similarity among organisms due to common ancestry.

another. And they confirm that Darwin's theory of descent with modification—which predicts such intermediate forms—is correct.

A Fin Is a Paw Is an Arm Is a Wing

In *On the Origin of Species*, Darwin asked, "What can be more curious than that the hand of a man, formed for grasping, that of a mole for digging, the leg of the horse, the paddle of the porpoise, and the wing of the bat, should all be constructed on the same pattern, and should include similar bones, in the same relative positions?" To Darwin, this uncanny similarity was evidence that all these organisms were related—that they share a common ancestor in the ancient past.

The fact that all tetrapods share the same forelimb bones, arranged in the same order, is an example of **homology**—a similarity due to common ancestry. Before Darwin, comparative anatomists had identified many such similarities in anatomy; what they lacked was a satisfactory explanation for why such similarity should exist. Darwin provided that explanation:

INFOGRAPHIC 16.4

Tiktaalik, an Intermediate Fossilized Organism

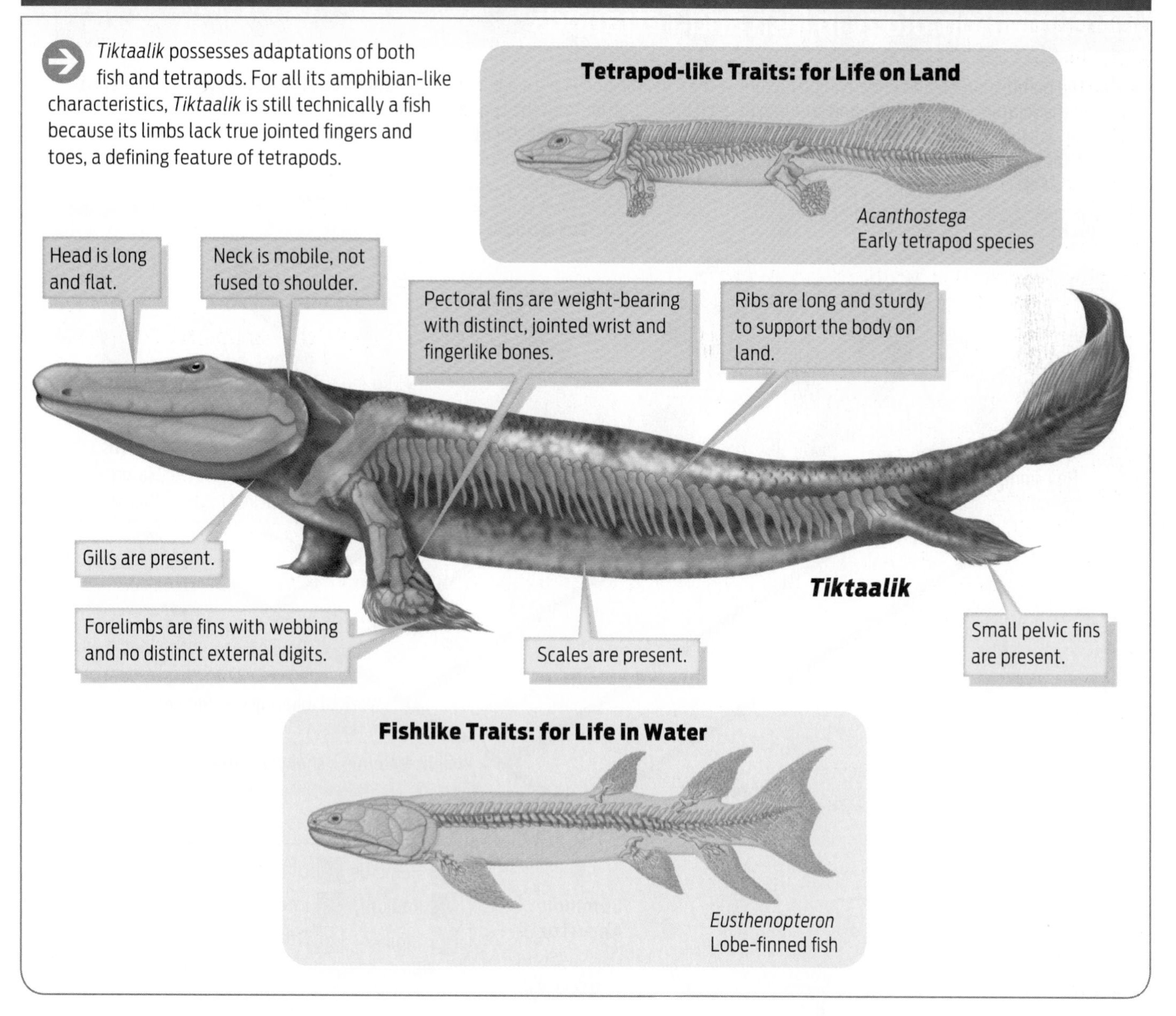

Tiktaalik possesses adaptations of both fish and tetrapods. For all its amphibian-like characteristics, *Tiktaalik* is still technically a fish because its limbs lack true jointed fingers and toes, a defining feature of tetrapods.
Tetrapod-like Traits: for Life on Land
Acanthostega
Early tetrapod species
Head is long and flat.
Neck is mobile, not fused to shoulder.
Pectoral fins are weight-bearing with distinct, jointed wrist and fingerlike bones.
Ribs are long and sturdy to support the body on land.
Gills are present.
Forelimbs are fins with webbing and no distinct external digits.
Scales are present.
Tiktaalik
Small pelvic fins are present.
Fishlike Traits: for Life in Water
Eusthenopteron
Lobe-finned fish

homologous structures are ones that are similar because they are inherited from the same ancestor–in this case, an amphibious creature like *Tiktaalik*. Why is this significant? Think of it this way: every time you bend your wrist back and forth–to swipe a paint brush or hold a cell phone to your ear, for example–you are using structures that first evolved 375 million years ago in fish. In many ways, you have a fish to thank for some of your most useful adaptations. As Shubin points out, "This is not just some archaic, weird branch of evolution; this is *our* branch of evolution." (For more on this, check out Shubin's book *Your Inner Fish*.) **(Infographic 16.5).**

If they have the same bones, why then do a human arm and a bird wing look so different? Remember that during the process of inheritance mutations are continually introduced into the DNA of genes. Such mutations can produce subtle changes in the proteins encoded by those genes–proteins involved in constructing the bones that make up an arm or a wing, for example. Changes in bone proteins can result in

INFOGRAPHIC 16.5

Forelimb Homology in Fish and Tetrapods

The number, order, and underlying structure of the forelimb bones are similar in all the groups illustrated below. The differences in the relative width, length, and strength of each bone contribute to the specialized function of each forelimb. This anatomical homology is strong evidence that these organisms all shared a common ancestor at some time in the distant past. The variations in bone shape and function reflect evolutionary adaptations to different environments.

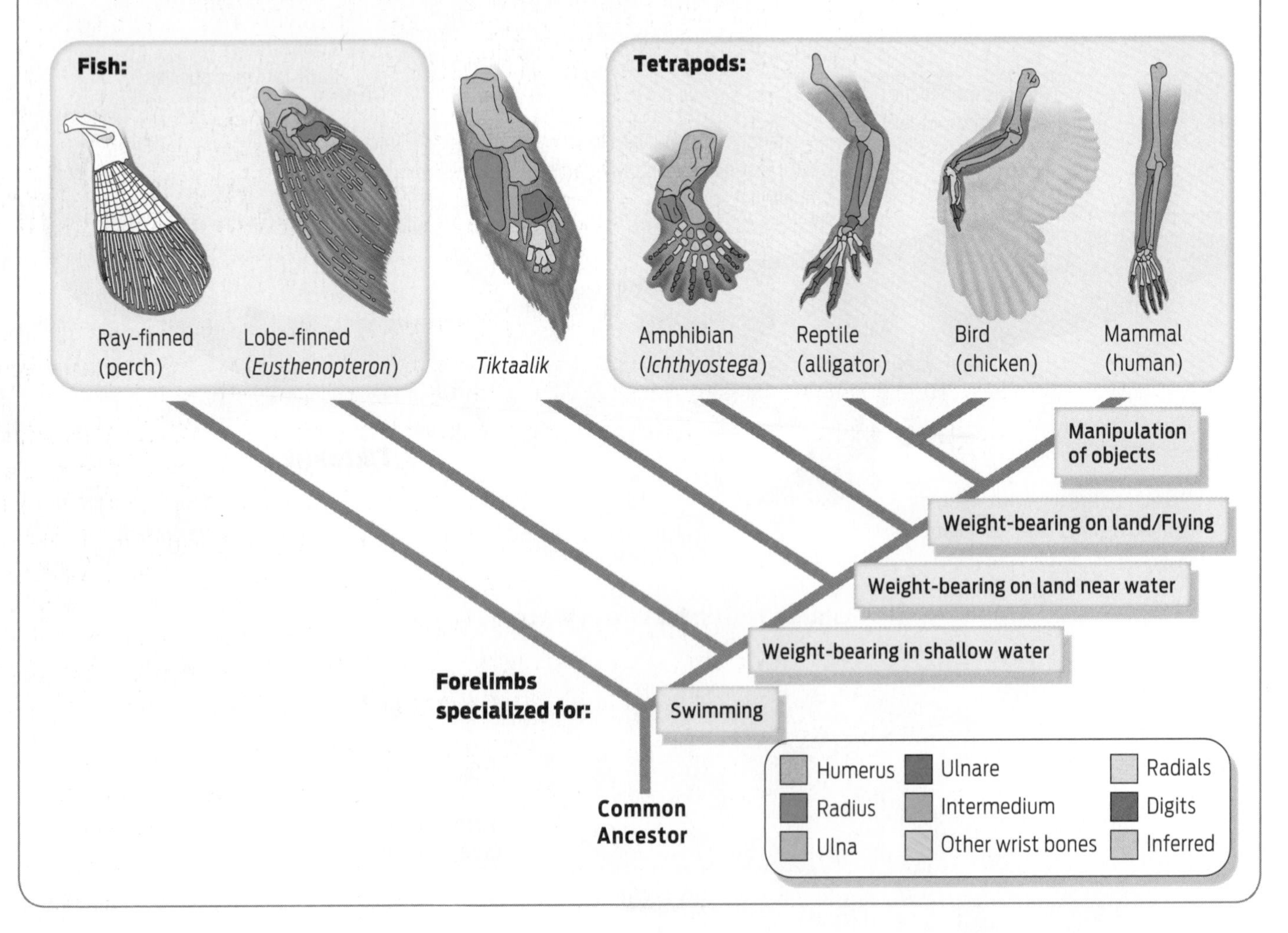

INFOGRAPHIC 16.6

Vertebrate Animals Share a Similar Pattern of Early Development

We can identify homologous structures by tracing their embryological development. Some of our middle ear bones, for example, are homologous with the jaw bones of reptiles and bones supporting gills in fish. We know this because all of these structures develop from the pharyngeal pouches that appear in all vertebrate embryos early in development. This developmental homology is strong evidence that all vertebrate animals are related by common ancestry. Genetic changes over time have introduced modifications in later stages that give rise to distinct species with vast physical differences.

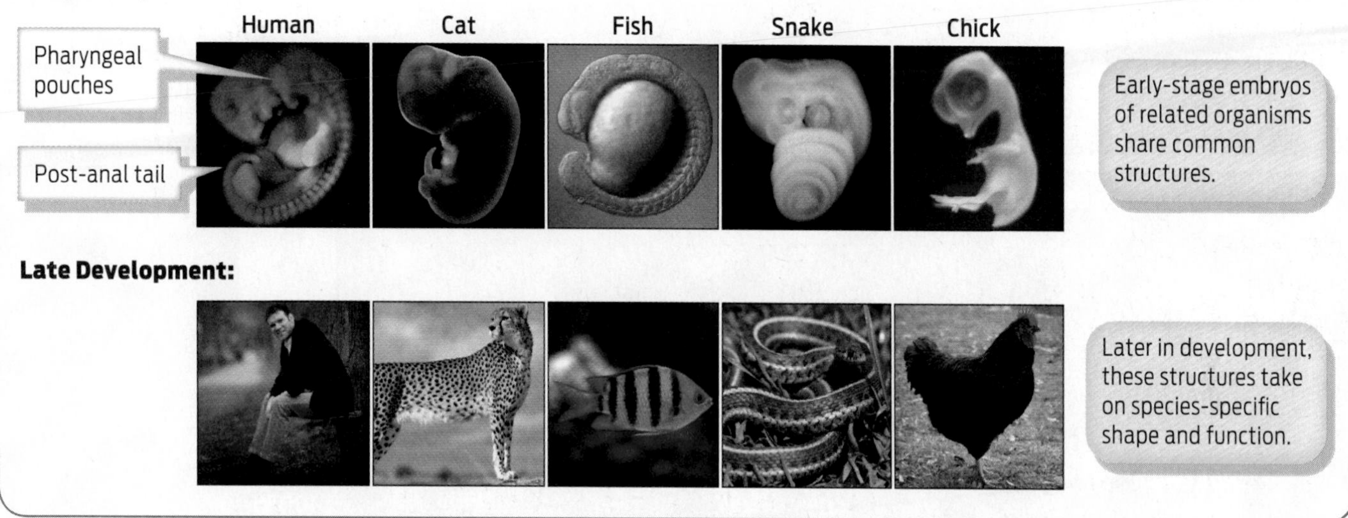

slightly altered bones, making them longer or thinner, for instance. When these modified bones are helpful to an organism's survival and reproduction, the advantageous traits are passed on to the next generation, and populations emerge that have these adaptations. This "descent with modification" (Darwin's phrase again) results in diverse organisms sharing common–homologous–structures and putting them to different uses.

> **"This is not just some archaic, weird branch of evolution; this is *our* branch of evolution."**
> —Neil Shubin

VESTIGIAL STRUCTURES
A structure inherited from an ancestor that no longer serves a clear function in the organism that possesses it.

We can see homology not only in adult anatomy, but in early development as well. Take a look at early embryos of vertebrate animals as diverse as humans, fish, and chickens and you'll see that they all look remarkably similar **(Infographic 16.6)**. Why should the embryonic stage of a human resemble the embryonic stage of a fish when the adults of each species look so different? Similar embryological structures are further evidence that all vertebrates shared a common ancestor.

Development helps us solve other evolutionary conundrums as well, such as why reptiles like snakes don't have legs like other tetrapods. In fact, snake embryos *do* possess the beginnings of limbs, but these limb "buds" remain rudimentary and do not develop into full-fledged limbs (although you can still see stubby hindlimbs in some species of snake today). Such **vestigial structures**, which serve no apparent function in an organism, are strong evidence for evolution: these "useless" features are inherited from an ancestor in whom they *did* serve a function.

Zooming in even further, to the molecular level, we find still more examples of homology–yet more evidence of common ancestry. Scientists have known since the 1960s that DNA is the molecule of heredity, and that it is shared by all living organisms on earth. Every molecule of DNA–whether from fish, maple tree, bacterium, or human–is made of the same four nucleotides

INFOGRAPHIC 16.7

DNA Sequences Are Shared among Related Organisms

Related organisms share DNA sequences inherited from a common ancestor. Over time, the sequence in each species acquires independent mutations. The more time that has passed, the greater the number of sequence differences that will be present. Thus, the percentage of nucleotides that differ between two species gives an indication of the evolutionary distance between them.

Sequence homology between species

Species A GGTATCGAGGTTCTACATTGCAACTTCTAC
Close relative GGAAACGAGGTTCTACATTGCCACTTCTAC
Distant relative GGAAACGAGGTTCGACATAGCCACTTCTAC

3 differences in 30 nucleotides
3/30 = 10%; or 90% similarity

5 differences in 30 nucleotides
5/30 = 17%; or 83% similarity

Similarity to human DNA sequences*

Pufferfish <2%
Mouse 40%
Chimpanzee 99%
Human 100%

Common ancestor of chimpanzees and humans, ≈5–7 mya.
Common ancestor of mice, chimpanzees, and humans, ≈60–100 mya.
Common ancestor of pufferfish, mice, chimpanzees, and humans, ≈420 mya.

*Data presented for the *CFTR* region. From Green et al., *Nature* 2003, 424:788–793.

(A, C, T, and G), and all organisms use the information encoded by those nucleotides to make proteins in the same basic way, using the universal genetic code (discussed in Chapter 8). Why should all living things use the same system of decoding genetic information? The best explanation is that this system was the one used by the ancient ancestor of all living organisms, passed on to all of its descendants, and preserved throughout billions of years of evolution.

DNA and Descent

While all living organisms share DNA and the genetic code, no two species will share the exact same sequence of DNA nucleotides. That's because (as described in Chapter 10) errors in DNA replication and other mutations are continually introducing variation into DNA sequences (and the proteins they encode). Over time, neutral and advantageous mutations will tend to be preserved, while harmful mutations will tend to be selected against and eliminated. In addition, much of our DNA is noncoding. Because mutations in noncoding DNA have no effect on an organism, they accumulate over time. As mutations are passed on to descendants, the number of sequence differences between the ancestor and its descendants grows—slowly in the case of sequences coding for critical proteins whose structures are well adapted to their functions, and more rapidly in the case of noncoding DNA. Closely related species will therefore have fewer DNA sequence differences than species that are more distantly related.

For example, when scientists looked at one specific region of DNA—the cystic fibrosis transmembrane region, which contains both coding and noncoding regions—they discovered that human DNA in this region is 99% identical to chimpanzee DNA. The fact that the DNA of the two species is nearly identical reflects the fact that humans and chimps share a common ancestor that lived relatively recently—just 5–7 million years ago. By contrast, human DNA is only 40% identical to the DNA of a mouse at this same region, which makes sense given that humans and mice share a common ancestor that lived between 60 and 100 million years ago. Even less sequence identity would be seen between a human and a toad, whose common ancestor—a lobed-finned fish— lived roughly 375 million years ago. The more distantly related two species are, the more sequence differences in DNA sequences you will see. In essence, DNA serves as a kind of molecular clock: each additional sequence difference is like a tick of the clock, showing the amount of time that has elapsed since the two species' common ancestor **(Infographic 16.7)**.

When combined with evidence from the fossil record, anatomy, and development, molecular data become a powerful tool for understanding evolution. As we'll see in Chapter 17, DNA evidence is often a more reliable clue to common ancestry than physical appearance, and can serve as a check on conclusions derived from the fossil record or anatomy. As well, DNA is deepening our knowledge of how limbs evolved. Scientists have discovered that even species that are only very distantly related share some of the same genes. Animals as seemingly different as humans and fruit flies, for example, use some of the very same genes to get their heads on straight and their limbs in the right place. Learning how these genes work and how changes in their DNA sequences can produce large-scale changes in body plan or limb structure is a hot area of biology right now, familiarly known as "evo-devo."

Filling in the Gaps

Asked what he thinks is most interesting about the discovery of *Tiktaalik,* Ted Daeschler homes in on what he sees as a popular misconception about the fossil record—that it's "spotty" and "chaotic." But that's simply not true, he says. Despite the fact that it does not record *all* past life, the fossil record is still "very good"—so good, in fact, that you can use it to make and test predictions. You can, for example, look at the fossil record of fish and tetrapods and—suspecting on the basis of anatomy that the two groups are related—hypothesize that an intermediate-looking animal must have existed at some point. Then you can go look for it. Daeschler refers to this process as "filling in the gaps," and it's exactly what he and Neil Shubin did with *Tiktaalik*. They knew, based on the existing fossil record, *when* such a creature was likely to have existed, so then it was just a question of *where* to look for it.

For Shubin and Daeschler, *Tiktaalik* is exciting mostly because it shows that our understanding of evolution is correct: "It confirms that we have a very good understanding of the framework of the history of life," says Daeschler. "We predicted something like *Tiktaalik,* and sure enough, with a little time and effort, we found it." ■

Summary

- The theory of evolution—what Darwin called "descent with modification"—draws two main conclusions about life: that all living things are related, sharing a common ancestor in the distant past; and that the species we see today are the result of natural selection operating over millions of years.
- The theory of evolution is supported by a wealth of evidence, including fossil, anatomical, and DNA evidence.
- Fossils are preserved remains or impressions of once-living organisms that provide a record of past life on earth. Not all organisms are equally likely to form fossils.
- Fossils can be dated directly or indirectly based on the age of the rocks they are found in, or on their position relative to rocks or fossils of known ages.
- When fossils are dated and placed in sequence, they show how life on earth has changed over time.
- As predicted by descent with modification, the fossil record shows the same overall pattern for all lines of descent: younger fossils are more similar to modern organisms than are older fossils.
- Descent with modification also predicts the existence of "intermediate" organisms, such as *Tiktaalik*, that possess mixtures of "old" and "new" traits.
- An organism's anatomy reflects adaptation to its ecological environment. Changed ecological circumstances provide opportunities for new adaptations to evolve by natural selection.
- Homology—the anatomical, developmental, or genetic similarities shared among groups of organisms—is strong evidence that those groups descend from a single common ancestor that lived many millions of years ago.
- Homology can be seen in the common bone structure of the forelimbs of tetrapods, the similar embryonic development of all vertebrate animals, and the universal genetic code.
- Many genes, including those controlling body plans, are shared among distantly related species, an example of molecular homology owing to common ancestry.
- More-closely related species show greater DNA sequence homology than do more-distantly related species.

THE FOSSIL RECORD

While incomplete, the record of past life preserved in the fossil record gives us valuable insight into evolutionary changes in organisms over time and the history of life.

HINT See Infographics 16.1–16.3.

KNOW IT

1. Generally speaking, if you are looking at layers of rock, at what level would you expect to find the newest—that is, the youngest—fossils?

2. Which of the following is most likely to leave a fossil?

a. a jellyfish
b. a worm
c. a wolf
d. a sea sponge (an organism that lacks a skeleton)
e. All of the above are equally likely to leave a fossil.

3. What can the fossil shown below tell us about the structure and lifestyle of the organism that left it? Describe your observations.

USE IT

4. You have molecular evidence that leads you to hypothesize that a particular group of soft-bodied sea cucumbers evolved at a certain time. You have found a fossil bed with many hard-shelled mollusks dating to the critical time, but no fossil evidence to support your hypothesis about the sea cucumbers. Does this find cause you to reject your hypothesis? Why or why not?

5. A specific type of oyster is found in North American fossil beds dated to 100 million years ago. If similar oyster fossils are found in European rock, in layers along with a novel type of barnacle fossil, what can be concluded about the age of the barnacles? Explain your answer.

TIKTAALIK AND ITS SIGNIFICANCE

Tiktaalik provides a glimpse into the adaptations of vertebrate animals as they moved from the water onto land.

HINT See Infographics 16.4 and 16.5.

KNOW IT

6. Which of the following features of *Tiktaalik* is not shared with other bony fishes?

a. scales
b. teeth
c. a mobile neck
d. fins
e. none of the above

7. *Tiktaalik* fossils have both fishlike and tetrapod-like characteristics. Which characteristics are related to supporting the body out of the water?

USE IT

8. *Tiktaalik* fossils are described as "intermediate" or "transitional" fossils. What does this mean? Why are transitional organisms so significant in the history of life?

9. *Tiktaalik* has been called a "fishapod"—part fish, part tetrapod. Speculate on the fossil appearance of its first true tetrapod descendant—what features would distinguish it from *Titkaalik*? How old would you expect those fossils to be, relative to *Titkaalik*?

10. If some fish acquired modifications that allowed them to be successful on land, why didn't fish just disappear? In other words, why are there still plenty of fish in the sea if the land presented so many favorable opportunities?

COMPARATIVE ANATOMY

Strong evidence for evolution can be seen in the anatomical and developmental structures shared among diverse species of vertebrate animals.

HINT See Infographics 16.5 and 16.6.

➔ KNOW IT

11. Compare and contrast the structure and function of a chicken wing with the structure and function of a human arm.

12. Vertebrate embryos have structures called pharyngeal pouches. What do these structures develop into in an adult human? In an adult bony fish?

➔ USE IT

13. What is the evolutionary explanation for the fact that both human hands and otter paws have five digits?

14. Could you use the presence of a tail to distinguish a human embryo from a chicken embryo? Why or why not?

MOLECULAR EVIDENCE

All living organisms use DNA as their hereditary molecule and make proteins using the same genetic code, a reflection of the fact that all life on earth shares a common ancestor that lived in the distant past.

HINT See Infographic 16.7.

➔ KNOW IT

15. You have three sequences of a given gene from three different organisms. How could you determine how closely the three organisms are related to one another?

➔ USE IT

16. If, in humans, the DNA sequence TTTCTAGGAATA encodes the amino acid sequence phenylalanine–leucine–glycine–isoleucine, what amino acid sequence will that same DNA sequence specify in bacteria?

17. Gene *X* is present in yeast and in sea urchins. Both produce protein X, but the yeast protein is slightly different from the sea urchin protein. What explains this difference? How might you use this information to judge whether humans are closer evolutionarily to yeast or to sea urchins?

SCIENCE AND ETHICS

18. Fossils allow us to understand the evolution of many lineages of plants and animals. They therefore represent a valuable scientific resource. What if *Tiktaalik* (or an equally important transitional fossil) had been found by amateur fossil hunters and sold to a private collector? Do you think there should be any regulation of fossil hunting to prevent the loss of valuable scientific information from the public domain?

Q & A: Evolution

What You Will Be Learning

Q & A: Evolution

The history, classification, and phylogeny of life on earth

The modern theory of evolution draws two main conclusions about life on earth: that all living things are related, and that the different species we see today have emerged over millions of years as a result of genetic change. We can use evidence from geology, chemistry, paleontology, biogeography, comparative anatomy, and genetics to reconstruct the details of that evolutionary history.

GEOLOGY

Q How old is the earth, and how do we know?

A When the *Apollo 11* astronauts Neil Armstrong and Buzz Aldrin returned to earth from their historic 1969 moon walk, they carried with them a cargo of lunar rock chipped from the moon's surface. Embedded within these hunks of shimmering anorthosite lay clues to the earliest history of our solar system, including the planet we call home.

According to the nebular hypothesis, the planetary objects in our solar system are the result of a single event: the collapse of a swirling solar nebula, which formed both the sun and the planets out of cosmic dust. Since all the planets were formed at roughly the same time, we can date the age of the solar system by dating any planetary object within it. Of the many moon rocks obtained over the course of the six *Apollo* missions, the oldest have been calculated to be some 4.4 to 4.5 billion years old, which means that the earth is at least that old as well.

Why go to the moon to date the earth? With the exception of a few meteorite battle scars, the moon's surface has remained largely intact over the course of its existence. In contrast, the earth is a swirling ball of molten lava that continuously churns and digests its rocky outer crust. Because of this perpetual churning, it is difficult to find original, undisturbed rocks from earth's earliest period. The oldest known intact piece of earth's land surface, the Acasta Gneiss in a remote region of northern Canada, dates from 3.9 billion years ago. Some of these ancient rocks contain minerals as old as 4.1 to 4.2 billion years.

Scientists estimate that the age of the earth—and of the solar system more generally—is 4.54 billion years.

While these values do not establish an absolute age of the earth, they do provide a lower limit: the earth is at least as old as the materials that make it up. From these earth minerals and moon rocks, as well as material from meteorites that have fallen to earth, scientists estimate that the age of the earth–and of the solar system more generally–is 4.54 billion years, give or take a few million years.

How are such rocks, extraterrestrial or earthly, dated? The most important method is **radiometric dating,** in which the amount of radioactivity present in a rock is used as a kind of geologic clock. When rocks form, the minerals in them contain a certain amount of **radioactive isotopes**–atoms of elements such as uranium-238, potassium-40, and rubidium-87–that are unstable and decay into other atoms.

Radioactive isotopes decay by releasing high-energy particles from the nucleus, a change that causes one element literally to transform into another. For example, an atom of the radioactive isotope uranium-238 decays in a stepwise fashion into a stable atom of lead-206. The time it takes for half the isotope in a sample to break down is called its **half-life**.

Different radioactive elements decay at different rates. Uranium-238 has a half-life of 4.5 billion years, whereas potassium-40 has a half-life of 1.3 billion years. The half-life of carbon-14 (which is used to date once-living, organic remains rather than rocks) is relatively short: it decays to nitrogen-14 in just 5,730 years.

The Genesis Rock, a sample of lunar crust from about the time the moon was formed, was retrieved by *Apollo 15* astronauts James Irwin and David Scott.

A visitor looks at a display in the Hall of Biodiversity at the American Museum of Natural History in New York City.

RADIOMETRIC DATING
The use of radioactive isotopes as a measure for determining the age of a rock or fossil.

RADIOACTIVE ISOTOPE
An unstable form of an element that decays into another element by radiation, that is, by emitting energetic particles.

HALF-LIFE
The time it takes for one-half of a substance to decay.

INFOGRAPHIC 17.1

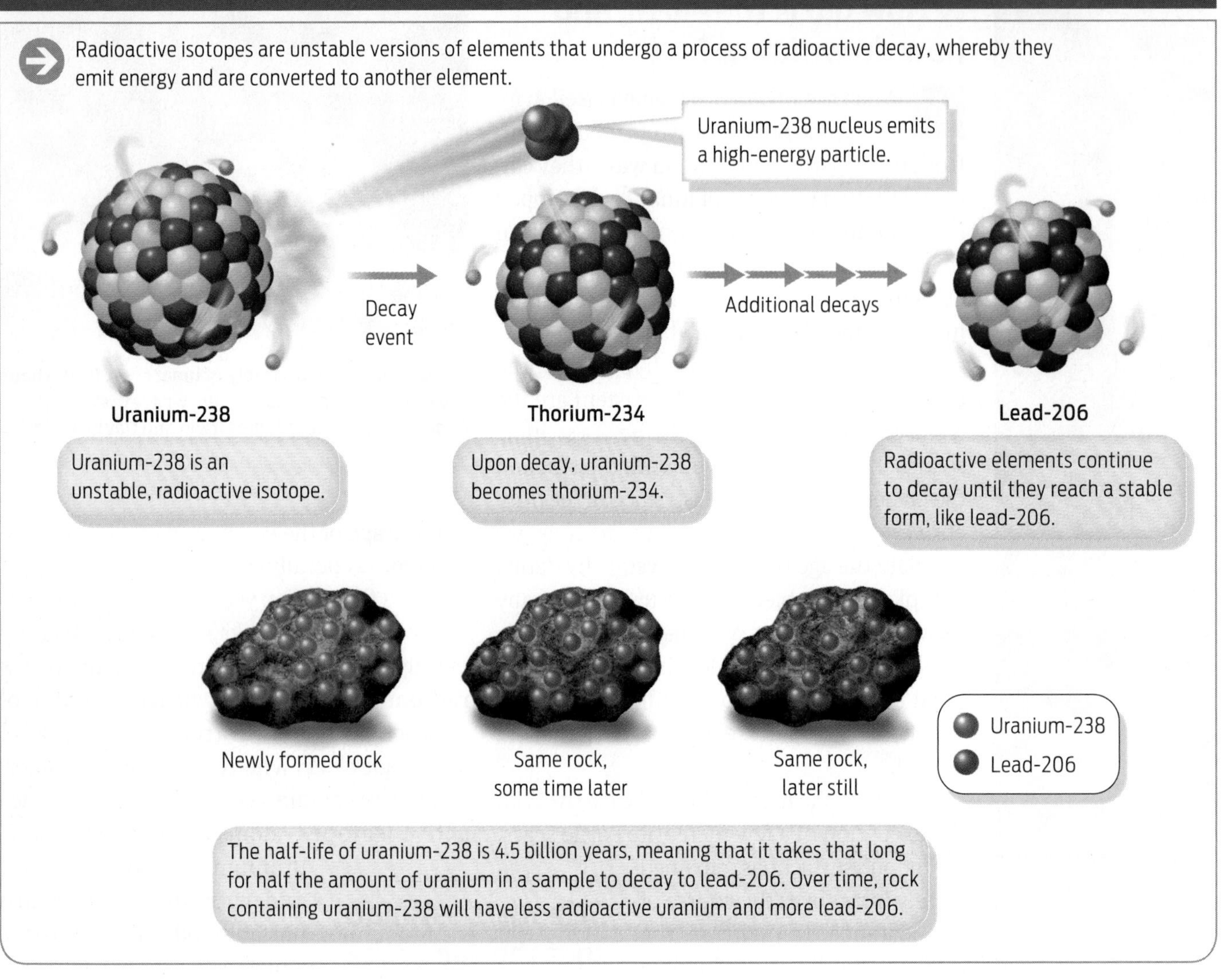

Unstable Elements Undergo Radioactive Decay
Radioactive isotopes are unstable versions of elements that undergo a process of radioactive decay, whereby they emit energy and are converted to another element.
Uranium-238 nucleus emits a high-energy particle.
Decay event
Additional decays
Uranium-238
Thorium-234
Lead-206
Uranium-238 is an unstable, radioactive isotope.
Upon decay, uranium-238 becomes thorium-234.
Radioactive elements continue to decay until they reach a stable form, like lead-206.
Newly formed rock
Same rock, some time later
Same rock, later still
Uranium-238
Lead-206
The half-life of uranium-238 is 4.5 billion years, meaning that it takes that long for half the amount of uranium in a sample to decay to lead-206. Over time, rock containing uranium-238 will have less radioactive uranium and more lead-206.

Because the isotopes decay at a known rate, they can be used to determine the age of the materials in which they're found **(Infographic 17.1)**.

As wind and water washed over rocks throughout earth's history, they stripped off, or eroded, particles and carried them to other places. Sometimes the deposited particles were compressed over many years into new rock layers by water or by additional particles. Such rock, called sedimentary rock, can be seen in the distinctive striations, or stripes, marking successive layers of sandstone and limestone found in former riverbanks like those surrounding the Grand Canyon in Arizona. Most fossils are found in sedimentary rocks.

Rocks can also form suddenly as erupting volcanoes spew lava and ash over an area. When this molten debris cools and hardens, it forms what is called igneous rock ("igneous" is from the Latin word for "fire"). Radiometric dating is performed on igneous rocks. When the rocks form, the radioactive clock is set to zero; no products of radioactive decay are present. Over time, more and more radioactive decay will occur, and more and more stable product will be present. By measuring the ratio of a radioactive isotope to stable product present in a layer

INFOGRAPHIC 17.2

Radioactive Decay Is Used to Date Some Rock Types

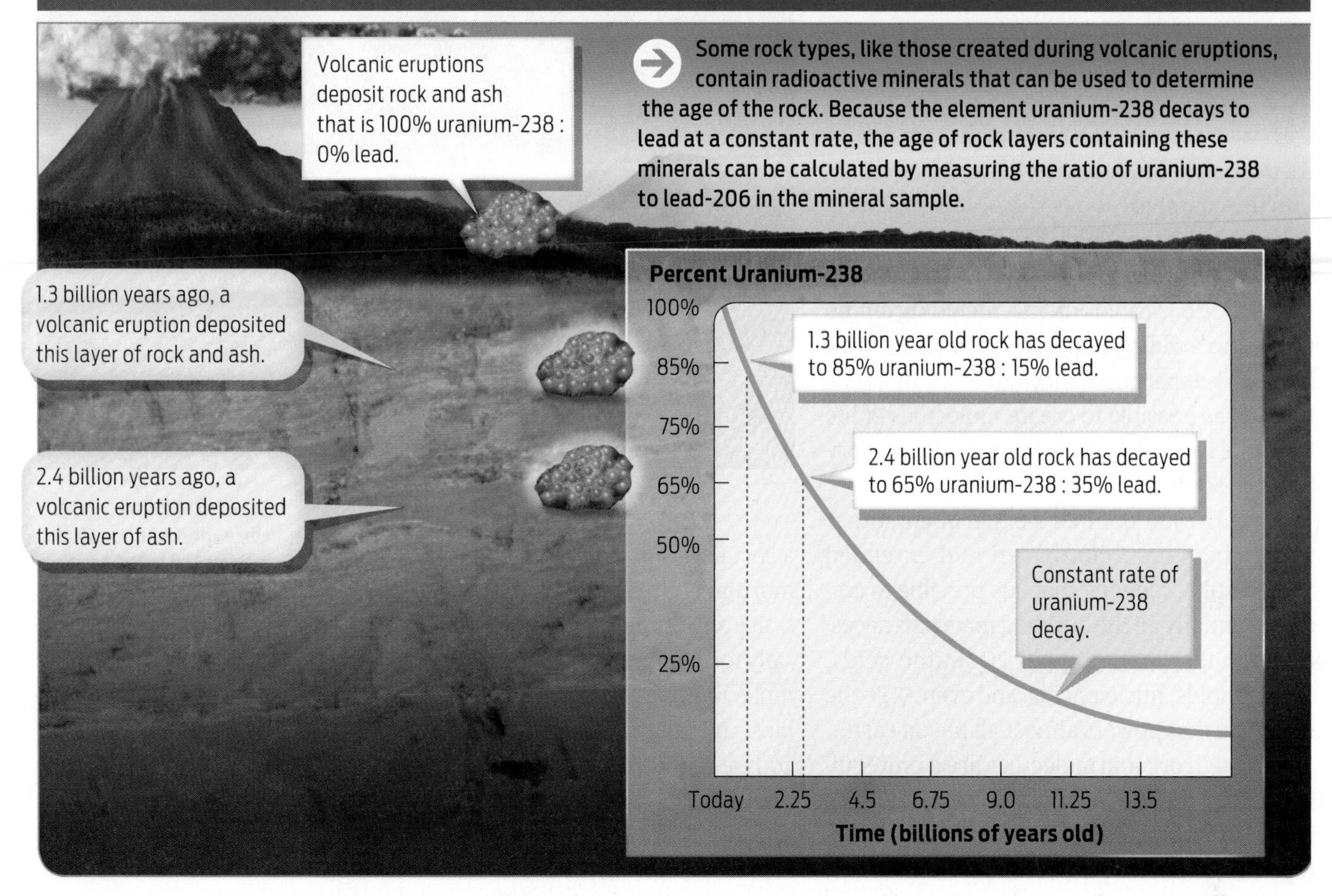

Volcanic eruptions deposit rock and ash that is 100% uranium-238 : 0% lead.
Some rock types, like those created during volcanic eruptions, contain radioactive minerals that can be used to determine the age of the rock. Because the element uranium-238 decays to lead at a constant rate, the age of rock layers containing these minerals can be calculated by measuring the ratio of uranium-238 to lead-206 in the mineral sample.
1.3 billion years ago, a volcanic eruption deposited this layer of rock and ash.
2.4 billion years ago, a volcanic eruption deposited this layer of ash.
Percent Uranium-238
100%
85%
75%
65%
50%
25%
1.3 billion year old rock has decayed to 85% uranium-238 : 15% lead.
2.4 billion year old rock has decayed to 65% uranium-238 : 35% lead.
Constant rate of uranium-238 decay.
Today
2.25
4.5
6.75
9.0
11.25
13.5
Time (billions of years old)

of igneous rock, scientists can calculate its age **(Infographic 17.2)**. Sedimentary rocks, on the other hand, cannot be dated by radiometric methods because they are made up of particles from rocks of various ages.

Dating rocks by radioactive isotopes is quite precise and can be confirmed by a number of methods. For example, minerals taken from layers of rock in Saskatchewan, Canada, were dated by three methods: the potassium-argon method yielded an age of 72.5 million years; the uranium-lead method, an age of 72.4 million years; and the rubidium-strontium method, an age of 72.54 million years.

BIOCHEMISTRY

Q When and how did life begin?

A At some point in the earth's distant past, life did not exist. Then, at a later point, it did. Where did this life come from? How did it start? The precise details of the transition from nonliving to living are lost in the mists of time. We can now only hypothesize how that transition might have occurred.

Scientists have offered a number of hypotheses to explain how life began on earth, including the idea that it arrived here fully formed on an asteroid or meteorite from outer space. Others hypothesize that life emerged in stages over time, as inorganic chemicals combined into successively more-complex molecules, including ones that were capable of self-replicating—that is, of copying themselves. A landmark experiment lending support to this hypothesis was performed by University of Chicago chemist Harold Urey and his 23-year-old graduate student Stanley Miller in 1953.

Urey and Miller hypothesized that they could synthesize organic molecules—the building blocks of life—by replicating the chemical environment of the early earth. To simulate the early atmosphere, they combined the gases hydrogen (H_2), methane (CH_4), ammonia (NH_3), and water vapor (H_2O) in a flask filled with warm water—the "primordial sea." They then mimicked lightning by discharging sparks into the chamber.

As the gases condensed and rained into the sea, a host of new molecules was produced from these basic ingredients, including amino acids, the building materials of proteins. This landmark experiment showed for the first time that it was possible to create molecules of life from the inorganic materials found in the primordial soup.

Since Urey and Miller's experiment, other researchers have confirmed and extended their results, showing that it is possible to create essentially all the building materials necessary for life, including all 20 amino acids, sugars, lipids, nucleic acids, and even ATP—the molecule that powers almost all life on earth.

Although organic molecules are a prerequisite for life, they are not themselves alive. To be alive, something must be able to grow, reproduce, and metabolize, among other things. Today, of course, cells carry out these life-sustaining functions. How then did living cells come about? Recall from Chapter 2 that one of the major components of cells are their lipid membranes. Researchers hypothesize that at a certain point the lipid molecules in the primordial soup formed bubbles—which makes sense since lipids are hydrophobic and naturally form bubbles in water. The other organic molecules were incorporated into these bubbles and, as researchers speculate, over the course of millions of years, such membrane-bound bubbles filled with self-replicating molecules eventually became cells, capable of reproducing. While these ideas are highly speculative, research on microorganisms living today in such unlikely places as hydrothermal vents at the bottom of the ocean are giving us concrete insights into how life might have begun (see Chapter 18).

Stanley Miller recreates the experiment he first performed in 1953 with Harold Urey.

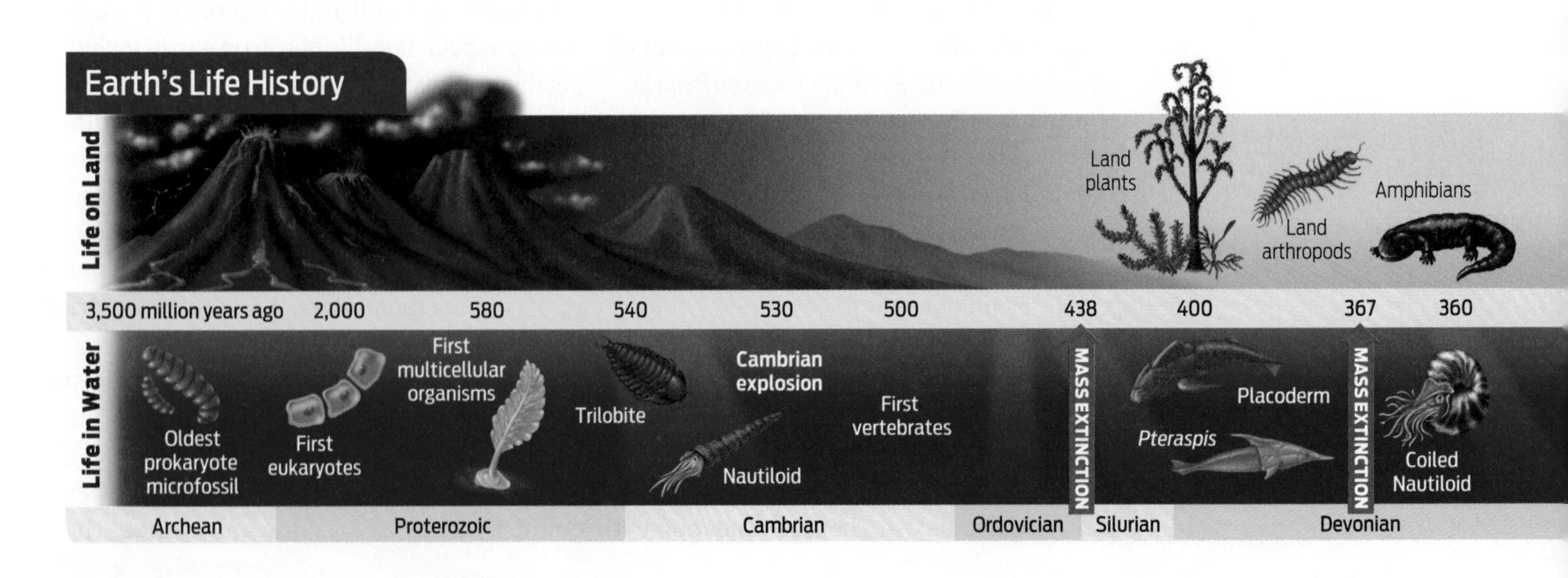

PALEONTOLOGY

Q What was life like millions of years ago?

A Humans weren't around millions of years ago, so we have no cave paintings or other records to help us picture what life on earth was like. Most of what we know about past life on earth comes from fossils—the preserved remains of once-living organisms, such as *Tiktaalik,* which is discussed in Chapter 16.

The oldest known fossils date from some 3.5 billion years ago.

While each fossil find is a treasure, any single specimen reveals only a tiny slice of geologic history. What paleontologists really want to understand is how each fossil fits into the larger story told by the fossil record. By dating the rock layers, or strata, near where fossils are buried, scientists can determine when different organisms lived on the earth. Combined with knowledge from geology, chemistry, and biology, the fossil record has enabled scientists to construct a geologic timeline of life on earth **(Infographic 17.3)**.

The geologic timeline shows that during the 4.6 billion years or so that the earth has been around, its geography and climate have gone through dramatic changes. For the first few million years or so it was a molten ball of lava continually bombarded by meteorites. Not until the surface cooled down about 3.8 billion years ago—to a balmy 45°C to 85°C (113°F to 185°F)—could it support life.

The oldest known fossils date from some 3.5 billion years ago, when earth's climate was very different from what it is today. The atmosphere lacked substantial oxygen (O_2), churning instead with ammonia, methane, and hydrogen. In this oxygenless world, the only organisms that could thrive were unicellular prokaryotes that used these other gases as a fuel source. Only with the emergence and proliferation of unicellular photosynthetic organisms, between 3.0 to 2.5 billion years ago, did oxygen begin to accumulate in the atmosphere, opening the door for more-complex eukaryotic organisms to evolve.

The first multicellular, eukaryotic organisms to make use of this oxygen were green algae, which appeared 1.2 billion years ago. Soft-bodied aquatic animals followed, about 600 million years ago, but it is only from 545 million years ago, during the Cambrian period, that we see fossil evidence of a truly diverse animal world. During the Cambrian explosion, as this event is known, ocean life swelled with a mind-boggling array of strange-looking creatures, including *Opabinia*, an organism with five eyes and a snout resembling a vacuum-cleaner hose, discovered in fossils from this period.

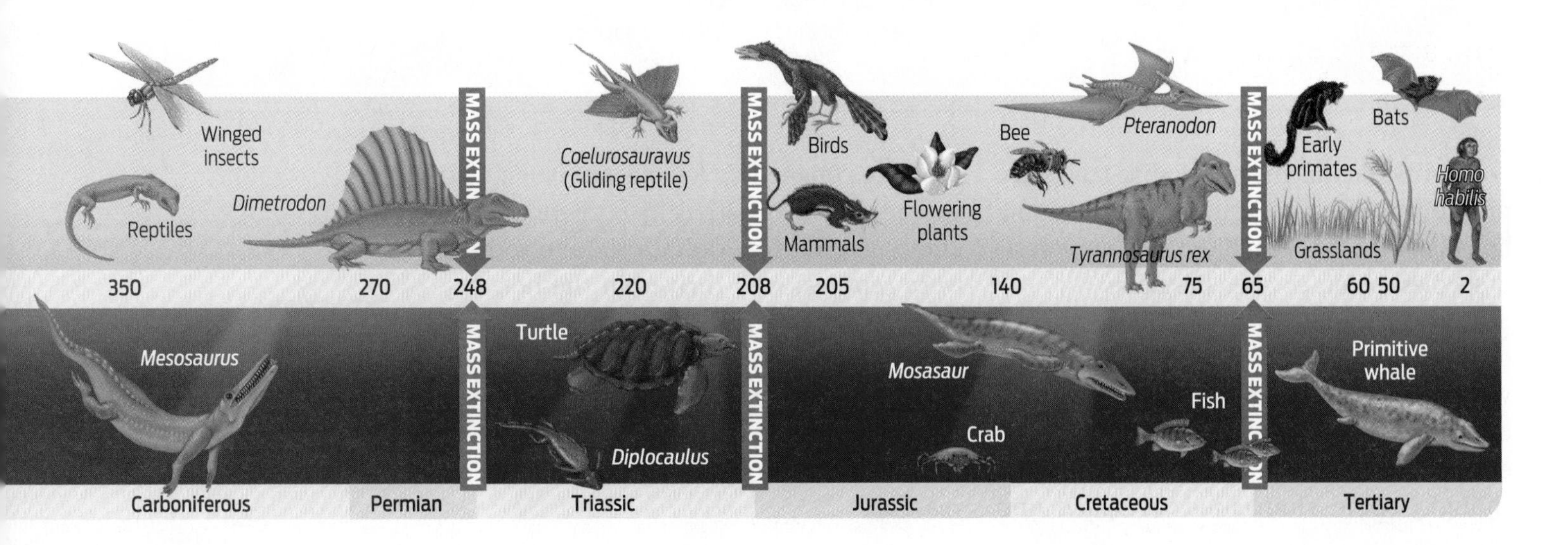

INFOGRAPHIC 17.3

Geologic Timeline of the Earth

Based on fossil evidence and radiometric dating of rock layers from around the world, scientists have produced a geologic timescale of the earth. The geologic timescale provides a chronological history of the main periods of the earth and its inhabitants.

Era	Period	Millions of Years Ago	Notable Events in the History of Life
Cenozoic	Quaternary	1.8	Many large mammals go extinct. Modern organisms are present today.
	Tertiary	65	Mammals, birds, and flowering plants diversify. Grasses appear. The first primates and early humans.
Mesozoic	Cretaceous	144	Dinosaurs diversify. Cone-bearing plants dominate, while flowering plants take over in many habitats. Era ends in mass extinction of dinosaurs.
	Jurassic	206	The first flowering plants and bird species appear. Large dinosaurs are plant-eaters.
	Triassic	251	Dinosaurs and mammals appear on land. Ocean life diversifies in recovery from Permian extinction.
Paleozoic	Permian	290	Reptiles appear on land. Oceans abundant in coral species. Era ends with mass extinction of 95% of living organisms.
	Carboniferous	354	Forests of seedless plants dominate land. Amphibians appear and begin to diversify.
	Devonian	408	Fish species diversify. The first insects and seed-bearing plants appear on land.
	Silurian	439	Seedless plants, primitive insects, and soft-bodied animals appear on land.
	Ordovician	495	Diverse plant and animal life in the oceans. The first fungal species appear.
	Cambrian	543	Expansion of ocean animal diversity. Ancestors of vertebrates appear.
Pre-Cambrian			Single-celled organisms in the ocean dominate life. Some soft-bodied invertebrates develop.

The first organisms to colonize land were primitive plants, appearing roughly 450 million years ago. By 350 million years ago, forests of seedless plants covered the globe.

Then, 290 million years ago, life was drastically cut down: roughly 95% of living species were suddenly extinguished in a mass die-off known as the Permian **extinction.** The extinction wasn't bad for all organisms, though; some flourished as space and resources opened up for the survivors, who spread and diversified in a phenomenon known as **adaptive radiation.** Among these were reptiles, who thrived in the hot, dry climate of the Triassic period. The most famous group of reptiles, the dinosaurs, dominated the land for nearly 200 million years, until they died out in another mass extinction at the end of the Cretaceous period, 65 million years ago.

EXTINCTION
The elimination of all individuals in a species; extinction may occur over time or in a sudden mass die-off.

The Burgess Shale, in the Canadian Rockies, is famous for the fine state of preservation of the soft parts of its fossils.

The reason for the extinction of the dinosaurs was a mystery for many years; evidence now suggests that what killed off the dinosaurs (and 60% of the other species living at the time) was a massive 6-mile-wide asteroid that plowed into earth with almost unimaginable force, sending a thick layer of soot and ash into the atmosphere and blocking out the sun for months. A crater 110 miles wide in Mexico's Yucatán peninsula, near the town of Chicxulub, is the likely impact site.

Evidence now suggests that what killed off the dinosaurs (and 60% of the other species living at the time) was a massive 6-mile-wide asteroid.

With the extinction of the dinosaurs, it was mammals' chance to spread and diversify on land and thus give rise to many of the species of organisms we see on the planet today. This pattern of sudden change—extinctions followed by adaptive radiations—is seen in the fossil record, and is an example of **punctuated equilibrium,** the theory that most evolutionary change occurs in sudden bursts related to environmental change rather than taking place gradually.

ADAPTIVE RADIATION
The spreading and diversification of organisms that occur when they colonize a new habitat.

PUNCTUATED EQUILIBRIUM
The theory that most species change occurs in periodic bursts as a result of sudden environmental change.

BIOGEOGRAPHY
The study of how organisms are distributed in geographical space.

PLATE TECTONICS
The movement of the earth's upper mantle and crust, which influences the geographical distribution of landmasses and organisms.

BIOGEOGRAPHY

Q Why are there no penguins at the north pole, and no polar bears at the south pole?

A In terms of habitat, the north pole and the south pole are quite similar: cold, snowy, and surrounded by ocean. Yet each place is home to different creatures. Why? We can get some clues from **biogeography,** the study of the natural geographic distribution of species. Biogeography seeks to explain why islands and isolated land areas—such as the Arctic and the Antarctic—have evolved their own distinct flora and fauna.

Penguins make their home in the southern hemisphere, especially in the coastal regions of Antarctica. According to fossil evidence, penguins first appeared about 65 million years ago near southern New Zealand and Antarctica. Polar bears, on the other hand, live only in the Arctic. From fossil and DNA evidence, polar bears likely evolved from brown bears roughly 150,000 years ago in an area in Siberia, when the region became isolated by glaciers. Both penguins and polar bears have thus lived in their respective northern or southern habitat for a long time, and it's easy to understand why they haven't migrated from one pole to the other—obstacles included great distance and the warmer oceans ringing their icy habitat **(Infographic 17.4)**. But how did they get to their homes in the first place?

Though today they are at opposite ends of the earth, the Arctic and Antarctic landmasses weren't always so far apart. In fact, 250 million years ago, the continents we currently see on earth were bound together in one large landmass that geologists call Pangaea. At that time it was theoretically possible for populations of land-dwelling animals to roam far and wide over the entire land surface. But because of a geologic process known as **plate tectonics,** over time this giant landmass split and split again, forming the continents of the northern and southern hemispheres. In the process, the ancestors of penguins and polar bears were

INFOGRAPHIC 17.4

The Geographic Distribution of Species Reflects Their Evolutionary History

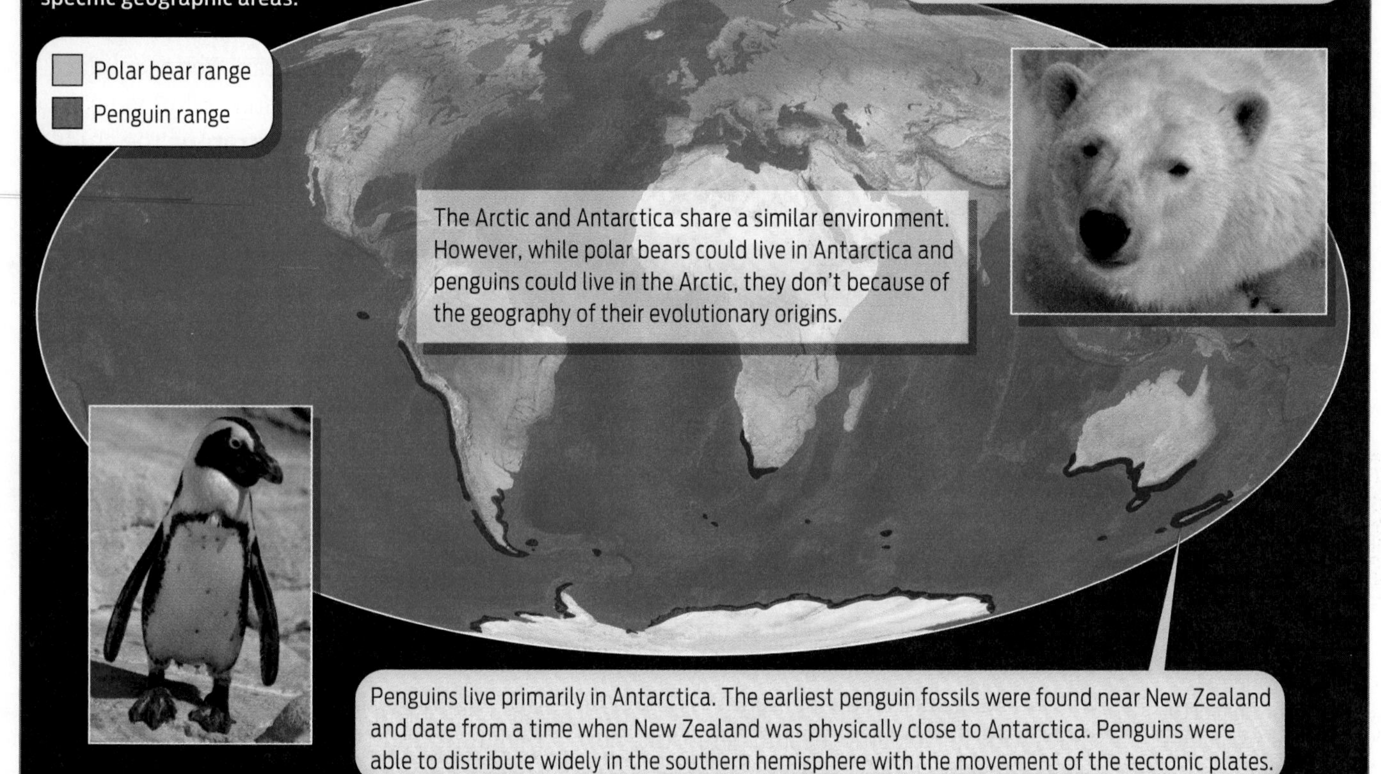

The study of the natural geographic distribution of species is known as biogeography. The geographic distribution of organisms on the earth today reflects both adaptations to the local environment as well as how organisms originally came to populate specific geographic areas.
Polar bears live in the Arctic. Their closest relatives are brown bears from Siberia. Polar bears have remained isolated on connected ice sheets of the northern hemisphere.
Polar bear range
Penguin range
The Arctic and Antarctica share a similar environment. However, while polar bears could live in Antarctica and penguins could live in the Arctic, they don't because of the geography of their evolutionary origins.
Penguins live primarily in Antarctica. The earliest penguin fossils were found near New Zealand and date from a time when New Zealand was physically close to Antarctica. Penguins were able to distribute widely in the southern hemisphere with the movement of the tectonic plates.

isolated from each other, as if on different lifeboats cast out to sea. Because the animals we know today as penguins and polar bears evolved from their ancestors after the split of the northern and southern landmasses, they are found today at different ends of the earth **(Infographic 17.5)**.

SYSTEMATICS

Q Are creatures that look alike always closely related?

A Polar bears share many traits with their brown-bear cousins–both species are recognizable as bears despite obvious differences in color. The fact that polar bears resemble brown bears in important respects is persuasive evidence that the two species share a recent common ancestor. But common ancestry is not the only reason that two species might appear similar. Even species that are not closely related may share similar adaptations as a result of independent episodes of natural selection, a phenomenon called **convergent evolution.**

Cold-dwelling fish provide a good example. In the frigid waters of the Antarctic Ocean, fish have a unique adaptation that keeps them from becoming ice cubes: their blood is pumped full of "antifreeze." Fish antifreeze is actually molecules called glycoproteins that lower the temperature at which body fluids would otherwise freeze by surrounding tiny ice crystals and keeping them from growing. Arctic fish, at the

CONVERGENT EVOLUTION
The process by which organisms that are not closely related evolve similar adaptations as a result of independent episodes of natural selection.

earth's other pole, also have antifreeze proteins, but the genes that code for them are different.

In the Antarctic Ocean, fish have a unique adaptation that keeps them from becoming ice cubes: their blood is pumped full of "antifreeze."

Arctic and Antarctic fish diverged from their common ancestor long before each species developed antifreeze proteins, which means these adaptations must have evolved more than once. In other words, at least two separate, independent episodes of evolution occurred with the same functional results. Sometimes similar environmental challenges will favor the same adaptations time and time again.

SYSTEMATICS

Q How many species are there on earth, and how do scientists keep track of them?

A Current estimates of the total number of species on earth range anywhere from 5 to 30 million, of which 1.8 million or so have been formally described. Many of these species are found in diversity hot spots such as rain

INFOGRAPHIC 17.5

Movement of the Earth's Plates Influences Climate and Biogeography

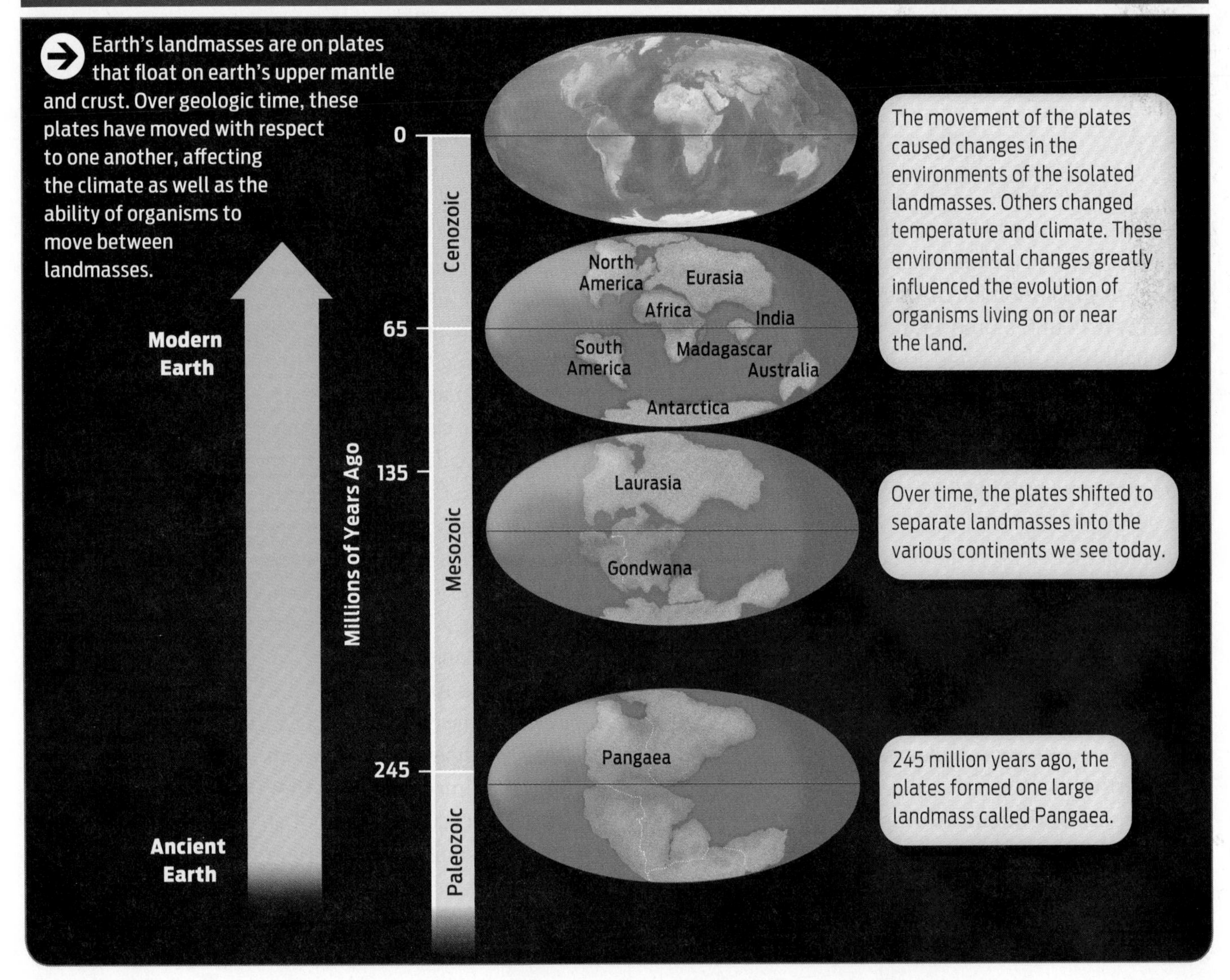

Earth's landmasses are on plates that float on earth's upper mantle and crust. Over geologic time, these plates have moved with respect to one another, affecting the climate as well as the ability of organisms to move between landmasses.
Modern Earth
Ancient Earth
Millions of Years Ago
0
65
135
245
Cenozoic
Mesozoic
Paleozoic
North America
Eurasia
Africa
India
South America
Madagascar
Australia
Antarctica
Laurasia
Gondwana
Pangaea
The movement of the plates caused changes in the environments of the isolated landmasses. Others changed temperature and climate. These environmental changes greatly influenced the evolution of organisms living on or near the land.
Over time, the plates shifted to separate landmasses into the various continents we see today.
245 million years ago, the plates formed one large landmass called Pangaea.

INFOGRAPHIC 17.6

How Many Species Are There?

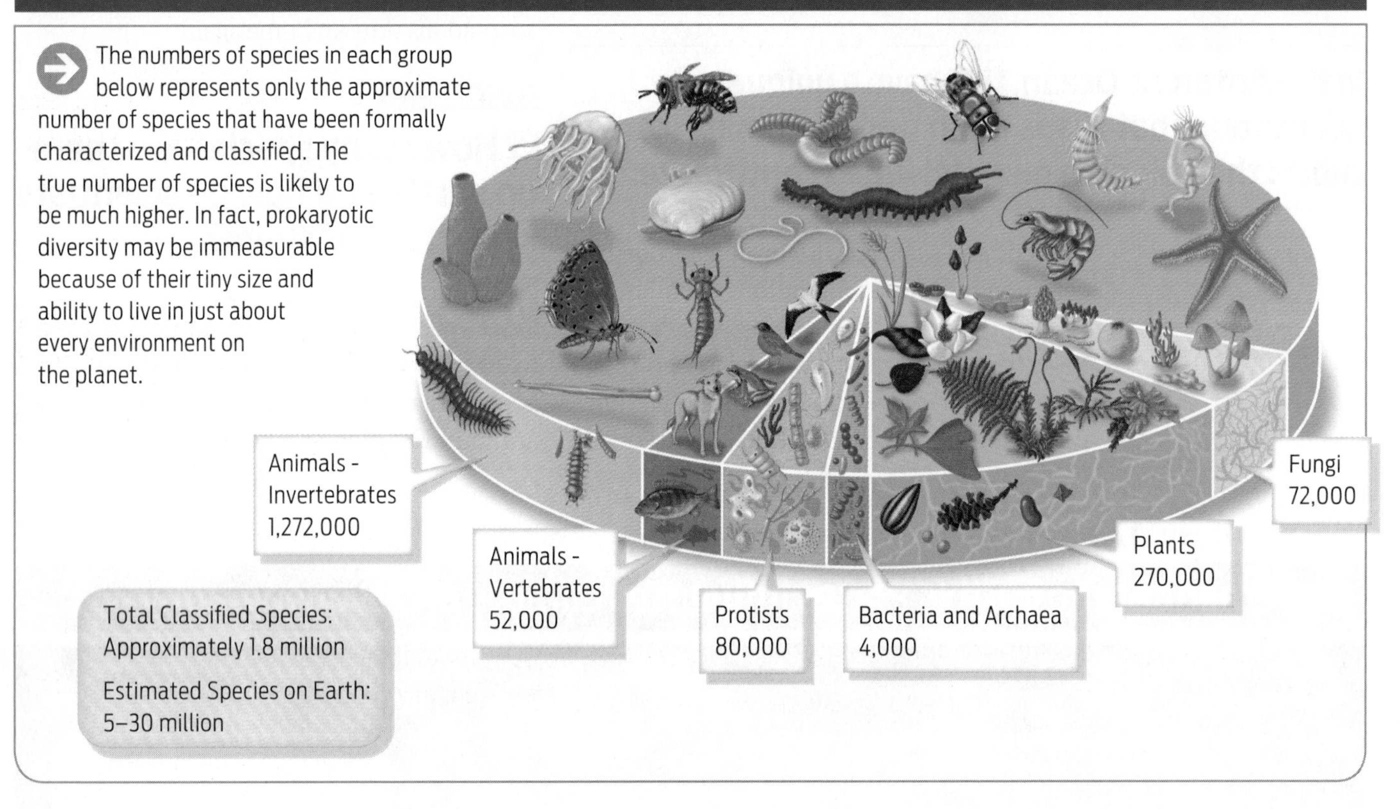

The numbers of species in each group below represents only the approximate number of species that have been formally characterized and classified. The true number of species is likely to be much higher. In fact, prokaryotic diversity may be immeasurable because of their tiny size and ability to live in just about every environment on the planet.
Animals - Invertebrates 1,272,000
Animals - Vertebrates 52,000
Protists 80,000
Bacteria and Archaea 4,000
Plants 270,000
Fungi 72,000
Total Classified Species: Approximately 1.8 million
Estimated Species on Earth: 5–30 million

forests. But as the wide range of the estimate implies, it's hard to put an exact number on the number of species on earth—there are simply too many to count. Moreover, new species are continually being discovered. In 2007, for example, scientists identified 11 new species of plants and animals in a remote part of Vietnam. And a recent study by researchers at Arizona State University found that 17,000 new eukaryotic species were discovered in 2006 alone, more than half of them insects **(Infographic 17.6)**.

With so many species out there, how do scientists keep track of them all? The process by which scientists systematically identify, name, and classify organisms is called **taxonomy.** (Taxonomy is part of the broader study of systematics, or the study of biological diversity of life on earth.)

Taxonomy is an attempt to impose a human sense of order on this vast array of species, categorizing them on the basis of features they have in common, such as whether their cells are eukaryotic or prokaryotic, whether they photosynthesize, whether or not they have four legs and fur.

By studying the many similarities and differences among organisms, taxonomists have come up with a system for sorting organisms into a series of eight progressively narrower categories: domain, kingdom, phylum, class, order, family, genus, species. As you move down the list, from domain to species, the categories get increasingly exclusive, until finally only one member is included. The genus and species names provide the scientific name for every living organism. Because the scientific name is in Latin, it can be easily recognized in many languages.

Take humans, for example. Humans are eukaryotes, members of the domain Eukarya. They are also animals, members of the kingdom Animalia. Within the animal kingdom, they belong to the phylum Chordata, a group that includes the **vertebrates**, animals with a rigid backbone. Further, humans are **mammals**, members of the class Mammalia; they share with all members of this class mammary

TAXONOMY
The process of identifying, naming, and classifying organisms on the basis of shared traits.

VERTEBRATES
Animals with a rigid backbone.

MAMMALS
Members of the class Mammalia; all members of this class have mammary glands and a fur-covered body.

glands and a body that is covered with fur. Humans belong to the Primate order, which also includes monkeys, apes, and lemurs. And humans are members of the Hominidae family, and so are closely related to their fellow hominids: chimpanzees, gorillas, and orangutans. Our scientific name—made up of our genus and species names—is *Homo sapiens* ("wise man") **(Infographic 17.7)**.

Classification would seem to be a simple matter—just observe, measure, and sort. But deciding which category an organism belongs

INFOGRAPHIC 17.7

Classification of Species

Organisms are classified into groups that are increasingly exclusive. In the broadest category (Animal Kingdom), all animals are included. Closely related organisms are grouped based on morphological, nutritional, and genetic characteristics. There are far fewer organisms in an order than in a phylum.

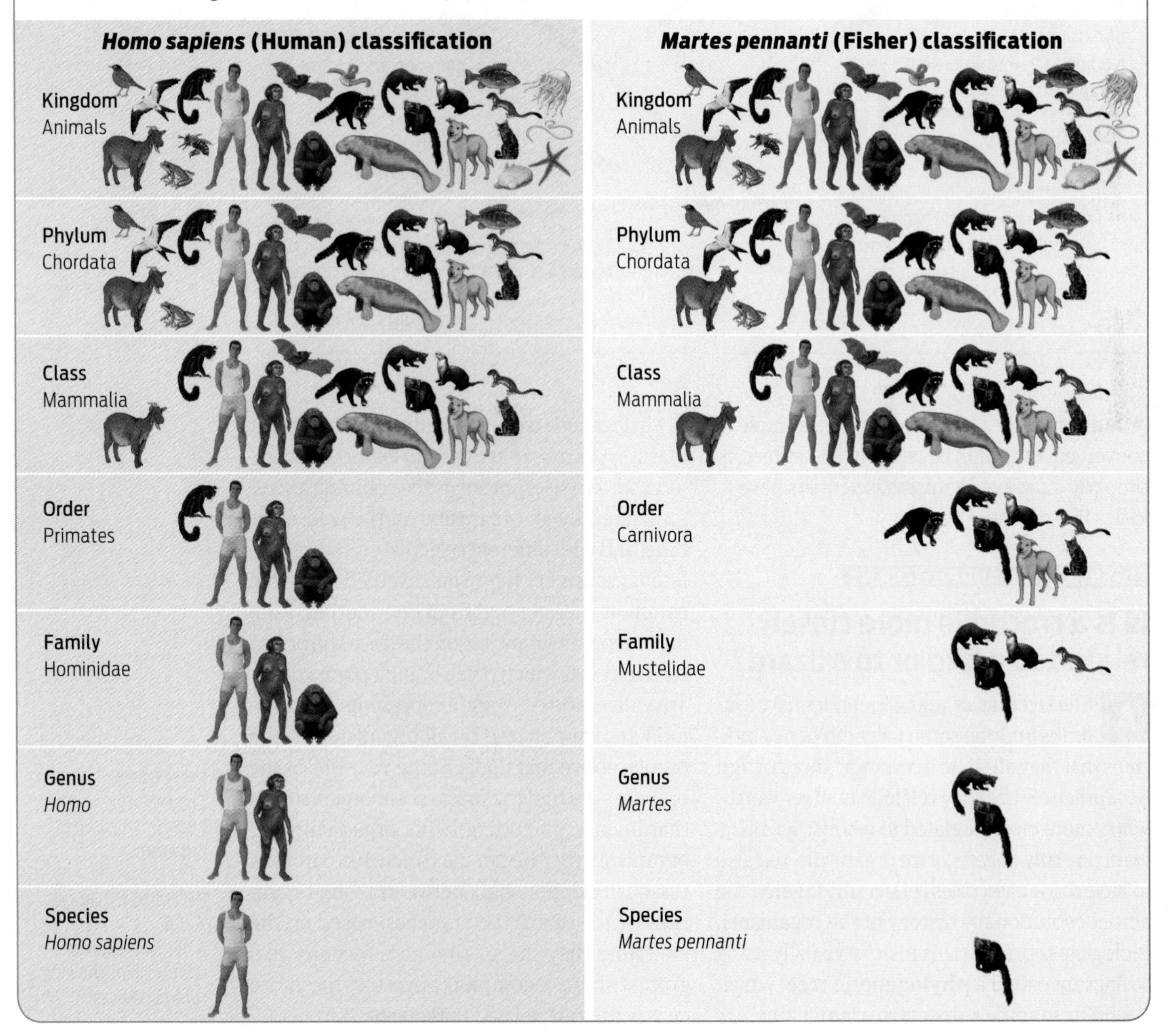

INFOGRAPHIC 17.8

How to Read an Evolutionary Tree

Evolutionary history, or phylogeny, is represented visually by an evolutionary tree. Trees have a common structure, with a root, nodes, and branch points. To determine evolutionary relationships among living or extinct organisms, consider the most recent common ancestors.

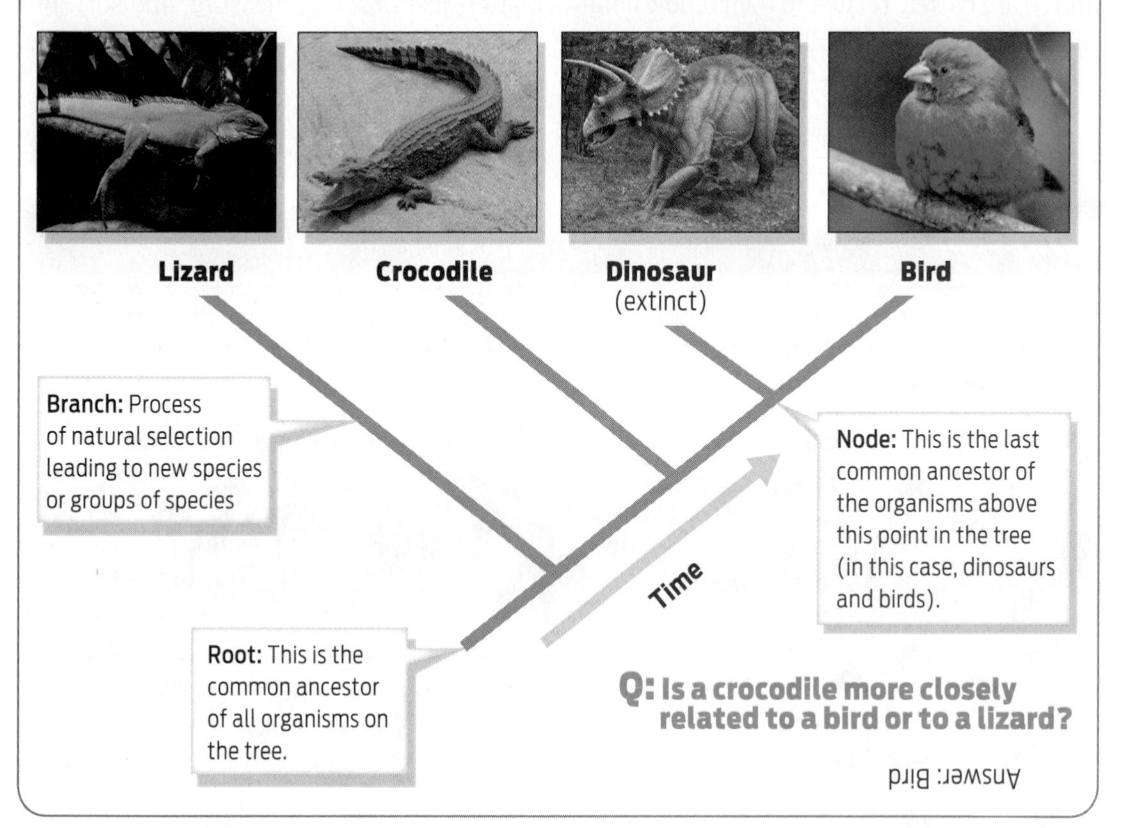

Lizard
Crocodile
Dinosaur (extinct)
Bird
Branch: Process of natural selection leading to new species or groups of species
Node: This is the last common ancestor of the organisms above this point in the tree (in this case, dinosaurs and birds).
Root: This is the common ancestor of all organisms on the tree.
Time
Q: Is a crocodile more closely related to a bird or to a lizard?
Answer: Bird

in can sometimes be tricky, as the example of convergent evolution has shown. Sometimes, to properly classify organisms, scientists have to look a little deeper.

CLASSIFICATION AND PHYLOGENY

Q Is a crocodile more closely related to a bird or to a lizard?

A The fact that all land vertebrates have four limbs and the same forelimb bones indicates that they all share a common ancestor. But how precisely are they related? In other words, who's more closely related to whom? Scientists want not only to categorize organisms, but also to have those categories reflect **phylogeny,** the actual evolutionary history of the organisms. Biologists represent this history visually using a diagram called a **phylogenetic tree,** which is similar in some respects to a family tree.

Phylogenetic trees can be drawn in a number of ways, but most have certain features in common. At the base, or root, is the common ancestor shared by all organisms on the tree. Over time, and with different selective pressures, different groups of organisms diverged from that common ancestor and from one another, leading to separate branches on the tree. The points on the tree at which these branch points occur are called nodes. A node represents the common ancestor shared by all organisms on the branch above that node. At the very tips of the branches we find the most recent organisms in that lineage, including living organisms and organisms that became extinct. We can thus establish relationships between living organisms (at the tips of the branches) based on the ancestors they share. The more recently two groups share a common ancestor, the more closely they are related **(Infographic 17.8)**.

PHYLOGENY
The evolutionary history of a group of organisms.

PHYLOGENETIC TREE
A branching tree of relationships showing common ancestry.

A phylogenetic tree is a visual representation of the best hypothesis we currently have for how species are related. The evidence for a phylogenetic tree comes from many sources, including the fossil record, physical traits, and shared DNA sequences. For many years, biologists relied solely on observable physical or behavioral features to construct evolutionary trees. But with the genetic revolution, it's become common to include DNA evidence. Typically, researchers compare sequence differences in a gene that is found in all living organisms, such as the ribosomal RNA (rRNA) genes.

Sometimes the new genetic information yields surprises. Modern genetic evidence shows, for example, that crocodiles are more closely related to birds than they are to lizards, appearances notwithstanding. Genetics, you might say, is shaking the evolutionary tree.

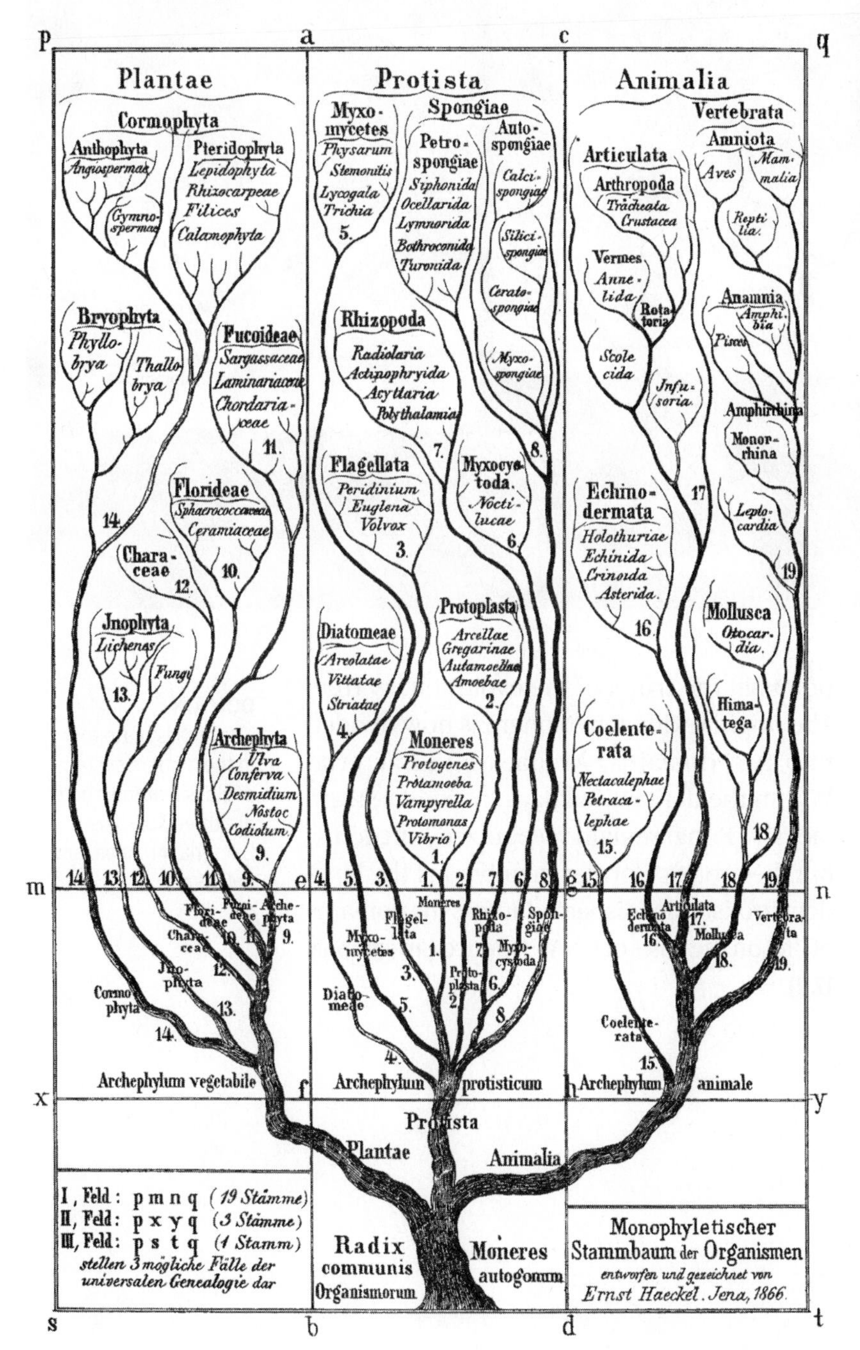

An early version of the tree of life drawn in 1866 by E. H. P. A. Haeckel

CLASSIFICATION AND PHYLOGENY

Q How many branches does the tree of life have?

A Since each living species sits on its own branch in a phylogenetic tree, the complete tree of life has as many branches as there are species in the world. Today's species are like thin twigs in the upper branches of an enormous tree. Closer to the bottom of the tree, nearer to the ancient trunk, however, we find significant forks. Just how many forks there are at the bottom of the tree is a question that has been debated for decades.

Before the 18th century, biologists divided living things into just two main categories: animals and plants. This classification was based on whether an organism moved around and ate or did not move around and eat. By the mid-19th century, use of the microscope had revealed a whole new world of microscopic organisms, and so a third branch was added to life's tree: protists.

By the 1960s, taxonomists realized that even three such branches did not fully capture the diversity of life; many organisms—such as fungi—didn't fit neatly into any of these groups, and so another classification scheme was proposed. This one grouped all living organisms into five large kingdoms on the basis of how they looked (both anatomically and microscopically) and how they obtained their food. The five kingdoms were Animalia, Plantae, Fungi, Protista, and Monera. Protista comprised mostly single-cell eukaryotic organisms (such as the amoeba), and Monera included all prokaryotic organisms (such as bacteria).

Yet even this revised classification scheme eventually had to be overhauled as more infor-

INFOGRAPHIC 17.9

DNA Defines Three Domains of Life: Bacteria, Archaea, Eukarya

All living organisms have evolved from a common ancestor. Based on DNA evidence, we can group living things into one of three domains of life, each with a distinct evolutionary history. While the Bacteria and Archaea both have prokaryotic cells, they have distinct evolutionary histories, with Archaea being more genetically related to Eukarya than Bacteria. The domain Eukarya encompasses protists, plants, fungi, and animals, including humans (see Chapter 19).

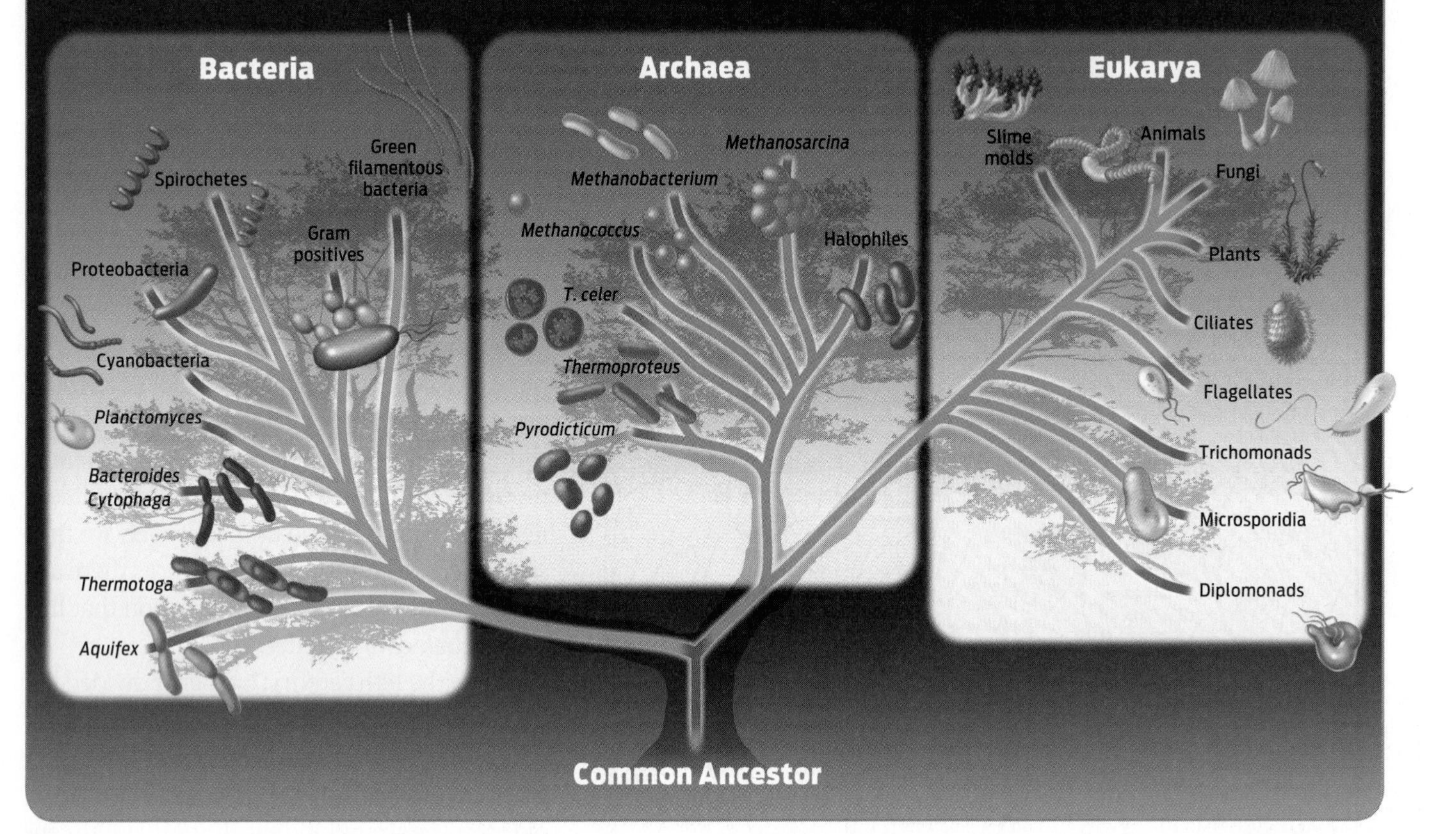

mation became available. In the 1970s, genetic studies by Carl Woese revealed that, on the basis of on genetic relatedness, not all prokaryotes could be lumped together; likewise, protists were too genetically diverse to be put in one category. Consequently, scientists now group organisms into one of three large **domains**–Bacteria, Archaea, and Eukarya–which represent three fundamental branch points in the trunk of the evolutionary tree. The original kingdom Monera is now divided into two domains, Archaea and Bacteria. Within the domain Eukarya, Animalia, Plantae, and Fungi remain recognized kingdoms, but the protists (former members of the kingdom Protista) are dispersed across the domains of life, on the basis of DNA evidence **(Infographic 17.9)**. ■

DOMAIN
The highest category in the modern system of classification; there are three domains—Bacteria, Archaea, and Eukarya.

Summary

- The age of the earth and its rock layers can be determined by measuring the amount of radioactive isotopes present in certain types of rocks, a method known as radiometric dating.
- Life on earth may have emerged in stages, as inorganic molecules combined to form organic ones in the primordial soup, and as these were incorporated into lipid bubbles to form cells.
- Using geological evidence and the fossil record, paleontologists have been able to construct a geologic timeline of life on earth.
- Earth's history can be divided into important eras and periods. Dinosaurs, for example, lived primarily from 250 to 65 million years ago, during the Mesozoic era, from the Triassic through the Cretaceous periods.
- The history of life on earth is marked by repeated extinctions and adaptive radiations, a phenomenon of intermittent rather than steady change known as punctuated equilibrium.
- Ancient movement of earth's major landmasses affected the eventual distribution of species around the globe, the study of which is known as biogeography.
- Convergent evolution is the evolution of similar adaptations in response to similar environmental challenges in groups of organisms that are not closely related.
- Life is astoundingly diverse. Current estimates of the total number of species on earth range anywhere from 5 to 30 million, of which 1.8 million have been formally described.
- Biologists sort organisms into a series of nested categories based on shared anatomical and genetic features: domain, kingdom, phylum, class, order, family, genus, species.
- The scientific name of an organism is given by its genus and species names (for humans it is *Homo sapiens*).
- Both physical evidence and genetic evidence are used to understand evolutionary history, or phylogeny. Branching trees of common ancestry are used to represent that history visually.
- On the basis of DNA evidence, all living organisms can be classified into one of three domains: Bacteria, Archaea, or Eukarya.

Chapter 17 Test Your Knowledge

HISTORY OF LIFE

Life on earth has changed dramatically since it first emerged. We can learn a great deal about the history of life on earth by studying rock layers and the fossils found within them.

HINT See Infographics 17.1–17.3.

KNOW IT

1. What do uranium-238, carbon-14, and potassium-40 have in common?

2. To date what you suspect to be the very earliest life on earth, which isotope would you use: uranium-238, carbon-14, or potassium-40? Explain your answer.

3. Place the following evolutionary milestones in order from earliest to most recent (numbering them from 1 to 7), providing approximate dates to support your answer.

the first multicellular eukaryotes ____________
the first prokaryotes ________________
the Permian extinction ________________
the Cambrian explosion ________________
the first animals ________________
the extinction of dinosaurs ________________
an increase in oxygen in the atmosphere ________________

USE IT

4. Consider a rock formed at about the same time as the earth was formed.

a. How old is this rock?
b. How much of the original uranium-238 is likely to be left today in that rock?

5. If an igneous rock contains 75% lead, how old is that rock? (Look at Infographic 17.2.)

6. Diverse animal fossils are found dating from the Cambrian period, not earlier. Why might these organisms have made their first appearance in the fossil record only then, even though their ancestors may have been living, and evolving, for a long time before the Cambrian? (Think about what kinds of new structures might have evolved during the Cambrian period that would have allowed these organisms to leave fossils.)

7. Along the banks of a river, some sedimentary rock strata have been revealed by erosion. If the sedimentary layers had been deposited in the Carboniferous period, would you expect paleontologists to find fossils of amphibians in these strata? Reptiles? Sharks? Justify your answers.

BIOGEOGRAPHY AND PLATE TECTONICS

The current distribution of organisms on earth reflects both evolutionary history and accidents of geology.

HINT See Infographics 17.4 and 17.5.

KNOW IT

8. If two organisms strongly resemble each other in terms of their physical traits, can you necessarily conclude that they are closely related? Explain your answer.

9. What did the arrangement of landmasses on earth look like between 135 and 65 million years ago? What happened to these landmasses, and how does this change help explain the distribution of organisms found on the planet?

USE IT

10. A cactus called ocotillo (*Fouquieria splendens*), which grows in New Mexico, looks very much like *Alluaudia procera*, a species of plant that grows in the deserts of Madagascar. These two plant species are not closely related—why then do they look so alike?

11. If penguins and polar bears had evolved before Pangaea split into northern and southern continents, what might you predict about their geographic distribution today?

12. Both bats and insects fly, but bat wings have bones and insect wings do not. Would you consider bat and insect wings to be a result of convergent evolution, or of homology—evolution based on inheritance of similar structures from a common ancestor? Explain your answer.

CLASSIFICATION AND PHYLOGENY

Categorizing the many species on earth and understanding how they are related is a challenging task, made easier by genetic information.

HINT See Infographics 17.6–17.9.

KNOW IT

13. Which of the following is *not* a domain of life?

a. Animalia
b. Eukarya
c. Bacteria
d. Archaea
e. Plantae
f. Neither a nor e is a domain of life.

14. Put the following terms in order, from most inclusive (1) to least inclusive (5).

domain_____
species_____
kingdom_____
genus_____
phylum_____

15. A phylogenetic tree represents

a. a grouping of organisms on the basis of their shared structural features.
b. a grouping of organisms on the basis of their cell type.
c. a grouping of organisms on the basis of their complexity.
d. a grouping of organisms on the basis of their evolutionary history.
e. a grouping of organisms on the basis of where they are found.

USE IT

16. Why was the classification of the kingdom Monera split into two domains? What are these two domains?

17. Which number on the tree below shows the most recent common ancestor of humans and corn?

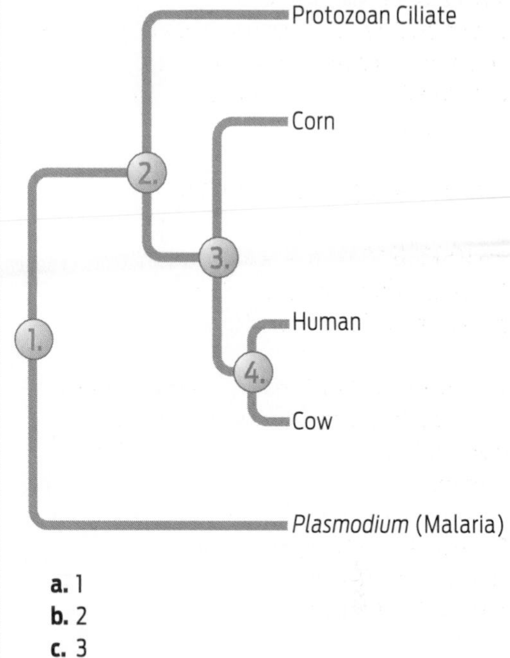

a. 1
b. 2
c. 3
d. 4
e. Humans and corn do not share any ancestors.

SCIENCE AND ETHICS

18. How might knowledge of the evolutionary history of organisms affect human health? How might such knowledge affect decisions regarding the environment?

Lost City

What You Will Be Learning

➔ USE IT

15. What is the significance of methane and other hydrocarbons at Lost City? (Think of both the origin of life and the sustenance of early life.)

16. If methane were not produced abiotically at Lost City, what would be the implications for early life?

17. Would you expect to find photosynthetic organisms at Lost City? Explain your answer.

SCIENCE AND ETHICS

18. Do you think that the scientists studying Lost City should be concerned about introducing microbial contaminants from their submersibles onto the towers of Lost City? How probable is this, given the conditions at Lost City and on the surface? If such an event could happen, what would be the implications?

19. Do you think that, instead of spending time studying microbes from extreme and remote environments, scientists should be studying microbes that are more apparently relevant to humans, such as ones that cause disease? In what ways might understanding the organisms at Lost City might be useful to humans?

Rain Forest Riches

What You Will Be Learning

area covered by forests shrank by 23 million acres a year during the 1990s–most of it cleared for agriculture. Some experts estimate that only half the acreage of the planet's original rain forest remains.

Forests are only one place biodiversity is in danger. Habitat destruction in ecosystems around the globe–wetlands, ice caps, coral reefs–poses a grave threat to countless species. If current rates of habitat destruction continue, we may witness levels of extinction rivaling the greatest extinction events of geological history. Can anything be done to reverse the trend of dwindling biodiversity around the globe?

Though the rapidly expanding human population (Chapter 24) is gobbling up resources faster than the earth can restore them, there are things we can do to mitigate the destruction. One conservation strategy is to protect those areas that are known to be especially diverse, ensuring that they remain so. That means safeguarding habitat and forbidding overhunting. Where possible, it also means taking efforts to restore missing diversity, in an effort to keep ecosystems whole. "We have an obligation to try to keep an ecosystem intact if we can," says Lewis, from Washington Department of Fish and Wildlife, noting that the best way to protect the environment may be to keep all of its parts in place.

"We have an obligation to try to keep an ecosystem intact if we can." –Jeffrey Lewis

For the fishers, at least, things seem to be looking up. They are dispersing and reproducing in the forest–at least seven females so far have had kits–and park manager Happe says she is "guardedly optimistic" about their chance of survival. Over the next few years, Happe and Lewis will continue to monitor the fishers, which have been equipped with radio collars, to make sure they are adapting to and surviving in their new home. Only then will they be able to label the restoration project a success. Fishers may have returned to the forest, says Happe, but "they're not out of the woods." ■

Summary

- Rain forests are sites of great biological diversity, as measured by the number and variety of different species present.
- The domain Eukarya encompasses all eukaryotic organisms—plants, animals, fungi, and the many types of protists.
- Plants are multicellular eukaryotes that carry out photosynthesis. All plants have cells with cell walls, but not all have a vascular system, not all produce seeds, and not all produce flowers.
- Plants can be subdivided into groups, including the bryophytes, ferns, gymnosperms, and angiosperms, on the basis of their terrestrial adaptations.
- Animals are multicellular eukaryotic heterotrophs that obtain nutrients by ingestion.
- Most animals are invertebrates (that is, they lack a backbone). The most abundant invertebrates by far are arthropods, and especially insects.
- Vertebrates (animals with a backbone) are members of the phylum Chordata. Common vertebrates include mammals such as the fisher, as well as amphibians, reptiles, birds, and fish.
- Fungi are decomposers, acquiring their nutrition and energy by breaking down dead organic matter and absorbing the results. There are unicellular and multicellular fungi.
- Protists are a diverse group of mostly unicellular eukaryotic organisms that do not cluster on a single branch of the evolutionary tree. They include photosynthetic plantlike algae and animal-like parasites.
- All eukaryotes are descendants of a unicellular protist that first emerged some 2 billion years ago as the result of endosymbiosis.

DOMAIN EUKARYA

The domain Eukarya encompasses all eukaryotic organisms, including plants, animals, fungi, and protists. Eukaryotic organisms display great diversity in their many evolutionary adaptations.

HINT See Infographics 19.1 and 19.3.

KNOW IT

1. How does the physical landscape diversity of Olympic National Park affect biodiversity in the park?

2. What are the defining features of the domain Eukarya?

3. What do a fisher and a fir tree have in common?

USE IT

4. How do you think the diversity of eukaryotic organisms in each of the following areas would compare to the diversity in Olympic National Park—would there be more or less? Explain the reasons for your answers.

a. Lake Michigan
b. the Sonoran Desert in Arizona
c. the prairies of Kansas

5. If a fungicide were applied throughout Olympic National Park, how might it affect eukaryotes in the park? Explain your answer.

PLANT DIVERSITY

Plants are photosynthetic, multicellular eukaryotes. Plants have evolved many different structures and adaptations for living and reproducing on land.

HINT See Infographic 19.6.

KNOW IT

6. Which group of plants was the first to live on land? Why do we find these plants only in particular environments (after all, if they were first, shouldn't they have spread everywhere by now)?

7. A major difference between a fern and a moss is

a. the presence of seeds.
b. the presence of flowers.
c. the presence of cones.
d. the presence of a vascular system.
e. the ability to carry out photosynthesis.

USE IT

8. What is an advantage of having seeds? (Think about spreading to new locations and whether or not reproduction relies on water.)

9. What type of seed plant is likely to rely on hungry animals to spread its seeds? Explain your answer.

10. How did the evolution of vascular systems in plants change the landscape?

ANIMAL DIVERSITY

Animals are multicellular eukaryotes that obtain nutrients through ingestion. Animals exhibit a wide variety of body shapes and structures.

HINT See Infographic 19.5.

KNOW IT

11. A sand dollar gets its name from its body shape—it resembles a large coin. What type of body symmetry does a sand dollar have?

a. bilateral
b. radial
c. none (sand dollars are amorphous)
d. hyphae
e. mycelium

12. What do a backbone and an exoskeleton have in common?

a. They are found in closely related groups of animals.
b. They are made of the same substance.
c. They both help provide support to an animal's body.
d. They both require an animal to molt in order to be able to grow.
e. all of the above

13. You and a fisher are both mammals; as such, what are some characteristics you and the fisher have in common?

14. Which of the following statement(s) is/are true about both cockroaches and lobsters?

a. They are invertebrate insects with bilateral symmetry.
b. They are mollusks with an exoskeleton.
c. They are arthropods with segmented bodies and no symmetry.
d. They are arthropods with an exoskeleton.
e. They are mollusks with a segmented body.

USE IT

15. Many characteristics are used to classify animals. Why do we need to use so many different characteristics? Consider the following five animals: woodpecker, human, wasp, ant, and fisher; and the following three characteristics: ability to fly, two-legged, bearing feathers

a. Which of the five animals could be grouped by each characteristic?
b. Would this grouping reflect their real taxonomic relationship?

c. What feature(s) would you use to put wasps and ants together in their own group? What about human and fisher?

16. Judging from their numbers, arthropods are a tremendously successful group. What traits do you think have enabled them to be so successful? Justify your answer with examples.

FUNGAL DIVERSITY

Fungi are unicellular and multicellular eukaryotes that obtain nutrients by secreting digestive enzymes onto organic matter and absorbing the digested product.

HINT See Infographic 19.7

KNOW IT

17. Consider the "eating habits" of fungi.
- **a.** Can fungi carry out photosynthesis?
- **b.** Can fungi ingest their food?
- **c.** How do fungi obtain their nutrients and energy?

18. Which of the following meals include fungi as food?
- **a.** a bread and blue cheese platter with fruit
- **b.** mushroom risotto
- **c.** a and b
- **d.** a fruit salad
- **e.** yogurt

USE IT

19. A very early classification scheme placed the fungi together with the plants. Why do you think fungi were grouped with plants? What features distinguish them from plants?

A PLETHORA OF PROTISTS

Protists are a diverse group of primarily unicellular eukaryotic organisms that are considered together only because they do not sort neatly into any other single evolutionary category.

HINT See Infographics 19.8 and 19.9.

KNOW IT

20. What do members of the informal group known as protists have in common?
- **a.** nothing
- **b.** They are all eukaryotic.
- **c.** They all carry out photosynthesis.
- **d.** They are all human parasites.
- **e.** They are all decomposers.

21. Endosymbiosis describes the process by which:
- **a.** protists diverged from plants.
- **b.** eukaryotic cells acquired certain organelles.
- **c.** protists became multicellular.
- **d.** eukaryotes diverged from prokaryotes.
- **e.** a and c
- **f.** b and d

USE IT

22. Why do scientists no longer consider protists a separate kingdom? How might scientists find new taxonomic "homes" for the protists? Do you think structural features (for example, chloroplasts) or genetic information will be more useful in their classification?

23. Many protists have an organelle called the contractile vacuole that pumps out water that enters the cell by osmosis. Why is this a useful adaptation for a protist? What might happen to a protist if its contractile vacuole stopped working? (Think about where many protists live, and what happens to bacteria whose cell walls are disrupted by antibiotics.)

SCIENCE AND ETHICS

24. Reintroducing species to their native habitats is sometimes controversial. One reintroduction effort in particular that has caused quite a stir is the reintroduction of the Mexican gray wolf (*Canis lupus baileyi*) into New Mexico and Arizona. You can read about this project at http://www.fws.gov/southwest/es/mexicanwolf/.
- **a.** Why might it be important to reintroduce species into their native habitats? Answer first in general terms, then specifically for the Mexican gray wolf.
- **b.** What factors could impede the success of such reintroductions? Again, answer in general terms first, then specifically for the Mexican gray wolf.

25. Many species reintroductions are being carried out across the United States. Do some research to learn about at least one such effort. For the species you research, address the following questions:
- **a.** What caused it to be lost from its native habitat?
- **b.** Is its reintroduction important?
- **c.** Are there are controversies about its reintroduction?
- **d.** What made you interested in this particular species and its reintroduction? Is it an "attractive" species? Is it being reintroduced near where you live?

Chapter 20 Human Evolution

What Is Race?

What You Will Be Learning

20.1 How Do We Define Race?

20.2 Melanin Influences Skin Color

20.3 Folate and Vitamin D Are Necessary for Reproductive Health

20.4 Human Skin Color Correlates with UV Light Intensity

20.5 Modern Human Populations Are Ancestors of “Eve”

20.6 Mitochondrial DNA Is Inherited from Mothers

20.7 Out of Africa: Human Migration

20.8 Traits of Modern Humans Reflect Evolutionary History

20.9 Natural Selection Influences Human Evolution

20.10 The Evolution of Skin Color

What Is Race?

Science redefines the meaning of racial categories

When Barack Obama was elected in 2008, he was hailed as America's first black president. When Tiger Woods won the Masters Golf Tournament in 1997, he was lauded as the first black man to win. When Halle Berry won an Oscar in 2001 for best actress, she was commended as the first black woman to win in that category.

Why was skin color so significant? A 250-year history of slavery and racial discrimination in the United States has left a bitter legacy. Almost 150 years after slavery was legally abolished in the United States, people of color are still underrepresented in positions of power and prestige. Although the reasons for this underrepresentation are complex, the recognition of the achievements of Obama, Woods, and Berry was important because it signaled a major change: barriers to social advancement were beginning to come down.

To shoehorn any of these three people into a simple racial category, however, is misleading: Barack Obama was born to a white mother and an African father; Tiger Woods's background includes African, Chinese, Dutch, and Thai forebears; Halle Berry was born to a white mother and a black father. So what does the term "black" mean?

Historically, racial categories were employed primarily by one group to maintain power over another and to justify forms of oppression, including slavery. In the United States, racial categories were reinforced by laws such as the "one drop" rule adopted by several states in the 1920s, which held that any American with one drop of African blood was to be considered black. People then continued to use these categories and their connotations to justify racial discrimination and, in some places, racial segregation.

Though social and political attitudes have changed, people continue to invoke racial categories like "black" or "white" for various reasons, including simple physical description. Regardless of the reason, it is increasingly clear that from a biological perspective racial categories are meaningless. Research on the evolution of humans increasingly shows that race is a

social, not a biological, category. Groups of people can and do share similar physical characteristics, such as skin color and other features, but all humans are members of a single biological species, *Homo sapiens*. The only thing that skin color might accurately identify is the geographical location where a person's ancestors lived (**Infographic 20.1**).

Humans are a recent species, first walking the earth a mere 200,000 years ago.

Why is there so much variation in human skin tone? And how did the geographical variation come about? The answers lie in the evolution and migration of our earliest ancestors. Humans are a recent species, first walking the earth a mere 200,000 years ago. The physical differences we see among people today have all emerged in the very recent past. And while the physical differences between, say, an African from Senegal and a European from Sweden may appear large, biologically speaking such differences are actually quite small. In fact, genetic studies comparing regions of the human genome from person to person show that each person's DNA is 99.9% identical to any other unrelated person. Nevertheless, this 0.1% difference holds clues that help explain how our varying skin tones and other physical traits evolved.

INFOGRAPHIC 20.3

Folate and Vitamin D Are Necessary for Reproductive Health

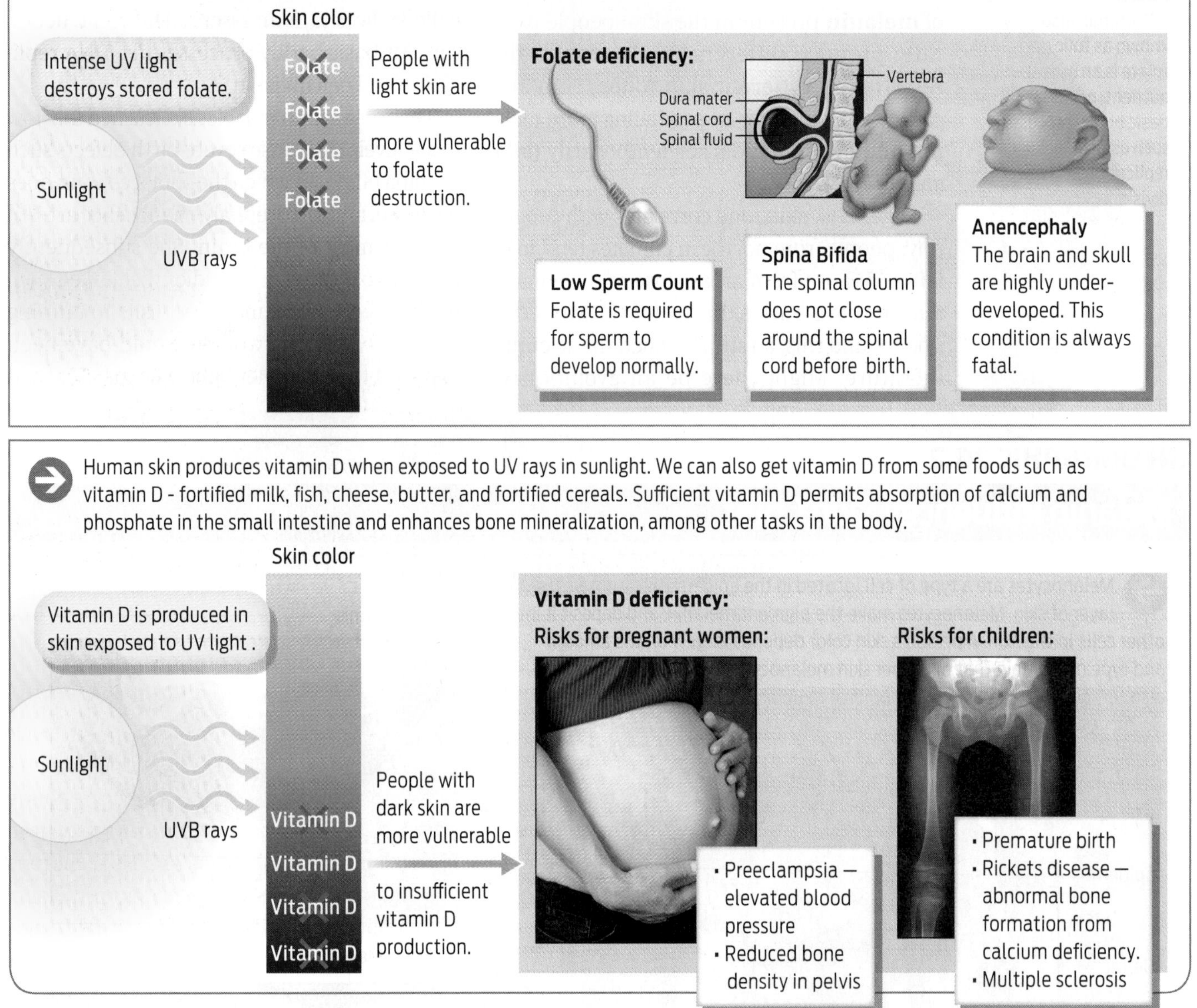

Folate, also known as vitamin B_9, is abundant in beans, citrus fruit, dark green leafy vegetables, whole grains, poultry, pork, shellfish, and liver. Folate is especially critical during periods of rapid cell division, such as during embryonic and fetal development. UV light destroys the body's folate stores.
Skin color
Intense UV light destroys stored folate.
Folate
Folate
Folate
Folate
Sunlight
UVB rays
People with light skin are more vulnerable to folate destruction.
Folate deficiency:
Vertebra
Dura mater
Spinal cord
Spinal fluid
Low Sperm Count
Folate is required for sperm to develop normally.
Spina Bifida
The spinal column does not close around the spinal cord before birth.
Anencephaly
The brain and skull are highly under-developed. This condition is always fatal.
Human skin produces vitamin D when exposed to UV rays in sunlight. We can also get vitamin D from some foods such as vitamin D - fortified milk, fish, cheese, butter, and fortified cereals. Sufficient vitamin D permits absorption of calcium and phosphate in the small intestine and enhances bone mineralization, among other tasks in the body.
Skin color
Vitamin D is produced in skin exposed to UV light.
Sunlight
UVB rays
Vitamin D
Vitamin D
Vitamin D
Vitamin D
People with dark skin are more vulnerable to insufficient vitamin D production.
Vitamin D deficiency:
Risks for pregnant women:
Risks for children:
• Preeclampsia – elevated blood pressure
• Reduced bone density in pelvis
• Premature birth
• Rickets disease – abnormal bone formation from calcium deficiency.
• Multiple sclerosis

that folate is necessary for sperm to develop normally.

Taken together, the results of Jablonski's literature search suggested that people with light skin are more vulnerable to folate destruction than are darker-skinned people–presumably because melanin absorbs and dissipates damaging UV light as heat. Could the need to protect the body's folate stores from UV light have driven the evolution of darker skin shades? The supporting evidence was compelling. But then what was the advantage of having light skin at all, as many populations today do? And why are there geographical differences?

Jablonski's work built on a hypothesis first proposed in the 1960s by biochemist W. Farns-

VITAMIN D
A fat-soluble vitamin required to maintain a healthy immune system and to build healthy bones and teeth. The human body produces vitamin D when skin is exposed to UV light.

worth Loomis, who suggested that vitamin D might play a role in the evolution of skin color. Unlike folate, which is destroyed by excess sunlight, the production of vitamin D *requires* ultraviolet light. **Vitamin D** is crucial for good health: it helps the body absorb calcium and deposit it in bones. During pregnancy women need extra vitamin D to nourish the growing embryo. In addition, since vitamin D is so important for healthy bone growth, too little might also cause bone distortion, and a distorted pelvis would make it difficult for a woman to bear children **(Infographic 20.3)**.

In 2000, Jablonski and Chaplin published a study in the *Journal of Human Evolution* that compared data on skin color in indigenous populations from more than 50 countries to levels of global ultraviolet light. They found a clear correlation: the weaker the ultraviolet light, the fairer the skin, a compelling suggestion that both dark and light skin are linked to levels of global sunlight. They now had a complete hypothesis: light skin evolved because in sun-poor parts of the world it helped the body produce vitamin D, while dark skin evolved because it helped protect the body's folate stores in people who lived in sunny climates. The body's need to balance levels of these two important nutrients explains why there is so much variation in skin tone around the globe **(Infographic 20.4)**.

Since the publication of Jablonski and Chaplin's work, many other scientists have tested their hypothesis, and it is now the most widely

The body's need to balance levels of vitamin D and folate explains why there is so much variation in skin tone around the globe.

INFOGRAPHIC 20.4

Human Skin Color Correlates with UV Light Intensity

Nina Jablonski and George Chaplin used NASA satellite measurements of UVB intensity to predict the amount of skin pigment that would best block harmful UV rays yet still enable the body to produce sufficient vitamin D in populations around the globe. Their predictions closely match actual skin color variations around the world.

Predicted pigmentation: skin-color prediction based on UVB intensity

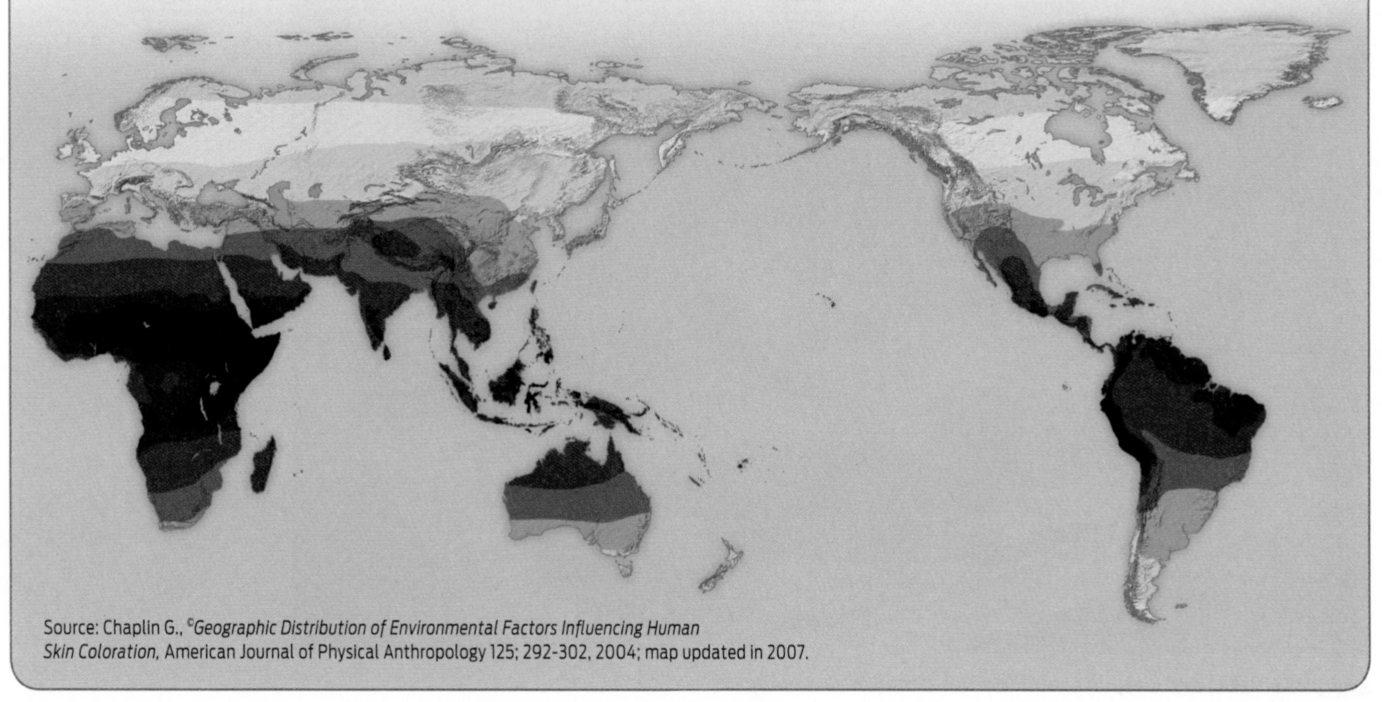

Source: Chaplin G., "*Geographic Distribution of Environmental Factors Influencing Human Skin Coloration*, American Journal of Physical Anthropology 125; 292-302, 2004; map updated in 2007.

accepted explanation for the evolution of human skin color. As Jablonski points out, "It synthesizes the available information on the biology of skin from anatomy, physiology, genetics, and epidemiology, and has not been contradicted by any subsequent data."

Out of Africa

If light and dark skin tones developed over time, then at some point in history, all humans likely had the same skin tone. This scenario, accepted by most scientists, is supported by genetic studies suggesting that anatomically modern humans first evolved in Africa.

In 1987, a team led by Allan Wilson of the University of California at Berkeley used **mitochondrial DNA (mtDNA)**–genetic material we inherit solely from our mothers–to construct an evolutionary tree of humanity. Wilson and his two colleagues Rebecca Cann and Mark Stoneking determined that all humans can trace their ancestry back to a single woman who lived in eastern Africa some 200,000 to 150,000 years ago, a woman he called the mitochondrial Eve.

In other words, if every person on the planet were to construct a family tree that listed every relative for thousands of generations back in time, they would all eventually converge at a single common female ancestor and a single common male ancestor; this female is the one Wilson dubbed Eve **(Infographic 20.5)**. Note that Eve wasn't the only female living at the time; she was merely one female in a population of many ancient humans. But her mitochondrial DNA is the only DNA that modern humans still carry today. In other words, other females living at the time also had descendants, but the lines of these descendants died off over time. Eve's descendants–and only her descendants–populate the earth today.

MITOCHONDRIAL DNA (mtDNA)
The DNA in mitochondria that is inherited solely from mothers.

Jablonski (right) and Chaplin examine a world map of skin color.

INFOGRAPHIC 20.5

Modern Human Populations Are Descendants of "Eve"

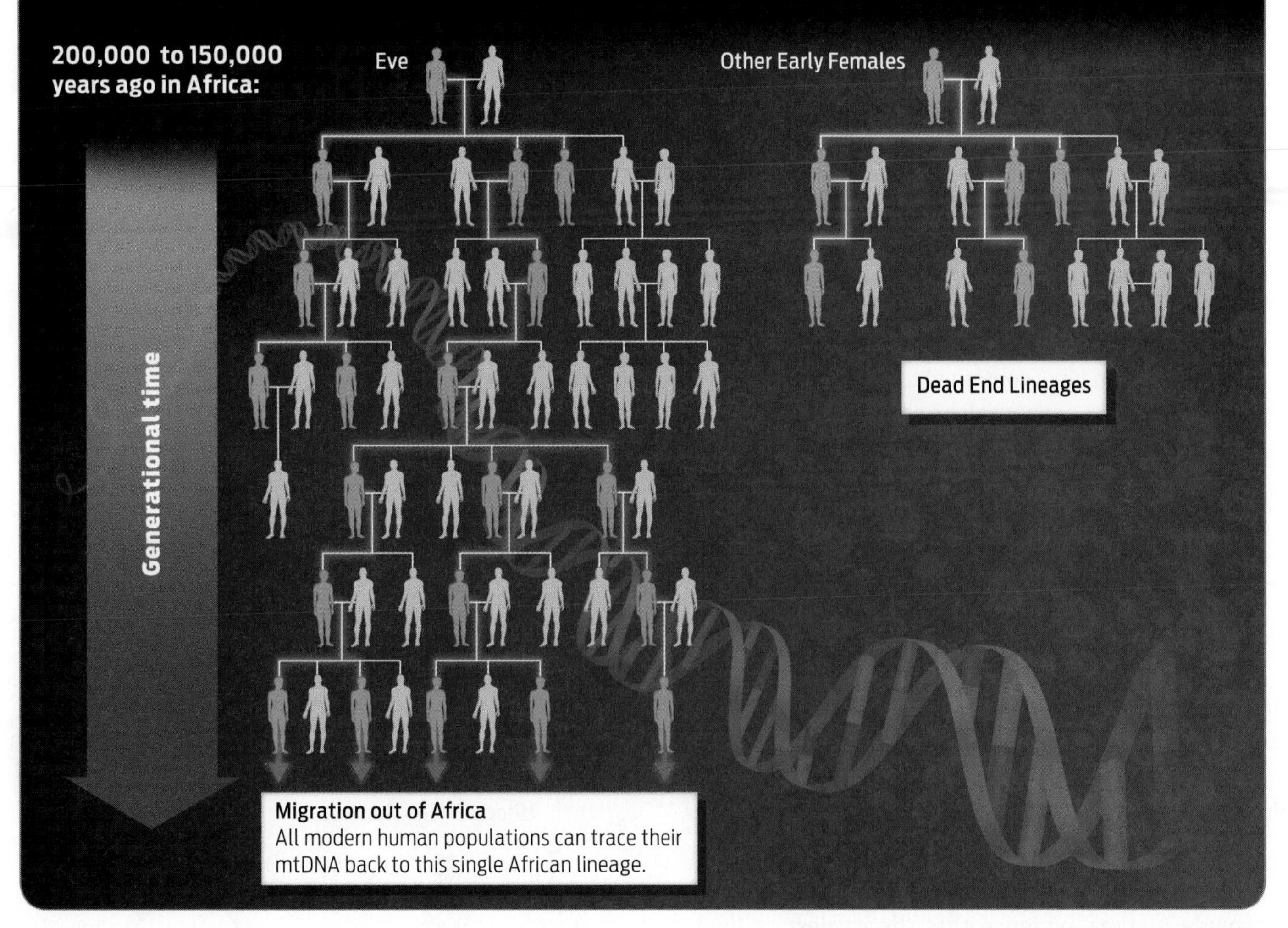

Many women of Eve's generation left descendants, but the mtDNA data suggest that Eve's descendants are the ones who went on to become the modern human populations across the globe we know today.
200,000 to 150,000 years ago in Africa:
Eve
Other Early Females
Generational time
Dead End Lineages
Migration out of Africa
All modern human populations can trace their mtDNA back to this single African lineage.

Mitochondrial DNA is DNA located in the mitochondria in all our cells. Unlike nuclear DNA, which is inherited from both parents in most multicellular organisms (including humans and other animals), and which undergoes recombination during meiosis, mtDNA passes from mothers to offspring essentially unchanged. That's because sperm do not contribute their mitochondria to the newly formed zygote **(Infographic 20.6)**.

All humans can trace their ancestry back to a single woman who lived in eastern Africa some 200,000 to 150,000 years ago.

Like nuclear DNA, mtDNA mutates at a fairly regular rate, although it appears to mutate faster than nuclear DNA. A mother with a mutation in her mtDNA will pass it to all her children, and her daughters will pass it to their children in turn. Because these mutations pass down without being combined and rearranged with paternal mitochondrial DNA, mtDNA is a powerful tool by which to track human ancestry back through hundreds of generations.

To conduct the Eve study, Wilson and his colleagues collected mtDNA from 147 contemporary individuals from Africa, Asia, Australia, Europe, and New Guinea. On the basis of the mtDNA sequence patterns, the researchers created an evolutionary tree. Branches of the tree from all five areas could be traced back to Eve. However, the tree had two major evolutionary branches: one that included the ancestors of populations now living in Asia, Australia, Europe, and New Guinea, and one that included the ancestors of modern-day Africans.

The mtDNA of people on the African branch had acquired twice as many mutations as the mtDNA of people on the rest of the tree. The most likely interpretation of these data, the scientists reasoned, was that the African mtDNA had had more time to accumulate mutations, and was consequently older, evolutionarily speaking. This would mean that humans likely originated in Africa, where they formed several ancestral populations. After some period of time, one group of Africans left the continent, and their descendants continued to migrate to other continents, eventually becoming the ancestors of modern-day Asians, Australians, and Europeans.

Since Wilson's study, additional evidence continues to back the "out of Africa" hypothesis. Fossil discoveries in Ethiopia in 2003 and 2005 represent the oldest known fossils of modern humans–160,000 and 195,000 years old, respectively–and plug a major gap in the human fossil record. Both sets of remains date precisely from the time when Wilson and his colleagues think that a genetic Eve lived in eastern Africa. The fossil discoveries provide evidence that anatomically modern humans were living in that region around the same time that Eve lived, and provide further evidence that the earliest humans originated in Africa.

This hypothesis is also supported by research that sampled genetic diversity from nuclear DNA. In a 2008 study, for example, Richard Myers, of the Stanford University School of Medicine, and colleagues found less and less genetic variation in people the farther away from Africa they lived–the same pattern of variation that scientists have found in human mtDNA sequences. This finding suggests that as each small group of people broke away to explore a new region, it took only a sample of the parent population's genes. Consequently, genetic diversity decreased in tandem with the distance people traveled away from Africa–a classic example of the founder effect described in Chapter 15 **(Infographic 20.7)**.

INFOGRAPHIC 20.6

Mitochondrial DNA Is Inherited from Mothers

When egg and sperm fuse during fertilization, sperm contribute only nuclear DNA to the nucleus of the newly formed zygote. The egg provides all other organelles, including mitochondria. Consequently, only mothers contribute mitochondrial DNA to their children.

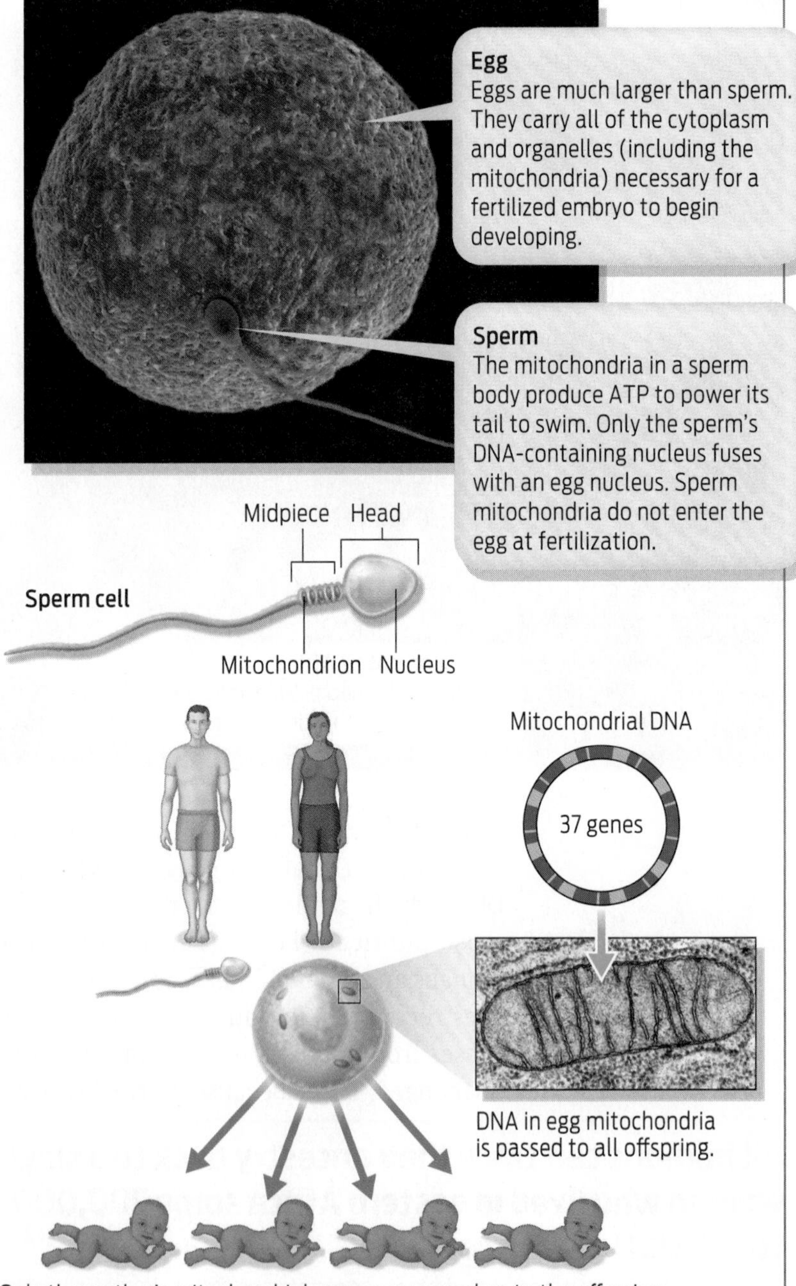

Egg
Eggs are much larger than sperm. They carry all of the cytoplasm and organelles (including the mitochondria) necessary for a fertilized embryo to begin developing.
Sperm
The mitochondria in a sperm body produce ATP to power its tail to swim. Only the sperm's DNA-containing nucleus fuses with an egg nucleus. Sperm mitochondria do not enter the egg at fertilization.
Midpiece
Head
Sperm cell
Mitochondrion
Nucleus
Mitochondrial DNA
37 genes
DNA in egg mitochondria is passed to all offspring.

Only the mother's mitochondrial genes are passed on to the offspring.

HOMINID
Any living or extinct member of the family Hominidae, the great apes—humans, orangutans, chimpanzees, and gorillas

Becoming Human

A number of lines of evidence peg Eve as the likely common ancestor of all humans living today. However, she represents merely one branch on the evolutionary tree that includes our species; this tree has several other branches representing other hominid species that came before her **(Infographic 20.8)**. A **hominid** is any member of the biological family Hominidae, which includes living and extinct humans and apes.

Humans and apes are grouped together because the fossil evidence shows that modern humans and present-day apes evolved from a common ancestor that lived 13 million years ago. Of the living primates, humans and chimpanzees are the most closely related, although it has been more than 6 million years since their shared ancestor lived. During those 6 million years, both humans and chimps have undergone a tremendous amount of evolutionary change, which is why living humans look and behave so differently from chimps—or any other primate species living today.

Scientists haven't yet discovered fossil remains of the last common ancestor between chimps and humans. However, in October 2009

INFOGRAPHIC 20.7

Out of Africa: Human Migration

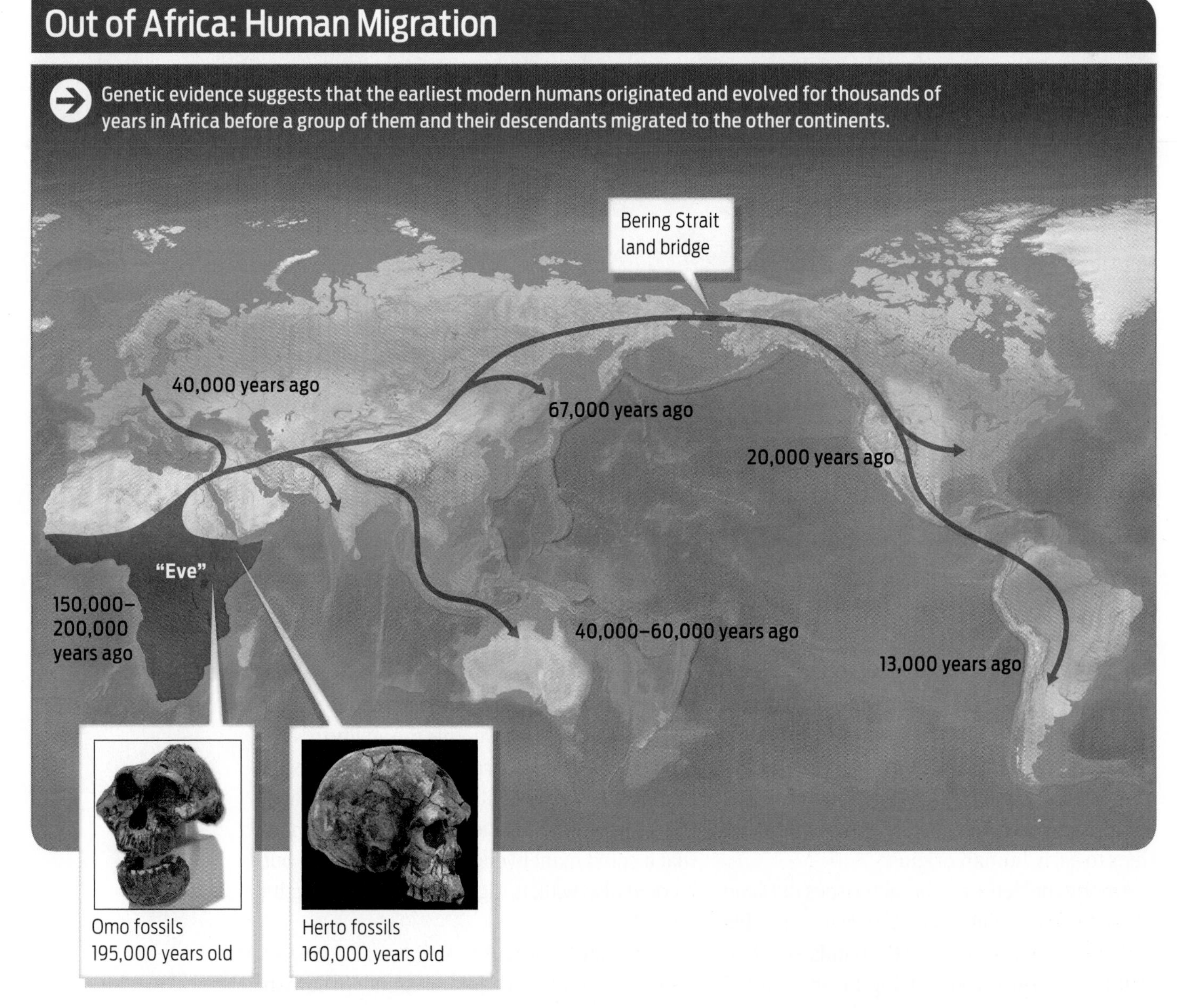

INFOGRAPHIC 20.8

Traits of Modern Humans Reflect Evolutionary History

Homo sapiens is the only surviving lineage in the evolutionary history of humans. In other words, several hominids have existed or coexisted as related but distinct species in the past. The physical traits that modern-day humans have, such as skin color and body hair, evolved in response to selective pressures. A species with less hair could better regulate body temperature in hot and sunny environments, for example, but would require darker skin to protect it from high UV light exposure.

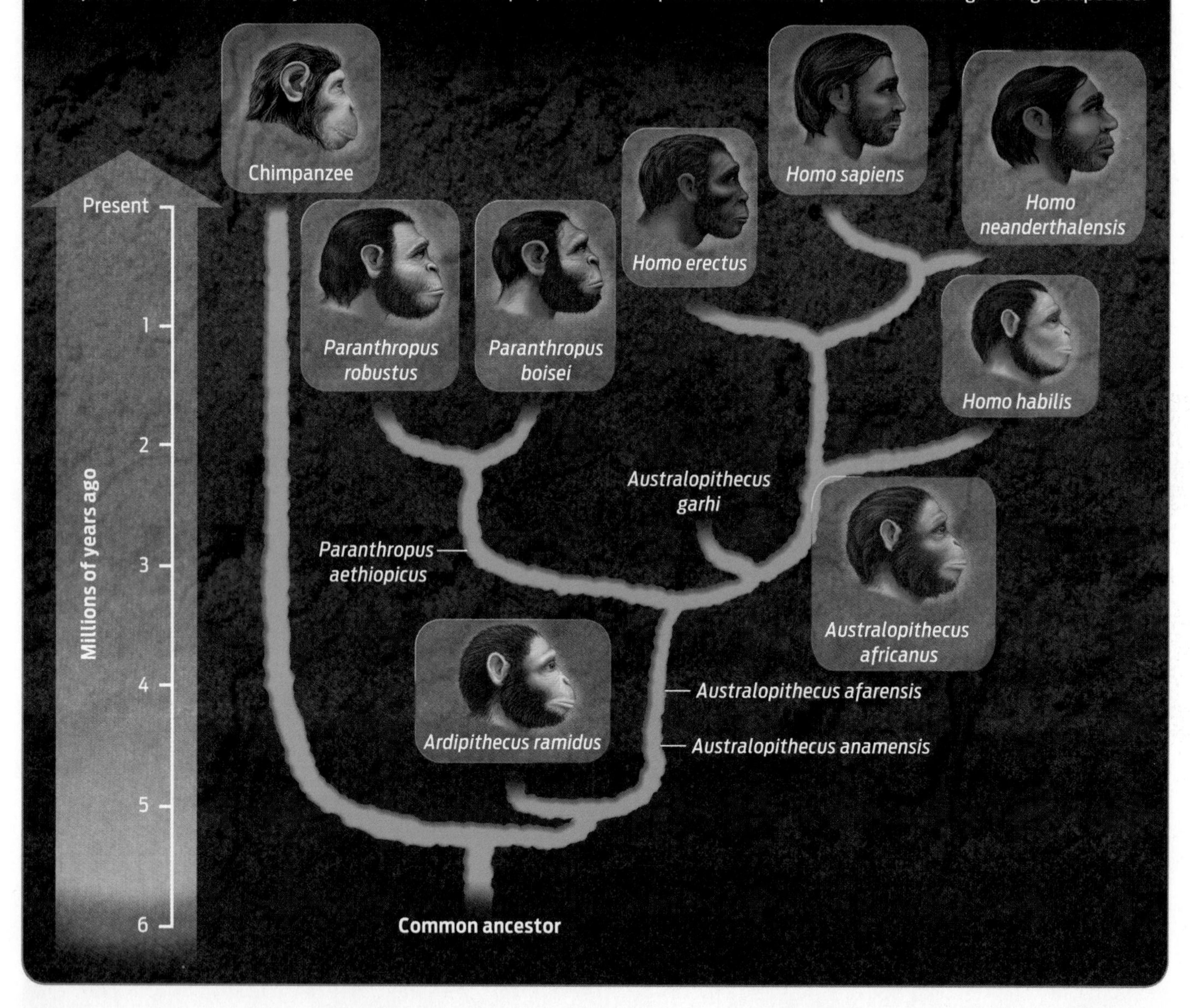

the first analyses of fossil remains of a 4.4-million-year-old hominid, *Ardipithecus ramidus*, nicknamed Ardi, were published. Ardi's remains are among the oldest hominid fossils so far discovered and, as such, give tantalizing clues to early human origins.

Among the defining characteristics of *Homo sapiens* are the ability to walk upright and a big brain. An upright gait meant the hands were free to make and use tools. A big brain enabled H. sapiens to develop complex language. Ardi helped scientists discover that the ability to walk upright evolved first. Ardi had a small brain, suggesting that it could not use complex language. By studying Ardi's bones, scientists also know that it could maneuver on all fours in trees, but it could also walk upright without dragging its knuckles.

The fossil record after Ardi has also helped show us some of the major milestones in human evolution. For example, artifacts found at various archeological sites indicate that simple tool use began approximately 2.6 million years ago, most likely when our hominid ancestors began eating meat from large animals. The first tool-users were members of the genus *Australopithecus*. This genus walked upright and appears to have lost the ability to live in trees, as evidenced by the lack of an opposable big toe, which had helped the early hominids grip branches.

Another milestone was the ability to use and control fire, which appeared about 800,000 years ago. Artifacts such as clay shards found at various fossil sites show that *Homo erectus* was likely the first species able to control fire. Fire use enabled *Homo erectus* to cook meat and bone marrow, to stay warm, and probably to fight off predators.

Finally, at some point between 800,000 and 200,000 years ago, hominid brain size began to expand rapidly. Geological studies show that this was also a time of rapid and dramatic climate change. Scientists hypothesize that a larger brain would have enabled better communication and problem-solving, which would have been very useful as our hominid ancestors had to cope with climate change. This was also around the time that anatomically modern humans like Eve and our own species, *Homo sapiens*, appeared.

Selection for Skin Color

That anatomically modern humans evolved in Africa suggests that the first humans likely had dark skin. But then how might varying skin tones have later developed? Nina Jablonski's research has revealed that environmental factors likely played a role in the evolution of different skin tones. Environment alone doesn't cause evolution, however. Rather, the environment acts on traits, or phenotypes, increasing or decreasing the frequency of alleles in a population by natural selection. Where did these alleles come from?

Recall that each time a cell replicates, mutations—errors in replication—can occur. If these mutations occur in germ cells during meiosis (Chapter 10), they will permanently change the genome of the next generation. This process continually introduces new alleles into the population. Some of these alleles can be negative or harmful, as in the case of hereditary cancer or cystic fibrosis. But new alleles can also be benign or even beneficial. Indeed, sometimes alleles can be so positive and confer such a survival advantage that they become more common in succeeding generations and can eventually become fixed in a population **(Infographic 20.9)**.

Sometimes alleles that are harmful in one environmental context may be beneficial in another. For example, the recessive allele responsible for cystic fibrosis (CF) can cause this serious disease when it occurs in homozygotes, who have two copies of the allele. However, research has suggested that being heterozygous for CF—that is, having only one CF allele—may have reduced the severity of diarrhea caused by cholera or some other infection. Consequently, carrying a CF allele provided an advantage during epidemics. This would help explain why the CF allele became relatively common.

Skin color is another example of a trait that likely conferred an advantage to humans and underwent natural selection at some point in human history. Otherwise, dark or light skin color wouldn't be so common among specific populations. In fact, the dark skin of those *Homo sapiens* who evolved in Africa was probably an early adaptation; it is likely that before evolving dark skin, our earliest ancestors had light skin, just as chimpanzees do today.

Fossil and genetic evidence suggests that about 2 million years ago hominids became "bipedal striders, long distance walkers and possibly even runners," according to Nina Jablonski. But to sustain such activities, they needed an effective cooling system, a feat they could have accomplished only by losing excessive body hair and gaining more sweat glands. In contrast, hairy chimpanzees, our closest living animal relatives, can sustain only short bouts of activity without getting overheated. "It's like sweating in a wool blanket," Jablonski explains. "After that blanket gets saturated, you can't lose very much heat."

At some point, some factor—food scarcity, perhaps—forced ancient hominids out of the forests

INFOGRAPHIC 20.9

Natural Selection Influences Human Evolution

The environment selects for specific genetically determined traits. Different environments will select for different traits, and therefore different alleles.

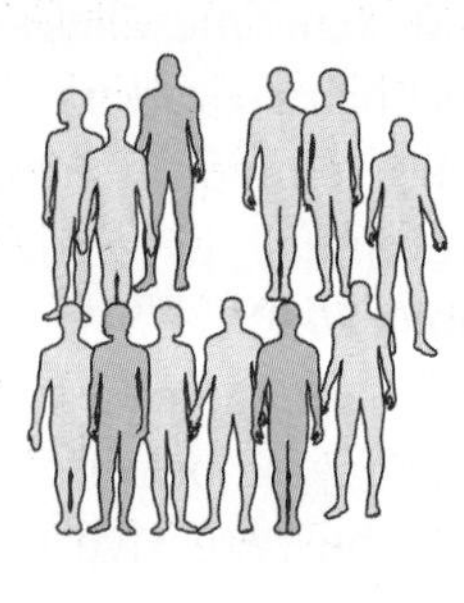

Positive selection

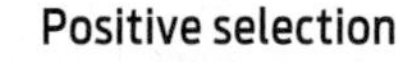

Green trait confers greater reproductive success in a particular environment.

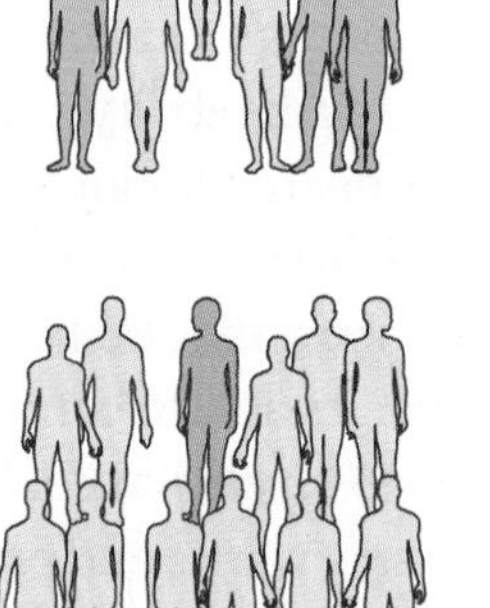

The frequency of the green allele increases in the next generation, so more people have the green trait.

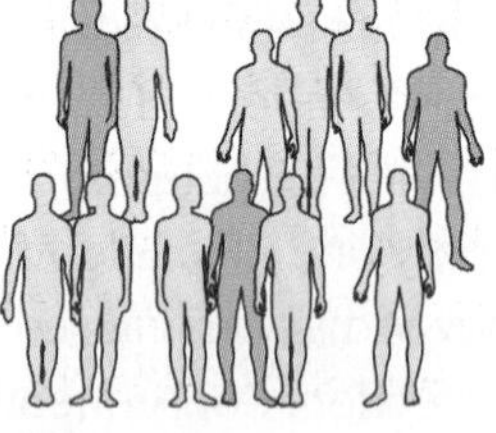

Negative selection

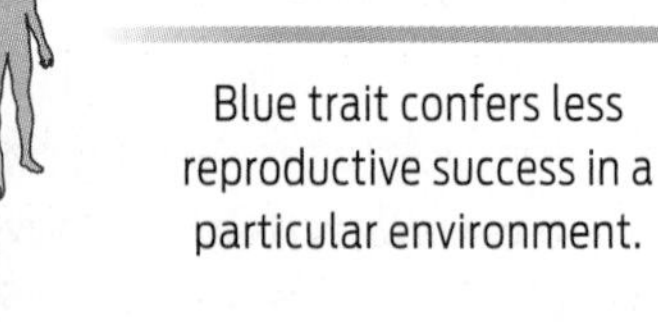

Blue trait confers less reproductive success in a particular environment.

The frequency of the blue allele decreases in the next generation, so fewer people have the blue trait.

Neutral (no selection)

Orange trait does not influence reproductive success in a particular environment.

The frequency of the orange allele does not change in the next generation, so the proportion of people with the orange trait remains the same.

and into the open savannahs to hunt for food. Hominids with less hair and more sweat glands were likely better hunters because they could sustain long bouts of activity without getting overheated. Like modern-day chimpanzees, these hominids likely had fair skin under their hair. Without hair to protect their light skin, they were exposed to the intense African sun. And, scientists hypothesize, exposure to the sun would have reduced their folate levels and thus their fitness in the sun-drenched environment.

Any of these ancient hominids that carried an allele or developed or inherited a mutation that increased their ability to produce more melanin would have been able to spend more time in the sun without the detrimental effects. Darker coloration would have protected their skin, and consequently their folate levels, from the sun, enabling these prehistoric humans to hunt and travel in the open fields.

Evidence to support this hypothesis comes from genetics. In 2004, Alan Rogers and his colleagues studied a gene that influences skin shade. They discovered that more than a million years ago, an allele that contributes to dark skin became fixed—that is, its frequency approached 100%—in the African population. "This is critical," Jablonski says. "It shows that darkly pigmented skin became extremely important to us" around the time that hominids became more humanlike.

The allele for darker skin was such an advantage in terms of survival and reproduction that hominids with darker skin left more offspring than their lighter-skinned relatives. Though some hominids were certainly born with rare mutations that gave them light skin, they weren't able to survive and reproduce in great enough numbers for the trait to persist in the population. The allele for darker skin eventually increased in the population until it reached 100%.

Populations that migrated north, away from the African sun, however, faced a different environment. Folate was not as easily destroyed in this lower-UV-light environment. But the high levels of melanin present in dark skin were a disadvantage; they prevented bodies from producing enough vitamin D. In this low-UV-light environment, fair skin allowed the body to soak up more ultraviolet light and produce essential vitamin D. In these environments fair-skinned people thus were more fit and left more descendants than dark-skinned people. Consequently, the frequency of light skin in northern climates increased with each generation.

"Skin tone is one of the best examples of human evolution." –Mark Stoneking

Genetic studies show that the frequency of alleles for light skin increased and swept through populations as they migrated north–most likely more than once. The fact that light skin in people from Northwestern Europe and light skin in people from Eastern Asia is determined by at least three different genes suggests that mutations for light skin arose independently and spread through those two populations separately.

There are a number of other hypotheses for the evolution of skin tone, but of all of them, the folate-vitamin D hypothesis has the most evidence supporting it and is consequently "the most reasonable," says Mark Stoneking of the Max Planck Institute for Evolutionary Anthropology in Leipzig, Germany. In fact, Stoneking says, "skin tone is one of the best examples of human evolution." It's an example in which we can see that genes have definite phenotypic effects on skin pigmentation and for which scientists can also see that the trait was selected, he explains.

There are at least a dozen different genes that interact to determine skin color, maybe more than 100 genes in total, says Stoneking. The dozen or so that have been identified are the ones that are known to have very strong effects.

Scientists also know that these skin color genes have been favored by specific environments because they carry genetic signatures of natural selection. To study whether natural selection favored any particular trait, scientists typically look at the amount of sequence variation that exists in a gene of interest. Less variation than average means that there was some environmental pressure that selected the alleles for that trait to be conserved over time.

Indeed, skin-color genes show this very pattern–they show less sequence variation than genes for other traits. Consequently, we know that the amount of melanin in the skin represents a compromise, or evolutionary trade-off, most likely between the need to protect folate levels from excess sunlight and the need to absorb sunlight to make vitamin D–and the way the trade-off was resolved depended on the environment **(Infographic 20.10)**. Skin color is thus a proxy for the geographic origin of our ancestors, but not much else.

Throughout human history, the lines between what we have come to call races have been fluid. Genetic studies show that hardly any population is pure in the way that many have thought. As people moved around the globe, they settled and often bore children with people they met along the way, introducing their alleles into the local gene pool. The particular environment people encountered favored some traits over others, and that is why populations that live in similar environments share similar features.

Though people tend to create racial groupings based on obvious physical characteristics, such features can be shared with other groups, says Jablonski. Not all Africans have equally dark skin and not all Europeans are fair-skinned, for example. And as humans travel more, settle in different areas, and intermarry, Jablonski says, "racial categories will get messier and messier." Perhaps the concept of race itself in time will disappear. ■

INFOGRAPHIC 20.10

The Evolution of Skin Color

Human skin color is an example of a trait that has undergone natural selection. Varying levels of UV light have selected for a range of skin tones around the globe. In each case, the amount of melanin represents a compromise between the need to protect folate and the need to make vitamin D.

2. **Migration into Low-UV-Light Environment**
Individuals with light skin reproduce more successfully:
- Lower melanin levels enable sufficient vitamin D levels even with low UV light levels.
- Folate is not destroyed in low-UV environments.

1. **High UV Light Environment**
Individuals with dark skin reproduce more successfully:
- More melanin protects folate even from high levels of UV light.
- High UV light intensity allows even those with more melanin to produce sufficient vitamin D.

3. **High UV Light Environment**
Individuals with dark skin reproduce more successfully:
- More melanin protects folate even from high levels of UV light.
- High UV light intensity allows even those with more melanin to produce sufficient vitamin D.

Summary

- Physical features shared by people within populations reflect adaptations to specific environments.
- Alleles can be harmful, beneficial, or neutral in their effect on survival and reproduction.
- Skin color most likely evolved in response to environmental UV levels, an example of human adaptation by natural selection. Alleles for darker skin conferred an advantage in sunnier environments, while alleles for lighter skin conferred an advantage in regions that receive weak sunlight.
- Skin color represents an evolutionary trade-off between the need for vitamin D, which requires adequate sunlight for its production, and the need for folate, which is destroyed by too much sunlight.
- Fossil evidence shows that humans and apes descended from a common ancestor and that walking upright preceded development of a big brain. There were many species that could walk upright before *Homo sapiens* appeared.
- Mitochondrial DNA evidence shows that modern-day humans first emerged in Africa, approximately 200,000 years ago, and subsequently spread to other continents.
- Humans evolved from apelike primate ancestors who likely had fair skin. Darker skin emerged in tandem with loss of body hair as our hominid ancestors ventured into the hot savannah.
- Biologically distinct human races do not exist. All humans are members of the same biological species.

THE EVOLUTION OF SKIN COLOR

Skin color most likely evolved in response to environmental UV levels, and represents an evolutionary trade-off between a need for vitamin D and folate.

HINT See Infographics 20.1 to 20.4, 20.9, and 20.11.

KNOW IT

1. In the course of human evolution, which of the following environmental factors likely influenced whether populations had mostly light-skinned individuals or mostly dark-skinned individuals?

a. average annual temperature
b. average annual rainfall
c. levels of UV light
d. the vitamin D content of the typical diet
e. mitochondrial DNA inheritance

2. As hypothesized by Jablonski and Chaplin, darker skin is advantageous in _________ UV environments because darker skin __________.

a. high-; reduces Vitamin D production
b. high-; protects folate from degradation
c. high-; increases the rate of folate synthesis
d. low-; allows more vitamin D to be produced
e. low-; allows more folate to be produced

USE IT

3. If folate is *not* destroyed by UV radiation, predict the skin color you might find in each of the following populations. Explain your answers.

a. populations living at the equator
b. populations living in Greenland

4. Which of the following would help darker-skinned people who live in low-UV environments remain healthy?

a. folate supplementation
b. sunscreen
c. reduced production of melanin
d. vitamin D supplementation
e. calcium supplements

5. Our closest primate relatives, chimpanzees, have light-colored skin yet live in tropical (high-UV) environments. How would the Jablonski-Chaplin hypothesis explain this observation?

a. Chimpanzees don't need folate for successful reproduction.
b. Chimpanzees are not susceptible to skin cancer.
c. The hair of chimpanzees protects their light skin from UV light.
d. Chimpanzees require much higher levels of vitamin D than humans do.
e. Chimpanzees use a light-colored pigment as their UV protection.

EARLY HUMAN ORIGINS

Fossil and mtDNA evidence suggests that humans and apes evolved from a single ancestor, likely in Africa.

HINT See Infographics 20.5 to 20.8, and 20.10.

KNOW IT

6. Why is mtDNA a useful tool in the study of human evolution?

7. According to the "out of Africa" hypothesis of human origins and migration, which group of people should show the highest level of genetic diversity?

a. Africans
b. Europeans
c. Asians
d. South Americans
e. Australians

USE IT

8. Of the following traits that are associated with being human, which evolved most recently?

a. upright walking
b. ability to control fire
c. social communication
d. tool use
e. big brain

9. Rank the levels of genetic diversity you would expect to find within the four populations listed in Question 7 from highest to lowest. Justify your ranking.

10. Why would individual Australopithicines who could make and use tools have had a selective advantage (that is, higher fitness) over individuals who could not make or use tools?

11. Ardi was partially arboreal (that is, the species could live in trees). The ability to move around in trees was facilitated by an opposable big toe that would help grip branches. Once ancient hominids moved permanently to a grounded lifestyle, would there have been any selective pressure to maintain an opposable big toe? Explain your answer.

12. Members of the genus *Australopithecus* walked upright, and their fossilized footprints show no evidence of an opposable big toe.

a. What foot structure and lifestyle might have been selected if early hominid evolution occurred in a forested environment? In a grasslands environment? Would you predict any differences because of the selective pressures in each environment? Why or why not?

b. What other traits would you expect to be favored in a forested environment? In open grasslands?

SCIENCE AND SOCIETY

13. Vitiligo is a disease in which melanocytes are destroyed, causing loss of pigmentation. If a dark-skinned person develops vitiligo and therefore lighter-colored skin, would his or her race change? What factors have led people to classify (or misclassify) themselves or others as members of one race or another?

14. Visit the 2010 Census Constituent FAQs page, http://2010.census.gov/partners/pdf/ConstituentFAQ.pdf. What is the U.S. Census definition of race? Why do you think the Census asks people to specify their race? What factors do you think go into a person's choice of a particular race on the U.S. Census form?

On the Tracks of Wolves and Moose

What You Will Be Learning

On the Tracks of Wolves and Moose

Ecologists are learning big lessons from a small island

"Teeth, hooves, blood, bruises, adrenalin, exhaustion. Romeo killed a moose. Very likely, this is the first moose he'd ever killed. He'd seen his parents, the alpha pair of Chippewa Harbor Pack, do it many times. He would have even helped his parents kill moose. He'd wounded moose a couple of times this winter, but never killed one. His pride heightened because he killed this moose with the help of a girlfriend. By early morning they slept with full bellies while a dozen ravens celebrated the accomplishment with a feast of their own."

That's an entry, made on February 20, 2010, in biologist John Vucetich's field journal, describing the exploits of a young gray wolf researchers have named Romeo. For almost 20 years, Vucetich has been shadowing wolves like Romeo and his kin on Isle Royale, a remote island about 15 miles off the Canadian shore in the northwest corner of Lake Superior. A 200-square-mile slice of roadless wilderness that is accessible only by boat and seaplane, Isle Royale may seem an unlikely place for a scientific laboratory, but that's exactly what it is for Vucetich and his colleagues. Every summer, and for a few weeks every winter, they investigate the island's packs of gray wolves (*Canis lupis*) and the herd of moose (*Alces alces*) that are their lifeblood.

Begun in 1958, the Isle Royale wolf and moose study is the longest-running predator-prey study in the world. For more than 50 years, researchers have studied how these two island inhabitants have interacted and co-existed. They are motivated by a simple goal: "to observe and understand the dynamic fluctuations of Isle Royale's wolves and moose, in the hope that such knowledge will inspire a new, flourishing relationship with nature." And all the effort may finally be paying off.

ECOLOGY
The study of the interactions between organisms, and between organisms and their nonliving environment.

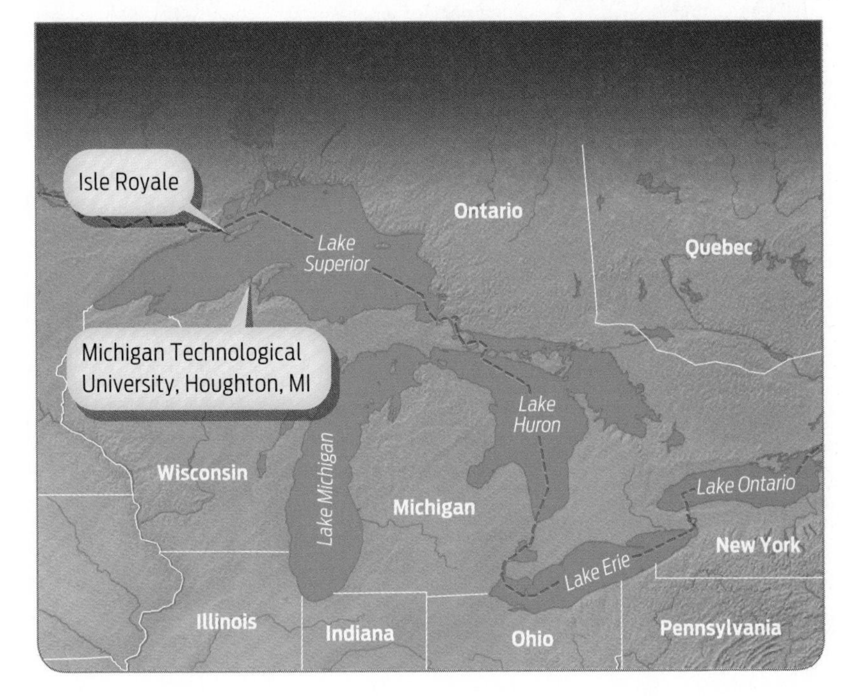

Vucetich began studying wolves as a college student at Michigan Technological University in the early 1990s. In 2001, he became coleader of the Isle Royale study, working alongside his former teacher and mentor, Rolf Peterson. It's challenging work at times, but Vucetich says he may have been destined for this career path: "*Vuk*"–the root of his last name–"is the Croatian word for wolf."

In Nature's Laboratory

Isle Royale essentially functions as a natural laboratory in which biologists can learn about **ecology,** the interactions between organisms and between organisms and their nonliving environment. A number of features make Isle Royale a good place for ecological research.

Because the island is uninhabited by humans and is protected as a national park, scientists can study moose-wolf interactions in a nearly natural environment, undisturbed by settlement, hunting, or logging.

Isle Royale is also an ideal distance from shore—close enough to the mainland for moose and wolves to have got there, but far enough away that other animals do not migrate easily to it. Because there are no other predators or prey on the island, the only things eating moose are wolves, and moose are just about the only thing wolves eat. These simplified conditions allow scientists a good look at the two residents' behavior and ecological impact.

Another thing that makes Isle Royale good for research is its size. The island is not so big as to have an unmanageably large population of moose, and not so small as to be unsupportive of a wolf population. "It's a little bit of the Goldilocks thing," says Vucetich, now a professor of ecology at Michigan Tech. "Isle Royale is not too big and it's not too small and it's not too close and not too far. It's just the right size to have a population of wolves and moose that we can study."

Ecologists study organisms at a number of levels: They can look at an individual organism, such as a single moose or wolf, studying how it fares in its surroundings. They may also look at a group of individuals of the same species living in the same place—a herd of moose, or a pack of wolves, for example—watching what happens to this **population** over time. Two or more interacting populations constitute a **community.** Isle Royale, for example, is home to a community of wolves, moose, and the plants the moose feed on.

Finally, ecologists may want to understand the functioning of an entire **ecosystem,** all the living organisms in an area and the nonliving components of the environment with which they interact. When moose eat trees, for example, they reduce the available habitat for other animals, such as birds. However, the heat of summer can reduce the ability of moose to feed, which in turn improves tree growth (**Infographic 21.1**).

Ecology draws not only on many areas of biology but also on many other branches of science, including geography and meteorology as well as mathematics.

A multidisciplinary science, ecology draws not only on many areas of biology but also on many other branches of science, including geography and meteorology as well as mathematics. Vucetich was initially drawn to ecology as a way to experience the outdoors, and being good at math was not something he could have

POPULATION
A group of organisms of the same species living and interacting in a particular area.

INFOGRAPHIC 21.1

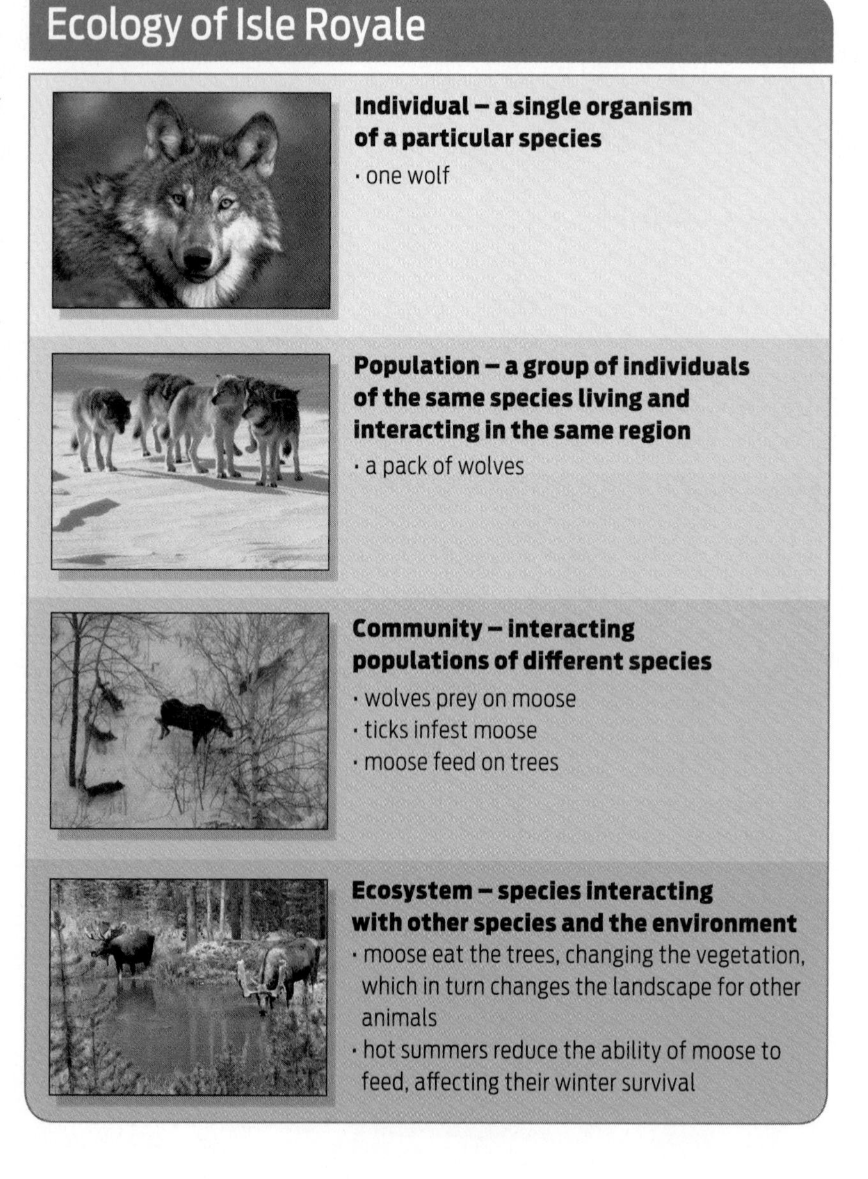
Ecology of Isle Royale
Individual – a single organism of a particular species
· one wolf
Population – a group of individuals of the same species living and interacting in the same region
· a pack of wolves
Community – interacting populations of different species
· wolves prey on moose
· ticks infest moose
· moose feed on trees
Ecosystem – species interacting with other species and the environment
· moose eat the trees, changing the vegetation, which in turn changes the landscape for other animals
· hot summers reduce the ability of moose to feed, affecting their winter survival

predicted. "As a high school student, I didn't like math at all," he says. Only when he saw that math allowed him to spend more time outdoors doing what he loved did he become "interested and inspired to learn a great deal about math."

Vucetich is a population ecologist, and population ecology is all about numbers. On Isle Royale, the main numbers the researchers are interested in year after year are the numbers of wolves and moose. "In any given season there are more or less of those species and we want to understand why," says Vucetich. Answering the "why" involves a lot of time, patience, and, of course, counting.

Much of the counting is done from the air. Sitting one in front of the other inside a tiny two-person plane, pilot and observer circle the island scanning for evidence of wolves and moose. Wolves are relatively easy to find and count because their tracks are easy to follow in the snow. "You follow the wolf tracks until you find the wolves," says Vucetich. The other thing that makes counting wolves easy is that they live in packs: if you find one wolf, you've generally found the others. And since there are usually only a couple dozen wolves on the island at any time, it's possible to count every one.

COMMUNITY
Interacting populations of different species in a defined habitat.

ECOSYSTEM
All the living organisms in an area and the nonliving components of the environment with which they interact.

DISTRIBUTION PATTERN
The way that organisms are distributed in geographic space, which depends on resources and interactions with other members of the population.

It's a different story with moose. There can be more than a thousand moose on the island—too many to count all at once. Besides, moose are relatively solitary creatures, and their brown coloring makes them harder to spot against the backdrop of dark evergreen trees. When moose are feeding in the forest—which is much of the time—counting them is, according to Vucetich, "like trying to count fleas on a dog from across the room." It's simply not possible to count them all.

Instead, the team uses a shortcut: they count all the moose in a series of square-kilometer plots representing about 20% of the island, average the number of moose per plot, and then extrapolate to the rest of the island. But even this shortcut requires many careful hours of study in the plane, straining to see the moose through the trees. To help himself concentrate, Vucetich recites a sort of mantra: "Think moose, think moose, look for the moose." That's the only way to make sure he doesn't miss one **(Infographic 21.2)**.

The somewhat random dispersion of individually roaming moose represents one type of **distribution pattern** found in nature. Distribution patterns generally reflect behavioral or ecological adaptation. For moose, being solitary and randomly distributed may help protect them from predation, since single moose are harder to spot in the forest than a large group would be. A random distribution may also allow individuals to maximize their access to resources. Pine trees, for example, have air-blown seeds that are spread far and wide by gusty winds, resulting in a random distribution of trees in the forests on the island.

When moose are feeding in the forest, counting them is "like trying to count fleas on a dog from across the room."
—John Vucetich

A truly random distribution is rare in nature; even wind-blown seeds must fall on fertile soil to grow, and this does not always happen. More common is a clustered, or clumped, distribution, which results when resources are unevenly distributed across the landscape, or when social behavior dictates grouping, as it does with the highly social wolf. Clumping has its advantages: for wolves, clumping helps them to gang up on moose; they circle their prey and close in for the kill. Clumping can also be a defense against predation, as it is for a school of fish.

A third distribution pattern found in nature is uniform distribution. In this case, individuals keep apart from one another at regular distances, usually because of some kind of territorial behavior. Birds such as penguins that nest

INFOGRAPHIC 21.2

Distribution Patterns Influence Population Sampling Methods

Wolves and moose have different distribution patterns on Isle Royale. Determining the size of each population requires a distinct counting strategy.

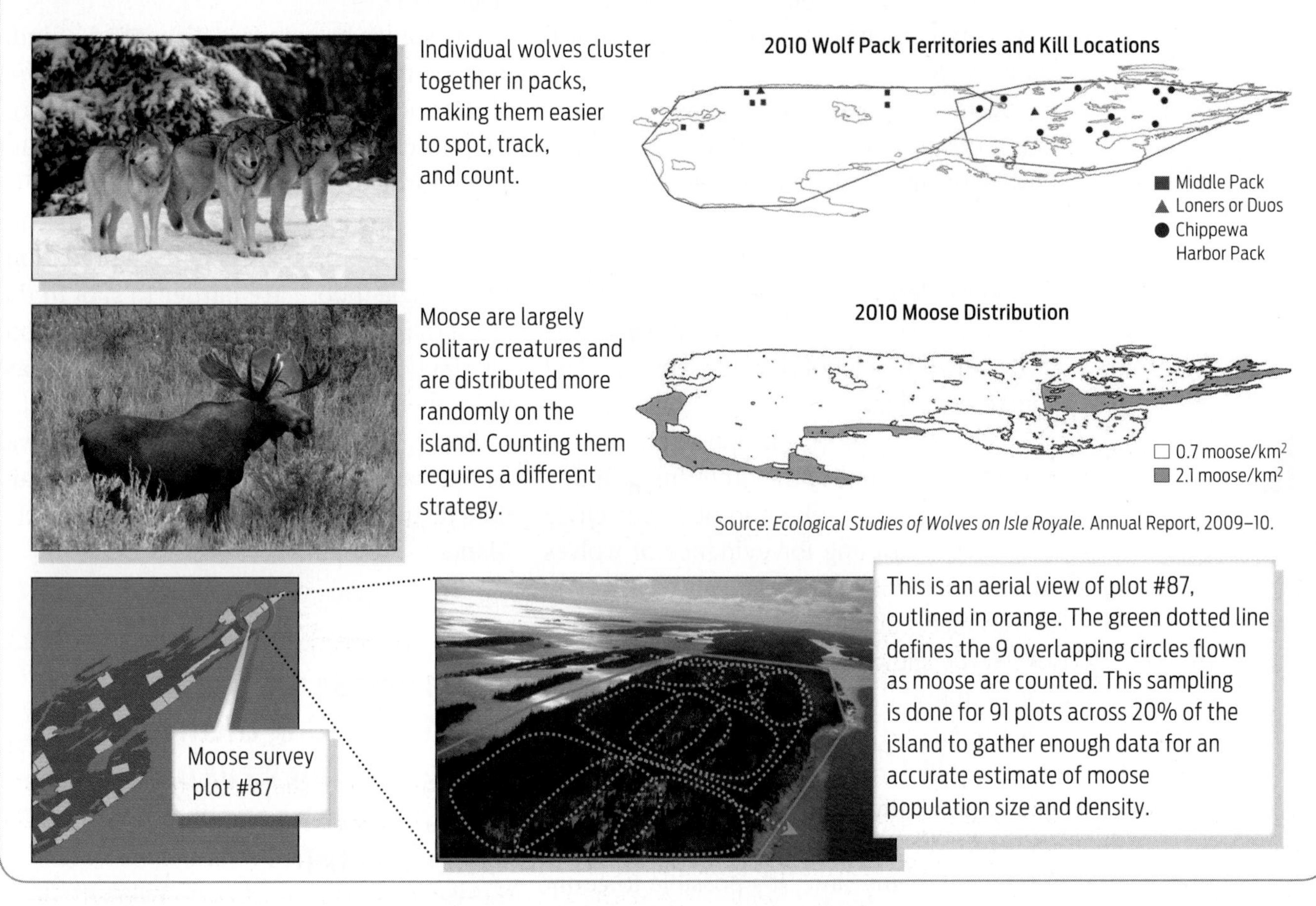

in defined spaces a few feet away from each other are a good example **(Infographic 21.3)**.

Individual wolves cluster together in packs, making them easier to spot, track, and count.
2010 Wolf Pack Territories and Kill Locations
Middle Pack
Loners or Duos
Chippewa Harbor Pack
Moose are largely solitary creatures and are distributed more randomly on the island. Counting them requires a different strategy.
2010 Moose Distribution
0.7 moose/km²
2.1 moose/km²
Source: Ecological Studies of Wolves on Isle Royale. Annual Report, 2009–10.
Moose survey plot #87
This is an aerial view of plot #87, outlined in orange. The green dotted line defines the 9 overlapping circles flown as moose are counted. This sampling is done for 91 plots across 20% of the island to gather enough data for an accurate estimate of moose population size and density.

Population Boom and Bust

Moose have not always roamed Isle Royale. The first antlered settlers likely arrived around 1900, when a few especially hardy individuals swam across the 15-mile-wide channel from Canada. With an abundant food supply and no natural predators on the island, the moose population exploded, growing from a handful of individuals around the turn of the century to more than a thousand by 1920.

This rapid increase reflected the population's high **growth rate,** a rate defined as the birth rate minus the death rate. Because it denotes the simple balance between birth and death, the growth rate is also known as the rate of natural increase. When the birth rate of a population is greater than the death rate, the population grows; when the death rate is greater than the birth rate, the population declines; and when the two rates are equal, the result is zero population growth.

In many populations, immigration and emigration make substantial contributions to population growth. But because the moose and wolves of Isle Royale are isolated, and individuals neither come to nor go from the island on a regular basis, their population growth rates are due only to births and deaths.

Ecologists describe two general types of population growth. The rapid and unrestricted increase of a population growing at a constant rate is called **exponential growth.** When a population is growing exponentially, it increases by a certain fixed percentage every generation. Thus, instead of a constant number

GROWTH RATE
The difference between the birth rate and the death rate of a given population; also known as the rate of natural increase.

EXPONENTIAL GROWTH
The unrestricted growth of a population increasing at a constant growth rate.

A bull moose walking across a waterway.

of individuals being added at each generation—say, the population going from 100 to 120 to 140 to 160—the increase is more like credit card interest, with each increase added to the principal (the population) before the percentage is applied. And so, with an exponential growth rate of 20%, that population would increase at each generation from 100 to 120 to 144 to 173 to 207. If the population continued to grow exponentially, it would quickly get out of control, not unlike a credit card bill you don't pay on time.

Such unrestricted growth is rarely if ever found unchecked in nature. As populations increase in numbers, various environmental factors such as food availability and access to habitat limit an organism's ability to reproduce.

INFOGRAPHIC 21.3

Population Distribution Patterns

Different organisms have different distribution patterns. There are three main types, but few organisms in nature fall into strictly one category.

Individuals are equally likely to be anywhere within the area.

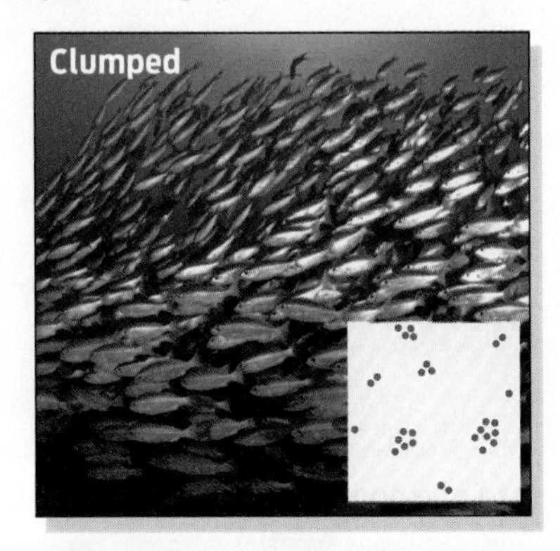

High-density clumps are separated by areas of low abundance.

Individuals maximize space between them by being uniformly spaced.

When population-limiting factors slow the growth rate, the result is **logistic growth**, a pattern of growth that starts rapidly and then slows.

Eventually, after a period of rapid growth, the size of the population may level off and stop growing. At this point, the population has reached the environment's **carrying capacity**–the maximum number of individuals that an environment can support given its space and resources. Carrying capacity places an upper limit on the size of any population; no natural population can grow exponentially forever without eventually reaching a point at which resource scarcity and other factors limit population growth. This is true even of the human population, as discussed in Chapter 24 **(Infographic 21.4)**.

Note that the size of a population may fluctuate around the environment's carrying capacity, briefly exceeding it and then dropping back. After an initial overshoot of carrying capacity, factors such as disease or food shortage will cause the population to shrink. This drop in turn may allow the environment time to recover its food supply, at which point the population may begin to grow again, briefly exceeding carrying capacity, and so on, in a cycle of boom and bust.

When moose first arrived on Isle Royale, their population grew exponentially. This unchecked proliferation of hungry mouths took a severe toll on the island; by 1929, the moose had munched their way through most of its vegetation. In turn, the reduction of the island's food supply caused the moose population to crash. The moose population had exceeded the island's carrying capacity, and by 1935, they had dwindled to a few hundred starving individuals.

The herd got lucky, though. The next summer, fire consumed 20% of the island, and the scorched areas provided space for new trees to

LOGISTIC GROWTH
A pattern of growth that starts off fast and then levels off as the population reaches the carrying capacity of the environment.

CARRYING CAPACITY
The maximum population size that a given environment or habitat can support given its food supply and other natural resources.

INFOGRAPHIC 21.4

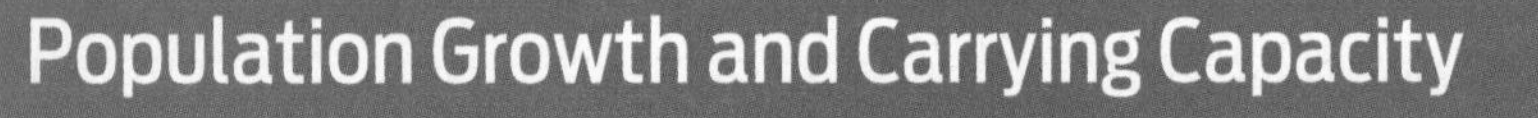

Population Growth and Carrying Capacity

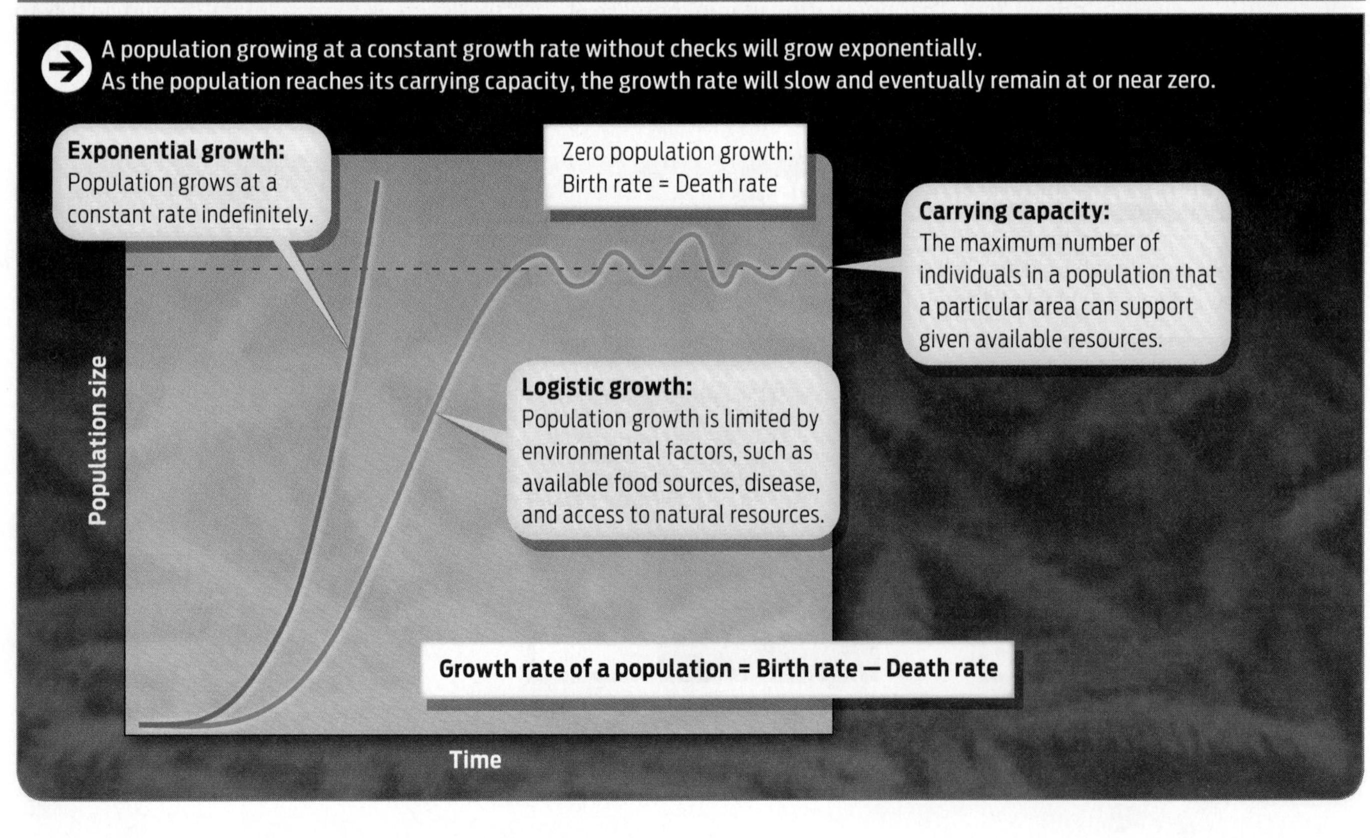

A population growing at a constant growth rate without checks will grow exponentially.
As the population reaches its carrying capacity, the growth rate will slow and eventually remain at or near zero.
Exponential growth: Population grows at a constant rate indefinitely.
Zero population growth: Birth rate = Death rate
Carrying capacity: The maximum number of individuals in a population that a particular area can support given available resources.
Logistic growth: Population growth is limited by environmental factors, such as available food sources, disease, and access to natural resources.
Growth rate of a population = Birth rate — Death rate
Population size
Time

grow. But as soon as the forest recovered, moose numbers again began to explode, ravaging the forests once more.

Then, around 1950, everything changed. One especially cold winter, a pair of gray wolves crossed an ice bridge connecting Canada to Isle Royale, forever altering the ecology of the island. Since then, the fates of the wolves and moose have been inextricably linked, with the size of one population influencing the size of the other.

Near the beginning of the Isle Royale study, in 1959, there were about 550 moose and 20 wolves. Moose numbers climbed for about 15 years, reaching a peak of approximately 1,200 animals in 1972, and then declined rapidly, to a low of approximately 700 moose in 1980. As moose numbers fell, wolf numbers rose—from a low of 17 wolves in 1969 to a high of 50 animals in 1980. These two trends were linked: the wolves were feeding themselves well enough to increase their own population, but by hunting and killing so many moose they caused the moose death rate to exceed the birth rate. With a negative growth rate, the moose population shrank.

What would happen next? Would the wolf predators simply drive their moose prey to extinction? No one knew. The only thing to do was watch and wait. Eventually, it became clear that the two populations were rising and falling together in a specific pattern, with the size of the wolf population peaking several years after the size of the moose herd and then dropping.

Why does the wolf population fall? Because even for wolves, there's no such thing as a free lunch: they pay a price for predation in the form of a declining food supply. The result is a repeating cycle in the number of predator and prey. Rather than growing exponentially and leveling off, the populations cycle through repeated rounds of boom and bust (Infographic 21.5).

John Vucetich with the carcass of a dead moose.

INFOGRAPHIC 21.5

Population Cycles of Predator and Prey

The wolf and moose populations are intimately linked. The wolf population peaks about 10 years after a peak in the moose population, and then declines, following the decline in the moose.

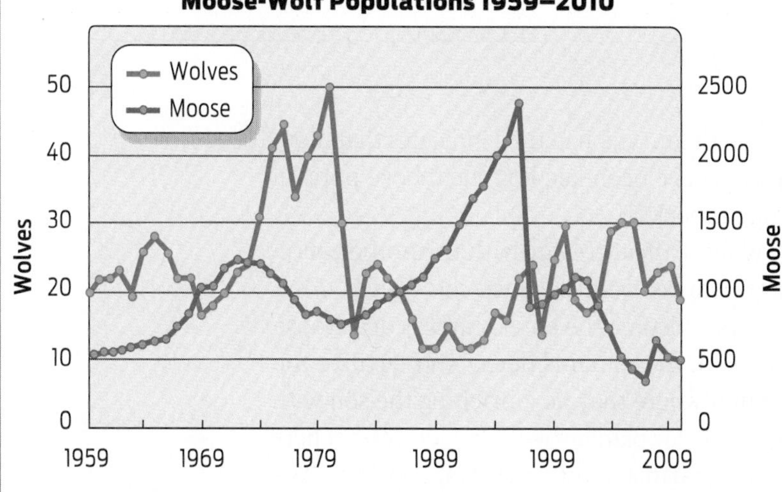

Isle Royale National Park, 1959–2010. Moose population estimates during 1959–2001 are based on population reconstruction from recoveries of dead moose. Estimates from 2002–2010 are based on aerial surveys.

Ecological Detectives

One pattern to emerge in the decades of data collected on Isle Royale is a correlation between

INFOGRAPHIC 21.6

Patterns of Population Growth

Wolf, moose, and tree populations are all interconnected. Trees provide food for moose, and moose provide food for wolves. Anything that impacts the size of one population will impact the size of the others.

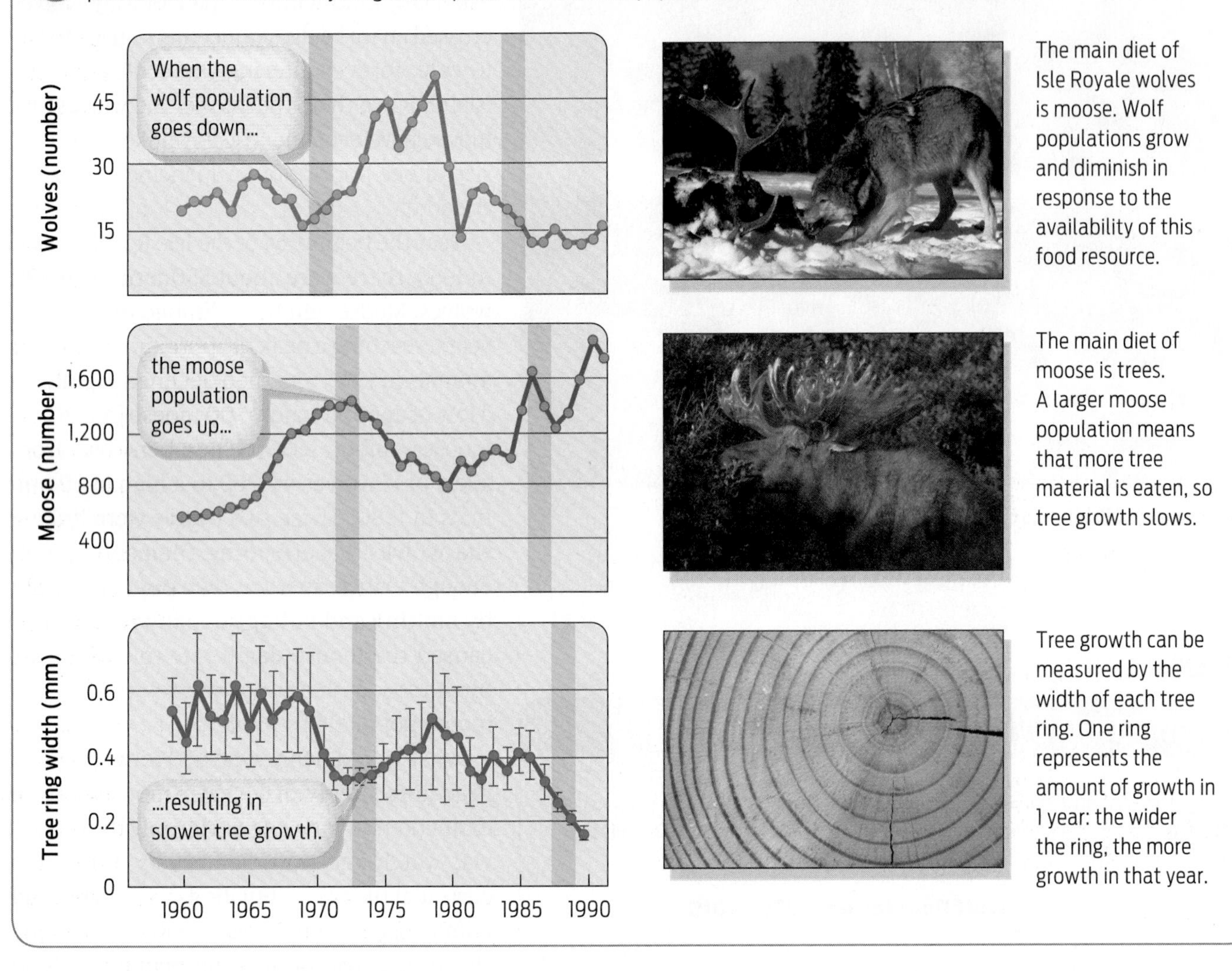

Wolves (number)
45
30
15
When the wolf population goes down...
Moose (number)
1,600
1,200
800
400
the moose population goes up...
Tree ring width (mm)
0.6
0.4
0.2
0
...resulting in slower tree growth.
1960
1965
1970
1975
1980
1985
1990
The main diet of Isle Royale wolves is moose. Wolf populations grow and diminish in response to the availability of this food resource.
The main diet of moose is trees. A larger moose population means that more tree material is eaten, so tree growth slows.
Tree growth can be measured by the width of each tree ring. One ring represents the amount of growth in 1 year: the wider the ring, the more growth in that year.

a large wolf population and vigorous tree growth. When wolves are plentiful, they keep the moose population in check. Because trees are the primary food source for moose, they grow more when fewer moose are eating them. It's therefore possible to follow the rise and fall of the wolf population by monitoring the state of the forest.

One way ecologists can determine forest growth and health is to count and measure the width of tree rings, which reflect how much trees have grown season by season. They also measure how tall the trees are. Taller and bigger trees mean that fewer moose have been foraging on them, which in turn indicates that more wolves have been keeping the moose population in check.

Wolves affect tree growth in another, indirect, way as well. Because the wolves don't always consume the entire carcass of a moose they kill, the remains decay and fertilize the ground where they lie, enriching the soil with nutrients for plant growth. In fact, researchers have found that nitrogen levels are between 25% and 50% higher in these hot spots compared to controls. This work shows that predators–in this case, wolves–are an important component of a balanced and healthy ecosystem, and "illustrates what can be protected or lost when predators are preserved or exterminated," according to Joseph Bump, an assistant professor at Michigan Tech's School of Forest Resources and Environmental Science **(Infographic 21.6)**.

INFOGRAPHIC 21.7

Moose and Wolf Health is Monitored Using a Variety of Data

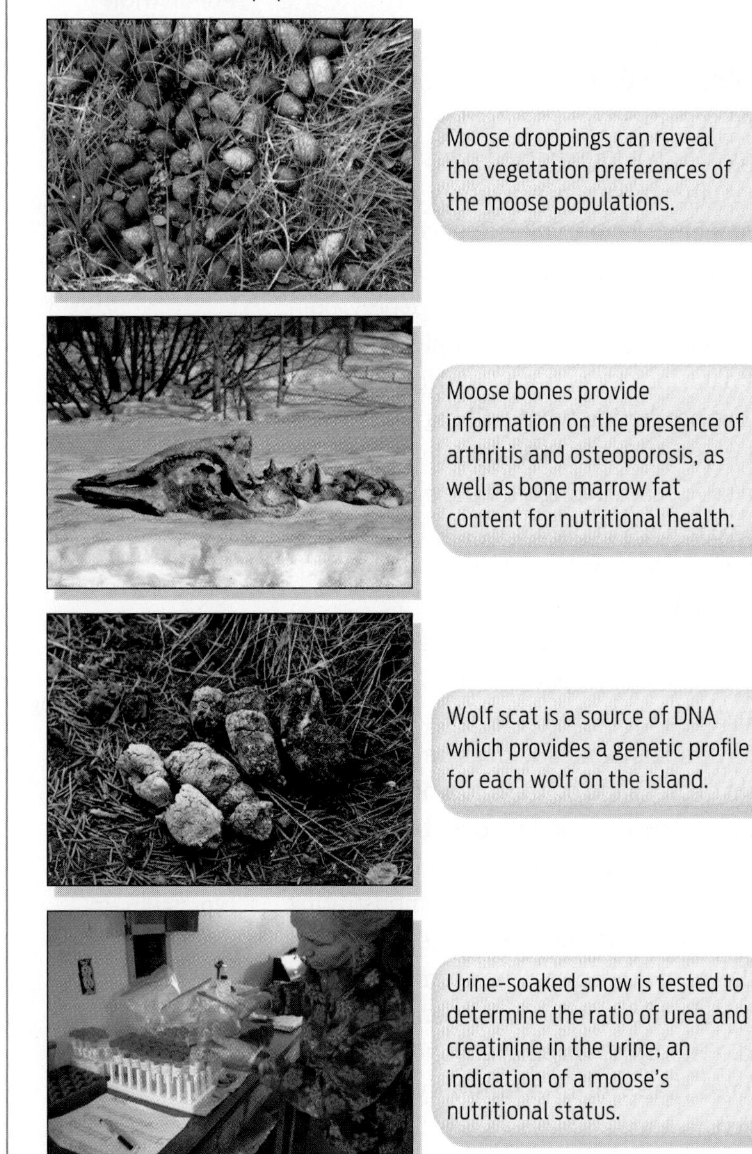

In addition to information about population size, researchers collect other data that are essential for monitoring the physical health of moose and wolf populations.
Moose droppings can reveal the vegetation preferences of the moose populations.
Moose bones provide information on the presence of arthritis and osteoporosis, as well as bone marrow fat content for nutritional health.
Wolf scat is a source of DNA which provides a genetic profile for each wolf on the island.
Urine-soaked snow is tested to determine the ratio of urea and creatinine in the urine, an indication of a moose's nutritional status.

Another clue the ecological detectives look at in determining the population patterns on Isle Royale is urine-soaked snow and droppings, also known as scat. Urine and scat may seem crude objects of scientific study, but to the trained eye aided by a microscope they reveal a host of information about the animal that produced them. For example, by analyzing scat samples under the microscope, researchers can tell exactly what moose have been eating. During the winter months, for example, moose eat mostly twigs from deciduous (leaf-shedding) plants and needles from balsam fir and cedar trees.

Scat also provides important information about an animal's genetics. It is used to obtain DNA profiles, for example, which can be used to confirm population counts and to track which wolves were involved in killing which moose. DNA can also be used to look for diseases or signs of inbreeding. "Through the DNA we can get a good sense of individual wolves—how they live and how they die," says Vucetich. (For more information on DNA profiling, see Chapter 7.)

Yet more clues can come from studying a moose kill site, which is a bit like analyzing a crime scene. Researchers can tell if a moose was killed by wolves because in that case there will often be blood spattered on nearby trees and signs of struggle in the form of broken branches. Wolves also typically scatter bones as they feast, whereas the carcasses of moose that die of starvation may be relatively intact.

At the kill site, researchers gather moose bones. From these bones, the researchers can tell how old a moose was when it died, as well as learn about other aspects of the animal's health, such as whether it had arthritis or osteoporosis. The value of this information goes beyond understanding individual animals. It allows researchers to know whether wolves are targeting healthy moose or sickly ones. Killing a healthy moose has a bigger effect on moose population dynamics than killing one that is already near death, because a young, healthy moose might have gone on to reproduce had it lived **(Infographic 21.7)**.

Too Close for Comfort?

Moose are formidable foes of their wolf predators. At 900 pounds and 10 times the weight of a wolf, an adult moose can successfully defend itself against an aggressive pack of wolves with its powerful front legs. For that reason, wolves often attack older and weaker or young moose. They typically target the nose and rear, where they bite and latch onto the flesh like a steel trap. When enough wolves are attached, their collective weight brings down the moose, and the feeding begins.

A number of factors can influence the likelihood that wolves will kill moose. One of the simplest is **population density,** the number of organisms per given area. Because the area of Isle Royale stays the same, as the size of the moose population increases so does its density. At high population density, moose are easier for wolves to locate and kill. Further, when the moose population is at high density, food scarcity can also be a problem, leaving moose hungry and weak and therefore more vulnerable to attack.

Because wolf predation and plant abundance have a greater effect on moose when the moose population is large, these are examples of **density-dependent factors** influencing population size. As living organisms, they are also examples of **biotic** factors influencing growth. Not all density-dependent factors are biotic. Nonliving, or **abiotic,** factors like weather, habitat, and breeding places can also influence population size in a density-dependent manner.

Some environmental pressures take a toll on a population no matter how large or how small it is. In an exceptionally cold winter with deep snow, for example, moose can die of cold or starvation. The weather can also weaken them so they are easier to hunt and kill. Since cold weather affects moose regardless of population size, it is considered a **density-independent factor**: whether 10 moose or 1,000, a harsh winter affects them all.

Conversely, harsh winters tend to benefit wolves, since moose are easier to catch in deep snow. "A mild winter is always tough on the wolves," says Peterson, who notes that moose can more easily escape wolves when snow cover is light. Other common density-independent factors influencing population growth include rainfall, drought, and fire. Density-independent factors can be nature's form of bad luck, often striking without warning. Most, but not all, density-independent factors are abiotic **(Infographic 21.8)**.

At 900 pounds and 10 times the weight of a wolf, an adult moose can successfully defend itself against an aggressive pack of wolves with its powerful front legs.

Watching and Waiting

For the scientists on Isle Royale, population ecology is full of unexpected twists and turns. There is often no sure way to know how various environmental factors will influence the growth of a population. Even on an isolated island with only one large predator and one large prey, population dynamics are never simple. Scientists gather data, look for patterns, and form hypotheses, but predicting what will happen next is much more difficult. "[What] Isle Royale has shown us . . . convincingly for the past 50 years, is that we're lousy at predicting the future," says Vucetich. "What we're a fair bit better at is explaining the past."

For example, beginning around 1980, a disease known as canine parvovirus (CPV) infected Isle Royale's population of wolves. The disease typically affects domestic dogs and was likely brought to the island on the boots of unsuspecting hikers. The disease killed all but 14 of the island's wolves, and over the next 10 years the moose population skyrocketed, demonstrating that wolves exert a strong influence on the abundance of their prey. The event was useful from a scientific standpoint–but entirely unexpected. "There's no way that anyone could have predicted that. Not in a million years," says Vucetich.

That wasn't the end of the surprises. In the last 15 years, it's become apparent that a

POPULATION DENSITY
The number of organisms per given area.

DENSITY-DEPENDENT FACTOR
A factor whose influence on population size and growth depends on the number and crowding of individuals in the population (for example, predation).

BIOTIC
Refers to the living components of an environment.

ABIOTIC
Refers to the nonliving components of an environment, such as temperature and precipitation.

DENSITY-INDEPENDENT FACTOR
A factor that can influence population size and growth regardless of the numbers and crowding within a population (for example, weather).

INFOGRAPHIC 21.8

Abiotic and Biotic Influences on Population Growth

Both living (biotic) and nonliving (abiotic) environmental factors influence the size and growth of populations.

Abiotic Factors:
Climate:
Harsh winter temperatures and snowfall stress moose. Food-seeking and hiding from predators become difficult.
Temperature:
High summer temperatures can cause heat stress in moose. Changes in climate may increase insect parasitism.
F
C
120
100
80
60
40
20
0
-20
-40
50
40
30
20
10
0
-10
-20
-30
-40
Biotic Factors:
Food:
The availability of trees as a food source influences moose population growth.
Predators:
The presence of a wolf predator limits moose population growth.
Disease:
Tick infestations weaken moose.

warming climate, not just predation by wolves, is influencing moose population size. The first decade of the 21st century was one of the hottest on record. Sweltering summer temperatures hit moose especially hard. The large herbivores get hot easily, and they don't perspire; they escape the heat by resting in the shade. A lot of time spent resting means less time for eating, and a moose who's been dieting all summer has less insulation for winter.

Warmer temperatures have affected moose in a more insidious way as well. About 10 years ago, Vucetich and his colleagues began to notice that a tick parasite was bothering the moose, and that warm weather seems to favor ticks. Ticks suck the moose's blood and cause them to itch. The moose scratch themselves against trees and chew their hair out trying to rid themselves of the itchy freeloaders. Since a single moose may host many thousands of ticks, the combination of tick-related blood loss and heat-induced weight loss can be deadly. In 2004, the average moose had lost more than 70% of its body hair, the result of carrying more than 70,000 ticks.

By 2007, the deadly combination of blood-sucking ticks, hot summers, and relentless predation from wolves had driven the moose population to its lowest point in at least 50 years—385, down from 1,100 in 2002. Predictably, the wolf population followed suit, declining from 30 individuals in 2005 to 21 in 2007. As of 2010, the moose population has remained low, at about 510 individuals—half their typical abundance—while the wolves declined to just 19 individuals **(Infographic 21.9)**.

Hunted by wolves, preyed on by ticks, dogged by oppressive heat, moose certainly do not have it easy. They can live to be 17 years old, but most moose die before reaching their tenth birthday. To paraphrase philosopher Thomas Hobbes, a moose's life is often nasty, brutish, and short.

It's no picnic for wolves, either. While the wolf lifespan—the longest they can live—is about 12 years, most die by age 4. The most common cause of death is starvation. With few available food sources, a wolf may go 10 days without eating. Obtaining a meal on the eleventh day may mean having to wrestle a 900-pound moose on an empty stomach.

As of 2010, the moose population has remained low, at about 510 individuals—half their typical abundance—while the wolves declined to just 19 individuals.

The difficulty of finding food is just one obstacle for wolves. They also have a very high incidence of bone deformities, which cause back pain and partial paralysis of the hind legs. In the early years of the Isle Royale study, such deformities were rare, but they've become more common in recent years, almost certainly as a result of inbreeding. For the last 12 years, every dead wolf on Isle Royale has had such deformities.

It's not the first time wolf populations have been in trouble. When colonial settlers first arrived in North America, the gray wolf roamed throughout all of the future 48 contiguous U.S. states. By 1914, hunting and trapping had greatly reduced the population, and survivors were limited to remote wooded regions of Michigan, Wisconsin, and Minnesota. The federal government officially listed the species as endangered

INFOGRAPHIC 21.9

Warming Climate Influences Moose and Wolf Population Size

In recent years, climate change has become a significant influence on moose and wolf populations on Isle Royale. Warmer temperatures lead to increased tick infestations of moose, resulting in a weakened and depleted population.

One moose may be home to tens of thousands of ticks at a time.

Ticks cause moose to lose their hair, their appetite, and a good deal of blood.

Ticks make moose weak and vulnerable to predation and starvation. So, while ticks have been increasing...
...the moose population has been decreasing.
Moose weakened by ticks are easier for wolves to catch. After an initial population increase in response to an abundance of moose, the wolf population begins to suffer (2007) as the moose population continues its decline.
% Hair loss
80
60
40
20
0
Ticks
2001
2003
2005
2007
Population size
1,200
800
400
0
Moose
2001
2003
2005
2007
Population size
30
20
10
0
Wolves
2001
2003
2005
2007

Chippewa Harbor, Isle Royale. © JOHN AND ANN MAHAN

in the early 1970s, when it seemed on the verge of extinction.

The wolves' latest plight poses an ethical dilemma: should scientists intervene on their behalf–say, by importing wolves from another population to reintroduce genetic diversity–or let nature take its course? It's a question that Vucetich thinks about a lot. The answer, he says, will require balancing a number of competing values–not just the value of individual animals, but the values of population and ecosystem health in addition to the values of scientific knowledge and the value of wilderness. Without wolves, for example, would the moose population once again explode and decimate the island's forest? Would healthier wolves be able to completely overwhelm moose, and drive them to extinction on the island? These are the sorts of difficult questions that wildlife managers will need to consider when debating whether and how to intervene.

If only one value mattered, Vucetich notes, it would be easier to make a decision, but here the values are often competing. The dilemma is a familiar one to conservation biologists. According to Vucetich, these competing values show up in varying degrees in almost any management question that we have in any part of the world. They represent, he says, "this grand question of How should humans relate to nature?" To this question, there are no easy or obvious answers. Nevertheless, he believes it is important for people to debate and discuss these issues–not just scientists and experts, but lay people, too, because "every citizen has a stake in this question of how we relate to nature." ■

Summary

- Ecology is the study of the interactions between organisms and between organisms and their nonliving environment.
- Ecologists study these interactions at a number of levels, including population, community, and ecosystem.
- Living organisms can have a clumped, random, or uniform distribution pattern, depending on ecological and behavioral adaptations. Few organisms fall into strictly one category.
- Population growth is an increase in the number of individuals in a population. The growth rate of a population is defined as the birth rate minus the death rate. When immigration and emigration are excluded, it is also known as the rate of natural increase.
- Exponential growth is the unrestricted growth experienced by a population growing at a constant rate. Logistic growth is the slowing of the growth of a population due to environmental factors such as crowding and lack of food.
- Carrying capacity is the maximum population size that an area can support, given its food supply and other life-sustaining resources. Populations cannot grow exponentially forever; eventually, they hit the carrying capacity for the region and stop growing.
- Population growth can be limited by a variety of factors, including biotic (living) and abiotic (nonliving) parts of the environment.
- Density-independent factors, such as a severely cold winter, can affect a population of any size.
- Density-dependent factors, such as the presence of predators, have different impacts on the population, depending on the size and crowding of individuals in the population.
- Populations in a community are interconnected, with the fate of one often influencing the fate of the others.

STUDYING ECOLOGY

Ecology is the study of the ways organisms interact with one another and the environment. Ecologists study these interactions at a number of levels.

HINT See Infographics 21.1–21.3.

KNOW IT

1. What is the difference between a community and a population?

2. An ecosystem ecologist might study
- **a.** plant populations.
- **b.** herbivores that eat the plants.
- **c.** predators in the population.
- **d.** the impact of precipitation patterns on the plant populations.
- **e.** all of the above

3. Why do the researchers collect scat as part of their study on Isle Royale?

USE IT

4. Which of the following is an example of population growth?
- **a.** The average weight of Americans has increased substantially in the past decade.
- **b.** Tropical fish have been found in more northern waters than their usual habitat.
- **c.** The number of people in a town has increased by 25% in the past 5 years.
- **d.** The number of butterflies in a region have stayed the same from 1950 to 2010.
- **e.** all of the above

5. How would you explain to a 10-year-old what ecologists do?

6. A local environmental group wants to determine the population size of squirrels in a nearby nature preserve. What are some methods you could use to estimate the size of the squirrel population? Would the same approaches be as useful in determining the population size of maple trees in the same area? Why or why not?

7. Why is it important for researchers to determine the cause of death of moose on Isle Royale? Can this information be used to help make predictions about moose and wolf populations? Explain your answer.

8. How would you use scat analysis to determine whether an herbivore had a preference for a particular type of vegetation? Be specific about both the type of analysis, and what the analysis would reveal for herbivores with or without a preference for a particular type of vegetation.

POPULATION GROWTH

Ecologists analyze the growth of populations with the help of concepts from population ecology. A number of different environmental factors can affect the growth of populations.

HINT See Infographics 21.4–21.9.

KNOW IT

9. Which of the following would cause a population to grow?
- **a.** identical increases in both the birth rate and the death rate of a population
- **b.** a decrease in the birth rate and an increase in the death rate of a population
- **c.** an increase in the birth rate and a decrease in the death rate of a population
- **d.** an increase in the birth rate and no change in the death rate of a population
- **e.** an identical decrease in both the birth rate and the death rate of the population

10. When a population reaches its carrying capacity, what happens to its growth rate?

USE IT

11. You are studying a group of predatory fish that live in a school in a large lake. If a parasite were introduced to the lake by a vacationing fisherman, would you expect it to have a greater impact on the population if the fish were at high density or low density (assume the parasite is passed from one fish to another through the water, but can only remain alive in the water for a very short period of time)? What would happen to this same population if there was a severe drought and very hot summer?

12. Classify each of the following as a biotic or an abiotic factor in an ecosystem. Then predict the impact of each on the moose population of Isle Royale. Explain your answers, keeping in mind possible interactions between the various factors and between the moose and wolf populations.
- **a.** hot summer temperatures
- **b.** ticks that parasitize moose
- **c.** declining numbers of balsam fir trees
- **d.** a parvovirus in wolves
- **e.** deep winter snowfall

13. Assume that a new herbivore is added to Isle Royale that is not a prey for wolves. Predict the effect of this introduction on

a. the populations of trees.
b. the moose population.
c. the wolf population.

14. If the moose population remains stable, what other factors could influence the wolf population on Isle Royale?

15. Population Q has 100 members. Population R has 10,000 members. Both are growing exponentially at a 5% annual growth rate.

a. Which population will add more individuals in 1 year? Explain your answer.
b. After 5 years, what will be the size of each population?
c. If the larger population reaches its carrying capacity at the end of the third year, what will its size be after 5 years?

SCIENCE AND ETHICS

16. The wolves of Isle Royale are suffering from bone deformities, probably as a result of inbreeding in their small population.

a. Do you think that humans should intervene to save the wolves? Would your answer be different if the wolves were near human populations or agricultural centers?
b. If humans were to intervene, what kinds of strategies might help stabilize or increase the wolf population? Explain your answer.

17. In the 1960s the Asian carp was introduced into U.S. waterways, where it now consumes massive amounts of plankton (including photosynthetic algae). This species is becoming a concern to ecologists and sport fishermen; why do you think this is the case? Think about possible consequences for the communities in which this species now resides. What if the Asian carp invades waterways with recreational or commercial fisheries? What management strategies can you suggest?

Chapter 22 Community Ecology

What's Happening to Honey Bees?

In three short
American be
have lost 3 m
colonies.

POLLEN
Small, thick
plant struct
contain cell
develop int

POLLINATI
The transfe
from male t
plant struct
that fertiliz
occur.

What You Will Be Learning

22.1 Bees Are Keystone Species
22.2 Commercial Crops Require Bees
22.3 Flowering Plant Reproduction Relies on Pollinators
22.4 Energy Flows Up a Food Chain
22.5 A Honey Bee Food Web
22.6 Organisms Live Together in Symbioses
22.7 Pollinators Have Different Ecological Niches
22.8 Bees Compete for Resources
22.9 What is Causing Colony Collapse Disorder?

> **"People need to take more time to understand where their food comes from, what it takes to produce food and have this incredible supply of food available to us."** –Maryann Frazier

launched a "Help the Honey Bee" campaign, noting that honey bee-dependent products are used in 25 of its 60 flavors. The company also introduced a new flavor called Vanilla Honey Bee, the proceeds of which are being used to fund CCD research. Even ordinary citizens are catching the bee buzz. From city-dwellers becoming amateur rooftop beekeepers to sub-urbanites letting more flowers grow in their yards, the ranks of people wanting to make the environment pollinator-friendly has swelled. And that's a cause that just about everyone can get behind–because, as more and more people are coming to realize, a world without honey bees just wouldn't be as sweet. ■

Summary

- An ecological community is made up of interacting populations of different species.
- Bees are keystone species in that they play a fundamental role in supporting the entire community, much like the keystone in an arch.
- Bees are the primary pollinators for many species of flowering plants, which depend on the pollinators to transfer pollen between plants of the same species.
- Flowers are the reproductive hub of a plant, containing male and female reproductive structures. Pollination, the transfer of pollen from male to female structures, results in fertilization.
- The organisms in a community are connected by a food chain. Each player in the chain is an important ecological link in the chain.
- Organisms at the base of the food chain are producers—they obtain energy directly from the sun and supply it to the rest of the food chain; organisms higher up the food chain are consumers—they obtain energy by eating organisms lower on the chain.
- In predation, one organism eats another. Herbivory—eating plants—is one type of predation.
- As energy flows through different trophic levels in the food chain, some of it is lost to the environment.
- Organisms can have different types of symbiotic relationships. In mutualistic symbioses, both members benefit; in parasitism, one member benefits while the other suffers; and in commensalism, one member benefits while the other is unharmed.
- The space and resources, including other members of the community, that a species uses to survive and reproduce define its ecological niche. Some species have overlapping niches, leading to competition for resources.
- Bees are not the only pollinators in peril. Human development and agriculture have decreased habitat and foraging areas for many natural pollinators, resulting in increased competition among them.

Chapter 22 Test Your Knowledge

KEYSTONE SPECIES AND POLLINATION

Keystone species are critical in community structure. Insect pollinators play key roles by ensuring reproduction of many species of flowering plants.

HINT See Infographics 22.1–22.3.

KNOW IT

1. How does a community differ from a population?

2. What are keystone species?

3. A rocky shoreline that is covered at high tide but exposed at low tide supports a community of mussels, algae, barnacles, and starfish. An ecologist systematically removes species from different areas of the beach. Removing the mussels doesn't substantially change the community, but removing the starfish dramatically changes the mix of species in the area. Which is the keystone species?

- **a.** mussels
- **b.** barnacles
- **c.** algae
- **d.** starfish
- **e.** all of the above

4. Bees transfer pollen from the ____________ to the ____________.

- **a.** anther; stigma
- **b.** stigma; style
- **c.** filament; ovary
- **d.** anther; ovary
- **e.** stigma; anther

USE IT

5. Think about a community of organisms that you are familiar with. From what you know about this community, choose what you think might be a keystone species and defend your choice.

6. If you have pollen allergies, are you more likely to be suffering from the effects of bee-carried pollen or wind-carried pollen? Explain your answer.

FOOD CHAINS AND ENERGY FLOW

Energy is initially captured by autotrophs and flows through organisms in food chains. As energy flows from producers through consumers, some of it is lost to the environment as heat.

HINT See Infographic 22.4.

KNOW IT

7. In relation to a food chain, what do plants and photosynthetic algae have in common?

- **a.** nothing
- **b.** they are both producers
- **c.** they are both first level consumers
- **d.** they are both top level consumers
- **e.** their numbers are limited by the energy they take in from heterotrophic food sources

8. A bear who eats both blueberries and fish from a river can be referred to as

- **a.** an omnivore
- **b.** a heterotroph
- **c.** a consumer
- **d.** a producer
- **e.** all of the above
- **f.** a, b and c
- **g.** a and c

USE IT

9. Describe a natural food web that includes a terrestrial food chain (including honeybees) and at least one aquatic organism from an aquatic food chain.

10. Explain how a cow can eat so many kilograms of grain but not produce the equivalent amount of energy in the form of meat. What happens to the energy stored in the grain once it is ingested by the cow?

11. Compare the diet of a human who is an herbivore with that of a human who is a top consumer.

COMMUNITY INTERACTIONS

Organisms in a community interact in many different ways, which are sometimes helpful to one another, sometimes not. The precise role that each species plays in a community defines its niche.

HINT See Infographics 22.5–22.7.

KNOW IT

12. What are some important features of a honey bee niche? How is it that other nectar-feeding organisms can coexist with bees as part of a community?

13. Competition is most likely to occur
  a. when one species eats another.
  b. when two species occupy different niches.
  c. when one species helps another.
  d. when two species occupy overlapping niches.
  e. when two species help each other.

USE IT

14. On a rocky intertidal shoreline (the area between the highest and lowest tidelines, so the intertidal zone is alternately exposed and covered by seawater), mussels and barnacles live together attached to rocks where they obtain food by filtering it from ocean water. Since these two species coexist in the same habitat, we predict that they do not have identical niches. What might be separating their niches enough to allow them to occupy the same rocky intertidal zone?

15. If a meadow of wildflowers were converted to a field of corn, would you predict the number and diversity of bees in the community to increase or decrease? Explain your answer.

KNOW IT

16. Which of the following characterizations best defines a symbiotic relationship?
  a. Both organisms benefit.
  b. The organisms live in close association.
  c. Only one organism benefits.
  d. The relationship is mutually harmful.
  e. Neither organism benefits.

17. Would you characterize the relationship between the bacteria that live symbiotically within bees and their bee hosts as a type of competition, parasitism, mutualism, or commensalism? Explain your answer.

USE IT

18. What is the evidence for and against varroa mites and IAPV being responsible for colony collapse disorder (CCD)?

19. We all have *E. coli* bacteria living in our intestinal tracts. Occasionally these *E. coli* can cause urinary tract infections. From this information, which of the following terms would you say describe(s) the relationship between us and our intestinal *E. coli*? Why did you choose the term(s) you did?
  a. competition
  b. mutualism
  c. parasitism
  d. symbiosis
  e. predator–prey

SCIENCE AND ETHICS

20. Many people consider bees a stinging nuisance. What could you say to such people to dissuade them from killing all the bees in their backyards?

21. Farmers often plant large acreage of a single crop in order to maximize yield and simplify harvesting. From what you have read in this chapter, what arguments can you make for *not* growing acre after acre of almonds, even for almond lovers?

Chapter 23 Ecosystem Ecology

The Heat Is On

What You Will Be Learning

The Heat Is On

From migrating maples to shrinking sea ice, signs of a warming planet

For more than two centuries, Burr Morse's family has collected sap from Vermont's maple trees and boiled it to sweetened perfection. If you pour maple syrup over your breakfast pancakes or eat maple-cured ham, you've likely enjoyed the results of their careful craft, or that of other Vermont sugar farmers. About one in four trees in the state of Vermont is a sugar maple (*Acer saccharum*), and each year the state produces between half a million and a million gallons of syrup, making Vermont the number one maple syrup producer in the United States. Yet what has been a proud family tradition and the economic lifeblood for generations of sugar farmers could very well be in jeopardy.

"In the last 20 years we have had a number of bad seasons and most of those I would attribute to temperature that is a little too warm," says Morse. "For maple sugaring to work right, the nights have to freeze down into the mid 20s, and the days have to thaw up into the 40s. And the nights for those 20 years, it seemed, were not quite getting cold enough."

Morse isn't the only one to notice the shift. Sugar farmers across New England have noted the changes in temperature and are leery about their long-term effects.

Warmer winters in New England could have a large economic impact on the region. As

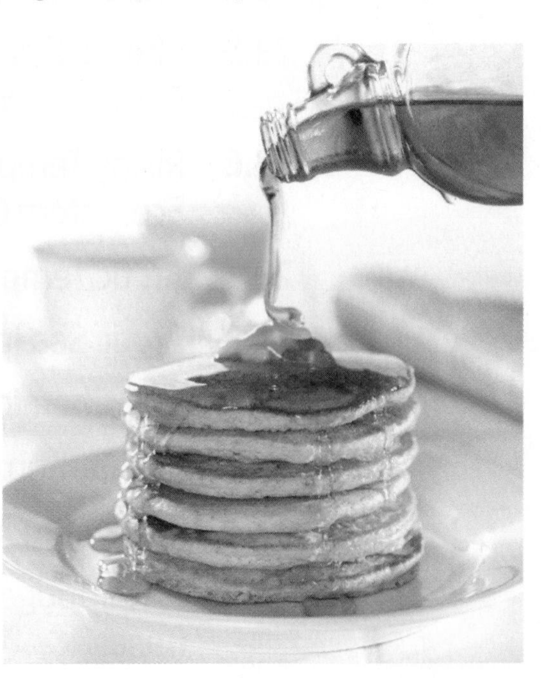

One out of 4 trees in Vermont is a sugar maple.

ecologist Tim Perkins, director of the Proctor Maple Research Center at the University of Vermont, testified to Congress in 2007, "If the northeast regional climate continues to warm as projected, we expect that the maple industry in the U.S. will become economically untenable during the next 50-100 years." This is not just icing on the cake; according to Perkins, "the total economic impact of maple in Vermont alone is nearly $200 million each year."

"The total economic impact of maple in Vermont alone is nearly $200 million each year."
—Tim Perkins

Before 1900, 80% of the world's maple syrup came from trees in the United States, the rest from Canada. Today, the pattern is reversed, with Canada greatly out-producing the United States. Canada now accounts for about 80% of world maple syrup production. While part of this reversal has to do with marketing, Canadian government subsidies, and improved technologies, scientists believe that climate change is a significant contributing factor, putting New England sugar farmers at a competitive disadvantage **(Infographic 23.1)**.

New England's maples are not the only ones feeling the heat. Plant and animal species throughout the world—from herbs in Switzerland to starfish in California—are being affected by rising temperatures. Some are shifting their ranges as a result: many historically subtropical aquatic animals, such as seahorses and turtles, are drifting toward the coasts of northern England and Scotland, where ocean temperatures are warmer than they used to be. And fish that were once wholly tropical are turning up in North Atlantic waters. Other organisms that cannot easily relocate, such as plants and mountain-dwelling animals, are being driven to extinction.

Climate change is a natural part of the environment, of course, and nothing new in the long history of earth. But scientists are finding increasing and compelling evidence that humans are accelerating the pace of change, with potentially dire consequences for life on our planet.

To Everything a Season

In nature, timing is everything. And for many species temperature is nature's clock, cueing their seasonally appropriate tasks such as mating or producing flowers in the springtime. Rising temperatures around the globe are interfering with these natural rhythms. Many plants are flowering earlier now than they once did; animals–an example is the yellow-bellied marmot–are emerging from hibernation earlier; and many bird and butterfly species are migrating north and breeding earlier in the spring than they did a few decades ago. It's a pattern of change that scientists are seeing around the globe **(Infographic 23.2)**.

So what, you might ask, if flowers bloom earlier or marmots shake off their wintry slumber earlier in the season? Because living things are exquisitely adapted to their environments, a change in one part of an ecosystem may upset others.

As the name implies, an **ecosystem** is a complex, interwoven system of interacting components. It includes both the community of living organisms present in an area and features of the nonliving environment–physical conditions such as temperature and moisture and chemical resources found in soil, water, and

ECOSYSTEM
The living and nonliving components of an environment, including the communities of organisms present and the physical and chemical environment with which they interact.

INFOGRAPHIC 23.1

Vermont Maple Syrup: A Thing of the Past?

The amount of maple syrup produced in Vermont has been declining, in part because of a shortening of the maple syrup season in Vermont. Meanwhile, Canadian production has been increasing because of increased marketing, government subsidies, improved technologies, and, likely, climate change.

Maple Syrup Production 1916–2000

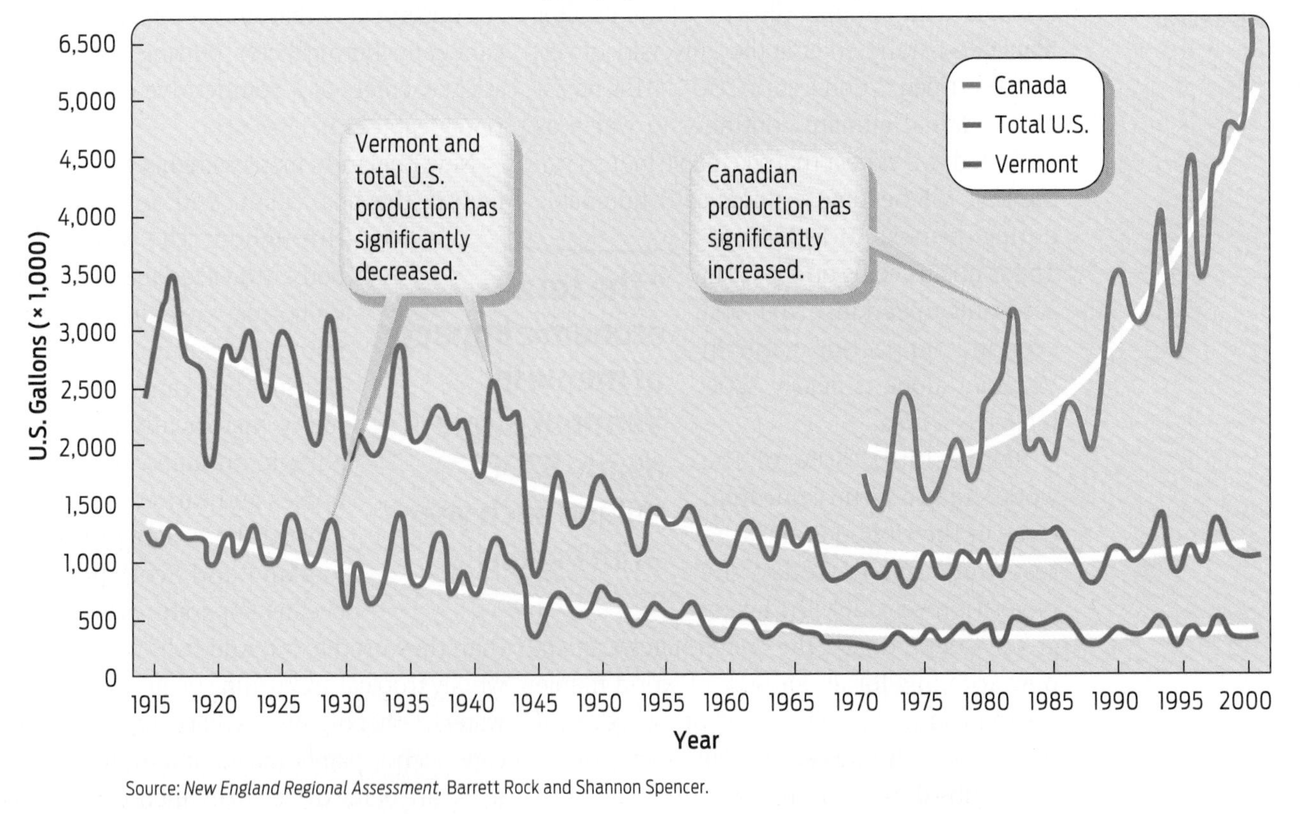

Source: *New England Regional Assessment*, Barrett Rock and Shannon Spencer.

INFOGRAPHIC 23.2

Rising Temperatures Affect Plant Behavior

Global warming is changing the seasonal behavior of plants and animals. Near Oxford, England, many plants are flowering earlier now than they did between 1954 and 1990. In fact, the average first date of flowering is now 4.5 days earlier than the long-term historic average (1954–1990). One species studied is now flowering 55 days sooner than it did in previous decades.

Plants are flowering earlier:

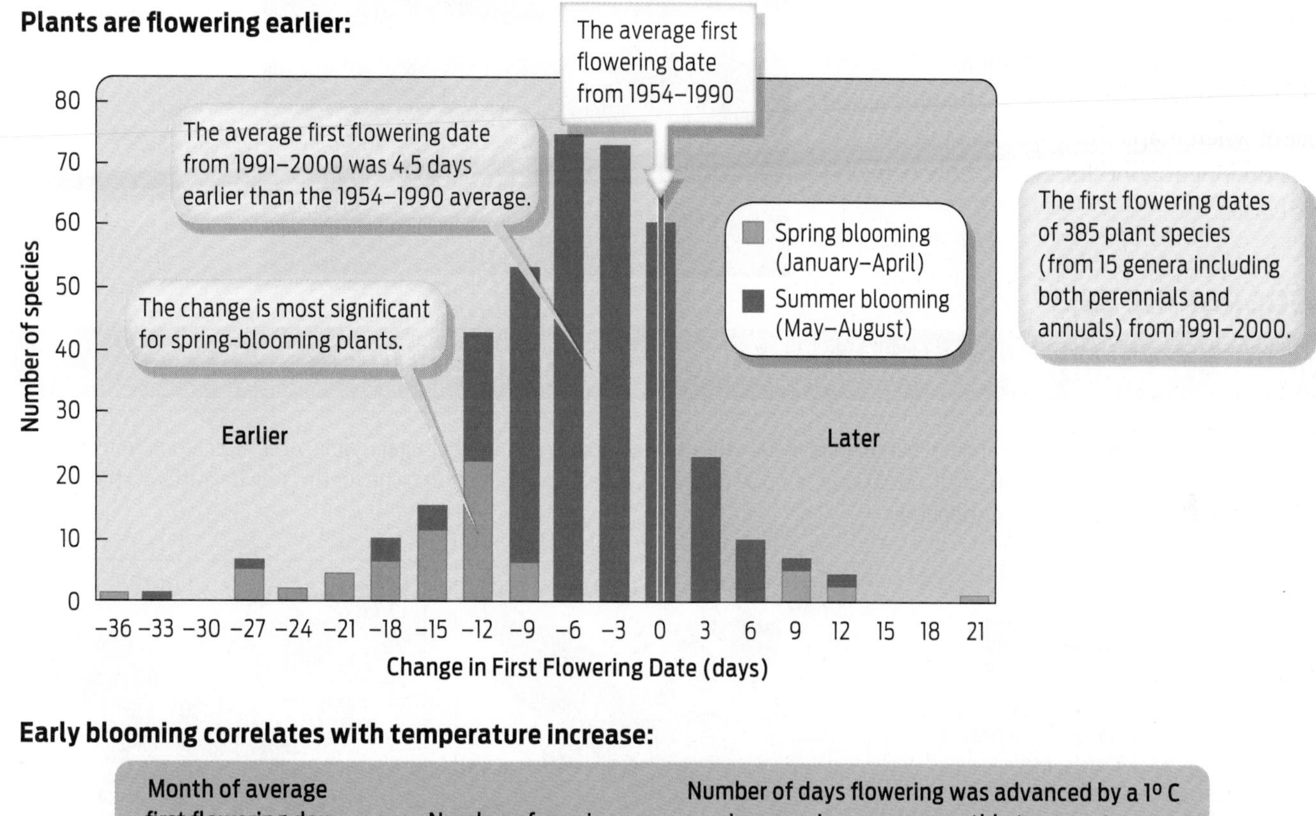

Early blooming correlates with temperature increase:

Month of average first flowering day	Number of species	Number of days flowering was advanced by a 1° C increase in average monthly temperature
February	12	6.0 days earlier
March	22	4.5 days earlier
April	63	4.3 days earlier
May	105	2.0 days earlier
June	108	1.7 days earlier
July	67	2.8 days earlier

Rapid Changes in Flowering Time in British Plants. A. H. Fitter and R. S. R. Fitter (2002) *Science*, Vol 296, p. 1689–1691.

air. Because the biotic and abiotic parts of an ecosystem can and do change, ecosystems are not static entities but dynamic systems. And because the parts of an ecosystem are so interconnected, a small change in one part of an ecosystem can have a domino effect.

No one knows this better than sugar farmers. "The flow of sap from maple trees during the spring season is controlled almost entirely by the daily fluctuation in temperature," explains ecologist Perkins. "Small changes in the day-to-day temperature pattern will have large consequences on sap flow."

Historically, trees have been tapped in early March when the sap began to flow; the sap was then collected for the next 6 weeks. But about 10 years ago, Perkins started getting calls from sugar producers saying that they were tapping earlier and making syrup earlier. Curious, he and his colleagues decided to investigate. They scoured historical records and surveyed thousands of maple sugar producers in New England. Their results were startling: over a mere 40 years, between 1963 and 2003, the start of the tapping season had moved forward by about 8 days. Even more significant, the end of the season, when maples begin to leaf out and the sap is no longer good for syrup, now comes 11 days earlier.

Collecting sap from a sugar maple tree (*Acer saccharum*) to make maple syrup.

INFOGRAPHIC 23.3

Maple Tree Range Is Affected by Increasing Temperature

Models including 25 environmental parameters predict the rapid disappearance of the sugar maple, *Acer saccharum*, from the United States with even small increases in temperature. As the ideal environmental niche for this tree migrates north into Canada, so does the tree population.

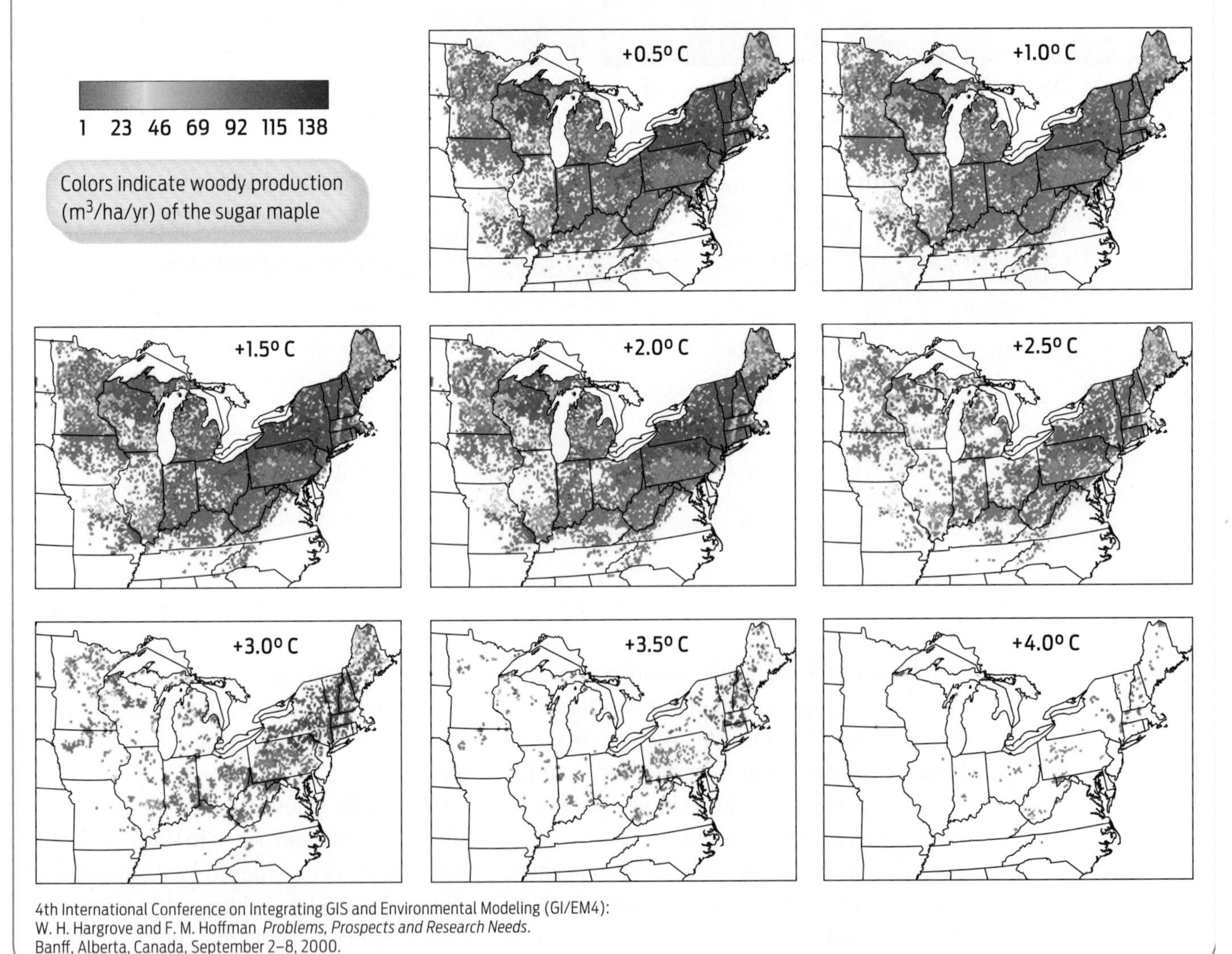

4th International Conference on Integrating GIS and Environmental Modeling (GI/EM4): W. H. Hargrove and F. M. Hoffman *Problems, Prospects and Research Needs.* Banff, Alberta, Canada, September 2–8, 2000.

HABITAT
The physical environment where an organism lives and to which it is adapted.

BIOME
A large geographic area defined by its characteristic plant life, which in turn is determined by temperature and levels of moisture.

"Over that 40-year time period we've lost about 3 days of the season," Perkins told Vermont Public Radio. "That doesn't seem like a lot until you realize that the maple production season averages about 30 days in length. So we've lost about 10% of the season."

To some extent, losses from a shortened tapping season have been offset by improved sap-removal technologies that make it possible to extract sap even under poor conditions. The bigger problem is what will happen if climate changes so much that New England no longer provides a suitable **habitat** for maple trees.

As Perkins testified to Congress, current climate models predict that by the end of the century New England's forests will more closely resemble those of present-day Virginia, North Carolina, and Tennessee, dominated by hickory, oak, and pine rather than maple, beech, and birch (Infographic 23.3). If that happens, not only maple syrup but the brilliant fall foliage New England is famous for will be a thing of the past.

New England's colorful foliage is part of a **biome** known as temperate deciduous forest. Biomes are large, geographically cohesive regions whose defining vegetation–its plant life–is determined principally by climatic factors such as temperature and rainfall (see Up Close: Biomes). Climate change is beginning to alter the boundaries and plant composition of

UP-CLOSE Biomes

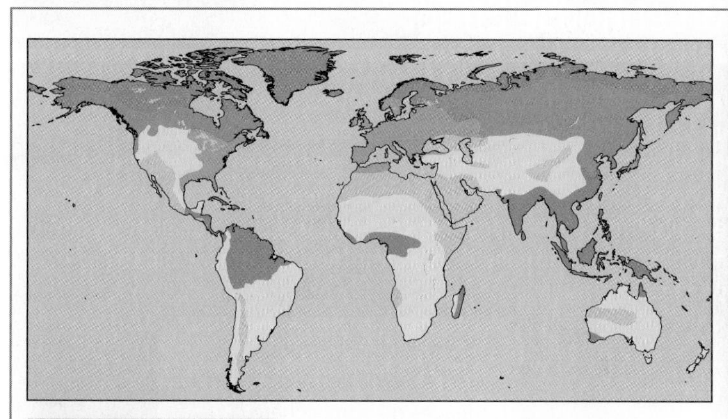

Temperate Deciduous Forest:
A biome characterized by trees that drop their leaves in winter. Winters are much colder than summers.

Tropical Forest:
Tropical forests are biomes characterized by warm temperatures and sufficient rainfall to support the growth of trees. Tropical forests may be deciduous or evergreen, depending on the presence or absence of a dry season.

Desert:
A biome characterized by extreme dryness. Cold deserts experience cold winters and hot summers, while hot deserts are uniformly warm throughout the year.

Grassland:
A biome characterized by perennial grasses and other nonwoody plants. In North America, the prairies are examples of grasslands.

Coniferous Forest:
A biome characterized by evergreen trees, with long and cold winters and only short summers.

Aquatic: Marine
This biome covers about three-fourths of the earth and includes the oceans, coral reefs, and estuaries.

Tundra:
A biome that occurs in the Arctic and mountain regions. Tundra is characterized by low-growing vegetation and a layer of permafrost (frozen all year long) very close to the surface of the soil.

Aquatic: Freshwater
A biome characterized by having a low salt concentration. Freshwater biomes include ponds and lakes, rivers and streams, and wetlands.

temperate deciduous forests, and may one day push sugar maples north into Canada.

Such changes are not uncommon—climate change is beginning to redraw the map of biomes around the world. In northern Alaska, where once there was only sparsely vegetated tundra, woody shrubs now grow. When Montana's Glacier National Park was opened in 1910, it held approximately 150 large glaciers; in 2010, there were only 25. As the vegetation in these landscapes changes, so will the community of organisms that rely on it for food and habitat.

When Montana's Glacier National Park was opened in 1910, it held approximately 150 large glaciers; in 2010, there were only 25.

Warming Planet, Diminishing Biodiversity

Although temperature swings and shifts in the ranges of organisms are natural phenomena, the amount of warming in recent years is unprecedented, and evidence suggests that the change is not merely part of a natural cycle. From 1880 until 2010, the earth's surface has warmed, on average, by about 0.8°C (1.4°F), according to a 2010 study by NASA's Goddard Institute for Space Studies. That may not sound like a lot. But consider this: the difference in global average temperatures between today and the last ice age—when much of North America was buried under ice—is only about 5°C (9° F). Where global temperatures are concerned, even a 1-degree change is significant.

The rate of warming has increased as well. Eighteen of the warmest years on record occurred in just the past 20 years. The last decade, from 2000 through 2010, was the hottest decade so far, with 2010 tying 2005 for the title of hottest year on record. Much of this warming is attributable to the **greenhouse effect,** the trapping of heat in earth's atmosphere. As sunlight shines on our planet, it warms the earth's surface. This heat radiates back to the atmosphere, where it is absorbed by **greenhouse gases** such as carbon dioxide. The heat trapped by greenhouse gases raises the temperature of the atmosphere, and in turn, the surface of the earth **(Infographic 23.4)**.

The greenhouse effect is a natural process that helps maintain life-supporting temperatures on earth. Without this greenhouse effect, the average surface temperature of the planet would be a frigid –18°C (0°F). In recent years, however, rising levels of greenhouse gases have increased the strength of the greenhouse effect, a phenomenon known as the enhanced greenhouse effect. As the amount of greenhouse gases in the atmosphere has increased, so have temperatures. The result is **global warming**, an overall

GREENHOUSE EFFECT
The normal process by which heat is radiated from the earth's surface and trapped by gases in the atmosphere, helping to maintain the earth at a temperature that can support life.

GREENHOUSE GAS
Any of the gases in earth's atmosphere that absorb heat radiated from the earth's surface and contribute to the greenhouse effect, for example carbon dioxide and methane.

GLOBAL WARMING
An increase in the earth's average temperature.

INFOGRAPHIC 23.4

The Greenhouse Effect

The greenhouse effect is a natural process that helps maintain steady and life-sustaining surface temperatures on earth. Sunlight heats the surface of the earth and that heat radiates back to the atmosphere. While some of the heat escapes to space, certain gases in the earth's atmosphere, known as greenhouse gases, trap heat within the atmosphere. This trapped heat warms the atmosphere and the earth's surface.

2. Some of the light and heat reflected off the earth's surface leaves the atmosphere.
1. The sun's energy enters the earth's atmosphere and heats its surface.
3. Greenhouse gases, including carbon dioxide, methane, and nitrous oxide, absorb some of the radiated heat, keeping the atmosphere and the surface of the earth warm.

INFOGRAPHIC 23.5

The Earth's Surface Temperature Is Rising with Carbon Dioxide Levels

As measured directly by thermometers, and as documented by historical records and other biological indicators (including tree rings, corals, and ice cores) the temperature on earth has increased rapidly in the past 140 years, along with increasing levels of carbon dioxide.

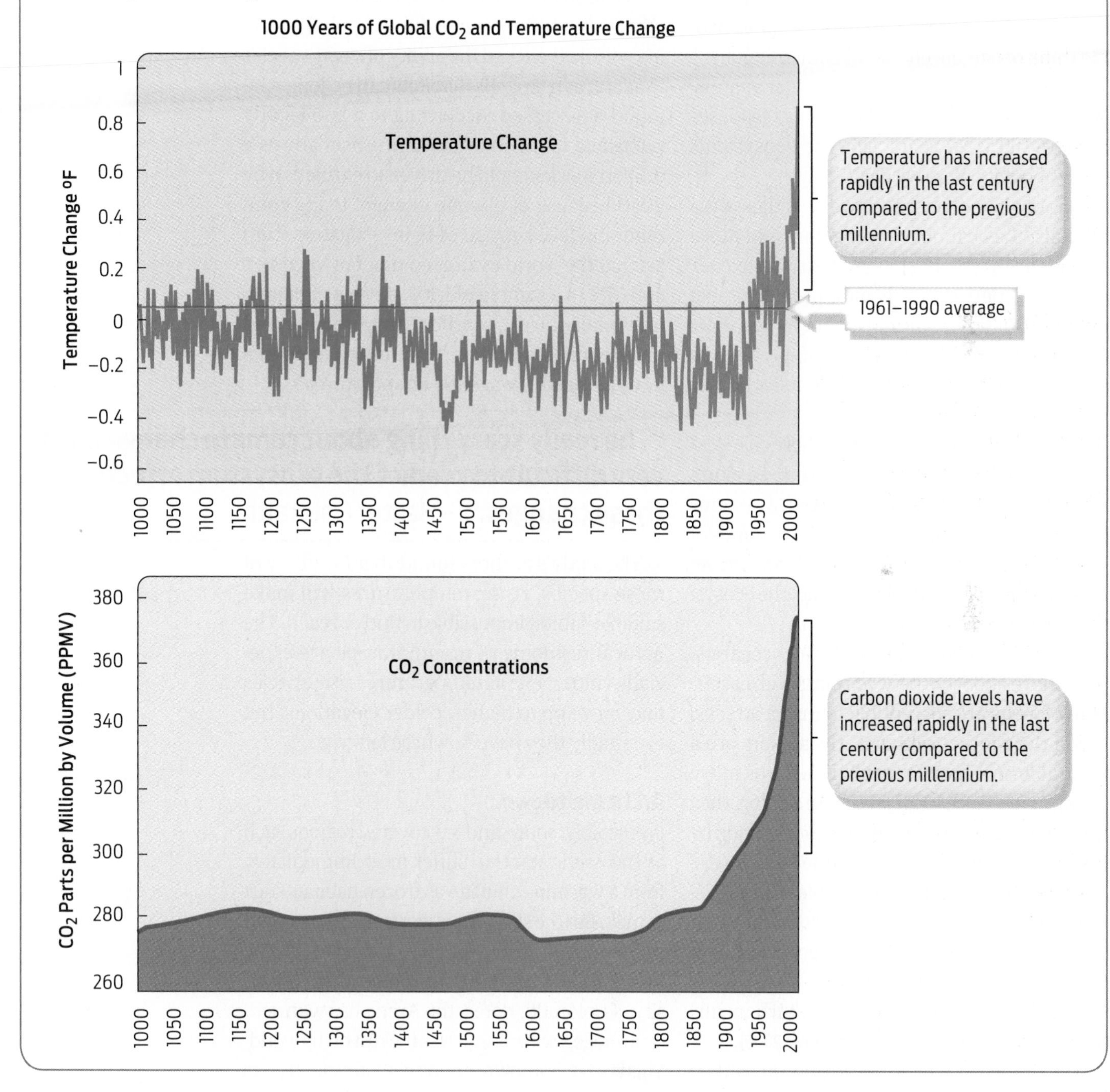

increase in the earth's average temperature **(Infographic 23.5)**.

For ecologist Hector Galbraith, director of the Climate Change Initiative at the Manomet Center for Conservation Sciences, in Massachusetts, one of the most worrying things about climate change is how quickly it is happening, and how sensitive species are to the changes. "Most people think of climate change as something that's 30 years out," says Galbraith. But that's simply not true, he notes. "We began seeing responses in ecosystems 20 years ago. The ecosystems knew about it before we did."

Plants, of course, are slower to adapt than animals; they cannot simply get up and move (although they may change their range over time by dispersing seeds into more favorable climes). But some animals can change their ranges quite quickly. "A bird can simply open its wings, and within two hours it's 50 miles farther north," says Galbraith.

What will be the outcome of all these changes? The answer, says Galbraith, is that we don't really know. "We're seeing changes to systems that have been relatively stable for thousands of years . . . . The really scary thing about climate change is it's very difficult to predict the ecosystem effects of these changes."

Nevertheless, there are disturbing scenarios. Take the relationship between birds and insects. Many forests are susceptible to insect attacks. Given their insect-rich diet, flycatchers are a natural form of pest control. If the birds move north, as evidence suggests they are doing, they leave behind a forest susceptible to predation by insects that might be less vulnerable to the changed climate or more adaptable. The maple-tree-loving pear thrip and the forest tent caterpillar are just two examples of insects that might be happy to see the flycatchers go. More insects means more dead trees, which in turn means more fuel for forest fires **(Infographic 23.6)**.

Not all species will be negatively affected by climate change–some may actually benefit. But one species' success in coping with climate change may contribute to another's demise. For example, the adaptable red fox (*Vulpes vulpes*), found throughout the northern hemisphere, is venturing into the range of the endangered Arctic fox (*Vulpes lagopus*), whose habitat–the Arctic tundra–has gotten warmer. When the two species share a range, the Arctic fox inevitably suffers because the red fox out-competes it for food and also preys on Arctic fox pups.

While some species can adapt to a changing climate by shifting range, future global warming will likely exceed the ability of many species to adapt, as hospitable habitats can no longer be found or accessed. According to a 2004 study published in the journal *Nature*, as many as a million species could be driven to extinction by 2050 because of climate change. Using computer models, a group of 15 investigators from around the world estimated that between 15% and 37% of a sample of 1,103 species of plants, mammals, birds, reptiles, amphibians, and invertebrates would be "committed to extinction" because of warming temperatures.

"The really scary thing about climate change is it's very difficult to predict the ecosystem effects of these changes." –Hector Galbraith

The study's authors found that for many of these species, rising temperatures will make suitable habitat impossible to find or reach. The natural residents of mountaintops are especially vulnerable: as temperatures rise, species may move up to higher, colder elevations, but eventually they have nowhere left to go.

Arctic Meltdown

Predictably, snow- and ice-covered regions such as the Arctic stand to suffer most immediately from a warming climate, as frozen habitats start to melt. But the situation is worse than one might imagine. As Mark Serreze, director of the National Snow and Ice Data Center at the University of Colorado, notes, the Arctic has warmed, on average, twice as much as the rest of the world. This is what is known among climate scientists as Arctic amplification, and it has to do with how sea ice affects temperature. As Serreze explains, sea ice both reflects solar radiation and insulates the ocean. As global temperatures rise, ice begins to melt. With less sea ice, more solar radiation is

INFOGRAPHIC 23.6

Rising Temperatures Mean Widespread Ecosystem Change

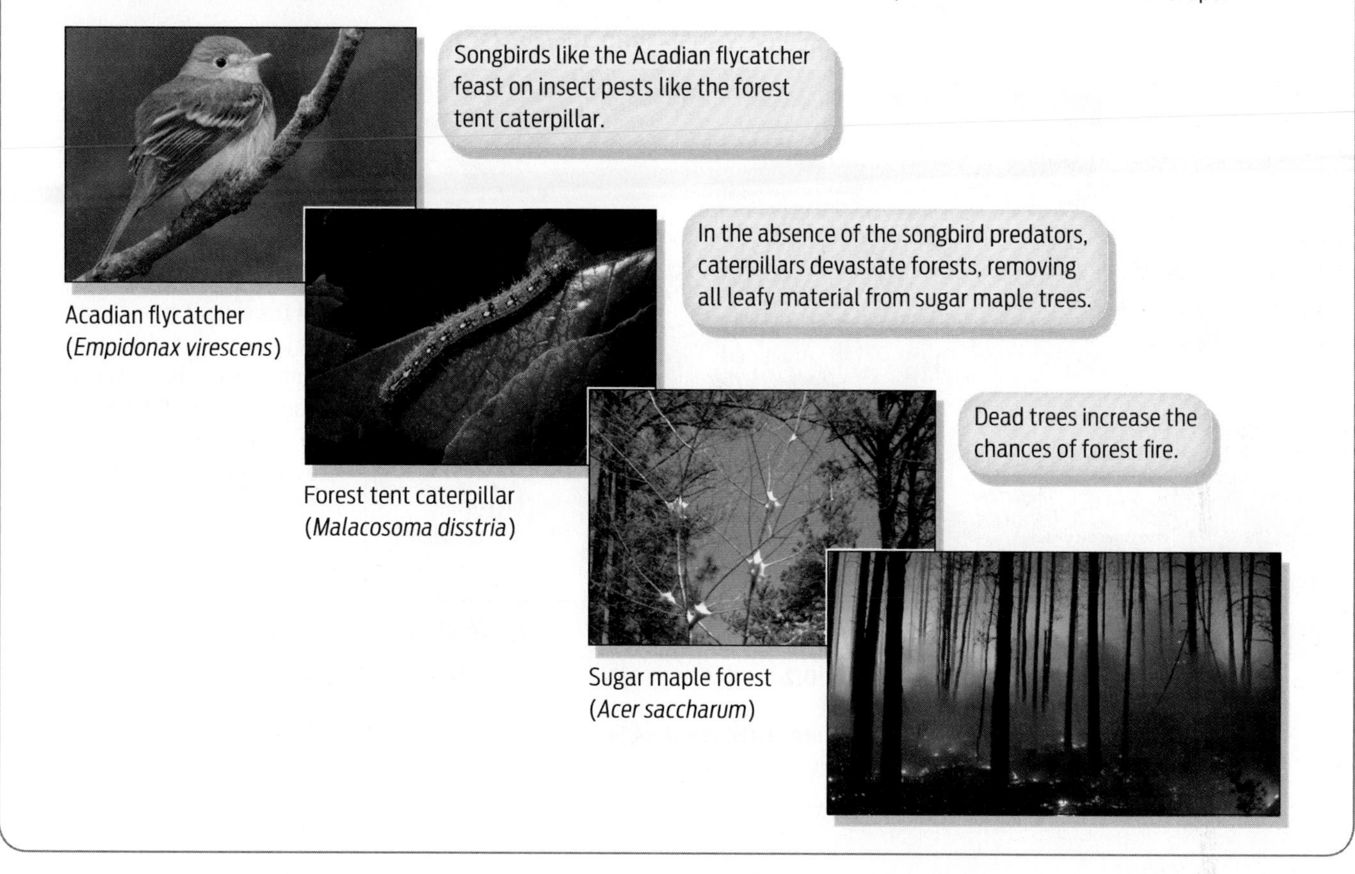

Climate change is having dramatic impacts on entire ecosystems. With warming temperatures, songbirds are expanding their habitat into more northern territories. As birds move northward, they leave behind their insect prey that are free to devastate the now unprotected forests. Dead trees, in turn, lead to more forest fires, which further alter the landscape.
Songbirds like the Acadian flycatcher feast on insect pests like the forest tent caterpillar.
Acadian flycatcher (*Empidonax virescens*)
In the absence of the songbird predators, caterpillars devastate forests, removing all leafy material from sugar maple trees.
Forest tent caterpillar (*Malacosoma disstria*)
Dead trees increase the chances of forest fire.
Sugar maple forest (*Acer saccharum*)

absorbed by the ocean and more of the relatively warm ocean is exposed to air, raising the air temperature even more. It's a positive feedback loop, which means that as additional ice is lost, temperatures will rise at an accelerated pace.

According to the extensive Arctic Climate Impact Assessment, the result of 4 years' work by more than 300 scientists around the world published in 2004, Arctic temperatures are projected to rise by an additional 4°-7°C (7°-13°F) over the next 100 years **(Infographic 23.7)**.

Warming temperatures could spell disaster for species that call the Arctic their home. Polar bears, for example, spend most of the year roaming the Arctic on large swaths of floating sea ice that blanket a good portion of the Arctic Ocean from September through March. The massive mammals use the sea ice to hunt for seals, which periodically pop up through "whack-a-mole"-like breathing holes in the ice and are nabbed by bears. Yet the size of this frozen habitat has been shrinking, greatly reducing the bears' ability to obtain food.

Moreover, over the past few decades the ice has been breaking up earlier and earlier in spring. The sea ice in Hudson Bay, Canada, for example, now breaks up nearly 3 weeks earlier than it did in the 1970s. In the absence of unbroken summer sea ice, the polar bears are stuck on land (where there are no seals), or are forced to swim long distances to reach sea ice. Some, exhausted by the journey, drown. Those that do survive have fewer opportunities to hunt. Canadian polar bears now weigh on average 55 pounds less than they did 30 years ago, seriously compromising their reproductive ability.

Scientists have monitored sea ice on a daily basis by satellite since 1979. Over the past three

INFOGRAPHIC 23.7

Arctic Temperatures Are Rising Fast

Current measurements suggest that the Arctic is warming faster than other parts of the earth. 2008 was the ninth warmest year on record (since measurements began in 1880). Much of the earth was warmer in 2008 than in the period between 1951 and 1980 (regions in yellow, orange, red, and brown). The Arctic, Antarctic, and Eurasia warmed more than the rest of the planet.

2008 Surface Temperature Change Compared to 1951–1980 Average (ºC)

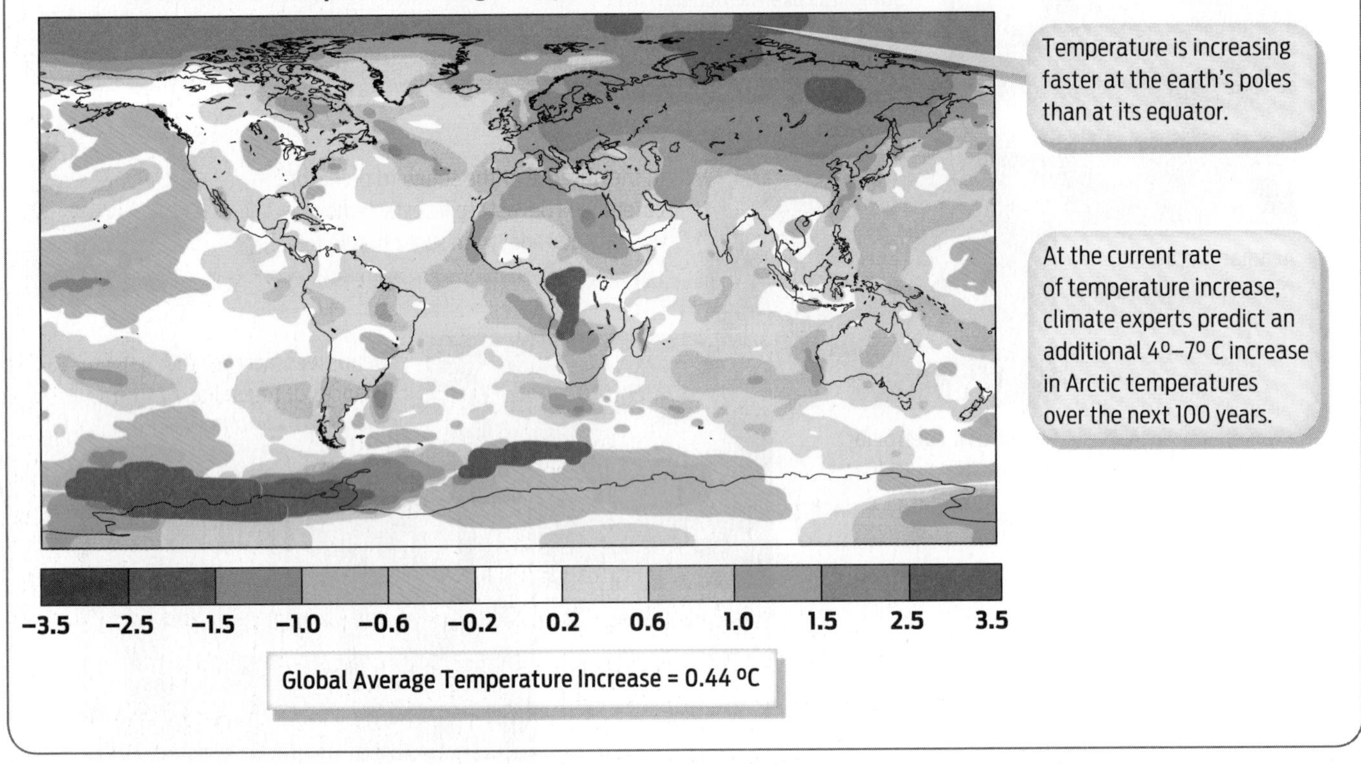

decades, the area of Arctic sea ice has shrunk by more than 1 million square miles, an area roughly four times the size of Texas, according to Walt Meier, a research scientist with the National Snow and Ice Data Center in Boulder, Colorado. Arctic sea ice hit a record low in September 2007, at the end of the summer melt season, shrinking to a level that climate change models had predicted wouldn't happen until at least 2050. Scientists now fear that nearly all of the polar bear's summer sea ice could vanish by 2040—possibly sooner **(Infographic 23.8)**.

Temperature is increasing faster at the earth's poles than at its equator.
At the current rate of temperature increase, climate experts predict an additional 4º–7º C increase in Arctic temperatures over the next 100 years.
−3.5 −2.5 −1.5 −1.0 −0.6 −0.2 0.2 0.6 1.0 1.5 2.5 3.5
Global Average Temperature Increase = 0.44 ºC

Warming temperatures are also causing glaciers and ice caps on land to melt. Unlike sea ice, which, like an ice cube in a glass of water, doesn't raise the water level as it melts, melting glaciers and ice caps do. How much will seas rise? "By 2100, you're looking at probably about a meter," says Serreze. "Here in Boulder we're at 5,400 feet, [so] we're not worried about that. But if you're living in Miami, this is something that should concern you."

Scientists now fear that nearly all of the polar bear's summer sea ice could vanish by 2040.

It's important to note that much of the data we have on climate change relates to global, long-term trends. From year to year, there may be slight variations—slightly warmer summers and less sea ice one year, slightly cooler summers and more sea ice the next. And indeed, from its all-time low in 2007, sea ice did indeed bounce back a bit in 2008 and 2009. But the trend is still unmistakably downward—toward less sea ice. By 2030 or 2040, says Serreze, there could be no summer ice to speak of. "You could take a ship across the north pole."

The evidence we have for global warming is clear, unmistakable, and alarming. Yet despite such evidence, recent surveys of public opinion show that large percentages of the American and British public do not believe in the reality of global warming. In response to such views, and media controversy, a group of roughly 250 scientists–all members of the U.S. National Academy of Science–signed a letter testifying to the legitimacy of climate data and climate science. Published in the May 7, 2010, issue of the journal *Science*, the letter concludes that the data we currently have establish with a 90% degree of confidence that the planet is warming.

Follow the Carbon

What's behind this planetary warming? The immediate cause is a fired-up greenhouse effect. And that, scientists argue, is the result of human activity. As they testified in their letter, "There is compelling, comprehensive, and consistent objective evidence that humans are changing the climate in ways that threaten our societies and the ecosystems on which we depend." How did we get to be the culprits in this situation? In short, by pumping more carbon dioxide into the atmosphere.

Carbon dioxide is the most notorious player in the greenhouse effect, and scientists believe it is responsible for most of the warming. In fact, atmospheric carbon dioxide concentrations are higher now than they have been in more than 700,000 years.

As we saw in Chapter 2, carbon is a natural ingredient in every living organism, part of the

INFOGRAPHIC 23.8

Arctic Sea Ice Is Melting

Rising temperatures have caused the polar ice cap to melt and break apart earlier in the season. The reduction in the extent of summer sea ice is threatening the survival of polar bears, which require the sea ice to hunt for seals.

Since 1979, more than 20% of the polar ice cap has melted away.

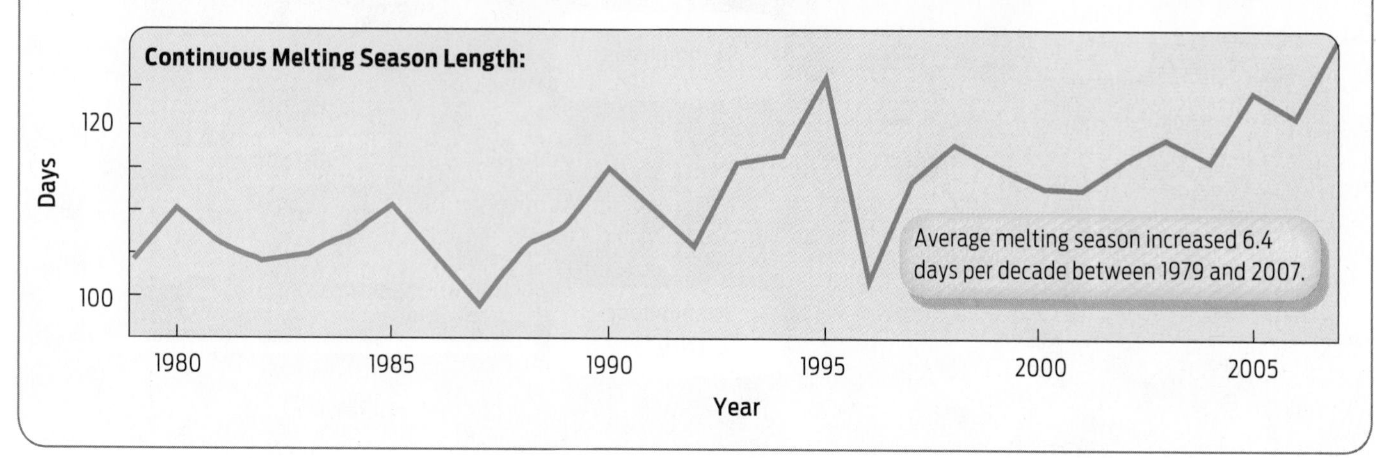

backbone of all organic molecules. Carbon also exists in inorganic forms: as carbon dioxide in the atmosphere, as carbonic acid dissolved in water, as calcium carbonate in limestone rocks. If dead organisms are fossilized before being digested by decomposers, the organic molecules contained within their bodies become trapped below the earth's surface or under the seas. Over time, these compressed organic molecules turn into **fossil fuels**–coal, oil, and natural gas.

Like other chemical elements, the total amount of carbon on earth remains essentially constant. In contrast to the way energy flows through an ecosystem in one direction (from the sun to producers to consumers and out to the universe as heat; Chapter 22), elements such as carbon move in cycles. Individual carbon atoms are recycled as carbon-based organisms die and new life is born, and as geological processes slowly reshuffle carbon in the nonliving environment. The movement of carbon through the environment follows a predictable pattern known as the **carbon cycle.**

At it cycles through the environment, carbon moves between organic and inorganic forms. For

FOSSIL FUEL
A carbon-rich energy source, such as coal, petroleum, or natural gas, formed from the compressed, fossilized remains of once-living organisms.

CARBON CYCLE
The movement of carbon atoms between organic and inorganic molecules in the environment.

UP CLOSE Chemical Cycles: Nitrogen and Phosphorus

Nitrogen atoms cycle between different chemical and biochemical compounds as they move from organisms to the soil, water, and air and back to organisms. A variety of natural processes as well as some human activities contribute to the transformation and movement of nitrogen through the ecosystem.

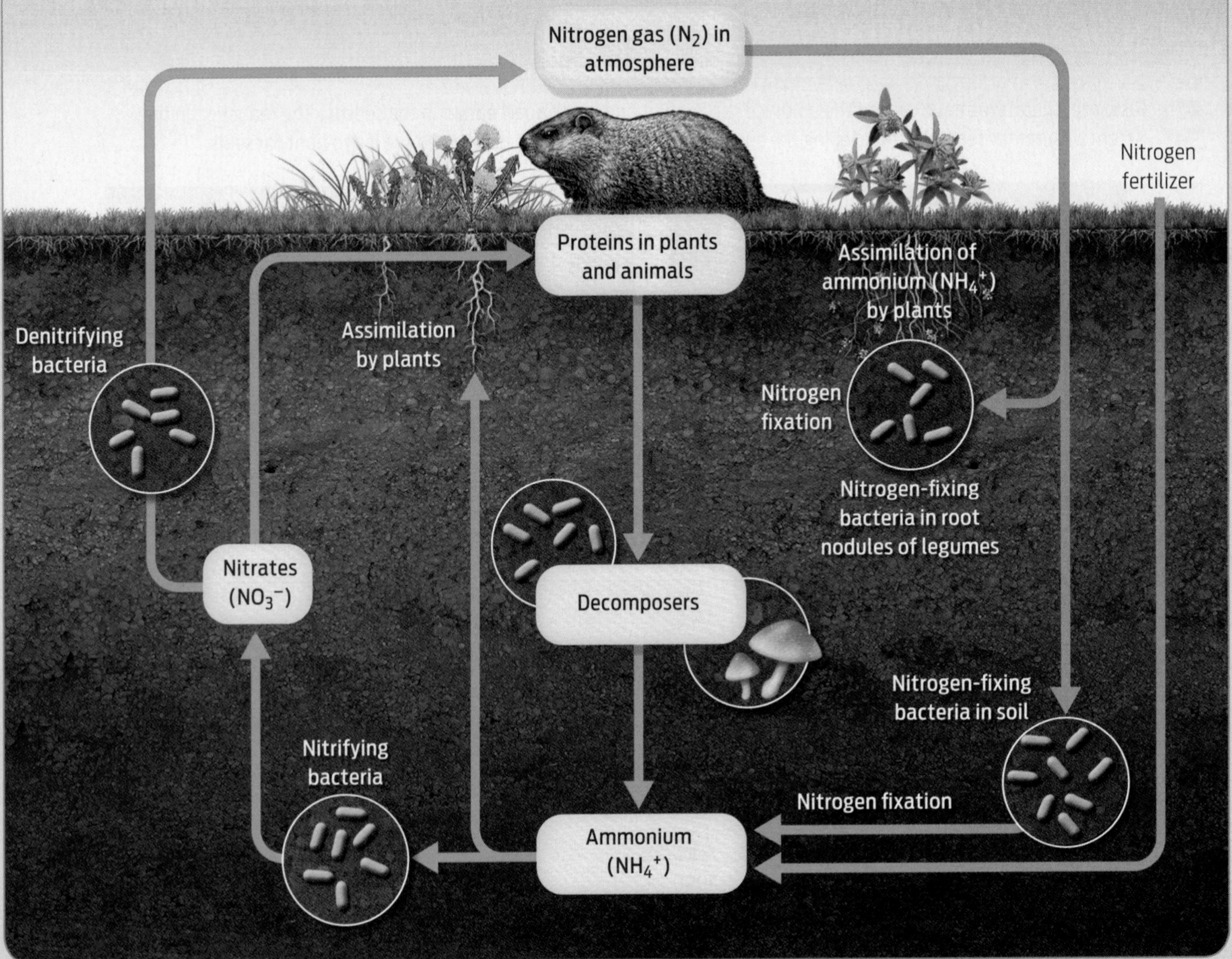

Nitrogen gas (N_2) in atmosphere
Nitrogen fertilizer
Proteins in plants and animals
Assimilation of ammonium (NH_4^+) by plants
Denitrifying bacteria
Assimilation by plants
Nitrogen fixation
Nitrogen-fixing bacteria in root nodules of legumes
Nitrates (NO_3^-)
Decomposers
Nitrogen-fixing bacteria in soil
Nitrifying bacteria
Nitrogen fixation
Ammonium (NH_4^+)

example, animals take in organic carbon when they eat other organisms and release inorganic gaseous CO_2 into the atmosphere as a by-product of cellular respiration. Similarly, when organisms die, decomposers in the soil use the dead organic material for food and energy, releasing some of the carbon during respiration as CO_2.

In turn, plants, photosynthetic bacteria, and algae take up CO_2 during photosynthesis and fix it into organic sugar molecules, thus reducing atmospheric CO_2 levels. Photosynthesis, respiration, and decomposition form a cycle that keeps carbon dioxide at a relatively stable level in the atmosphere. But human actions, such as deforestation and the burning of fossil fuels, can inject additional carbon dioxide into the cycle **(Infographic 23.9)**.

Note that carbon isn't the only element that cycles through ecosystems. Other elements, such as nitrogen, phosphorus, and sulfur, as well as water (Chapter 24), also follow natural cycles (see **Up Close: Chemical Cycles**). But it's the carbon cycle that is most relevant to the phenomenon of global warming.

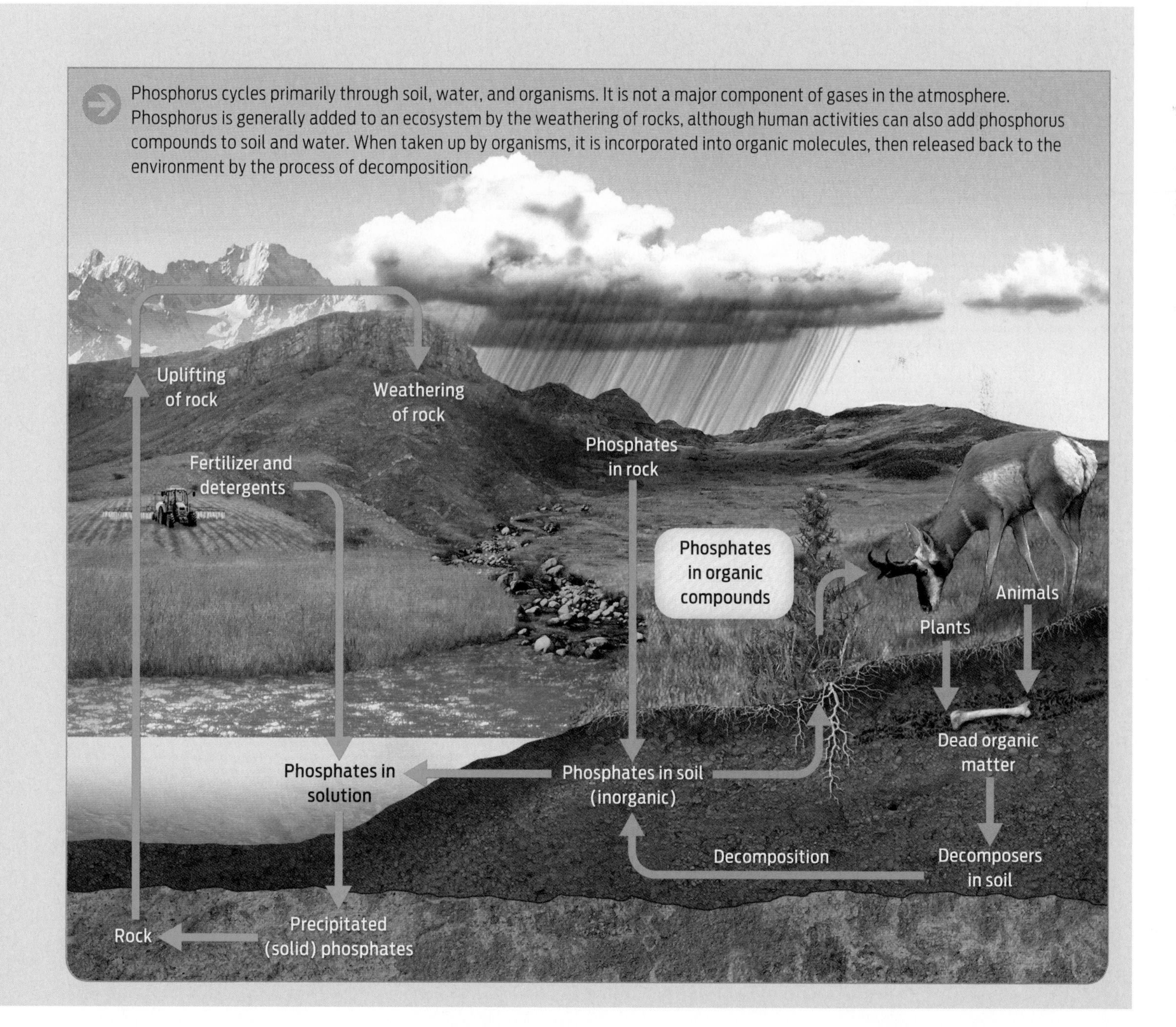

Phosphorus cycles primarily through soil, water, and organisms. It is not a major component of gases in the atmosphere. Phosphorus is generally added to an ecosystem by the weathering of rocks, although human activities can also add phosphorus compounds to soil and water. When taken up by organisms, it is incorporated into organic molecules, then released back to the environment by the process of decomposition.
Uplifting of rock
Weathering of rock
Phosphates in rock
Fertilizer and detergents
Phosphates in organic compounds
Animals
Plants
Dead organic matter
Phosphates in solution
Phosphates in soil (inorganic)
Decomposition
Decomposers in soil
Rock
Precipitated (solid) phosphates

INFOGRAPHIC 23.9

The Carbon Cycle

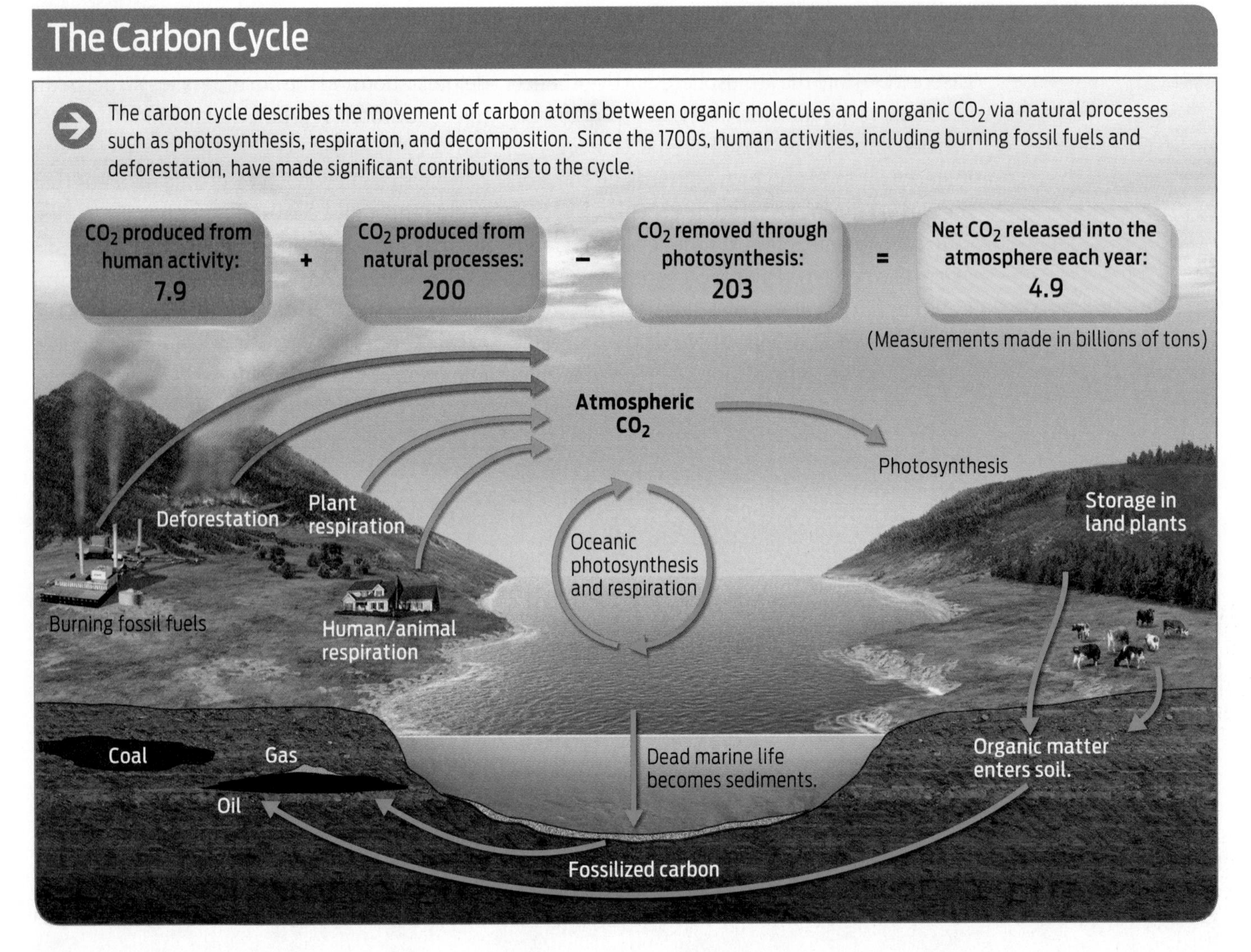

The carbon cycle describes the movement of carbon atoms between organic molecules and inorganic CO_2 via natural processes such as photosynthesis, respiration, and decomposition. Since the 1700s, human activities, including burning fossil fuels and deforestation, have made significant contributions to the cycle.
CO_2 produced from human activity: 7.9
+
CO_2 produced from natural processes: 200
–
CO_2 removed through photosynthesis: 203
=
Net CO_2 released into the atmosphere each year: 4.9
(Measurements made in billions of tons)
Atmospheric CO_2
Photosynthesis
Storage in land plants
Deforestation
Plant respiration
Oceanic photosynthesis and respiration
Burning fossil fuels
Human/animal respiration
Coal
Gas
Oil
Dead marine life becomes sediments.
Organic matter enters soil.
Fossilized carbon

For the most part, the amount of carbon present in the atmosphere as carbon dioxide has remained fairly constant. But since the late 1700s, with the rise of industry and the internal combustion engine, people have begun to alter the carbon cycle, adding increasing amounts of CO_2 to the atmosphere.

Before the industrial revolution, the carbon trapped in fossil fuels was not easily accessible, and therefore it wasn't cycling as part of the carbon cycle. But modern drilling and mining methods have unlocked the deep reserves of this ancient planetary energy. The CO_2 released when humans burn fossil fuels is the largest source of the carbon being added to the atmosphere by humans and is a major contributor to the enhanced greenhouse effect.

Air bubbles trapped in ice cores from the glacial ice of Greenland—an indirect measure of carbon dioxide in the atmosphere—show relatively constant amounts of CO_2 until 300 years ago. Since direct measurement of atmospheric CO_2 began late in the 19th century, its concentration has increased about 35% **(Infographic 23.10)**.

Virtually all climate scientists agree that greenhouse gases emitted by human activities—primarily driving gasoline-powered cars and burning coal to generate electricity—have caused most of the global rise in temperature observed over the past 50 years. In 2001, an international group of scientists and policymakers known as the Intergovernmental Panel on Climate Change concluded that the global

INFOGRAPHIC 23.10

Measuring Atmospheric Carbon Dioxide Levels

Ice cores provide a way to measure biological and atmospheric conditions from the distant past. Cylinders of ice representing a time frame covering thousands of years can be extracted from glaciers. Gas bubbles present in the ice reveal the atmospheric composition thousands of years ago.

Present day carbon dioxide levels are measured directly from the air:
Direct measurements of carbon dioxide are currently taken from the Mauna Loa Research Station in Hawaii.

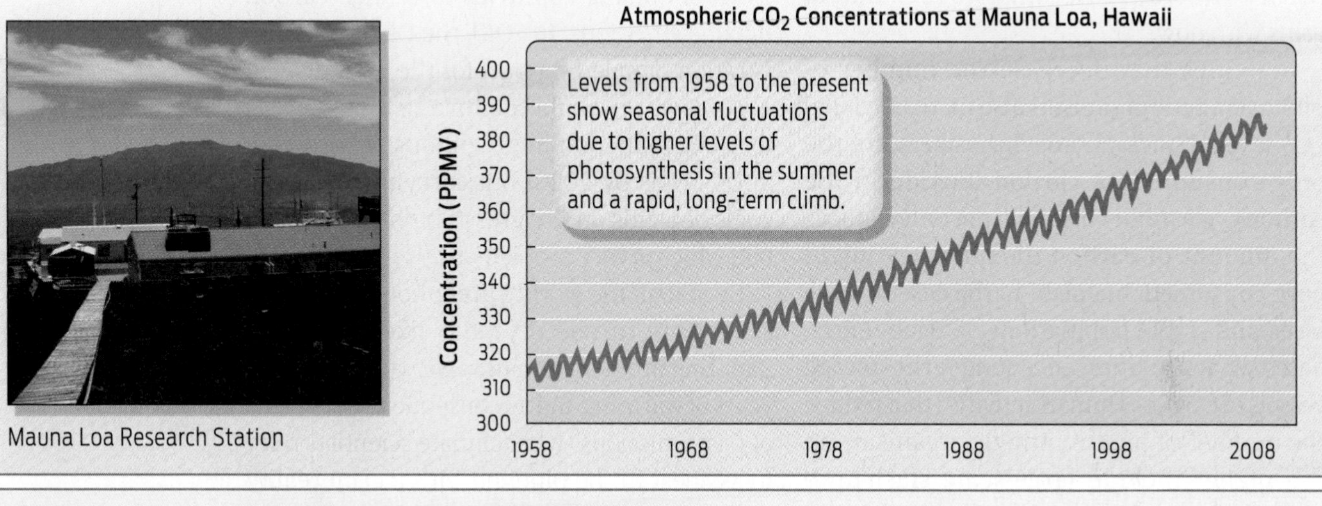

Mauna Loa Research Station

Historic carbon dioxide levels are measured in glacial ice cores:
Ice cores are long tubes of ice removed from a glacier. As each annual layer of ice was deposited on a glacier, it trapped the gas present in the earth's atmosphere at that time.

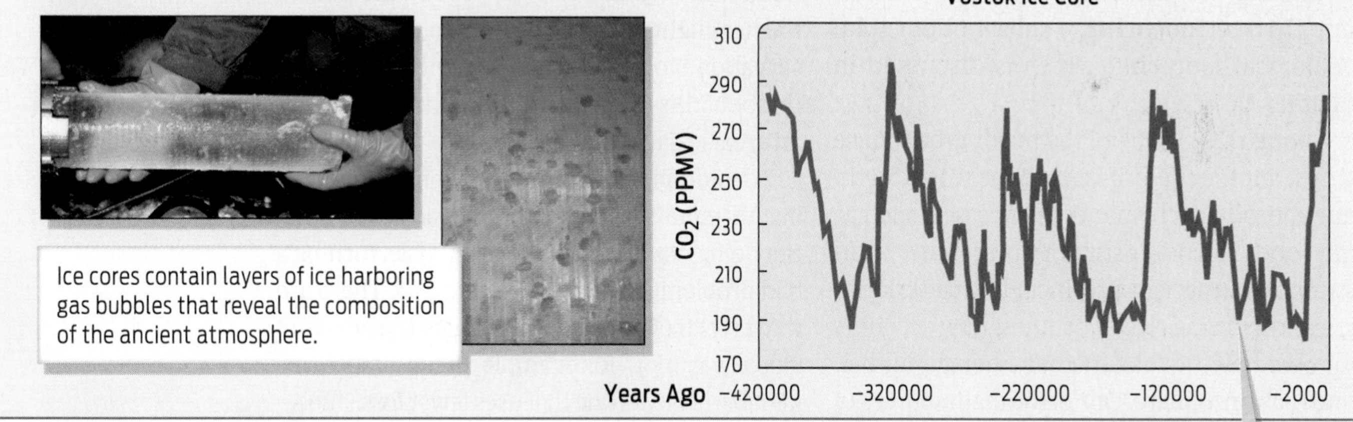

Ice cores contain layers of ice harboring gas bubbles that reveal the composition of the ancient atmosphere.

Some ice cores contain layers of ice from hundreds of thousands of years ago, and can be used to measure historical levels of CO_2. These measurements show that CO_2 levels cycle in patterns that correlate with major ice ages.

Together, these data provide a complete picture of atmospheric CO_2 levels over time:

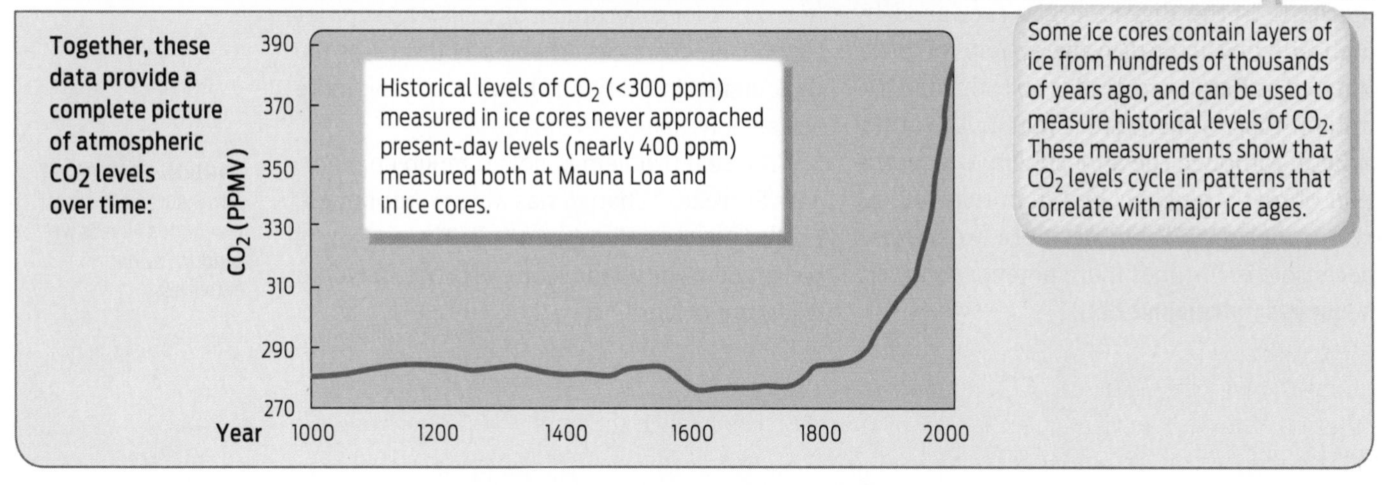

rise in average yearly temperature over the past 50 years was primarily anthropogenic—that is, caused by humans.

"Fossil fuels are incredibly efficient sources of energy," says Serreze, from the University of Colorado. "We've built our whole infrastructure around that. But what we didn't realize is that it's a trap, and that's what we're coming to grips with now."

Activities that decrease the number of photosynthetic organisms also increase global CO_2 levels. Since photosynthesizers are the only consumers of carbon dioxide in the carbon cycle, removing them not only reduces the amount of carbon dioxide they might have consumed, but also—in the case of large trees and stable populations of algae—eliminates what are in essence long-term storage vessels of carbon. Human activities that reduce the number of photosynthetic organisms on the planet include large-scale slash-and-burn agriculture, development that leads to deforestation, and various forms of pollution. Together, these activities contribute to our **carbon footprint,** a subset of our total ecological footprint, which is discussed in Chapter 24.

Though CO_2 is one of the major greenhouse gases, another culprit is methane (CH_4). Methane is produced by natural processes, such as microbes decomposing organic material in swamps, but agriculture, including cattle farming and growing rice in paddies, now accounts for over half the total methane being pumped into the atmosphere. One of the main sources of methane is the digestive gas produced by archaea living in the digestive systems of cattle. Emitted as flatulence, it adds an estimated 100 million tons of methane a year to the atmosphere. Although the atmospheric concentration of methane is far less than that of CO_2, atmospheric methane is more worrisome because it is 30 times more potent as a greenhouse gas **(Infographic 23.11)**.

No Time for Fatalism

The United States is among the world's biggest emitters of greenhouse gases, yet it has been, for political reasons, reluctant to make significant reductions. It is one of the few countries that refused to ratify the Kyoto Protocol, a United Nations agreement adopted in 1997 that obligates endorsing countries to reduce carbon dioxide emissions. In 2010, the Obama administration signaled support for the Copenhagen Accord, which would commit the United States to a 17% reduction in greenhouse gases from 2005 levels by 2020, but carrying out those goals depends on Congress's passing a climate bill, which is very problematic.

Even if all the world's greenhouse gas emissions were turned off today like a faucet, a daunting problem remains: we would still face years of warming and its consequences because of past emissions—what climate scientists refer to as "heat in the pipeline." It's a grim reality that could lead some to take a fatalistic attitude. That would be a dangerous mistake, says Hector Galbraith. "We've got to get beyond the deer in the headlights stage and begin to think as conservation biologists about what we're going to do about this to help to mitigate the impact." It's an area he calls "adaptation."

Adaptation will not be easy. For many species, like Vermont's maples, it may already be too late. But doing nothing, say scientists, risks turning a bad problem into a catastrophic one. There are things each of us can do to mitigate the effects of global warming—for example, living a more sustainable lifestyle, one that uses fewer fossil fuels—and voting for government officials who support sustainable practices. Chapter 24 discusses the topic of sustainability, and how you can live more sustainably, in more detail.

"The real problem is not so much change," says Serreze. "Change has always happened; change always will happen . . . . The real key is we've got to get a handle on the problem before it gets out of hand." ■

CARBON FOOTPRINT
A measure of the total greenhouse gases we produce by our activities.

INFOGRAPHIC 23.11

Anthropogenic Production of Greenhouse Gases

Human activities are increasing the levels of greenhouse gases in the atmosphere.

Burning Fossil Fuels
Burning fossil fuels (coal, natural gas, petroleum) liberates the carbon that was once stored as organic molecules in the earth and releases it into the atmosphere as carbon dioxide gas.

Methane from Cattle
The cattle that we raise have methane-producing microbes in their guts which help them digest the plant matter they eat. Cattle release large amounts of methane gas as flatulence.

Deforestation
Destroying and burning forests liberates carbon that was stored as organic molecules in trees and releases it into the atmosphere as carbon dioxide. In addition, it diminishes the capacity for carbon dioxide–capturing photosynthesis.

Rice Agriculture
Methane is released from rice paddies because of the methane-producing bacteria that live in the flooded, and therefore anoxic (oxygen-free), soil. Preparing land for growing rice may also destroy forests that formerly stored carbon in organic form.

Concrete Production
The production of concrete results in the release of large amounts of carbon dioxide. As we continue to develop cities on the earth, the impact of this process grows.

Summary

- Ecosystems are made up of the living and nonliving components of an environment, including the communities of organisms present and the physical and chemical environment with which they interact.
- Temperature is an important physical feature of any ecosystem and serves as a clock to time many biological events, such as breeding and hibernation.
- Biomes are large, geographically cohesive ecosystems, defined by their characteristic plant life, which in turn is determined by temperature and levels of moisture.
- Global climate change is a persistent pattern of change in the climate of the earth. Global warming is an increase in earth's average temperature over time.
- Global climate change, and especially global warming, is having widespread effects on plant and animal life on the planet—altering seasonal life cycles, shifting ranges, and contributing to species loss by extinction.
- The greenhouse effect is a natural process by which heat from the earth's surface is radiated to heat-trapping gases in the atmosphere, maintaining a global temperature that can support life. Rising levels of greenhouse gases have led to the enhanced greenhouse effect.
- Elements cycle through ecosystems. The carbon cycle is the movement of carbon atoms through living and nonliving components of the environment by the biotic processes of photosynthesis, cellular respiration, and decomposition, as well as by long-term geological processes.
- Global warming is the result of an increase in the amount of carbon dioxide and other greenhouse gases in the atmosphere, due primarily to human activities, such as burning fossil fuels and deforestation.
- Global warming is leading to melting sea ice in the Arctic, which is diminishing habitat for the organisms that rely on it. Melting of glaciers and ice caps is leading to rising sea levels.
- Methane is a significant greenhouse gas whose levels have increased because of human activities, including raising cattle and farming rice in paddies.

Data References

Infographic 23.5: (*Top*): IPCC Fourth Assessment Report; (*bottom*): http://cdiac.ornl.gov/trends/co2/lawdome.html; http://www.esrl.noaa.gov/gmd/ccgg/trends/

Infographic 23.7: http://data.giss.nasa.gov/gistemp/2008/

Infographic 23.8: (*Top left*): Arctic Climate Impact Assessment, 2004; 2 Intergovernmental Panel on Climate Change 4th Assessment Report, 2007; (*bottom*): http://earthobservatory.nasa.gov/IOTD

Infographic 23.10: (*Top*): http://www.esrl.noaa.gov/gmd/ccgg/trends/; (*middle*): Petit J.R., et al. 2001. *Nature*, 399, pp.429-436; (*bottom*): http://cdiac.ornl.gov/trends/co2/lawdome.html

ECOSYSTEMS AND CLIMATE CHANGE

Species are adapted to the ecosystems of which they are part. Climate change can alter their natural patterns and therefore change the dynamics of entire ecosystems.

HINT See Infographics 23.1–23.3 and 23.6.

KNOW IT

1. Which of the following are parts of an ecosystem:

a. the plant life present in an area
b. the animals living there
c. the amount of annual rainfall
d. soil chemistry
e. none of the above
f. all of the above

2. From what you've read in this chapter, list several examples of species that changed their geographic distributions or the timing of events in their life cycle as a result of global climate change.

3. If you were asked to identify a biome, which of the characteristics below would be most important to have data on (select all that apply):

a. monthly rainfall
b. temperatures throughout the year
c. plant life
d. animal life
e. human population size in the area

4. Which biome is characterized principally by evergreen trees?

5. Looking at Up Close: Biomes, where in North and South America do you find temperate deciduous forest? Tropical forest?

6. If global warming causes Arctic sea ice to melt, what will be the effect on sea levels in a low-lying region like Miami? What about if large parts of the Antarctic polar ice cap melted—what would be the effect on sea level?

USE IT

7. Although trees may not be able to walk away from increasingly warm regions, evolutionary adaptations may allow trees to survive in warmer regions. Discuss each of the adaptations listed below and decide if it is likely to be helpful or harmful in a warming environment. (Think about water—water is taken up by the roots of plants, and lost through pores in the leaves; CO_2 levels—CO_2 is taken up by plants through pores in leaves, then used by leaves for photosynthesis; and the movement of other species, for example insects, in response to global warming.)

a. having smaller leaves
b. having a larger number of pores on each leaf
c. having thicker and waxier bark

8. What is a possible risk for humans if insects that carry pathogenic bacteria or viruses expand their range northward?

GREENHOUSE EFFECT

Certain gases in the atmosphere act to trap heat. This heat trapping is essential for life on earth, but it can be altered by human activities in ways that harm life.

HINT See Infographics 23.4, 23.5, 23.7, 23.8 and 23.11.

KNOW IT

9. Which greenhouse gas is emitted every time you breathe out?

a. oxygen
b. carbon dioxide
c. methane
d. nitrogen
e. water vapor

10. Which of the following organisms will contribute to reducing atmospheric CO_2 levels?

a. maple trees
b. most algae
c. polar bears
d. pear thrips
e. a & b
f. a, b and d

11. Fossil fuels are most immediately derived from

a. organic molecules.
b. CO_2.
c. methane.
d. melting ice caps.
e. photosynthesis.

12. Could we live in the absence of the greenhouse effect? Explain your answer.

USE IT

13. Describe the evidence that increasing levels of greenhouse gases are responsible for global climate change. What if someone suggested to you that global climate change was due to increased intensity of solar radiation? What kind of evidence would you ask them to provide to support their hypothesis?

CARBON CYCLE AND GREENHOUSE EFFECTS

Carbon cycles through the environment, moving between organic molecules and inorganic carbon dioxide gas. Human activities can change the dynamics of the carbon cycle.

HINT See Infographics 23.9–23.11.

KNOW IT

14. Fill in the blanks in the image below.

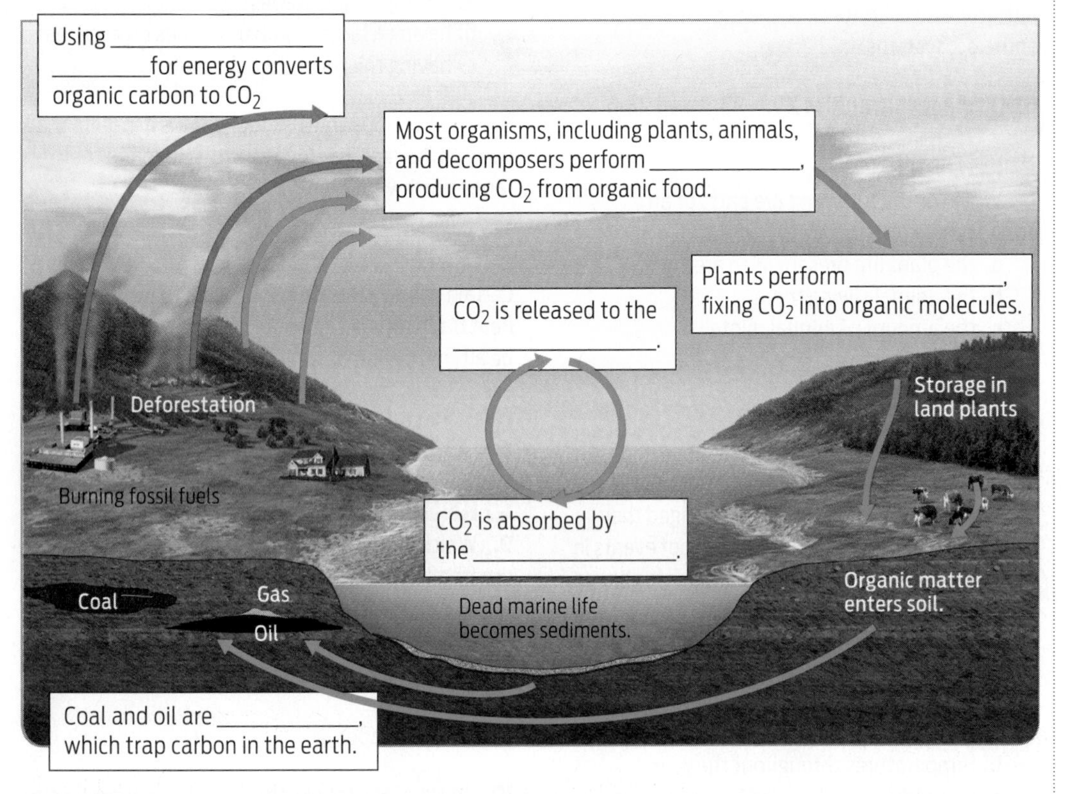

Using ________________ ________for energy converts organic carbon to CO_2
Most organisms, including plants, animals, and decomposers perform ____________, producing CO_2 from organic food.
Plants perform ____________, fixing CO_2 into organic molecules.
CO_2 is released to the ________________.
Deforestation
Storage in land plants
Burning fossil fuels
CO_2 is absorbed by the ________________.
Coal
Gas
Oil
Dead marine life becomes sediments.
Organic matter enters soil.
Coal and oil are ____________, which trap carbon in the earth.

15. Decomposers _______ CO_2 by the process of ________.

- **a.** emit; photosynthesis
- **b.** take up; photosynthesis
- **c.** emit; cellular respiration
- **d.** take up; cellular respiration
- **e.** store; cellular respiration

USE IT

16. How is ice useful in measuring atmospheric levels of CO_2?

17. Explain how each of the following contributes to an elevation of levels of greenhouse gases:

- **a.** large-scale slash-and-burn agriculture
- **b.** driving gasoline-fueled cars
- **c.** producing cattle for beef and dairy products
- **d.** rice production

18. Which of the following data would you use to determine the levels of atmospheric CO_2 in 1750? Justify your choice, including an explanation of why the other alternatives would not be as effective.

- **a.** historical weather records of daily temperatures
- **b.** archives of the Manua Loa observatory (to examine 1750 records)
- **c.** tree-ring analysis (to look for evidence of extreme fires)
- **d.** ice cores from ice formed in 1750

SCIENCE AND ETHICS

19. Visit an online carbon footprint or carbon emissions calculator (for example, http://www.epa.gov/climatechange/emissions/ind_calculator.html) and calculate your total carbon emissions.

- **a.** What is your largest source of emissions?
- **b.** What steps can you take to decrease your carbon emissions?
- **c.** Explain how line-drying (that is, air-drying) your laundry rather than drying it in the dryer can decrease your carbon emissions.

20. Using the carbon footprint calculator, design a low-carbon footprint menu for 1 day. Explain the basis for your food choices. Do you think that a low-carbon menu would be different in different parts of the country? Why or why not?

Eco-Metropolis

What You Will Be Learning

24.1 Human Population Growth

24.2 The Human Ecological Footprint

24.3 Countries Differ in Their Ecological Footprint

24.4 The Human Ecological Footprint Is Greater Than Earth's Biocapacity

24.5 Fossil Fuels Are Non-Renewable

24.6 Eco-Cities: Sustainable by Design

24.7 Renewable Resources Reduce Our Ecological Footprint

24.8 Water Is a Renewable Resource

24.9 Depletion of Fresh Water By a Growing Population

24.10 Water Availability Is Not Equally Distributed

24.11 Species Loss Increases as Human Population Grows

24.12 What You Can Do To Live More Sustainably

Eco-Metropolis

Designing the city of the future

Bumper-to-bumper traffic, a noxious cloud of gray smog, towering skyscrapers that seem to be straight out of the futuristic movie *Blade Runner*: welcome to Shanghai, China's largest city. With a population of 19 million and growing, it's not exactly a place you'd call environmentally friendly. But just 15 miles from this concrete jungle, on the island of Chongming at the mouth of the Yangtze River, something unprecedentedly green is in the works: the world's first eco-metropolis built completely from scratch. About three-fourths the geographic size of Manhattan, the eco-city known as Dongtan will be an urban oasis of green-roofed buildings, tree-lined streets, and pedestrian-friendly neighborhoods—the polar opposite of its dystopian neighbor.

More than just a nice place to live, Dongtan is being designed to incorporate lessons of cutting-edge ecological science. According to its designers, Dongtan will be entirely self-sufficient in food, water, and energy. It will produce no net carbon emissions and zero pollution. Agriculture will be entirely organic and local. All trash will be recycled, composted, or used to generate electricity. Vehicles will be powered entirely by renewable energy. In short, Dongtan will be a model of **sustainability** for the rest of the world to emulate.

Cities occupy just 2% of the terrestrial surface area of the earth and have only half the world's population, yet they consume more than 75% of its natural resources.

Many urban planners and environmentalists would agree that it's a model the planet badly needs. According to the United Nations, cities occupy just 2% of the terrestrial surface area of the earth and have only half the world's population, yet they consume more than 75% of its natural resources. The reason for this imbalance? Our cities are flawed in their very design, say urban planners—built as if natural resources like land and water were unlimited, and waste was something that would magically disappear.

SUSTAINABILITY
The use of the earth's resources in a way that will not permanently destroy or deplete them; living within the limits of the earth's biocapacity.

Will Dongtan ever be built?

Consider London, which imports more than 80% of its food from other countries. That's a population of 7.5 million people unable to feed itself. Or the mega-metropolis of New York City, which produces some 16,000 tons of garbage every day, sending it by truck, rail, and barge to landfills as far away as Virginia and South Carolina. Cities, it seems, are bursting at the seams.

China has roughly 20% of the world's population but only 7% of the world's land area.

Yet they continue to grow. The United Nations estimates that the ratio of people living in cities versus the country is the highest it's ever been. As of 2008, for the first time more people lived in cities than in rural areas. And the mass migration from countryside to urban center shows no signs of abating: by 2050, 70% of the world's population will live in a city. Compare this figure to 1900, when only 10% of the world's population lived an urban life. That's a huge shift, and one that poses significant environmental challenges, which urban planners are increasingly being asked to address. In China, where a population of 1.3 billion people represents an ecological force to be reckoned with, the task is both daunting and urgent.

China has roughly 20% of the world's population but only 7% of the world's land area, and its population is increasing by more than 10 million people each year. Every year the resources that each person uses increase as well, thanks to a rising standard of living. "China is one of the first countries in the world to realize this is an unsustainable direction and therefore is desperately trying to improve its energy efficiency and reduce its carbon intensity," says Peter Head, Director of Arup, the London-based engineering firm hired to design Dongtan.

But for Head and his colleagues, Dongtan was a chance to demonstrate to the world that urban growth can happen in a sustainable way, and that ecological challenges can be met with creative design solutions. While the plan does indeed look good on paper, at this point the eco-city is closer to fantasy than reality. Building has not yet begun on Dongtan, and it remains to be seen whether it will ever get off the drawing board. Arup is no longer working on the project

An aerial view of Dongtan.

ECOLOGICAL FOOTPRINT
A measure of how much land and water area is required to supply the resources a person or population consumes and to absorb the wastes it produces.

NATURAL RESOURCES
Raw materials that are obtained from the earth and are considered valuable even in their relatively unmodified, natural form.

(now that its design is complete), and the city's future prospects now rest with the Chinese government. But even if it remains just a twinkle in an architect's eye, Dongtan will have already achieved something important, shining a city-size spotlight on one of the most pressing issues facing humanity today.

Our Expanding Footprint

Judging by our numbers, humans are an extraordinarily successful species. Two thousand years ago, we numbered just 300 million globally—less than the current population of the United States. In 2010, there were 6.8 billion of us on the planet. Much of that growth occurred since 1950, thanks in large part to antibiotics and other advances in public health that have allowed people to live longer. And each hour more than 10,000 new people are added to the planet—roughly 3 per second, or 90 million per year. By 2050, demographers estimate, we'll hit the 9 billion mark. The human population is growing exponentially **(Infographic 24.1)**.

As the human population grows, so does our environmental impact. Ecologists measure that impact with a tool known as the **ecological footprint**, which calculates the amount of land and water area a population requires to supply the resources it consumes and to absorb the wastes it generates. Humans require a vari-

INFOGRAPHIC 24.1

Human Population Growth

Since the advent of agriculture, the human population has been following an exponential growth pattern, and is approaching 7 billion people. Some estimates predict that the human population will number 9 billion by 2050.

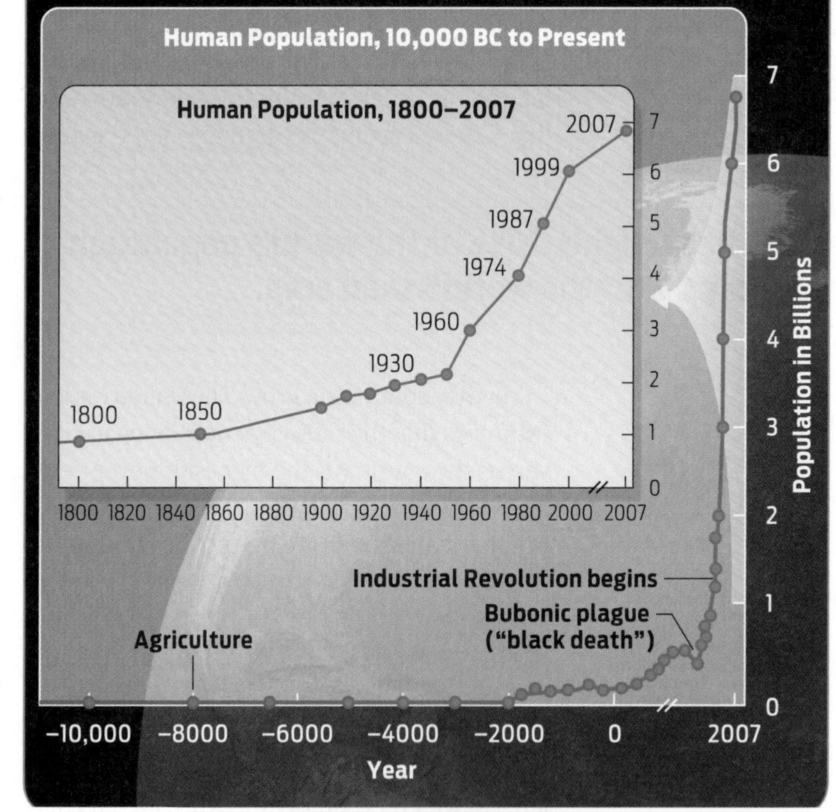

INFOGRAPHIC 24.2

The Human Ecological Footprint

How much of the earth's resources does your lifestyle require? The ecological footprint is a measure of people's demand on nature. It uses 5400 different measures gathered from government agencies and scientific publications to calculate a footprint in global hectares, a measure of how much biologically productive land and water area (cropland, forests, grazing lands, fishing area, and built-up land) a human population requires to produce the resources it consumes and to absorb the waste it produces.

How do you live?

Energy: Do you drive a car? Use a computer? Cook your food?

Buildings: Do you live in a house? Work in an office? Eat at restaurants?

Timber and Paper: Do you read magazines? Print your assignments? Use a textbook?

Food and Fiber: Do you eat three meals? Wear clothing? Have furniture?

Seafood: Do you eat fish? Take a fish oil supplement? Like shrimp cocktail?

What natural resources does it require?

Stored Carbon: Photosynthetic organisms and fossil fuel consumption

Built-Up Land: Once was biologically productive, now is space we live on.

Forest Land: Cut for consumer products.

Grazing and Crop Land: Requires a lot of land, water, and other resources like fertilizer.

Fishing Grounds: Depletion of wild stock or resources used in farming.

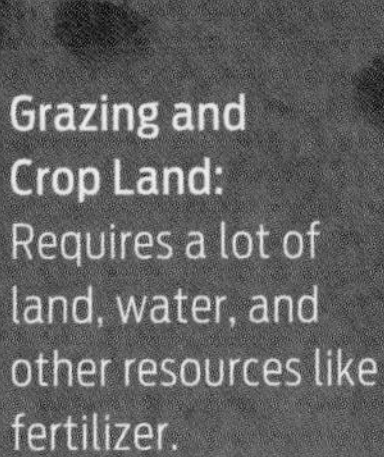

ety of **natural resources** to live: farmland to grow crops or raise cattle, gasoline to power cars, oxygen to fill our lungs, to name just a few. All these resources come, directly or indirectly, from the earth.

In addition to providing us with natural resources, the earth also acts like a sponge, absorbing our wastes: the carbon dioxide we emit, for example, and the garbage we produce. By quantifying the amount of biologically productive earth area it takes to sustain our lifestyles, the ecological footprint puts a number on our environmental impact **(Infographic 24.2)**.

Ecological footprints are expressed in units called **global hectares,** with 1 global hectare representing the biological productivity (both the resource-providing and waste-absorbing capacity) of an average hectare of land. A hectare is 10,000 square meters—about the size of a soccer field. As of 2006, the global average ecological footprint was 2.6 global hectares per person per year. In other words, it takes

GLOBAL HECTARE
The unit of measurement of the ecological footprint, representing the biological productivity of an average hectare of land.

that much land and water area to support one average human for 1 year.

But, of course, not everyone uses resources to the same extent. Patterns of consumption vary greatly from region to region and country to country. An average American, for instance, has an ecological footprint of about 9 global hectares, while the average Haitian uses just 0.48 global hectares. These are per capita figures, averages for one resident in each of those countries. It's also possible to calculate the ecological footprint of a whole country. For example, China has a per capita footprint of 1.85 global hectares, much smaller than that of the United States, but because China's population is so large, its *total* footprint is only slightly smaller than that of the United States, which has many fewer people **(Infographic 24.3)**.

INFOGRAPHIC 24.3

Countries Differ in Their Ecological Footprint

Per capita, the United States has a larger footprint than China. However, the enormous size of China's population means that China's total footprint is approaching that of the United States.

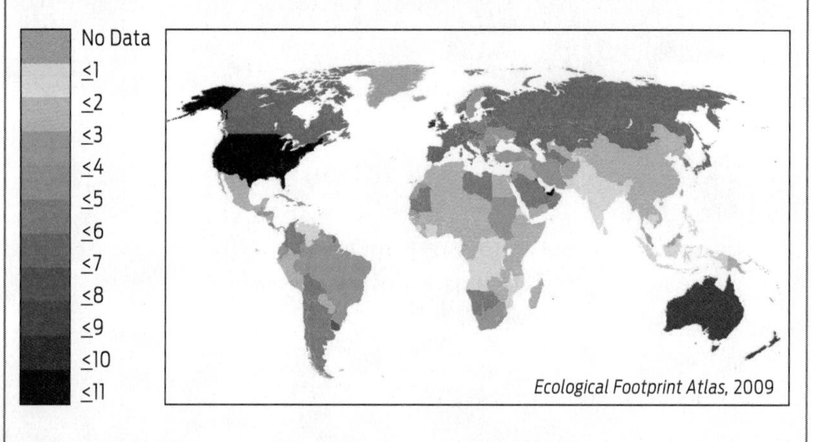

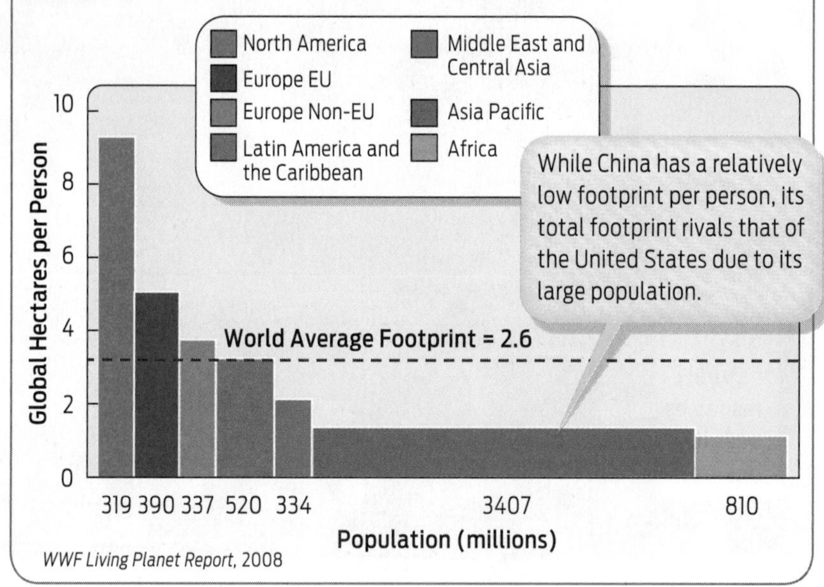

Moreover, China's footprint is expanding rapidly–about 3% a year. "Three percent a year doesn't sound like very much," says Head, "but it means that China ... needs to find about 90 million hectares of new land every year for all the resources needed to support the growth and footprint of urbanization."

The human ecological footprint is often compared with the earth's **biocapacity**–its ability to sustain human demand given its available natural resources and its ability to absorb waste. If we think of the footprint as our demand on the earth, the biocapacity is the amount of supplies that the earth can produce to meet that demand.

The earth's biocapacity can't always keep up with our demand. Currently, there are 6.8 billion people living on the earth. As of 2006, the earth had approximately 11.9 billion hectares of biologically productive land and sea (which doesn't include areas like deserts, glaciers, and open ocean), which works out to about 1.8 hectares available per person. Since our current average ecological footprint is 2.6 global hectares per person, we are clearly exceeding the earth's biocapacity, using resources faster than the earth can rejuvenate them. In other words, our current lifestyles are unsustainable **(Infographic 24.4)**.

"The human appetite for resources may be unlimited, but the planet's ability to sustain these needs is finite," says Mathis Wackernagel, co-creator of the ecological footprint concept and executive director of the nonprofit Global Footprint Network. "As our rising demand on ecological services pushes our natural systems to the breaking point, we are not only putting other species at risk, we are jeopardizing our own livelihoods and well-being."

According to Wackernagel, if everyone on the planet were to live like the average resident of the

BIOCAPACITY
The amount of the earth's biologically productive area—cropland, pasture, forest, and fisheries—that is available to provide resources to support life.

United States, it would take about five earths to support us. By contrast, if everyone in the world lived like the average person in India, we would need less than half an earth to satisfy our demands.

"The human appetite for resources may be unlimited, but the planet's ability to sustain these needs is finite." –Mathis Wackernagel

What is it about the U.S. lifestyle that leaves such a heavy footprint? Energy consumption, by far, is the largest culprit. The cars and SUVs we drive, the computers we work (and play) with and televisions that entertain us, the washers and dryers that clean our clothes, the air conditioners that cool our homes, the food we truck across the country or fly around the world–all these require energy. Globally, the energy component of our ecological footprint increased roughly 700% between 1961 and 2006, accounting for roughly half our total ecological footprint by 2006.

In the United States, as in most parts of the world, most of this energy comes from fossil fuels–oil, coal, and natural gas. As Chapter 23 discussed, burning fossil fuels releases carbon dioxide to the atmosphere, contributing to global warming. Therefore, the ecological footprint takes into account the amount of land and water area needed to absorb CO_2. This, combined with increased consumption, is what makes our energy footprint so large.

Because they take millions of years to form naturally, fossil fuels are considered **nonrenewable resources**, meaning that once depleted, they are essentially gone for good.

NONRENEWABLE RESOURCES
Natural resources that cannot be replaced.

INFOGRAPHIC 24.4

The Human Ecological Footprint Is Greater Than Earth's Biocapacity

When comparing our biological demand, or ecological footprint, with the earth's biocapacity, it is clear that our footprint has been exceeding biocapacity since the mid-1970s. Our greatest demand relates to energy use, indicated by our large carbon footprint.

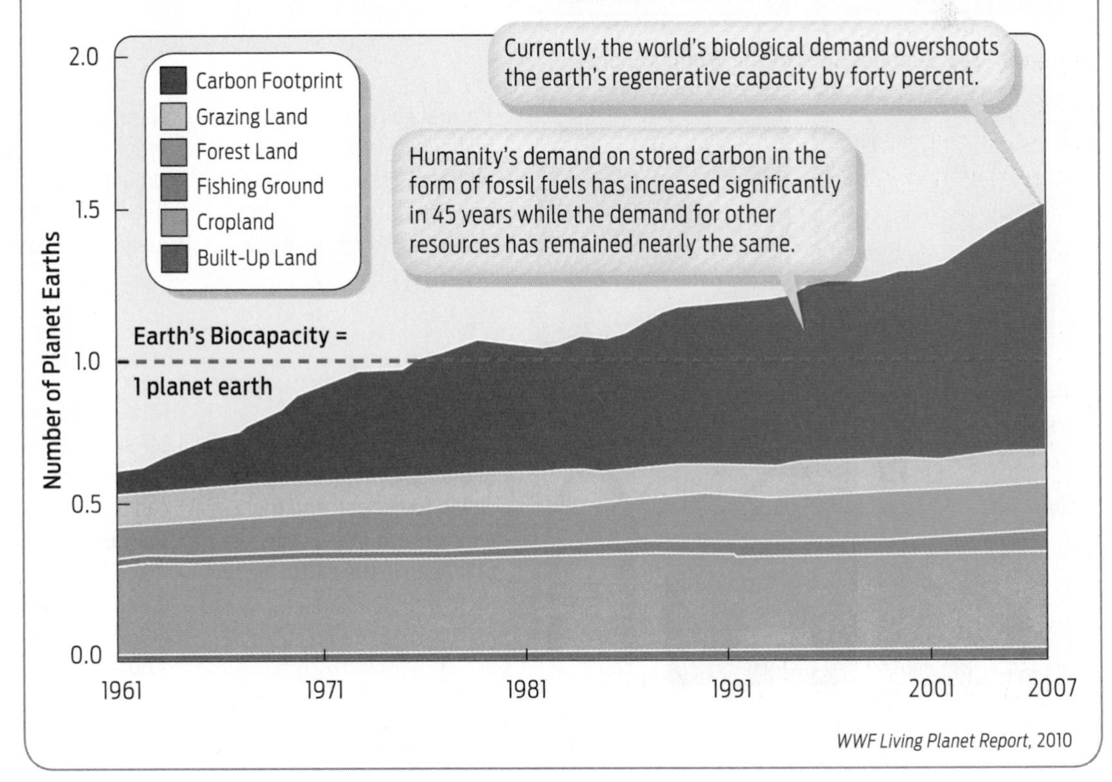

Carbon Footprint
Grazing Land
Forest Land
Fishing Ground
Cropland
Built-Up Land
Currently, the world's biological demand overshoots the earth's regenerative capacity by forty percent.
Humanity's demand on stored carbon in the form of fossil fuels has increased significantly in 45 years while the demand for other resources has remained nearly the same.
Earth's Biocapacity =
1 planet earth
Number of Planet Earths
2.0
1.5
1.0
0.5
0.0
1961
1971
1981
1991
2001
2007

WWF Living Planet Report, 2010

Besides contributing to our carbon footprint, burning these nonrenewable resources releases pollutants such as sulfur dioxide and nitrogen dioxide. Both of these emissions can combine with water in the atmosphere to form acid rain, which damages both terrestrial and aquatic life. Coal also contains toxic elements such as arsenic and fluorine, which can cause bone and lung disease when inhaled or consumed **(Infographic 24.5)**.

Can anything be done to reduce our ecological footprint? The most significant changes may ultimately have to come from government energy policy, but there are things that individuals can do as well.

Already, individuals and communities around the globe are devising creative ways to live more sustainably. At Carleton College in Minnesota, a wind turbine supplies 40% of the school's electricity. Energy-conscious residents of Calgary,

INFOGRAPHIC 24.5

Fossil Fuels Are Non-Renewable

Most of the natural resources we use to supply our energy needs are non-renewable. Coal, oil, and natural gas are fossil fuels that take millions of years to form as organic material is compressed by layers of sedimentary rock. While plentiful today, and relatively cheap to obtain, fossil fuels come with significant environmental and human costs.

	Why Do We Use It?	Environmental Impact
Coal: Supplies 23% of the World's Energy	Coal is burned in power plants to produce steam to turn turbines that generate electricity. Coal is relatively cheap to mine, and there is currently an abundance of it in the earth.	Mining coal from the earth often damages the habitat on large tracts of land. Greenhouse gases and pollutants like arsenic, nitrogen dioxide, and sulfur dioxide are released when coal is burned to make electricity. Coal miners have increased risk of respiratory illness.
Natural Gas: Supplies 24% of the World's Energy	Natural gas is burned to heat buildings and water. It is relatively cheap to extract, and there are currently large reservoirs of it deep in the earth.	Natural gas is extracted from underground and off-shore reservoirs. Drilling platforms can disrupt ocean habitat. Burning natural gas releases greenhouse gases to the atmosphere.
Petroleum Oil: Supplies 37% of the World's Energy	Oil is used to produce gasoline, petroleum products, and plastics. It is relatively cheap to extract, and there are currently large reservoirs of it deep in the earth.	Oil is drilled from undergound and off-shore reservoirs. Drilling platforms can disrupt ocean habitat. Oil spills can devastate ocean ecology and the seafood economy. Burning products made from oil produces pollution and emits greenhouse gases. Plastics do not biodegrade and therefore create a huge amount of landfill waste.

Unsustainable Mega-City Practices:

Fossil Fuel-Based Transportation
Sprawl means a daily commute, burning fossil fuels, and creating greenhouse gas emissions.

Dongtan-Style Alternatives:

Solar-powered water taxis

Living needs are within biking distance.

Taxis are electric or hydrogen fuel-cell powered.

Canada, can ride on a light rail transit system that obtains all its power from wind turbines. Outside London, England, the community of 100 residences known as BedZED satisfies all its energy needs from renewable sources such as solar panels and locally grown firewood. Apartments in Stockholm, Sweden, come equipped with stoves that burn gas extracted from organic waste generated in the community. In San Diego, garbage trucks run on methane gas captured from decaying garbage in landfills, while residents of Vermont can purchase "cow power"–energy obtained from cow manure–from their local utility company. The residents of Vienna and Paris bike freely around the city on municipally owned bicycles, greatly lowering their footprint.

The plan is for Dongtan to support 80,000 people by 2020, and 500,000 people by 2050.

But a whole city that is entirely self-sufficient in terms of energy and environmentally neutral in terms of carbon emissions and pollution? It sounds too good to be true.

Sustainable by Design

The plan is for Dongtan to support 80,000 people by 2020, and 500,000 people by 2050–the latter being about the population of central Atlanta. And yet, Dongtan's per capita ecological footprint will be a fraction of Atlanta's: 2.6 global hectares for a Dongtan resident versus 13 for an Atlantan.

How will Dongtan achieve a lower footprint? For one, the designers are not at all focused on superficial aesthetics, such as the decorative ornamentation of buildings, and are instead concentrating on what architects call performance-based design.

Focusing on performance and efficiency means rethinking the way cities work from the ground up–starting with transportation. "[T]the essential character of a city's land use comes down to how it manages its transportation," write Peter Newman and Jeff Kenworthy in their book *Sustainability and Cities: Overcoming Automobile Dependence*. When it comes to sustainability, car-based transportation is just about the worst thing that can happen to a city. Yet in many modern cities, such as Los Angeles and Houston, people have few alternatives to driving.

Not so in the eco-city. To eliminate the need for cars in Dongtan, all residential neighborhoods will be within 7 minutes' walking distance of public transportation, which will provide easy access to schools, hospitals, and businesses. Solar-powered water taxis and hydrogen fuel cell buses will provide the primary means of public transport. Bicycle paths and pedestrian walkways will crisscross the city. Cars will not be banned in the city, but car parks placed outside Dongtan will encourage leaving them behind. Curbing car culture will greatly reduce Dongtan's energy consumption.

The city will also be built in such a way that less energy is required to heat and cool it. Conventional cities are essentially "heat islands"–on average, 1°C (1.8°F) warmer than the countryside during the day, and up to 6°C (10.8°F) warmer at night. That's because concrete and asphalt absorb solar radiation. On a hot summer day, air

Unsustainable Mega-City Practices:

Concrete Heat Islands
Heat is generated, raising energy consumption required to cool buildings. Fresh rainwater is polluted as it runs off into drains.

Dongtan-Style Alternatives:

Plant trees along streets to cool cement neighborhoods.

The green roof at Chicago City Hall cuts air-conditioning costs.

conditioning can consume more energy in a city than any other single activity.

One very simple way to beat the heat is to plant more trees. Trees cool cities by providing shade. They also intercept solar radiation that would otherwise generate heat if it were absorbed by concrete or asphalt. In the eco-city, tree-lined streets, rooftop gardens, and green roofs will all temper the heat-island effect.

The buildings themselves will be constructed differently in Dongtan, with walls and windows designed to provide natural insulation and ventilation. To discourage the overuse of electricity, easy-to-read meters placed in obvious locations inside homes and offices will allow residents to see how much they use. Cost will be commensurate with usage.

Most important, Dongtan will generate all of its electricity and heat entirely from **renewable resources**–those that can be naturally replenished as long as the rate of consumption is not greater than the rate of replacement. Wind turbines and solar panels, for example, will provide the bulk of electricity. In addition, a combined heat and power plant will turn biomass such as leftover rice husks–the region has plenty–into valuable energy for human use. Heat given off during the process will in turn be piped into homes and businesses. Even human waste won't go to waste: treated sewage will be composted to fertilize crops. With such measures, designers estimate that Dongtan will use 65% less energy than a conventional city of the same size **(Infographic 24.6)**.

While cities get a bad rap for being resource hogs, they do have a key advantage over more-spread-out ways of living–they operate as economies of scale. In other words, the density of

RENEWABLE RESOURCES
Natural resources that are replenished after use as long as the rate of consumption does not exceed the rate of replacement.

Unsustainable Mega-City Practices:

Waste
Cities produce tons of solid waste per day, which ends up in landfills. Sewers work to capacity to remove human waste in high-density populations.

Dongtan-Style Alternatives:

Dump trucks fill their tanks with methane waste collected from landfils.

Batch reactors digest human waste to use for fertilizer.

INFOGRAPHIC 24.6

Eco-Cities: Sustainable by Design

An ecologically sustainable city would minimize the need for individual transport in cars, and promote cooling with trees and buildings positioned to take advantage of breezes. It would also recycle waste and use renewable energy like solar and wind power.

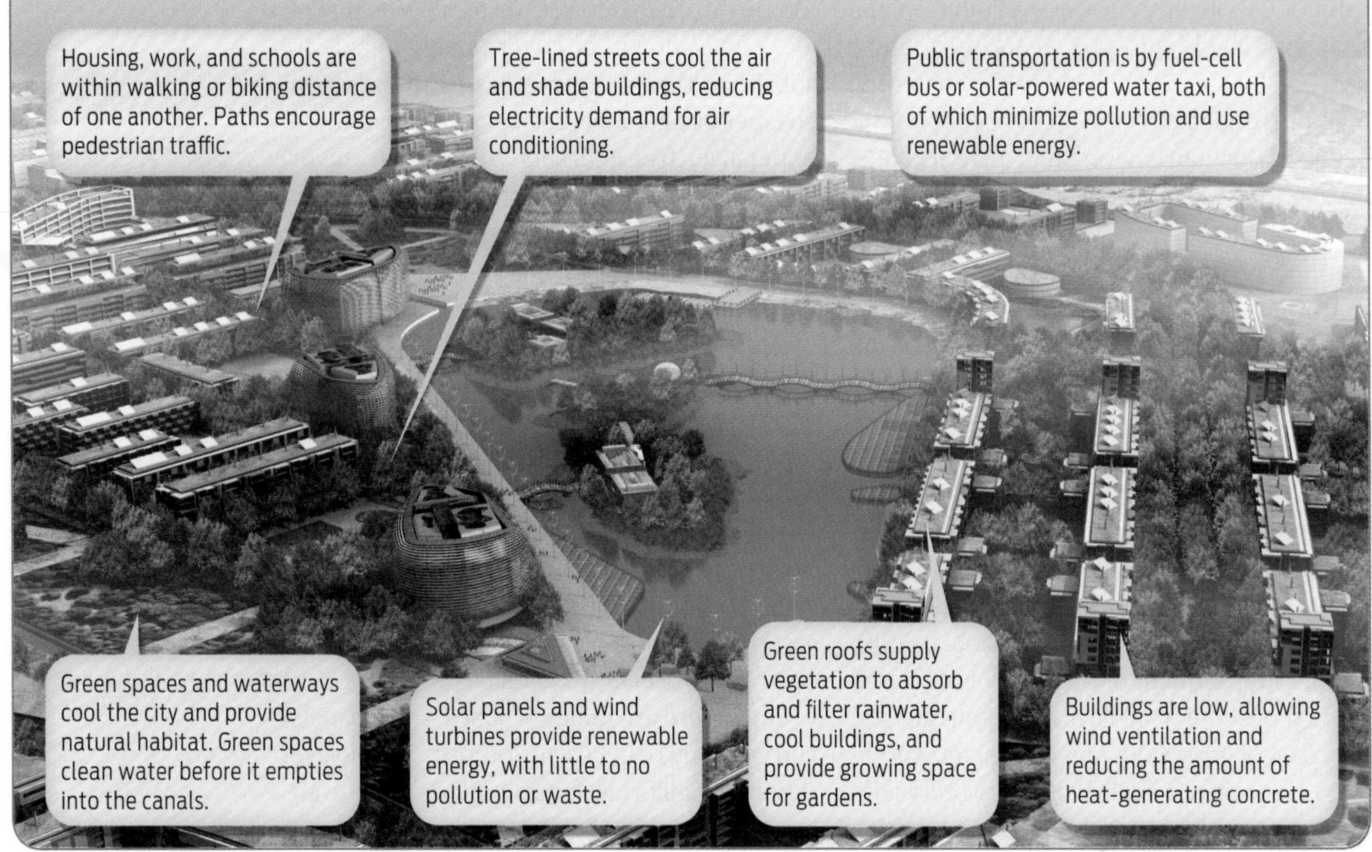

Housing, work, and schools are within walking or biking distance of one another. Paths encourage pedestrian traffic.
Tree-lined streets cool the air and shade buildings, reducing electricity demand for air conditioning.
Public transportation is by fuel-cell bus or solar-powered water taxi, both of which minimize pollution and use renewable energy.
Green spaces and waterways cool the city and provide natural habitat. Green spaces clean water before it empties into the canals.
Solar panels and wind turbines provide renewable energy, with little to no pollution or waste.
Green roofs supply vegetation to absorb and filter rainwater, cool buildings, and provide growing space for gardens.
Buildings are low, allowing wind ventilation and reducing the amount of heat-generating concrete.

people makes possible a more efficient and affordable utilization of resources, which is ultimately more sustainable. For instance, a public transportation system that transports tens of thousands of people who might otherwise be driving gas-guzzling cars can reduce the carbon footprint significantly.

New York is a good example. The average New Yorker who takes the subway 1 mile to work uses much less fossil fuel energy than a suburbanite who commutes 10 miles to work in a car. In fact, according to David Owen, author of *Green Metropolis: Why Living Smaller, Living Closer, and Driving Less are the Keys to Sustainability,* New Yorkers have the smallest per capita carbon footprint in the United States, a statistic that has even led some commentators to refer to New York as the Big Green Apple.

The solution to the problems of urbanism, then, is not to de-urbanize, but to make cities part of the solution rather than part of the problem.

Moving Forward

Construction of Dongtan was supposed to begin in 2007, with the first phase of development–including arrival of the city's first 10,000 residents–completed by 2010, in time for the Shanghai Expo. That didn't happen. Development has stalled, and it's unclear whether the ambitious plans will ever be fully realized. Critics of Dongtan have long held that the city was a utopian fantasy, more useful as a public relations ploy than a place to live.

Peter Head of Arup emphatically challenges that view. While the future of Dongtan itself is uncertain, he says China has plans to make the whole of Chongming an eco-island, using specifications developed for Dongtan. "In many ways, all of the ideas and thinking [are] alive and well," says Head. He acknowledges, though, that there is much to be done and that Dongtan is only a start. The eco-city's hypothetical footprint of 2.6, for instance, is still more than a truly sustainable one of 1.8. Reducing resource use even further will depend partly on the will of future residents.

Even if Dongtan is never built, there are hopeful signs that urban sustainability is catching on around the world. Eco-cities are currently being planned or built in countries as diverse as Argentina, Australia, Finland, Vietnam, and the United States. And key elements of sustainability—such as finding alternatives to fossil fuels—are increasingly being recognized as an issue of national and global importance. For example, as of 2009, the United States obtained 17% of its energy from renewable energy sources (including nuclear energy). While wind and solar constitute a small fraction of these renewable sources, their contribution is growing. In 2010, the U.S. Department of the Interior approved the first offshore wind farm, to be operated off the coast of Cape Cod, Massachusetts. And in 2008, the United States became the world leader in wind power investments—though in terms of the proportion of energy it obtains from wind, it is still greatly outperformed by many European countries. Highest marks for use of wind power go to Denmark, which in 2009 obtained 20% of its electricity from wind.

The appeal of renewable sources of energy such as wind and solar is undeniable: they are plentiful, powerful, and environmentally neutral in terms of their carbon emissions. Solar power alone could theoretically provide more than enough clean energy to supply the needs of everyone on the planet many times over—assuming we could adequately and inexpensively harvest it.

The technologies to harness wind and solar power are currently much more expensive to build and operate than coal-fired power plants, for example. What makes fossil fuels such convenient and inexpensive sources of energy is the fact that the difficult work of harvesting the energy of sunlight has already been done, by the fossilized photosynthetic organisms that have been compressed over millions of years into oil, coal, and gas. (In a way, we are already using the energy of sunlight to power our lifestyles, but indirectly.) Unless the price of the new technologies comes down, or governments decide to subsidize these alternatives to make them cheaper or tax fossil fuels to make them more expensive, free markets will tend to favor cheaper options.

Of course, when you consider the environmental and human costs of obtaining and burning fossil fuels—from coal-mine explosions, to air pollution, to oil spills—they aren't actually that cheap; think of the 2010 Gulf oil disaster, which is estimated to cost $40 billion and counting. These downstream costs of fossil fuels, which are not reflected in their market price, are known among economists as externalities. If externalities were included in the price, as some economists and environmentalists suggest they should be, then the playing field with other forms of energy would be more level.

Then there are issues of space: solar panels and windmills can take up lots of it. The Mojave Desert, in California, for example, is home to an increasing number of solar power plants. As of 2010, plans have been approved for projects that, when completed, will cover some 39 square miles of land. To some environmentalists, this represents a threat to local wildlife.

And while many people support the idea of renewable energy in theory, many would also prefer not to have the technology located in their backyards. Wind turbines, especially, are seen by many as a kind of "sight pollution," cluttering the landscape. (In fact, this was a controversial aspect of Cape Wind, the Cape Cod wind farm: residents didn't want to look at it.)

And of course, wind does not always blow and sun does not always shine, so they are less reliable than other forms of energy. Given the limitations of our current technologies, it is not yet possible to satisfy our energy demands with only the existing infrastructure of wind turbines, solar panels, biofuels, and the other renewable energy sources so far developed. At least for the next decade, we cannot take fossil fuels out of our energy mix **(Infographic 24.7)**.

INFOGRAPHIC 24.7

Renewable Resources Reduce Our Ecological Footprint

While many renewable resources are available to us, economic, technological, and environmental considerations currently limit their use as alternatives to fossil fuels.

Solar: 0.07% of World's Energy

Why Don't We Use It More?

Solar power is currently much more expensive to produce than non-renewable options. Producing solar panels involves using toxic chemicals, and the resulting waste must be properly disposed of.

Environmental Impact:

Solar energy traps energy from the sun and converts it into electricity and heat with little impact on the environment. As nothing is burned to make the electricity there are zero polluting emissions from this process.

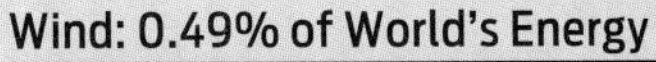

Wind: 0.49% of World's Energy

Wind power is currently much more expensive to produce than non-renewable options. Wind generators take up space, either on land or in the water, and must be located in windy areas. Some people don't want a visible wind farm near their homes.

Wind energy is used to turn wind turbines, producing electricity with little impact on the environment. In the absence of combustion, no pollutants are released to the environment. Bird species may be affected as turbines encroach on their air space.

Nuclear: 9% of World's Energy

Nuclear reactors are expensive to design and build, and a reactor has a limited life span. Extracting uranium from mines has an environmental impact, and the waste from uranium mines is radioactive. Most important, the waste from nuclear reactors is highly radioactive, making storage complicated. As well, weapons-grade plutonium can be made from reactor waste, which poses a security threat.

Nuclear energy uses radioactive elements harvested from the earth and concentrated. As these elements decay, they give off tremendous heat, which is used to produce electricity. As nothing is burned in the process, there are no polluting emissions.

Biofuels: 3.7% of World's Energy

While being intensively researched, biofuels have not yet become feasible replacements for fossil fuels. In some cases, significant emissions are associated with their production. In other cases, more research and investment is required to optimize the production process.

Biofuels are made from plant material. When burned, the only CO_2 released to the atmosphere is what the plants and algae took in through photosynthesis, so fossil deposits of carbon are not used. Biofuels can be made directly from plant material, or from energy-rich oils that algae make. In some cases, growing plants for biofuels competes with growing crops.

Hydroelectric: 2.4% of World's Energy

Hydroelectric power uses dams to block rivers, creating lakes with immense amounts of potential energy. Building dams destroys habitat, impacts local fish populations, and can force human populations to relocate.

Hydroelectric power relies on the conversion of potential energy (stored in the position of accumulated water behind a dam) to kinetic energy, which can turn a generator. There are no emissions associated with hydroelectric power. Hydro plants have long life spans, and hydro power can potentially power half the projected energy demands of the planet.

Geothermal: 0.35% of World's Energy

Geothermal energy is used extensively in Iceland and in some areas of California. However, it has yet to be fully developed in other areas, primarily because optimal technologies require further development.

Geothermal energy relies on naturally occurring heat from the magma layer beneath the earth's crust. This is a sizeable and sustainable resource that can be tapped to drive generators or directly heat homes and businesses. In some cases, noxious pollutants are released with the steam from geothermal resources.

tion. Canada, for example, hosts just 0.5% of the world's population, but 20% of the global freshwater supply is within Canada's borders. China, on the other hand, has 20% of the world's people but only 7% of the world's water. The United Nations estimates that at least a billion people in the world currently lack access to clean and safe drinking water, and by 2025, two-thirds of the world's population will live in areas of moderate to severe water stress. And climate change, if it changes precipitation patterns, may also affect the global availability of water in unpredictable ways **(Infographic 24.10)**.

Finding the Limits

When the designers at Arup were first approached about designing an eco-city from scratch, they were intrigued but skeptical. Surely there must be a catch, they thought. And indeed, there was a big one: the city, developers stipulated, must not disturb the migration path of the rare birds that use the wetlands on Chongming Island as a stopping point along their way between Siberia and Australia.

To protect the birds and their flyway, the Dongtan master plan calls for a buffer zone between the city and the bird resting area. The zone will be more than 2 miles wide and will help prevent pollutants–including light, sound, air, and water pollution–from reaching the surrounding wetlands.

These efforts to protect bird species illustrate a final point about our expanding footprint: it takes a toll not only on humans, but on other species as well. Every few years, the World Wildlife Fund's Living Planet report documents the health of nearly 1,700 species of vertebrates around the world. Between 1970 and 2005 the number of individuals in the populations studied declined 30% overall, reflecting not only increased hunting and fishing by humans but also the degradation of habitat as humans

INFOGRAPHIC 24.10

Water Availability Is Not Equally Distributed

Fresh water is not evenly distributed across the globe, and its availability does not always follow international borders. In addition, access to even a sufficient water supply may be limited by economic, social, and political circumstances, such as war and ethnic conflict. As the human population continues to grow, and access to clean fresh water continues to decline, these problems are likely to intensify, particularly in areas with existing scarcities of water.

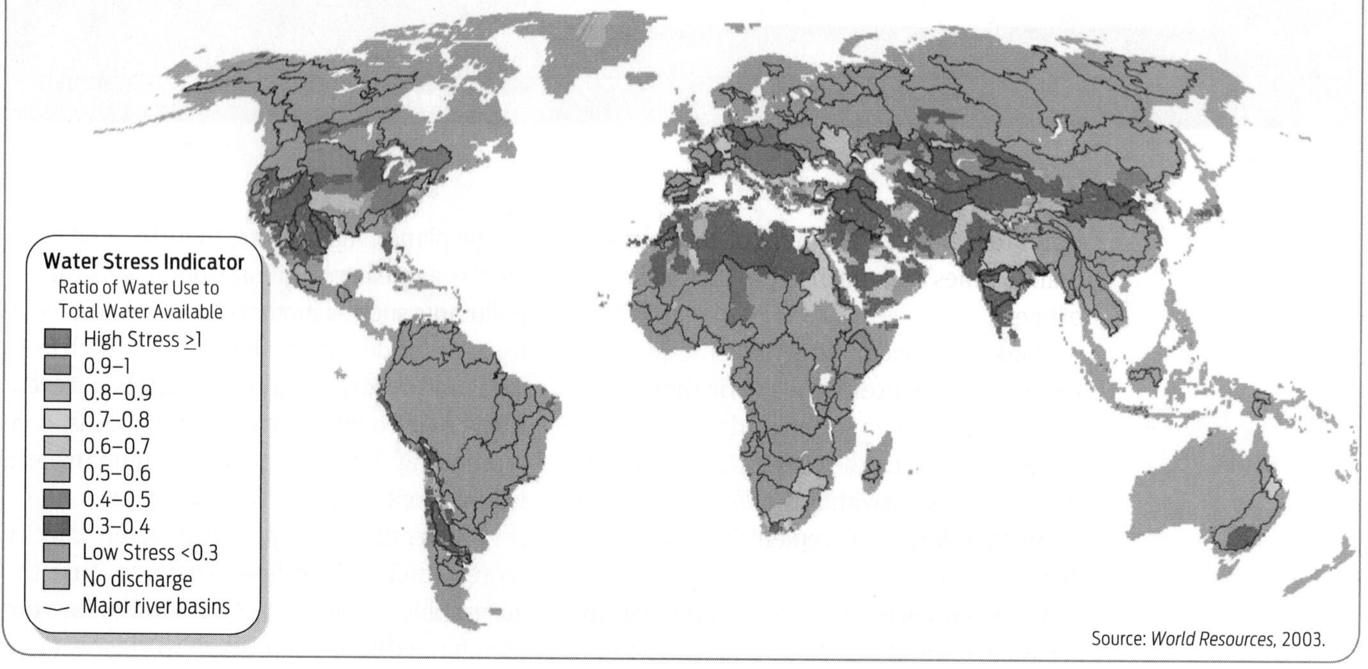

Source: *World Resources*, 2003.

INFOGRAPHIC 24.11

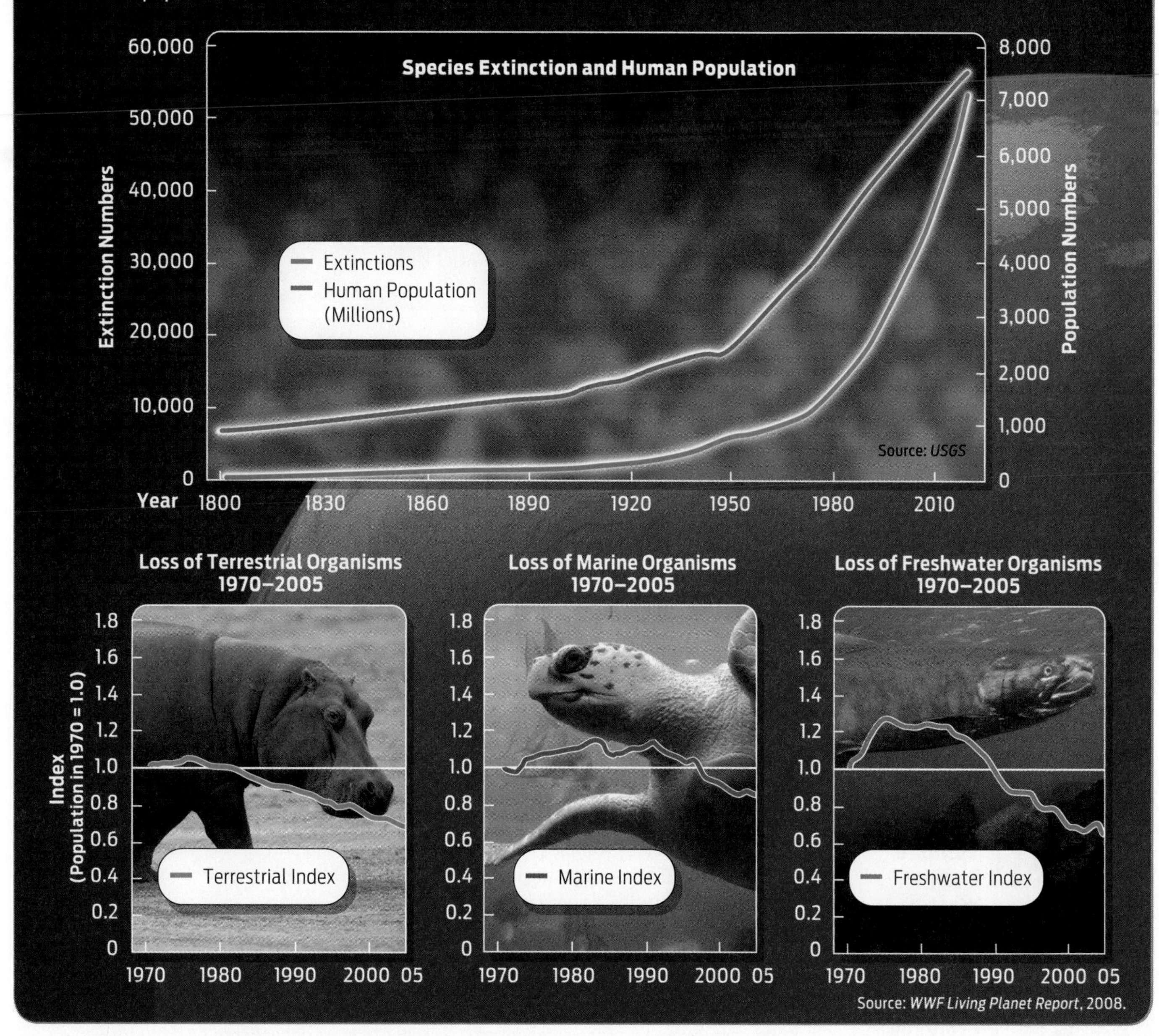

expanded into areas once occupied only by wild creatures **(Infographic 24.11)**.

Common sense tells us that the human population cannot continue to grow unchecked indefinitely—otherwise, within a few hundred years people would cover every square foot of the globe and many other species would be long gone. Experience with other species also tells us that the size of the human population will eventually reach an upper limit, at which point lack of resources and available space will curb growth.

What that precise limit is remains up for debate. The United Nations has estimated that the earth's carrying capacity (Chapter 21) is between 7 and 13 billion people; other researchers put the number much higher. Why is it difficult to pin down an exact figure? In part because the carrying capacity of the earth can change. The use of fertilizer and genetically modified crops, for example, has greatly increased the amount of food that can be produced from a given size plot of land. Likewise, modern sewer systems and advances in public health, which have helped prevent communicable diseases such as cholera, have allowed people to live in closer proximity, at much higher densities, than ever before. Quality of life concerns also muddy the calculation of carrying capacity. The earth may theoretically be able to support more than 13 billion people, but the question is, would anyone want to live in such a place?

Demographers tell us that there is little we can do in the short term to stem the human

INFOGRAPHIC 24.12

What You Can Do to Live More Sustainably

Take Action	Why?	Your Impact!
Reduce Home Water Use	The average U.S. household uses over 22,000 gallons of water per year for showers and baths. Water is almost always heated, resulting in increased fossil fuel consumption and greenhouse emissions.	If only 1,000 of us install faucet aerators ($2 – $5) and efficient showerheads (<$20), we can save nearly 8 million gallons of water and prevent over 450,000 pounds of carbon dioxide emissions each year!
Install Compact Fluorescent Lightbulbs (CFLs)	Electricity production is the largest source of greenhouse gas emissions in the United States, and lighting accounts for about 25 percent of American electricity consumption.	By replacing just four standard bulbs with CFLs, you can prevent the emission of 5000 pounds of carbon dioxide and reduce your electricity bill by more than $100 over the lives of those bulbs.
Eat Less Feedlot Beef	Feedlot beef is particularly wasteful. Producing one pound of feedlot beef in California, for example, requires five pounds of grain and over 2,400 gallons of irrigation water.	Eat more veggies. More of the energy in plants will go directly to you if it doesn't have to pass through a cow first. Plant food goes farther to support you than it does to raise feedlot beef.
Reduce Vampire Energy Waste	Electronics use energy even when they are turned off! This standby "vampire energy" accounts for 5 to 8 percent of a single family's home electricity use per year.	When you plug your electronics into a power cord that you turn off each night, you will save the equivalent of one month's electric bill each year.
Drive Less and Invest in Fuel Economy	With less than 5% of the world's population, America consumes a quarter of the world's oil and emits a quarter of the greenhouse gases, largely from automobiles.	Driving smaller vehicles and those with more fuel efficiency cuts carbon dioxide emissions and reduces dependence on non-renewable fossil fuels.
Recycle	75% of our trash can be recycled. The U.S. population discards each year enough glass bottles and jars to fill 12 giant skyscrapers. Recycling materials uses fewer non-renewable resources, saves energy, results in less air and water pollution, and creates more jobs than making new materials.	Recycling one aluminum can saves enough energy to run a TV for 3 hours. To produce each week's Sunday newspapers, 500,000 trees are cut down. Recycling a single run of the Sunday *New York Times* would save 75,000 trees. Taking reusable bags to do your weekly grocery trip reduces our demand for petroleum for plastic bags.

population climb. World population will continue to grow for years because of past growth: so many young people are alive today that even if they have only two children to replace themselves, the population will continue to expand for at least the next 50 years. This is true even in China where, since 1979 official government policy has limited couples to having one child.

For the longer term, demographers say that the best way to limit population growth is to raise the education level and quality of life for the world's women. According to Robert Engelman, Vice President for Programs at the Worldwatch Institute, there is a direct correlation between education level and the number of children women have. When women have more opportunities available to them, they tend to limit the size of their families.

"[T]he evidence suggests," Engelman writes in *Scientific American*, "that what women want–and have always wanted–is not so much to have *more children* as to have *more for* a smaller number of children they can reliably raise to healthy adulthood. Women left to their own devices, contraceptive or otherwise, would collectively 'control' population while acting on their own intentions."

Cities will also play a role. "In global terms, population growth is actually slowed by the growth of cities," writes Peter Newman. Far from being the harbingers of catastrophe, "[c]ities could indeed be helping to save the planet."

In fact, in many countries today, the birthrate is actually declining and may eventually stabilize as standards of living and educational levels of women rise around the world and as the world becomes more urbanized. Before it does, however, we will have to learn to adjust to a world occupied by at least a billion more people. And that means using all our resources, both renewable and nonrenewable, more wisely, more responsibly, and–indeed–more sustainably **(Infographic 24.12)**. ■

Summary

■ As of 2010, the human population totaled 6.8 billion people. Some demographers say the number could hit 9 billion by 2050.

■ As the human population grows, so does our ecological footprint, a measure of our demand on nature. Ecological footprint is measured in units called global hectares, the number of average hectares of land it takes to supply us with resources and to absorb our wastes.

■ The ecological footprint of the current human population is greater than the earth's biocapacity, its total natural resources and ability to absorb our wastes.

■ Natural resources include renewable resources, such as sunlight, wind, and water, and nonrenewable resources, such as fossil fuels (oil, coal, and gas).

■ Burning fossil fuels generates harmful wastes (for example, greenhouse gases and pollutants) and increases our ecological footprint.

■ Sustainability refers to the ability of humans to live within earth's biocapacity, without depleting nonrenewable resources. Sustainable practices minimize the consumption of nonrenewable resources by using renewable resources like wind and solar power instead of fossil fuels to generate electricity and heat.

■ At their current level of development, technologies to harvest renewable energy cannot meet our total energy demands. Fossil fuels cannot yet be taken out of our energy mix.

■ Although freshwater is a renewable resource, the world's supply is not distributed equally, and many people around the world suffer from water scarcity, a problem exacerbated by a rising population and the demands of agriculture.

■ City dwellers have a high per capita ecological footprint compared to people living in rural areas, in large part because of more intensive fossil fuel energy use linked to driving cars and rising consumption.

■ Cities can be more efficient than nonurban areas and can reduce their ecological footprint by limiting car use and incorporating sustainable technologies such as green roofs, public transportation, and renewable energy sources.

■ Individually, we can decrease our ecological footprint by driving less, reducing water use, eating less meat, and recycling.

HUMAN POPULATION GROWTH AND ECOLOGICAL FOOTPRINT

The human population has grown exponentially, and this growth has substantial impacts on the planet. Ecologists measure our environmental impact by calculating ecological footprints.

HINT **See Infographics 24.1–24.4.**

KNOW IT

1. From what you've read in this chapter, explain some of the advances that have permitted the human population to grow exponentially.

2. Describe an ecological footprint.

3. From your understanding of an ecological footprint and what you read in this chapter, which of the following places likely has a population with the greatest ecological footprint?

a. Dongtan, China
b. a rural village in China
c. Calgary, Canada
d. Houston, Texas
e. New York, New York

USE IT

4. For each place listed in Question 3, characterize its footprint as relatively high or relatively low. Justify your characterization by describing some of the factors that contribute to its footprint. (Refer to Infographics 24.4, 24.6, and 24.12 and the Global Footprint Network, http://www.footprintnetwork.org/en/index.php/GFN/.)

5. On the outskirts of a small town, a farmer has just sold his 5 acres of cropland to a developer who is planning to build 20 single-family condominium units on that land. Discuss the ways that this transaction will affect the size of the nearby town's population and the ecological footprint of the residents of the town.

6. What building considerations could the developer in Question 5 take into account to minimize the impact of this development on the ecological footprint of the town?

NATURAL RESOURCE USE

Natural resources occur naturally and cannot be produced by industrial processes. Some natural resources are renewable, while others are essentially irreplaceable.

HINT **See Infographics 24.5 and 24.7–24.11.**

KNOW IT

7. Which of the following waste products is/are associated with the burning of fossil fuels?

a. water
b. carbon dioxide
c. nitrogen dioxide
d. all of the above
e. b and c

8. Mark each of the following natural resources as renewable (R) or nonrenewable (N):

Freshwater _____
Coal _____
Codfish populations in the North Atlantic _____
Wind ____
Sunlight _____

9. If oil is formed from fossilized remains of once-living organisms, and if organisms keep dying, why is oil considered to be a nonrenewable resource?

USE IT

10. The renewability of some resources can depend on human choices and activities. List some such resources, and explain how human activities may lead a renewable resource to become essentially nonrenewable. (Look at Question 8 for some ideas.)

11. Think about your local region—for example, do you live in the desert southwest or on the northeast ocean shore? Describe the nonrenewable and renewable energy resources that are available in your region or that your region can harvest. What are some of the challenges that must be overcome in order to tap into the renewable energy resources in your region?

SUSTAINABILITY

Sustainability means living within the biocapacity of the earth. This includes using resources at a sustainable rate and not generating wastes faster than they can be decomposed or absorbed by the earth.

HINT **See Infographics 24.6 and 24.12.**

KNOW IT

12. The plans for Dongtan include many ideas that will contribute to sustainability. For each of the plans listed below, describe its impact on resource consumption and/or waste production.

a. schools and shops to be located near residences
b. buildings to have green roofs
c. solar panels to be mounted throughout the city

13. What are some ways in which waste can be used as a productive resource?

➔ USE IT

14. Dongtan would be an entire city with sustainable practices pre-engineered into its design. If you live in a traditional city, what practices can you adopt to reduce your ecological footprint and embrace the philosophy of sustainable living? For each practice that you think of, explain how it would contribute to sustainability and the reduction of your ecological footprint.

15. Many cities have been developed in the hot and dry southwestern states of the United States. What are some of the sustainability implications of living in the desert?

SCIENCE AND ETHICS

16. Infographics 24.9 and 24.10 provides a dramatic illustration of some of the choices we face in resource management. Water from the Colorado River is being used to irrigate crops in Arizona. How might this use affect the use of the water for recreation (for example, swimming, fishing, boating) in the downstream regions? Water currently destined for agriculture could instead be retained in the river to help preserve the endangered silvery minnow. How would you balance the competing agricultural, recreational, and ecological concerns involved in this choice?

17. In general, how do you think the ecological footprint of the United States compares to that of Bangladesh? If footprint expansion accompanies economic development, what will happen to the carrying capacity of the global human population as developing countries continue to develop? Do you think that developed countries such as the United States have an obligation to reduce their ecological footprint in order to make room for the development of other countries?

Answers

Chapter 1

1. c
2. e
3. Peer review means that the study has been reviewed by other scientists who are considered experts in that particular field. Ideally, both the results of the study and the methods used to conduct the study are reviewed and the article is rejected or revised accordingly. Peer review is important to ensure that sloppy studies are not reported and to ensure that scientists draw only appropriate conclusions from their results.
4. b
5. c
6. Ideally, the characteristics of both the control and experimental groups should be as close to identical as possible, meaning that there should be no intentional significant differences between the two groups with regard to age or breast cancer status. Because the study is intending to look at risk of developing breast cancer, both groups should be composed of women who do not currently have breast cancer.
7. The group of participants should be large and include frequent, infrequent, and non-coffee drinkers. The group should be composed of members of both sexes and multiple nationalities, ethnicities, ages, and socioeconomic groups. Randomly divide participants into two groups and give one group (the experimental group) caffeine in a drink and the other group (the control group) a placebo (no caffeine in the drink). Participants should not know whether they are part of the experimental or the control groups (the study is "blind"). Additionally, the scientist conducting the study should not know who is receiving which treatment until all the data are collected (the study is "double blind"). Multiple tests for brain function should be given. A larger study could include multiple experimental groups which receive differing amounts of caffeine to test dosage.
8. b
9. a
10. No. The study shows a correlation between abstaining from caffeine consumption and developing Parkinson disease but does not show that caffeine consumption can prevent the development of the disease.
11. a: No. The opinion expressed is anecdotal at best, meaning that it is reporting the experience of only a single person. Further, there is little reason to believe that the testimonial is even truthful; it is likely a paid endorsement, and there are no regulations regarding the truth of such endorsements. b: Results of either an experimental or epidemiological study reported in a peer-reviewed scientific journal.
12. Answers may vary. News organizations, and the corporations that own these organizations, are ultimately responsible for reports in the media. Scientists currently have little control over such reports.
13. Answers may vary. Items for the checklist may include the following:
    Was there a study done?
    Who conducted it?
    Was it peer reviewed?
    Was it reported in a respectable scientific journal?
    Who were the subjects in the study?
    Was it an experimental study or an epidemiological study?
    How many subjects?
    If the study was experimental, was it randomized?
    Do the authors of the study think that the results merit a behavioral change?

Chapter 2

1. c
2. Homeostasis is the ability of a cell or an organism to maintain a stable internal environment, usually in terms of pH, temperature, and chemical makeup, even when the external environment changes. The processes and molecules of life are delicate, and therefore homeostasis is generally important because small changes can destroy these molecules or disrupt the processes of life.
3. b
4. e
5. a
6. A polymer is a molecule is composed of smaller, sometimes repeating, subunits called monomers. Examples include complex carbohydrates (polysaccharides), which are composed of smaller sugar molecules called monosaccharides; proteins, which are composed of subunits called amino acids; and nucleic acids (DNA and RNA), which are composed of subunits called nucleotides.
7. a
8. Answers may vary. The characteristics of life are described in the text. Additional ideas include looking for common molecules representative of life on earth, including complex organic molecules, macromolecules, and water.
9. Answers may vary. However, dead organisms will generally still have a cellular structure and will not be reproducing, sensing and responding to the environment, or using energy.
10. Arguments for: Viruses reproduce. Viruses are generally able to maintain a stable internal chemical environment and make up for a lack of complex homeostatic mechanisms by being more resistant to harsh environments. Viruses can sense and respond to certain stimuli, such as detecting the presence of living host cell to invade. When inside living cells, viruses utilize the resources inside the cell for energy. Arguments against: Viruses require other living organisms in order to reproduce. Viruses generally do not grow, that is, they do

not increase in size, but only replicate. Viruses do not obtain or use energy on their own. Unlike living organisms, viruses are not made up of cells.

11. Life would be likely, but there is not sufficient evidence to conclude that life is present. There are nonliving (abiotic) means of producing glucose from carbon dioxide and water.

12. a: Sterols and triglycerides differ both in structure and in function. Sterols are composed of four carbon rings and function as color-producing pigments, hormones, or components of membranes (such as the cell membrane). Triglycerides are composed of three fatty acids covalently linked to the molecule glycerol. Triglycerides commonly function as energy storage molecules which in animals also serve to thermally insulate the organism from the environment. b: Phospholipids and triglycerides both have a glycerol backbone to which two fatty acids are attached. The difference is that in the third position either a third fatty acid is attached (triglycerides) or a phosphoryl-containing group (containing oxygen and the element phosphate) is attached (phospholipids). Because of this difference in structure, these molecules have different chemical properties and biological functions.

13. Olive oil is made up of triglycerides, which are nonpolar molecules that are hydrophobic ("water-fearing"). Salt is made up of Na^+ and Cl^- ions that are charged and interact with the polar ends of water molecules. Therefore salt is considered hydrophilic.

14. e

15. aqueous. a: The solvent is water. b: The solute is sugar. Additionally, both coffee and tea contain many other organic molecules that become dissolved in the water when tea or coffee is brewed. These molecules are produced by the tea leaves or coffee bean, and are extracted and dissolved into the hot water when brewed.

16. c

17. a

18. Hydrogen bonds and ionic bonds are both electrostatic attractions between charged atoms of two different molecules. Both are noncovalent interactions. Hydrogen bonds are between partially charged atoms (usually atoms of a water molecule), whereas ionic bonds are between ions, which have fully positive or negative charges.

19. Oil does not dissolve in the aqueous vinegar because the oil is hydrophobic. Water molecules are strongly attracted to one another (via hydrogen bonds) but not to the oil. Water excludes the oil in favor of interacting with itself or other hydrophilic molecules. Because salt is hydrophilic, it will dissolve in the vinegar but not in the oil. The ions in salt will remain together in oil because they are attracted more strongly to each other (by ionic interactions) than to the nonpolar triglycerides that make up the oil. In the vinegar, the ions that make up salt will become dissolved because they are attracted to the polar water molecules.

20. Liquid water can absorb and store large amounts of heat without evaporating and lose large amounts of heat before freezing. Therefore, seaside towns are buffered against changes in temperature because heat is absorbed by the ocean during the day and is transferred to the air from the relatively warm ocean water at night. In contrast, the desert sand is made primarily of silicon dioxide, which cannot absorb nearly as much heat during the day as can water, and rapidly cools at night.

21. b

22. Answers will vary. Some considerations if Martian dirt samples are brought to earth are disease; competition with native species (the invasive species effect); ecological effects, public fear. If an earth life form is released on Mars, effects may be destruction of the Martian environment and competition with Martian life forms if any are present, and possible extinction of those Martian life forms. Ethical implications include effects on religious views, cultural effects, the possibility of creating a disease-causing organism, and possible medical breakthroughs.

Chapter 3

1. Cell theory posits that the fundamental units of life are cells. All living organisms are made up of one or more cells. Cells arise only from other living cells.

2. d

3. prokaryotic cells and eukaryotic cells

4. a: No. According to cell theory, neither viruses nor prions (the protein aggregates that cause mad-cow disease) are considered to be living organisms. b: No. Nonliving agents can cause disease.

5. All living organisms contain genetic instructions in the form of DNA. All living organisms also synthesize the four classes of biological molecules (proteins, carbohydrates, lipids, and nucleic acids). Additionally, all known living organisms have a cell membrane and ribosomes.

6. e

7. c

8. a: Both involve the movement of a solute moving down a concentration gradient (that is, moving from a higher concentration to lower concentration). In both cases, additional energy is not required. b: Both facilitated diffusion and active transport require the function of a protein embedded in a membrane. However, the solute is moving in the opposite direction (relative to its concentration gradient) in these two cases.

9. Facilitated diffusion is necessary for molecules that cannot cross the phospholipid bilayer easily by themselves. Generally this is true of larger molecules and charged molecules. Small, nonpolar molecules, such as molecular oxygen (O_2) can pass freely through membranes, whereas polar molecules (like water) cannot.

10. c

11. Although the bacterial cell wall protects bacteria cells from lysis by keeping the cell from swelling, water can still escape from a cell placed in an environment that is high in salt. Because the concentration of salt outside the cell is high, the concentration of water is low. Water rushes out of the bacteria cell toward the lower concentration. As a result, bacterial cells shrink and die through desiccation. Some bacteria are more tolerant to high salt concentrations than others.

12. b

13. d
14. e
15. a: A mitochondrion (plural: mitochondria) is a rod-shaped organelle, approximately the size of some bacteria. Mitochondria are thought to have arisen by endosymbiosis, in which a bacterium is engulfed within another, larger cell. They are bound by a double membrane of two phospholipid bilayers. Mitochondria are considered the powerhouse of the cell; many of the reactions that extract energy from nutrients are housed within the mitochondria. b: The nucleus is a spherical membrane-bound organelle, bound by a double membrane composed of two lipid bilayers called the nuclear envelope. The nucleus houses the DNA of the cell. c: The endoplasmic reticulum (ER) is a network of single-membrane-bound tubes, actually an outgrowth of the outer layer of the nuclear envelope. The ER acts as a transport system within the cell. Proteins synthesized by ribosomes that coat part of the ER (the rough ER) are transported to other parts of the cell, often by first passing to the Golgi apparatus. d: The chloroplast is a membrane-bound organelle found in plants and other photosynthetic eukaryotes; also believed to have arisen from bacteria (again, by endosymbiosis). Reactions that capture energy from light are housed within the chloroplast, including the protein chlorophyll, which is responsible for the green color of plant tissues.
16. No. If you took an antibiotic that stopped bacterial reproduction, there would be no need for the bacteria to synthesize more peptidoglycan. So penicillin would not be able to interfere with new peptidoglycan production, as none would be synthesized in this situation.
17. Assuming that the concentration of solutes in the solution is the same as that inside the cell, then the cells will not burst, despite having weakened cell walls, because the osmotic pressure will not change–there will be no net movement of water into or out of the bacterial cells.
18. It is more challenging because both fungi and humans are eukaryotic organisms. Therefore the cell components of a fungal cell are much more similar to those of human than are those of bacteria. It is much harder to find a chemical that will selectively kill fungal cells without causing harm to human cells.
19. Muscles and nervous tissues are much more metabolically active than other tissues. Thinking and moving take a lot of energy. These tissues expend more energy and thus require more mitochondrial activity to produce energy. Tissues such as skin can rely more on alternative, but less effective, mechanisms for cellular energy production (which will be discussed later in the text).
20. Answers may vary. Physicians might explain that antibiotics will not have any effect on the viruses that cause flu or the common cold. Additionally, a physician might point out that using antibiotics decreases their effectiveness because of the development of antibiotic-resistant bacterial strains, which can be fatal. Because of the use of antibiotics over the past 60 years, there are now strains of the bacterium *Staphylococcus aureus* (and of other bacteria) that are resistant to all available antibiotics. For this reason, antibiotics should be used only when they are necessary to treat a bacterially caused disease.

Chapter 4

1. c
2. catabolic
3. c
4. e
5. b
6. c
7. a
8. A meal heavy in starch and fiber would cause less of a spike in blood sugar than a meal of only starch.
9. A diet rich in fiber will result in lower blood-glucose levels because the body cannot break the covalent chemical bonds holding the glucose monomers in fiber together. Therefore, it cannot be digested and only serves to slow the breakdown and absorption of other sugars. Fiber also has additional health benefits beyond controlling blood-sugar levels.
10. Because phospholipids are part of the cell membranes of both plants and animals, they can be directly acquired by consuming any food that contains cells, which includes all meats, vegetables, and fruits. However, phospholipids are not typically acquired directly from the diet in this way, but are "constructed" by anabolic reactions from fatty acids and other molecules. Fats and oils are the dietary sources of fatty acids.
11. Insulin causes cells of the body to take up glucose from the blood. Therefore, for a type I diabetic, who does not produce insulin naturally, insulin will be most effective if taken with, immediately before, or immediately after a meal, when blood-glucose levels are the highest or are rising. Blood-glucose levels steadily decline between meals.
12. a: calcium; b: calcium, vitamin D; c: vitamin C, very small amounts of calcium (not significant as a dietary source); d: vitamin C, calcium; e: vitamin D, very small amounts of calcium (not significant as a dietary source). None of the foods listed contains all three of these nutrients
13. e
14. Both cofactors and coenzymes are accessory molecules that enzymes use to accomplish their function. All coenzymes are cofactors, but not all cofactors are coenzymes. A coenzyme is a cofactor that is an organic molecule. Most vitamins are organic molecules that are consumed and are either coenzymes themselves or are converted into coenzymes.
15. b
16. c
17. The shape of the active site is very important in both the function of an enzyme and in specifying the substrates upon which the enzyme can act. Depending on how the shape of the active site is altered, the enzyme could act on different substrates, the enzyme could be completely nonfunctional, or (less likely) there could be no effect at

all. Generally, however, most enzymes are highly evolved and delicately tuned molecular machines, and so most changes will result in a complete loss of enzyme activity.

18. High fever can cause the enzymes of the body to malfunction. This is the main danger of hyperthermia, in which the body becomes overheated. However, this same mechanism is believed to be the reason why fever is a common response by the body to infection: the increased body temperature causes the enzymes in bacteria or other infectious organisms to malfunction, slowing their growth and aiding their removal from the body.

19. Osteoporosis is a disorder characterized by the thinning of the bones through loss of bone density. It is common with increasing age and more common in women than in men. Diet and exercise can both help to reduce the effects of osteoporosis.
Exercise reduces risk by stimulating bone growth. Dietary calcium reduces risk by preventing calcium from being reabsorbed from bone when it is needed elsewhere by the body. Dietary vitamin D regulates calcium metabolism and prevents bone loss due to calcium reabsorption. Dietary vitamin C is required for new bone deposition.

20. Vitamin D is important for regulating calcium levels in the body, and vitamin D deficiency can lead to loss of bone mass. Vitamin D is not found in many foods and is mainly synthesized in the skin. A critical reaction in this synthesis requires UV light (from sunlight). Typically only a few minutes in intense summer sunlight are necessary, but during the winter, or at latitudes far from the equator, the sunlight is not intense enough to adequately produce vitamin D. Vitamin D deficiency is a growing concern in the United States and around the world.

21. a: Perhaps the primary advantage to a nutritionally engineered diet is that it requires little time to prepare, and may involve consuming foods that are less expensive than whole-food alternatives. Additionally, with such a diet it can be easier to keep track of both macro- and micronutrient consumption because the exact amounts of each of these components in the food is known. Finally, such a diet could allow people with certain dietary restrictions such as food allergies to consume foods rich in nutrients that would otherwise be rare in their restricted diet. Disadvantages include the possibility that people do not understand enough about nutrition to adequately construct diets in this way. Certain micronutrients that are found only in unprocessed food may provide health benefits that are not yet appreciated, and therefore would not be included in the engineered diet. b: Engineered diets provide a particular advantage to those individuals for whom it is especially important to control diet.

22. Answers may vary. Responses should consider what evidence is necessary to make such a claim, and whether it is worth delaying the release of a potentially helpful food alternative to wait for extensive scientific studies to be done. Another important factor to consider might be whether the studies that demonstrated the ability of this particular fiber to reduce blood-sugar levels have been repeated with type I diabetics.

Chapter 5

1. b

2. a

3. Algae appear green because the chlorophyll within algae absorbs red and blue wavelengths and reflects the green wavelengths of sunlight. Our eyes perceive the reflected green wavelengths.

4. c

5. Photosynthetic algae obtain energy by using sunlight to create sugars by photosynthesis. The energy stored in these sugars can be used by the algae to carry out essential life processes. Animals are not able to use the sun's energy directly. Instead, to obtain energy animals eat plants or other organisms that have eaten plants. The energy the animals use ultimately comes from the sun.

6. oxygen (O); carbon dioxide (I); photons (I); glucose (O); water (I)

7. Increasing carbon dioxide levels should increase photosynthesis because carbon dioxide is an input for the reaction. If forests become immersed in water the plants will not be healthy because they are not adapted to living in water; thus this aspect of climate change will produce a negative effect.

8. Lipids harvested from algae are more "useful" because there are fewer steps required before the lipids can be used. To produce fuel from carbohydrates found in plants the plants need to be broken down and fermented to produce ethanol. A lot of energy is lost in these processing steps.

9. c

10. c

11. You should eat the algae directly to gain the most energy. If you feed the algae to a cow first, the cow will use some of the energy, and thus the burger made from the cow will contain less of the energy from the algae.

12. a

13. g

14. Biodiesel from algae requires the transfer of energy from sunlight to organic molecules (including oils). There is one major energy transfer (sun to organic molecule). Lipids from animals are only produced by the animal after the animal has eaten a plant or another animal that ate a plant. There are more steps between sunlight and animal lipids than between sunlight and algal lipids, and some energy is lost at each step.

15. One major advantage of growing algae in enclosed tubes is that the conditions can be controlled to maximize oil production. Disadvantages are the need for providing sufficient CO_2 for photosynthesis and the cost of maintaining an elaborate growing system.

16. Algae may be the perfect solution to the food versus fuel debate. Algae can be grown on land that is not suitable for crops, like the desert, and algae is not a major human food source and therefore algae production does not compete with food production. A conflicting factor is that algae still require water for growth, but much less than a crop would require.

Chapter 6

1. d
2. d
3. Her BMI would be ~27, which would make her overweight according to the CDC.
4. e
5. a
6. Your lifestyle would need to be modified to compensate for the additional Calories, regardless of the type of food eaten. Ultimately, the only ways to avoid weight gain are either to reduce Calories eaten from other food sources or to increase the number of Calories burned, preferably by additional exercise.
7. a: 884g; b: 3,536 Calories (assuming perfect aerobic metabolism); c: 442g; d: She will run for 2.91 hours at this pace and burn an estimated 2,575 Calories in that amount of time. Again assuming perfect aerobic metabolism, it will take her 4 hours at this pace to burn through her glycogen stores. She can run 36 miles in this time at this pace, which is well beyond the 26.2 required to complete a marathon. e: Her body will need to extract energy from another fuel source. Fat (from adipose tissue) and protein (from muscle) are the next sources, in that order. However, it is important to note that in cases of complete carbohydrate starvation (such as this one), protein from muscle will still be broken down to provide glucose (blood sugar) to the brain. This happens because humans are incapable of making glucose from fats. This process of protein wasting continues until the brain adjusts to another energy source (ketone bodies), which can be produced from fats.
8. Many possible reasons have been suggested, among them smaller portions, less energy-rich but nutrient-poor foods, longer meal times, less snacking, and greater self-control at meal times.
9. c
10. d
11. e
12. b
13. See Infographic 6.10. The carbon atom (in CO_2) will be taken up into the spinach leaf, into the plant tissue, into the plant cell cytoplasm, and eventually into the chloroplast. There it will be converted, along with water (H_2O), into carbohydrate (glucose). This glucose may be used to construct the plant cell wall (fiber) or stored for later use by the plant (starch). If it is stored as starch, then when it is consumed by a human, the glucose will be broken down again into CO_2 by glycolysis and the citric acid cycle. Thus, the carbon atom will have come full circle. (If the glucose had been converted to fiber, other organisms, particularly fungi and bacteria, will also break down the fiber to glucose and then to CO_2, so the atom of carbon can always return to CO_2.).
14. a
15. In the presence of oxygen, aerobic respiration is the preferred method by which glucose is utilized by most cells. The process involves catabolism of the glucose molecule in a process called glycolysis. Following glycolysis, each carbon atom originally found in the glucose molecule is converted to CO_2 in the citric acid cycle. Electrons removed from glucose during this process are transferred to the electron transport chain, by which they are eventually transferred to oxygen to form water (along with two hydrogen atoms). The energy from the electrons transported in this way is captured by the cell as ATP. The entire process produces ~36 ATP from each glucose molecule. In the absence of oxygen, aerobic respiration is not possible. Cells utilize fermentation to acquire a small amount of ATP, but only two ATP per glucose molecule are produced by fermentation.
16. c
17. A common hypothesis is that by eating longer meals, the French eat more slowly, and therefore allow time for stomach to signal the brain to quit eating. Often we do not realize that we are full until several minutes, or dozens of minutes, after eating. Therefore, by eating quickly, we may consume far more food than is necessary to feel full. Because of this difference, longer meal times may result in the French eating less food (and fewer Calories) at every meal.
18. Answers may vary. See the discussion in "The culture of eating" section of the text.
19. Answers may vary. They should be accompanied by a justification of how a particular intervention might result in reduced obesity. Examples: (1) Taxing high-fat foods might encourage people to purchase foods that contain higher amounts of carbohydrates and proteins. Theoretically these foods might be lower in Calories, given the lower energy density of these other nutrients. (2) Financial incentives might encourage people to exercise more, thereby increasing total Calories burned. (3) Fruits and vegetables, particularly vegetables, are high in nutrients but relatively low in Calories. Encouraging their consumption might decrease overall Calorie consumption in the American public. (4) Providing incentives to teachers to encourage teaching about nutrition might increase awareness of nutrition and its effect. The idea is that those who understand the dangers of poor nutrition and/or obesity will be motivated to change eating and exercise habits.

Chapter 7

1. e
2. c
3. b
4. (1) The two original strands of the DNA molecules are separated by means of *heat*. (2) The enzyme *DNA polymerase* "reads" each template strand and adds complementary nucleotides to make a new strand.
5. Step 1. The strands separate:
   ATCGGCTAGCTACGGCTATTTACGGCATAT

   TAGCCGATCGATGCCGATAAATGCCGTATA

   Step 2. DNA polymerase adds complementary nucleotides, forming two new double helices:.

ATCGGCTAGCTACGGCTATTTACGGCATAT
TAGCCGATCGATGCCGATAAATGCCGTATA

ATCGGCTAGCTACGGCTATTTACGGCATAT
TAGCCGATCGATGCCGATAAATGCCGTATA

6. a: F; b: T; c: F; d: F; e: T
7. Statement b in Question 6 is true because DNA is the molecule of heredity that is passed from parents to offspring. Statement e in Question 6 is true because all body fluids, including saliva, contain DNA that can be used for DNA evidence. Unfortunately, in the case of Roy Brown, the technicians were not able to extract enough DNA from the bite marks for PCR.
8. Statement a in Question 6 should be, "G pairs with C and A pairs with T." Statement c in Question 6 should be, "DNA consists of coding sequences, which encode information to produce proteins, and noncoding sequences, which regulate when coding regions are turned on or off." Statement d in Question 6 should be, "The number of STR repeats on your maternal chromosome can be different from the number of STR repeats on your paternal chromosome."
9. a
10. d
11. d
12. Lane B
13. a: Suspect B's profile matches the profile from the blood collected at the crime scene for all of the markers tested. b: Suspect A is most likely unrelated to the victim since they share very few STR bands. Suspect B is likely either the victim's parent or child since they share at least one band at each of the markers tested.
14. d
15. c
16. AMELY and AMELX
17. For AMELY: If the perpetrator is female, you would expect no bands, and if the perpetrator is male, you would expect one band. For AMELX: If the perpetrator is female, you would expect two bands or one thick band if the repeat is the same length on both X chromosomes. If the perpetrator is male, you would expect only one band.
18. a.

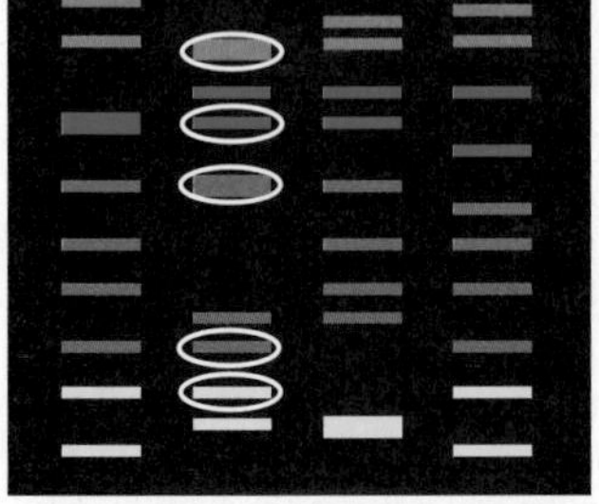

    b: Lane M1 identifies the father of the child.
19. The advantages of having a DNA bank include clearing individuals who have been wrongly convicted and helping to identify suspects in unsolved cases. Banking DNA could be problematic, depending on who has access to the information and how it is used. For example, people's DNA markers can indicate their susceptibility to certain diseases, enabling insurance companies with access to this information to discriminate against those individuals. This is particularly troubling because the markers indicate a correlation and not a definitive likelihood of developing the disease.

Chapter 8

1. A protein's function is determined by the shape, which is determined by the interaction of the amino acids that make up the protein–specifically, the order of the amino acids and how their side chains interact.
2. c
3. The protein will not be able to function because the heat will cause the protein shape to change and shape is critical for function.
4. The insulin gene is made up of DNA nucleotides that provide information about how to make the insulin protein. The insulin gene is transcribed into mRNA in the nucleus; then the mRNA is transported into the cytoplasm and translated into insulin protein.
5. d
6. e
7. The problem is likely in the regulatory sequence because the patient has reduced levels of normal antithrombin. The regulatory sequence controls how much mRNA is made and therefore how much protein is made. Alternatively, the patient could carry one allele with changes in the coding sequence that make a nonfunctional protein. The patient would still have low levels of functional protein made from the allele with the normal coding sequence.
8. To increase the level of antithrombin, the regulatory sequence should be modified. The regulatory sequence controls the timing and location of transcription of antithrombin. The amount of mRNA made is directly related to the protein produced.
9. To express a gene in skin cells, combine the regulatory sequence from the melanin gene with the coding sequence of the gene of interest. The regulatory sequence of melanin is specific to skin cells, so the gene of interest will only be produced there. To express melanin in yeast cells, use the regulatory sequence from a yeast gene and the coding sequence from melanin. The coding sequence is necessary to produce the correct melanin protein.
10. The beta casein regulatory sequence was used to express antithrombin in milk because the beta casein gene is expressed only in the mammary glands. This was important to ensure that the goats were not harmed by the production of antithrombin.
11. RNA polymerase (N); ribosome (C); tRNA (C); mRNA (C) (mRNA is transcribed in the nucleus and transported to the cytoplasm for transcription, so it is active and carries out its main function in the cytoplasm.)
12. a: The complementary DNA strand is TCTATGCTTTGT. b: The complementary mRNA strand is UCUAUGCUUUGU. c: The mRNA sequence contains four amino acids–Ser, Met, Leu, Cys–but only three will be translated into protein. Met is the start codon where translation will occur, so Ser

will not be translated. The final protein will include three amino acids: Met, Leu, Cys.

13. a: If RNA polymerase cannot bind to the regulatory region, the gene will not be transcribed; thus neither mRNA nor protein will be produced. b: A change in the coding sequence will not have an effect on transcription of the mRNA. Depending on the change in the sequence, the protein structure and levels could be unchanged if the change were to an amino acid with similar properties; or the protein could become nonfunctional because the new amino acid causes the protein to be shaped differently. A third possibility is that the change will create a stop codon in the middle of the protein, creating a truncated, nonfunctional protein. c: A change in the regulatory region that increases transcription will increase the amount of mRNA and protein that is produced. The function and structure of the protein will be normal. d: The change to the regulatory region will increase the level of transcription creating higher levels of mRNA. The change to the coding region may result in a nonfunctional protein resulting from changes in the shape of the protein. The combination of these changes will lead to an increase in nonfunctional protein which the cell will degrade; and the phenotype will be similar to that observed if the protein is not produced.
14. a: proline (Pro); b: proline (Pro); c: leucine (Leu)
15. The benefit of producing insulin in either pigs or bacteria is that more diabetics are able to live a long and healthy life because of the greater availability of insulin. The ethical question is whether or not we should modify organisms to produce unnatural proteins. One might argue that producing insulin in pig pancreas or bacteria is invasive and will harm the pig or bacteria. On the other hand, these human-insulin-producing animals were produced for the purpose, so harming the pig or bacteria for production of insulin is serving the greater good. In some ways it is easier to accept using bacteria to produce insulin: there are millions of bacteria, they have a short life cycle and rapid regeneration, and we can't see a single bacterium with the naked eye.

Chapter 9

1. e
2. c
3. Embryonic development, wound healing, and replacement of blood cells all require mitosis to create more cells.
4. c
5. Pregnant women should not take drugs that interfere with cell division because the cells of the developing embryo are rapidly undergoing cell division. If a woman were to take these drugs, the developing embryo would cease to grow and would die or have major defects.
6. b
7. Chemotherapy interferes with cell division to kill cancer cells. However, the drugs target any actively dividing cells such as intestinal cells, blood cells, and hair follicles. The side effects of killing these cells may include nausea and diarrhea (by interfering with normal cell division in the intestinal tract), and hair loss (by interfering with cell division in hair follicles).
8. a
9. c
10. Chemotherapy targets actively dividing cells like the lining of the digestive tract. When these cells are killed, side effects such as nausea may occur. Cognitive symptoms are not a side effect because neurons rarely (if ever) divide and so are not affected.
11. c
12. a: Irinotecan slows the growth of the tumor–the rate of cell division–by interfering with cell division. Thus the tumor grows more slowly in the presence of irinotecan than in the absence of chemotherapy. b: PHY906 plus irinotecan slows the growth of the tumor more than irinotecan alone, as can be seen in the graph by the very limited growth of the tumor in the presence of both irinotecan and PHY906. PHY906 enhances the effect of irinotecan.
13. Eating whole foods that are rich in beta-carotene may have more benefits than a beta-carotene supplement because there may be interactions between beta-carotene and other molecules in the whole food that will make the beta-carotene more potent.
14. anaphase
15. interphase (S phase)
16. If a cell does not complete cytokinesis there will be one cell with twice the number of chromosomes relative to the parent.
17. A drug interfering with spindle fiber shortening would be an effective cancer drug because anaphase cannot occur if the spindle fibers cannot shorten. Disrupting anaphase will lead to apoptosis and therefore slow the growth of the tumor.
18. Over-the-counter herbal supplements are not regulated by the FDA and may have varying amounts of the effective compound. Additionally, these supplements may contain other, potentially harmful, compounds.
19. Establishing the efficacy of PHY906 in humans will require a clinical study. The study should include cancer patients that are given chemotherapy alone or chemotherapy with PHY906. The patients should be closely monitored and allowed to stop taking PHY906 if negative effects occur at any point in the study. The patients should be informed of all possible risks before being asked to agree to participate in the study.

Chapter 10

1. Mutations in both tumor suppressor genes and in oncogenes increase the risk of developing cancer. Tumor suppressor genes cause cancer when the proteins the genes code for become nonfunctional; oncogenes cause cancer when the proteins become permanently activated, or "turned on." Both types of genes play important roles in cell division and its regulation. Tumor suppressors typically signal the cell to pause cell division in order to fix errors; oncogenes tend to promote cell division.
2. *BRCA1* is a tumor suppressor gene that produces a DNA repair protein that helps detect and repair mutations.

3. b
4. e
5. b
6. c
7. At birth, all of Lorene Ahern's cells–including her breast cells and her liver cells–were genetically identical and carried a mutation in one of her two *BRCA1* alleles. For cancer to develop, some of her breast cells must have accumulated additional genetic mutations, which would make these cells genetically different from her normal breast cells.
8. If there is no family history of breast cancer it is unlikely that the niece has a mutation in *BRCA1*. Therefore, she should be able to reduce her risk of developing cancer by adopting lifestyle changes like not smoking, using sunscreen, and minimizing exposure to carcinogens, which will decrease her chance of accumulating cancer-causing mutations.
9. e
10. a
11. The normal BRCA1 protein acts as a tumor suppressor to halt cell division and promote DNA repair. This means that it will take only one additional mutation in *BRCA1* (in the other allele) for them to lose all BRCA1 function. Nonfunctional alleles of *BRCA1* encode proteins that do not act properly to detect and repair damaged DNA.
12. There are many possible answers, as well are concerns about privacy. Nellie's doctor might advise Nellie to tell both her sister, Anne, and her brother that she carries the *BRCA2* mutation, but ultimately it is up to the sister to decide whether or not she wants to be tested for the mutation. Anne may not want to live with the burden of knowing that she has a higher risk but not a guarantee of developing breast cancer. A counselor might suggest that Anne be tested because there is evidence to suggest that there are treatment options available to carriers of the mutation, including prophylactic surgery. The brother should also be advised since men with mutations in *BRCA2* are also at higher risk for developing breast and prostate cancers. Another consideration is that their children's risk is affected if a parent carries the mutation.

Chapter 11

1. 46 (23 pairs)
2. 23
3. A person with CF is homozygous recessive at the *CFTR* gene and carries two of the CF-associated alleles in all of his or her lung cells. A heterozygous carrier for CF has one CF-associated allele and one normal allele at the *CFTR* gene. Someone who is homozygous dominant carries two of the normal alleles at the *CFTR* gene.
4. a: A heterozygous genotype will have a normal phenotype (like Emily's parents). b: A homozygous dominant genotype will have a normal phenotype. c: A homozygous recessive genotype will have cystic fibrosis.
5. Two individuals with different phenotypes may have different mutations at the *CFTR* gene or different alleles in other modifier genes that may affect the severity of the disease.
6. c
7. f
8. Maternal Paternal

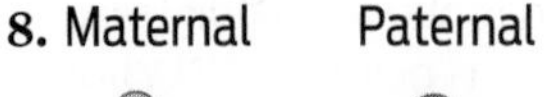

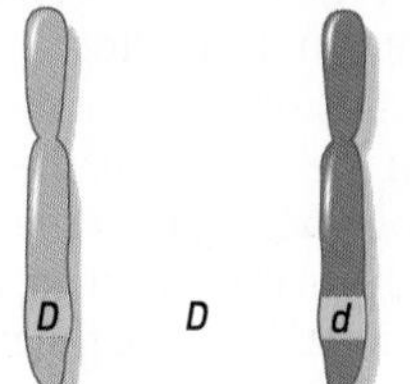

Each haploid gamete could contain one of the following chromosomes:

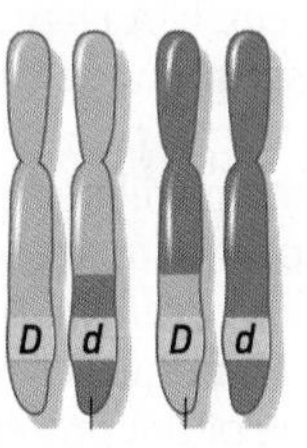

9. Since meiosis halves the total number of chromosomes, 41 unpaired chromosomes would be present in one of the alien's gametes.
10. Mitosis starts with a diploid cell and produces two genetically identical diploid cells. Meiosis also starts with a diploid cell but, because of recombination, results in the formation of four haploid cells containing new genetic combinations of alleles.
11. If meiosis I is skipped, there would be no recombination. The 46 replicated chromosomes would each separate their chromatids during meiosis II, resulting in gametes with 46 chromosomes (instead of 23).
12. b
13. recessive inheritance pattern
14. c
15. a: All of Emily's gametes will carry the allele that is associated with CF (*a*). The man's gametes will all carry the allele that is wildtype (*A*).

b:

	A	*A*
a	*Aa*	*Aa*
a	*Aa*	*Aa*

c: 0%

d: 100%

16. Since Huntington disease is a dominant disorder, the friend has a 50% probability of developing it, as shown in the Punnett square below:

	t	*t*
T	*Tt*	*Tt*
t	*tt*	*tt*

17. If you take a genetic test for a disease for which there is no cure, you do so knowing you would not be able to undergo treatment to improve your prognosis; thus, even if you are presently asymptomatic, you may become anxious about developing the disease. Knowing if you will

develop a disease may affect your choices about having children; it may help ongoing research; and may be helpful knowledge for your doctor in monitoring your overall health.

Chapter 12

1. c
2. Males have only one X chromosome, whereas females have two X chromosomes. Recessive traits occur when only recessive alleles are present. Males have only one allele of each X-linked gene because they have one X chromosome; therefore, if they have the recessive allele they will develop the recessive genetic disease. Females are less likely to be affected because their recessive allele can be masked by a dominant allele.
3. If a male has an X-linked recessive disease he cannot pass it to his son because the father must pass the Y chromosome to his son. The X chromosome in males will always come from the mother.
4. c
5. a: XX, female; b: XXY, male; c: XY, male; d: X, female
6. a: The brother and son of a female will not have identical Y chromosomes because the Y chromosome is inherited from the father. The exception is if the brother is the son's father. b: The brother and son of a male will have essentially identical Y chromosomes because the Y chromosome is inherited from the father. The two brothers will have received essentially the same Y chromosome from their father and then pass that Y chromosome to their sons.
7. a: DMD is X-linked recessive. 50% of the sons will have DMD and 0% of the daughters will have DMD. b: Rickets is X-linked dominant. 50% of the sons will have rickets and 100% of the daughters will have rickets.
8. There are many genes that contribute to the phenotype of height, so it is a polygenic trait.
9. d
10. Polygenic inheritance is primarily due to the influence of effects from multiple genes. Multifactorial inheritance includes an interaction with the environment.
11. Incomplete dominance describes traits in which heterozygous individuals have an intermediate phenotype between that of the homozygous dominant and homozygous recessive phenotypes. Traits that are codominant produce heterozygotes that display both the dominant and recessive phenotypes.
12. The only possible recipient for an A+ donor is type A+. The possible donors to an A+ recipient are type A+, type A-, type O+, and type O-.
13. Environment influences overall height. The two women may have had different diets while they were growing.
14. The hypothesis that genes and environment influence phenotype is supported by the significantly higher number of people diagnosed with depression who have two copies of the short allele and have had four or more stressful life experiences. If depression were controlled only by the number of short alleles, then all people with short alleles would have the same propensity for depression. Similarly, if depression were based entirely on environment, then all people who experience four or more stressful events should have a high propensity for depression.
15. From these data, the probability of people with two short alleles becoming depressed increases after three or more stressful life events.
16. Phenotype is the result of both genotype and environment. Even if two people have the same genotype for a predisposing allele, their environments may be different and thus change their probability of developing the disease.
17. a: 23 chromosomes (human egg); b: 23 chromosomes (human sperm); c: 46 chromosomes (zygote)
18. Genotypes cannot be deduced from karyotype analysis. Karyotype analysis is used to determine the number of chromosomes present.
19. c and d
20. Research supports a correlation between some of the most obvious birth defects and the age of a woman's eggs, but there are findings that the age of the male can also influence the frequency of cognitive disorders. It is more likely that the egg cells will include chromosomal abnormalities.
21. Factors for considering genetic counseling include age, family history, and medical history. The value of having this information is to be better prepared to support a child regardless of his or her abilities.

Chapter 13

1. A five-year-old child does have adult stem cells. The "adult" stem cells are somatic cells that are still able to divide to regenerate specific cell types.
2. a
3. d
4. Tissues are made up of different specialized cell types that work together. Neurons and glial cells are different cell types that work together to allow the firing of electrical impulses.
5. It would not be sufficient to replace only the neurons as the nervous tissue is made up of neurons and glial cells, and glial cells help the neurons with sending rapid signals.
6.

	Photoreceptor cells of the retina	Heart muscle fibers	Helper T cells
Myosin gene present?	X	X	X
Myosin mRNA present?		X	
Myosin protein present?		X	
Retinal gene present?	X	X	X
Retinal mRNA present?	X		
Retinal protein present?	X		
CD4 gene present?	X	X	X
CD4 mRNA present?			X
CD4 protein present			X

7. c
8. Advantages of using one's own cells include the following: There is no need to wait for a donor match because the cells will come from the recipient. Cells will not be rejected by the recipient's immune system: the cells come from the recipient so the immune system will recognize them as self. The recipient will not have to be on immune-suppressant medication, which can lead to other illness: the cells come from the recipient so the immune system will recognize them as self.
9. Embryonic stem cells can differentiate into almost any cell type and are found in early embryos. Adult stem cells are more limited as to the types of cells they can differentiate into; they are found in tissues.
10. c
11. It is more challenging to engineer a bladder because there are several types of cells (including muscle cells and nerve cells) that are required to make the organ, whereas skin is made up only of skin cells.
12. a: heart muscle; b: none; c: pancreas; d: neurons
13. Embryonic stem cells have a wider utility than adult stem cells in that embryonic stem cells can differentiate into almost any other cell type. These cells do not have an identity, so there is the promise that they could be put into a patient's body to stimulate regeneration in the damaged tissue. Adult stem cells are useful for a narrow range of cell types that are similar to the adult stem cell. For example, blood stem cells can make other blood cells, but could not make a liver cell. Embryonic stem cells are derived from early embryo cells and thus there are ethical concerns with using this cell type. Adult stem cells come from tissues and so there is less controversy about using them.
14. The genes that were inserted functioned to de-differentiate the cells back to an embryonic state. The genes had to be added into the cell because the endogenous genes had been shut off during development.
15. There are many possible opinions: one is to choose to allocate funds to all types of stem cells because each type will serve a unique purpose. Additionally, this field is relatively new and it is important to continue exploring all research avenues for the best solutions. Funding should also be allocated to those researching the ethical questions related to this research so that we don't cross a line that can't be undone.
    The technologies to create cloned embryos for "reproductive" or "therapeutic" cloning do not differ except that in reproductive cloning the embryo is implanted into a woman's uterus. There are many opinions and also regulations regarding reproductive cloning of humans. Many people think that reproductive cloning of humans should be illegal because they are concerned that the humans born from this process would not have the same rights as humans born from traditional means.

Chapter 14

1. a
2. Colonization means the bacteria are carried on or in the body without causing disease; infections are associated with disease.
3. c
4. MRSA can be passed from person to person by direct skin contact or touching contaminated surfaces. If the bacteria can find their way into a wound, they can get into the body and cause an infection. Athletes with cuts or scrapes can get MRSA from contact with other people or contaminated objects (for example, towels).
5. Once I confirmed that the infection is really caused by a MRSA strain (methicillin-resistant–also resistant to other beta-lactam antibiotics such as penicillin), I would try non-beta-lactam antibiotics. If these were not successful, then I would consider prescribing vancomycin as this antibiotic is reserved for severe MRSA infections that don't respond to other types of antibiotics. I would recommend that the teammates increase their hand washing, decrease contact when possible, and that the locker room be thoroughly cleaned to cut down the frequency that the other players will come in contact with MRSA.
6. Beta-lactam antibiotics work by interfering with the bacteria's ability to synthesize cell walls. Our eukaryotic cells are not affected because there is no cell wall in animal cells.
7. c
8. d
9. "Fitness" describes the ability of an individual to survive and reproduce in a given environment. An organism that has a higher fitness will be able to reproduce and pass its genes to the next generation at a higher frequency than a less fit individual.
10. a
11. c
12. Asexual reproduction occurs when an organism replicates its own genome and divides into two daughter cells. The daughter cells are a copy of the mother cell. The daughter cells could be different from the mother or the other daughter cell if mutations occur during replication.
13. Evolution is the result of change in allele frequency over generations (time). Bacteria evolve at a high rate because the generation time is minutes or hours compared to years for other organisms.
14. c
15. Some of the cells from the population will grow because during replication their DNA has accumulated mutations that make them resistant to vancomycin.
16. The genotype determines the phenotype, so if the genotype frequency is changed then the phenotype frequency will change.
17. a: The snails will be greenish in color to blend in with the grass. The snails that do not blend into their surroundings have a higher probability of being eaten by birds and therefore cannot pass their genes to the next generation.
    b: Individual snails will not be able to mutate to change their color in response to the environmental selective pressure. Snails that are brown will be selected for and will reproduce, resulting in more brown snails. The green snails will have a low fitness in this environment because they will stand out and have a higher probability of being eaten before they reproduce. Over time the color

phenotypes will shift to brown because these snails have a higher fitness in this environment and thus will reproduce more, leaving their brown alleles in the next generation of snails. The new color phenotype is the result of random mutation and recombination, leading to changes in the genome that are passed to the gametes and resulting in the brown phenotype.

18. a: It is troublesome to hear this story because antibiotics will not kill the virus and the increase in antibiotics in the environment will increase the chance that bacteria in the environment will become resistant. b: No. c: The risk to the friend is that bacteria in the friend's body are being exposed to the antibiotic and a few bacteria in his body may be resistant to that antibiotic. Those resistant bacteria will continue to replicate, leaving him with a population of bacteria that are resistant to that antibiotic. If those bacteria should infect his bloodstream (for example, through a break in the skin), that infection will be hard to treat (as the bacteria are already resistant to at least one class of antibiotic). The risk to the community is that if those bacteria should be transferred to others, they could cause an antibiotic-resistant infection in those affected.

Chapter 15

1. 0.34

2. b

3. Yes, evolution has occurred. The genetic definition of evolution is a change in the allele frequencies within a population.

4. Population B would be the most likely to survive a sudden environmental change, because it has the greatest allele and genotype diversity. This diversity increases the chances that some individuals will be better adapted to survive changes in their environment.

5. Populations 1 and 4 are the most threatened. Population 1 has both a low total size and a single gene for which there is only one allele in the population. However, population 1 has good genetic diversity with regard to the other two genes being studied. Population 4 has the least genetic diversity of all the populations but has a 20-fold higher total population. Either of these populations could be of great concern to conservationists.

6. If PKU occurs in 1 in 15,000 people, then q^2 is 1/15,000, and q is the square root of 1/15,000 (= 0.008). Therefore $p = 1 - q$ (= 0.992), therefore the carrier frequency = $2 \times p \times q$ (= 0.016, or 1.6% of the population).

7. $p = 0.45$, $q = 0.55$; predicted frequency of homozygous dominant (*AA*) is 20.25%, actual percentage is 5%; the population is not in Hardy-Weinberg equilibrium.

8. d

9. c

10. a: The frequency of alleles *A* through *L* has changed drastically. *A-L* but not *B* frequencies are reduced to zero. *B* frequency is now 1.0. b: This is an example of a bottleneck effect.

11. Genetic drift is an example of evolution because genetic drift changes the frequency of alleles in a population (the definition of evolution). Genetic drift differs from natural selection in that genetic drift does not necessarily lead to adaptation, since the changes in allele frequencies are not due to selection for a particular beneficial trait but rather are due to random events.

12. In this example, the descendant population's allele frequencies might remain similar to the founder's because the descendants are more likely to mate with other members of this same population rather than with members of the population around them.

13. d

14. Geographic isolation prevents gene flow. Due to natural selection or genetic drift, allele frequencies of the two separated populations will diverge over time. Without being able to exchange genetic material, the two populations may eventually experience changes in allele frequencies and evolve traits which prevent successful mating, and thus will have speciated.

15. According to the biological species concept, these populations are still the same species if they can mate and produce fertile offspring. You would need to observe mating between members of the previously separated populations and then follow their progeny to see if they are fertile–that is, if they can mate and produce offspring.

16. Inbreeding is detrimental because it decreases the number of heterozygotes in a population, increasing the proportion of individuals which are homozygous for recessive alleles. Many recessive alleles are mutations that are detrimental but which do not confer a phenotype in heterozygous individuals because the nonmutated gene is a dominant allele; however, in a homozygous recessive individual, these traits are expressed, usually with extremely negative consequences including decreased fitness, fertility, or viability (that is, the trait is lethal). This phenomenon is called inbreeding depression.

17. Over time the gene pools of these groups would converge, becoming more similar to one another.

18. *A* and *E* frequencies will increase; all other frequencies will decrease.

19. a: Answers will vary depending on date and sources used. In general, endangered species are at risk for becoming extinct, whereas threatened species are at risk for becoming endangered in the near future. b: Because the genetic diversity of the population has already been reduced, and further habit preservation will not restore genetic diversity in the short term. Other interventions are necessary to create gene flow and restore genetic diversity, such as the introduction of pumas from another area as described in the text. c: There are a number of possible answers, which include but are not limited to: genetic testing to identify any individuals with new alleles followed by intentional breeding of those individuals to increase rare allele frequencies; attempts to breed cheetahs with other closely related cats in the hope that some combinations may produce fertile offspring; genetic engineering to introduce new traits artificially; separation of the species into multiple distinct environments in the hope that each new founder population will evolve new traits distinct to that environment.

Chapter 16

1. the shallowest layers (those closest to the surface)
2. c
3. This fossilized skeleton appears to be most similar to extant (that is, currently living) bony fish. Specific characteristics noted could include the presence of dorsal, pectoral, pelvic, and tail fins, which are all clearly visible and are useful for swimming, indicating the organism likely lived in the water. Additionally, the connected skull and shoulder bones indicate protected gills, again indicating that this is a marine organism related to bony fish. The presence of teeth and large mouth indicate that this organism was likely predatory.
4. No. The hypothesis that these sea cucumbers existed does not predict that their fossils should be present in the fossil record, as organisms which had only soft-tissues (such as a sea cucumber) rarely produce fossils.
5. The barnacles are at least as old as the oysters, which are at least as old as the surrounding rock, which is dated at 100 million years. Fossils must be at least as old as the surrounding rock which encased them. Likewise, organisms found to coexist in the fossil record must have lived at approximately the same time.
6. c
7. elongated bony ribs and weight-bearing pectoral fins that include ankles
8. Transitional fossils represent midpoints between two groups of organisms. Often they are extinct organisms that represent a transitional form between the ancestors of two groups of extant (that is, currently living) organisms. In this case, *Tiktaalik* represents a transitional form between bony fish and amphibians (or all tetrapods). Often transitional fossils help scientists understand how organisms changed morphologically over time.
9. See Infographic 16.4. The first real tetrapod would likely have more distinguishable hindlimbs that would be capable of supporting weight, rather than pelvic fins. This would likely be the defining characteristic of the first tetrapod and would be the primary difference distinguishing it from *Tiktaalik*. Given the trends observed in the *Tiktaalik* fossil, you might also expect the tetrapod fossil to also have longer and thicker ribs, a less defined or smaller gill slit, longer neck, more developed fore-limbs with more defined digits ("fingers").
10. The land represented a new ecological niche into which life could expand, but this does not mean that the oceans and freshwater environments are not places in which life could thrive. Fish were already well adapted to surviving and proliferating in these marine environments, and therefore descendents of those ancient fish exist today.
11. The skeletal anatomy of a chicken wing and a human arm are very similar. All major bones are present and in the same locations relative to other bones. The primary difference is found in the fine bones that make up the digits. In the human hand, these bones are longer, more numerous, and arranged in a way that allows independent movement, an important feature in human evolution required to elegantly manipulate objects.
12. Middle ear bones in humans; gills in fish
13. The presence of five digits indicates that having five digits may have provided an evolutionary advantage to the ancestors of both otters and humans. Both humans and otters have evolved to utilize fine motor movements of their hands or paws. It is likely that having five digits may have improved motor skills, providing a reproductive advantage that is reflected in the complexity that is the human hand. Because otters are known to use their paws for grasping and, in simple tool use, using stones to crack open shellfish, we might predict that otters will continue to evolve paws that are more and more functionally complex. Because humans and otters share an ancient ancestor, both humans and otters use the homologous bones in their hands and paws.
14. It depends on the stage of development at which the embryos are observed. At early stages of development, both human and chicken embryos have post-anal tails, so the presence of a tail at that stage cannot be used to distinguish the two. Later, the post-anal tail disappears in the human embryo.
15. By comparing the three sequences to one another in a pairwise fashion. By counting the number of differences between each pair of sequences, a quantitative measure of similarity (% similarity) can be established. The more similar the sequences (the higher the % similarity between them), the more likely it is that they are closely related. Sequence evidence from a single gene is often combined with evidence from other genes to establish relationships between organisms. DNA sequence data is just one means by which to determine such relationships. Comparisons of morphological traits, such as the arm bones, are another way to gather evidence to establish such relationships.
16. With only a few rare exceptions, the genetic code is universal for all living organisms known on this planet. Therefore, the same piece of DNA encodes the exact same amino acids in bacteria as it does in humans.
17. The two proteins differ in the specific amino acid sequence that makes them up. This is encoded by the specific nucleotide sequence of the gene. Although the code is the same, the specific sequence of nucleotides differs between distantly related organisms. Because the more closely related two organisms are, the more similar their gene sequences will be, it is possible to determine relationships from the similarities between these genes.
18. Answers will vary. Considerations may include the importance of protecting important scientific finds, the role of government in such protection, and the extent to which the government may act to protect such finds at the cost of limiting personal freedom. There are many things which people do not have a right to collect because the act of collecting, trading or selling such items causes harm to others or the environment, and therefore there are bans on the collection and trade of pelts, tusks, teeth, horns, feathers, or other animal parts of protected species. Likewise, visitors to many protected geological sites are prohibited from taking rocks or other items.

Chapter 17

1. They are all radioisotopes that decay at steady and predictable rates, changing into other elements.
2. You would use uranium-238 because it has the longest half-life (4.5 billion years). Other isotopes with shorter half-lives will be barely detectable in a sample that is extremely old, having long ago decayed to levels that are below detectable limits.
3. (1) the first prokaryotes (~3.0 billion years ago), (2) an increase in oxygen in the atmosphere (~2.5 billion years ago), (3) the first multicellular eukaryotes (~1.2 billion years ago), (4) the Cambrian explosion (545 million years ago) (5) the first animals (~540 million years ago), (6) the Permian extinction (248 million years ago), (7) the extinction of dinosaurs (~65 million years ago)
4. a: approximately 4.5 billion years old; b: approximately half
5. ~9 billion years old
6. Many of the ancestors of these organisms may have been evolving for a long time without appearing in the fossil record because not all organisms leave fossils behind. The sudden appearance of numerous organisms in the fossil record of the Cambrian explosion may be largely due to the development of shells and other hard-body parts, which are more likely to leave behind a fossil.
7. Amphibians were certainly present. Early reptiles and sharks might also be found in these layers.
8. No. They may indeed be closely related, but similar morphology does not necessarily indicate homology. The two organisms may share common characteristics because of convergent evolution rather than homology.
9. See Infographic 17.5. The continents were then generally closer together than they are now. Since that time, because of plate tectonics–the movements of independent continental plates in the earth's mantle or crust–the continents have largely drifted apart. One major exception is the Indian continent, which has since collided with the Asian continent, forming the Himalayan mountain range. As landmasses moved and separated, so did the organisms that lived on those landmasses.
10. The two species may look alike because they have evolved similar traits independently because they are adapting to similar environments (the desert climate). This is an example of convergent evolution.
11. They would have migrated to both the north and south polar regions, in which case both penguins and polar bears would be found at both regions today.
12. Bats are mammals and hence share a common ancestor with all mammals. This common ancestor is not thought to have possessed any structures homologous to insect wings that could have been inherited. Therefore, it is more likely that bat wings evolved separately. This is another example of convergent evolution.
13. f
14. domain, kingdom, phylum, genus, species
15. d
16. Monera was divided because DNA evidence showed that it was made up of two distinct groups of organisms. These were later separated into the current domains of Bacteria and Archaea.
17. c
18. Answers will vary. For instance, because organisms are related, we can study possible cancer treatments in yeast or mice to determine if these treatments interfere with, for example, cell division in these organisms. If they do, they may also stop the division of cancerous cells in humans. If we understand the evolutionary adaptations of organisms to their environments, we may be able to identify organisms that will do well in habitats that have been degraded by human activities, or that will do well in environments that are changing as the result of global climate change.

Chapter 18

1. c
2. The fundamental difference between the two groups is that prokaryotes lack internal membrane-bound organelles. The lack of a nucleus enclosing the chromosomal DNA is the defining characteristic of a prokaryotic organism.
3. d
4. They were originally grouped together because of their similar size and morphology. Both are prokaryotic organisms.
5. "Archaebacteria" literally means "ancient bacteria." The name was originally used because archaea were then seen as a particular sort of bacteria, one that might be very old. The strength of this term is that it emphasizes the structural similarity of archaea to bacteria; the great weakness of this term is that it implies that archaea are a subset of bacteria. We now understand this to not be the case–bacteria and archaea are separate but related groups of organisms.
6. a
7. e
8. c
9. No. Many archaea live in environments that are difficult to replicate in the lab. Therefore, there are many archaea that scientists are unable to culture, but they are nonetheless present in the environment.
10. No. Bacteria and archaea generally look similar, and therefore DNA sequence evidence is usually used to distinguish between these two groups of prokaryotes.
11. These processes are important because they convert CO_2 and N_2 gases into forms that humans can use.
12. No. *N. gonorrhoeae* use pili to attach to human cells and evade host defenses. Without the pili, *N. gonorrhoeae* would not be a very effective pathogen.
13. High temperatures: Most organisms cannot survive outside a narrow temperature range. Temperatures outside this range lead to protein denaturation and membrane instability, resulting in cell death. High pressure: Most organisms are evolved to live within a

specific pressure range. The high pressures found at the bottom of the ocean would crush many other organisms. Alkalinity: Most organisms are evolved to live within a specific pH range. The high pH (basic) conditions at Lost City would kill most other organisms. Toxic gases: Many other organisms have not evolved the ability to tolerate the high concentrations of certain toxic gases found at Lost City.

14. c

15. Their presence supports the idea that these compounds can be produced abiotically (that is, without life). This supports the idea that the molecules that make up living organisms could have been produced by the harsh conditions present on the early earth. The organisms that live inside the vents may be more similar to the first organisms on earth than any other extant organisms on the planet.

16. It might modify the hypothesis that life may have started at specifically this type of thermal vent; however, if no methane was produced abiotically at this particular vent, that does not establish that methane cannot be produced at thermal vents or abiotically in other ways.

17. No. You would not expect to find photosynthetic organisms at Lost City because there is no appreciable sunlight at that depth.

18. Because of the harsh conditions present at the vents, it seems unlikely that the scientists working there might contaminate the area with surface organisms that would be able to survive and compete with the natural microbes. However, microbial life is so diverse and unpredictable that such contamination is certainly possible. Should such microbes survive, they might out-compete natural organisms (as an invasive species) and could disrupt or totally destroy the current ecosystem. This could lead to the extinction of a species, and so this threat should be taken seriously by the scientists studying the vents.

19. Answers will vary. There are countless examples of how breakthroughs in basic science (that is, science for the sake of understanding the world, rather than for a specific purpose, such as treating a disease) have led to breakthroughs which benefit human society. By studying the thermal vents, scientists hope to better understand the diversity of life on earth and its origins. The vents have already led to changes in our understanding of evolution and the mechanisms by which organisms harness energy on earth. For example, this ecosystem is one of the only known ecosystems that does not ultimately rely on the energy of the sun, as photosynthetic organisms do not make up the base of this ecosystem's food chain.

Chapter 19

1. Olympic National Park contains many species not found anywhere else, like the Olympic torrent salamander and the Olympic gopher, because of the park's isolation and topography. The park is an ecological island with saltwater on three sides and during the last ice age was separated from the rest of the United States. A large diversity of animals is able to survive in the various habitats found in the park, which include glacier-topped mountains, temperate rain forest, lakes, rivers, and the Pacific coastline.

2. The domain Eukarya encompasses all eukaryotic organisms, including plants, animals, fungi, and protists, which contain membrane-bound organelles.

3. Both the fisher and the Douglas fir are eukaryotes and inhabit the low-elevation rain forests of the Olympic National Park.

4. a: One would expect less diversity in Lake Michigan because there is less variation in habitat and Lake Michigan has been subjected to repeated glaciations, decreasing the time for evolution of new species. b: One would expect less diversity in the Sonoran Desert in Arizona because the climate limits the diversity of organisms to those that can survive in a dry habitat. There are fewer protists and bryophytes because these organisms must avoid drying out. c: One would expect less diversity in the prairies of Kansas because there are fewer trees, so there are fewer potential habitats.

5. A fungicide would kill the fungi, which are critical for decomposition. The lack of fungi would lead to an accumulation of nonliving organic matter and there would be fewer nutrients available for other eukaryotes.

6. The first group of plants to live on land was the bryophytes, which lack roots and tissue for transporting nutrients and water. The lack of a vascular system limits these plants to damp environments. So even though they were first, they don't have all the adaptations necessary to live in a variety of terrestrial habitats.

7. d

8. The evolution of seeds allowed plants to survive harsh conditions and spread to new locations. Seeds are protected within cones or fruit, and can be spread relatively easily.

9. A hungry animal is more likely to disperse the seeds of an angiosperm than a gymnosperm since the seeds are enclosed in fruit, a tasty treat for a hungry animal.

10. Ferns were the first true vascular plants. They were able to grow taller and overran the landscape during the Carboniferous period. Since then, vascular plants such as trees have dominated many landscapes.

11. b

12. c

13. Both the fisher and the human are predators and have a backbone, mammary glands, and hair on their bodies.

14. d

15. a: Using flight as a criterion, woodpecker and wasp would be grouped together and the nonflying group would include human, ant, and fisher. The "two-legged" group includes human and woodpecker; the "more-than-two-legged" group includes wasp, ant, and fisher. The only animal having feathers is the woodpecker; the "nonfeather" group includes human, wasp, ant, and fisher. b: These groupings do not accurately reflect the taxonomic relationship since, for example, wasps and ants are arthropods and woodpecker, human, and fisher are chordates. Thus, it is necessary to use molecular biology and multiple characteristics to characterize the relationship between organisms correctly. c: Wasps and ants share characteristics of arthropods, including segmented bodies with jointed appendages and a hard exoskeleton, and characteristics of insects, including three pairs of jointed legs and a three-part body consisting of head, thorax, and abdomen. The human and fisher are grouped together on the basis of the presence of an endoskeleton, the production of milk, and the presence of hair.

16. All arthropods have an exoskeleton for protection from predators, to prevent them from drying out, and to support movement. Some arthropods, like spiders, have evolved the ability to produce venom for protection and predation. Beetles have been extraordinarily successful because of the development of wings and specialized mouthparts. Wings allow beetles to escape predators and to access habitats and remote food sources. Ants have evolved complex social behavior that allows them to coordinate the behavior of the group.

17. a: Fungi are heterotrophs and cannot carry out photosynthesis. b: Fungi do not ingest their food–they perform external digestion instead. c: Fungi digest food by secreting enzymes onto their food, which break down molecules into smaller organic compounds that can be absorbed by the fungi.

18. c

19. Both fungi and plants are eukaryotes containing cell walls. Neither are mobile. The key difference between the two is how they obtain their nutrients. Plants are autotrophs, producing their own food through photosynthesis; fungi are heterotrophs, obtaining their nutrients by decomposing organic matter.

20. b

21. f

22. Protists are no longer considered a separate kingdom because protists do not form a cohesive evolutionary group: some members undergo photosynthesis like plants, other members eat other organisms like animals, and still other members are decomposers like fungi. Genetic information will be the most useful basis for creating new taxonomic "homes" for the protists.

23. If the protist is living in a freshwater environment, water will enter it by osmosis, decreasing the concentration of solutes in the protist and potentially causing the protist to swell and burst. The contractile vacuole prevents this from happening by removing some of the excess water.

24. a: The Mexican gray wolf is an endangered species, so its reintroduction may prevent them from becoming extinct. In addition, there may be benefits for the habitat by reintroducing wolves. For example, the wolves may perform a necessary evolutionary function by removing unfit individuals from the prey population. b: The reintroduction of species may negatively affect the habitat and other human activities. Negative effects on the environment include changes in other species that may increase the likelihood of certain diseases in the area or alter the biodiversity of the area. With the Mexican gray wolf, some of the considerations relating to humans include concern that the wolves may attack pets, livestock, or humans, may affect military activity in the area, and may decrease prey of interest to hunters.

25. Answers will vary depending on locality.

Chapter 20

1. c

2. b

3. Folate is normally destroyed by UV light, and darker skin evolved as a mechanism to maintain folate. a: If this were no longer true, and folate levels were unaffected by UV light, levels of melanin would likely decrease in populations living at the equator because the role of UV light in producing vitamin D would be beneficial. b: There would be no change expected in this population. The levels of UV are low in this environment, so there is no pressure for dark skin to protect folate, and light skin allows vitamin D production.

4. d

5. c

6. Mitochondrial DNA is inherited solely from the mother. Mitochondrial DNA mutates at a fairly regular rate. A mother with a DNA mutation in her mitochondrial DNA will pass it down to all her children, and her daughters will pass it to all their children. Because these mutations are passed on intact, without recombination, mtDNA is a useful tool to track human ancestry back through generations.

7. a

8. e

9. See Infographic 20.7. Africans have the highest levels of genetic diversity, as they are the descendants of the most ancient populations. The next highest levels would be expected in Asian populations (the migration from Africa to Asia took place ~67,000 years ago), followed by Europeans (~40,000 years ago) and then South Americans (13,000 years ago).

10. Tools would have aided in hunting and food preparation, allowing Australopithecines to have better nutrition and less risk of starving to death.

11. If there were another benefit to having an opposable big toe (faster movement, better walking stance) it would have been maintained by selection, and the number of individuals with this feature would increase. Since humans no longer have an opposable big toe, there must not have been a benefit to this feature and it was lost from the population.

12. a: If there was better hunting or mate selection on the ground, an early hominid in a forested environment might have moved to the upright walking position and lost an opposable big toe. In grassland, where there is nothing to climb, the arboreal traits would confer no selective advantage. b: Other traits that might be favored in a forested environment include the ability to see well in low-light conditions (for example, under a dark canopy of leaves) and a good sense of balance (necessary to walk on branches).

13 & 14. Answers will vary. Students should address race as a construct that is often produced by looking at genetics, physical traits, geography, history, and cultural traditions.

Chapter 21

1. A population is a group of organisms living in a particular geographic region. A community is all the organisms (species) in a geographic area. Communities differ from populations in that they include interactions between populations of different species.

2. e

3. Scat (feces) reveals information about the organism that produced it. Studying scat samples can elucidate what an organism has been eating, as well as providing a source of the organism's DNA. DNA analysis can be used to identify individuals, as well as to look for genetic diseases and the effects of inbreeding.

4. c.

5. Ecology is the study of organisms, interactions between organisms, and between organisms and the nonliving components of the environment. Ecologists look at (1) the way individual organisms respond to the environment (temperature, pH, light, etc.); (2) how a population of organisms grows, breeds, and changes genetically over time; (3) how populations interact with each other–such as the effect of predators or symbiosis between populations; (4) how elements like rain, disease, fire, etc., affect communities and populations.

6. You could travel and track the squirrels by their tracks and count them at their nests, but this would be very time consuming. Alternatively, you could set up a series of square plots that represent 20%-25% of the nature preserve. You could then count the number of squirrels in that area, and extrapolate to find an estimate of how many squirrels are likely to inhabit the whole nature preserve. These approaches could also be used to determine the population size of maple trees, but since the trees do not move, they could be counted more individually more easily than squirrels.

7. The populations of moose and wolves are linked–the moose provide virtually all of the wolf's food. By knowing the changes to the size of the moose herd, researchers would be able to predict changes to the size of the wolf pack. It is important to know the moose's cause of death because if other (nonwolf) factors are causing the moose to die, researchers would have to take that additional factor into consideration when assessing the future moose population size as well the wolf population size. For example, if researchers find that many moose are dying because of tick infestation or starvation, then the moose population will be affected independently of wolf predation, and this in turn will affect the wolf population.

8. Scat (feces) reveals information about the organism that produced it. Studying scat samples can reveal what an organism has been eating. If the scat from a particular type of herbivore has a variety of vegetation in it in approximately the same proportion as vegetation in the local environment, we could infer that those herbivores tend to eat what is available, without a strong preference. On the other hand, if a herbivore has a strong preference for a particular type of vegetation, then that vegetation would be present in the scat at a higher proportion than the proportion of the plant in the local environment, suggesting that the herbivore is eating that plant preferentially.

9. c and d

10. Carrying capacity is the maximum number of individuals that an environment can support given its space and resources. At carrying capacity the population will level off, fluctuating slightly but maintaining a relatively constant size around the carrying capacity.

11. Because the parasite passes from fish to fish through the water, but can't survive for long periods of time alone in the water, the parasite will have the biggest impact at a high fish density. This is because at high density, the fish will be crowded, and so the parasite will likely find new host in a short amount of time. At low population density, it might take a parasite a long time to encounter a new host, and it may die before reaching a new host. A hot summer and drought may cause a lake to start to dry up (or at least lose a lot of its volume). This would leave less volume for the same number of fish, thereby increasing their population density, and permitting greater opportunities for density-dependent factors to affect the population.

12. a: Abiotic. Hot summer temperatures increase the ticks on moose and also may weaken the health of the moose, eventually causing death, decreasing the size of the moose population, and affecting the size of the wolf pack in subsequent years. b: Biotic. Ticks weaken the health of the moose, eventually causing death, decreasing the size of the moose population, and affecting the size of the wolf pack in subsequent years. c: Biotic. The lack of trees would mean a lack of food for the moose, which would weaken the health of the moose, eventually causing death, decreasing the size of the moose population, and affecting the size of the wolf pack in subsequent years. d: Biotic. The illness in the wolves would make them less able to hunt the moose or result in fewer wolves in the population. This will allow the moose population to grow. e: Abiotic. Deep snowfall is likely to trap the moose, making it easier for the wolves to catch and kill. This will cause death, decrease the size of the moose population, and affect the size of the wolf pack in subsequent years.

13. a: The introduction of a new herbivore will cause a decrease in the population of trees on Isle Royale. This is because the new herbivore and the moose will both be eating the trees. b: The moose population will decline with the introduction of a new herbivore that is not preyed on by wolves. This is because there will now be competition for food between the organisms, and some moose will die from starvation. c: The wolf population will decrease as the moose population decreases after the introduction of the herbivore. However, if the wolves can prey on the new herbivore, the wolf population may increase, because of this new food source.

14. If the population of moose remains stable on the island, the wolf population could be influenced by factors such as weather, disease, and genetic inbreeding.

15. a: Population R (10,000) would add more individuals at the end of the first year. This is because 5% of 10,000 is 500 individuals added to the population,whereas 5% of 100 (the size of population Q) is an increase of only 5 individuals.

    b: After 5 years the size of each population would be:

Year	Population Q	Population R
0	100	10,000
1	105	10,500
2	110	11,025
3	116	11,576
4	121	12,155
5	127	12,763

    c: If the populations reached carrying capacity at year 3, the level of resources would stop them from growing any larger. The populations would then remain very close to the carrying capacity (with minor fluctuations above and below the carrying capacity). The larger population would level off at approximately 11,576 individuals, which would remain its size in subsequent years.

16. a: Answers will vary. Some people may value the wolves when they are at remote locations, but less so when they are near (and potential predators of) pets and herds of cattle or flocks of chickens. b: A number of strategies could be considered. (1) Introduction of wolves from another population would not only increase the number of wolves but also provide new alleles to increase the genetic diversity in the inbred population of wolves. (2) Wolf pups could be protected in a refuge where they would be fed and have adequate shelter in a cold winter to increase their survival rate. Once the wolf pups were judged to be healthy and strong, they could be released to the pack. If more wolf pups survived each year, that would increase the size of the population (although it wouldn't help with the problem of inbreeding). (3) Instead of introducing both male and female wolves from another population, only females could be introduced. This would increase the number of mates for males and contribute to increasing the genetic diversity of the population.

17. If the carp are eating huge amounts of algae, there will be fewer algae for other organisms that rely on algae as their major food source. Many of these organisms will be other fish, whose populations will suffer as a result. If these fish are commercial or recreational sport fish, the fishing industry will be negatively affected. Furthermore, if the carp are seriously devastating the algae population (much as the moose can do to the tree population on Isle Royale), there may be local impacts on CO_2 levels. Photosynthetic algae take up CO_2 as they photosynthesize, and this helps mitigate climate change caused by elevated levels of CO_2. Management strategies are certainly challenging. One is to essentially try to overfish the Asian carp, rewarding the capture and removal of the carp. Another (and riskier) strategy might be to introduce a parasite that will attack the carp but not native fish. Unfortunately, this strategy tends to backfire, in that the introduced parasite has unintended negative consequences on other species.

Chapter 22

1. A population is a group of organisms of the same species that live in the same area and can mate with one another and produce fertile offspring. A community consists of populations that interact and are connected by their actions. Communities contain more than one species.
2. Keystone species are those that are very important to a community because of their central role in supporting all the species in the community.
3. d
4. a
5. There are many possible answers. An example: Phytoplankton are a keystone species in the ocean, providing food for many organisms in the ocean and producing a large amount of oxygen (via photosynthesis) that is essential for life on earth.
6. Those suffering from pollen allergies are most likely allergic to pollen that is wind-carried because that pollen is in the air and can be inhaled.
7. b
8. f
9. The answer to Question 8 provides an example. A bear is part of a terrestrial food chain. The bear eats blueberries, which are pollinated by bees. Humans and birds also eat blueberries. Those birds in turn can be eaten by predatory birds, such as hawks. The bear can also "cross over" to an aquatic food chain and eat salmon. Those salmon eat organisms lower on the food chain, for instance algae.
10. The energy stored in the grain is released as the cow digests the grain. Energy is used in digestion and some energy is lost to heat. Energy is also used to sustain the life of the cow. When the meat of a cow is eaten it contains less energy than the grain because much of the energy stored in the grain has been lost or used.
11. The herbivore eats only producers (plants). Herbivores receive ~10% of the energy that is stored in the producer; the rest is burned as fuel or given off as heat. The top carnivore would eat only meat, and depending on the size of the food chain would have access to 1% or less than 1% of the energy. This is because at each level up the food chain an organism can extract only 10% of the energy in the level below.

    Producer (100%) → herbivore (10%) → carnivore 1 (1%) → carnivore 2 (> 1%)
12. Bees are attracted to flowers with yellow, blue, or purple petals; other nectar-seeking organisms are attracted to other colors of flowers. Bees and other pollinators have also coevolved with some plants in such a way that the shape of the flower and the shape of the bee fit together to maximize pollen pick-up and release and provide the bee with abundant nectar and pollen. Although bees and other pollinators both feed on nectar, they do not try to feed on the same flowers and thus are not in direct competition.

13. d

14. Although mussels and barnacles are both filter feeders, they might be able to coexist because the sizes of their filters are different. The mussel may be able to eat larger organisms than the barnacles, so there is not competition for the same food.

15. The number and diversity of bees in the area would likely decrease drastically because corn is wind pollinated and does not make nectar or have flowers flowers to the bees. The bees would have to leave the area in search of a food source.

16. b

17. The relationship between bees and the bacteria that live within them can be characterized as mutualistic symbiosis. The bacteria get nutrients and a safe place to live from the bees, and the bees benefit from the bacteria because the bacteria help the bees to defend themselves from disease.

18. Researchers studying colony collapse disorder (CCD) noticed that the bees were very sick and seemed to have weakened immune systems. The virus IAPV was a good candidate for causing weakened immune response, and it was found in 96% of the hives with CCD. However, further research showed that not all colonies that are infected with IAPV have CCD, so there must be another factor causing CCD. A second hypothesis was that the parasitic varroa mite was feeding on the bees' blood, causing a weakened immune system and making them more susceptible to disease. However, research has shown that the levels of mite infection in colonies with CCD are no higher than the levels seen in previous years when CCD was not a problem.

19. The relationship between *E. coli* and humans is a symbiosis. Most of the time *E. coli* is beneficial to humans because these bacteria can prevent other pathogenic bacteria from colonizing (competition), and some types of *E. coli* can produce vitamin K_{12}. *E. coli* benefits from living in the human intestine because of the available nutrients and environment that is conducive to growth.
    This type of relationship is called mutualism. Some strains of *E. coli* are parasitic: the bacteria get nutrients and a place to live and the humans get sick.

20. Point out the importance of bees in helping plants to reproduce. These plants are used as food sources by humans or by other animals which humans eat. Without the bees the plants would not be fertilized and no seeds would form.

21. Planting a single crop over a wide area decreases the number of pollinators that can be supported because there will be competition for the common resource. Also, there might not be enough variety in food sources for the pollinators to maintain a healthy diet. Similarly, a single crop will flower (and produce pollen and nectar) all at once, creating a situation of feast at the time of flowering and famine for the rest of the year. These periods of famine can cause the loss of pollinator species, affecting many other crops.

Chapter 23

1. f

2. Species discussed in this chapter that might be affected by global climate change include maple trees, seahorses, turtles, fish, yellow-bellied marmots, the Arctic fox, the red fox, and polar bears.

3. a, b, and c.

4. The coniferous forest biome is characterized principally by evergreen trees.

5. Temperate deciduous forest–eastern North America; tropical forest–Central America and northern South America; tundra–northern North America

6. Melting sea ice does not cause sea levels to rise. However, when ice caps melt, they cause sea levels to rise, thus putting low-lying areas at risk of flooding.

7. a: Smaller leaves would decrease the surface area that can lose water to the environment, and that means that less water would be lost by evaporation. Less surface area also means a decreased ability to take up CO_2 from the environment. However, given that increased global temperatures are caused by increased CO_2 levels, smaller leaves might not be a negative factor. Overall this could be a *useful adaptation* for the increased temperatures. b: More pores on each leaf would increase the amount of water lost because there would be more exposed area from which to lose water. Having more pores would also increase the amount of CO_2 that could be taken into the plant and used for photosynthesis. Overall the loss of water would be more detrimental to the plant than the increased CO_2, so this would *not be a useful adaptation to global climate change.* c: Thicker, waxier bark would serve to maintain water in the trunk of the plant, since water won't evaporate out of these surfaces. This waxy layer would not affect CO_2 uptake since this tissue is not photosynthetic. Similarly, thicker and waxier bark would be harder for insects to munch on, so this feature would be protective in the face of migrating populations of plant-eating insects. Overall this would be an *effective adaptation* for global climate change.

8. People in northern climates could contract new and different diseases if insects that carry diseases expand their ranges northward because of global climate change.

9. b

10. b

11. a

12. In the greenhouse effect, the heat trapped by greenhouse gases raises the temperature of the atmosphere and in turn raises the temperature of the surface of the earth. Without the greenhouse effect, the temperature on earth would be -18°C; we could not survive at that temperature.

13. The evidence that increasing levels of greenhouse gases (particularly CO_2) are responsible for global climate change includes these points: (1) Carbon dioxide concentrations are higher than they have been in 700,000 years. (2) Since direct measurement of atmospheric carbon dioxide began in the late 19th century, its concentration has increased ~35%. Figure 23.5 shows

how CO_2 levels increased with temperatures over the last 1,000 years. Both temperature and CO_2 stayed relatively level until the start of the industrial revolution, when CO_2 levels increased.
Data from the past 50+ years showing that the level of solar radiation had increased over the entire planet would support the hypothesis. Also supportive would be a graph and data analysis that show a correlation between increased levels of solar radiation and the increasing temperatures of the planet.

14. Using *fossil fuels* for energy converts organic carbon to CO_2. Most organisms, including plants, animals and decomposers, perform *respiration,* producing CO_2 from organic food. CO_2 is released to the *atmosphere.* CO_2 is absorbed by the *photosynthetic organisms in the ocean.* Plants perform *photosynthesis,* fixing CO_2 into organic molecules. Coal and oil are *fossil fuels* that trap carbon in the earth.

15. c

16. Surface ice is not useful, but ice cores provide a way to measure atmospheric conditions from the distant past. Cylinders of ice can be extracted from glaciers, and the composition of gas bubbles within them can be analyzed. This analysis reveals the atmospheric conditions of thousands of years ago, when those bubbles were initially trapped in the ice.

17. a: Large-scale slash and burn agriculture releases carbon that was stored in trees directly to the atmosphere as carbon dioxide. Additionally, as the land does not generally support long-term crop production, the overall levels of photosynthesis (which removes CO_2 from the atmosphere) are reduced relative to those of the original forest. b: Driving gasoline-fueled cars (in fact, burning any fossil fuel) releases carbon (as CO_2) that had been stored for a very long time in the earth. This carbon is essentially "new" carbon being introduced into the atmosphere, adding to the amount of carbon already cycling. c: Cattle raised for beef and dairy products have methane-producing microbes in their guts that help them digest food, but the cattle release large amounts of methane gas as flatulence. Methane is a potent greenhouse gas. d: Rice production also releases methane gas into the environment. Methane is a potent greenhouse gas.

18. d. Ice cores that contain ice formed in 1750 would be the most effective way to determine the CO_2 concentration at that time. The Mauna Loa observatory has been continuously monitoring and collecting data related to atmospheric change only since the 1950s–it would not have data from 1750. Neither tree rings nor historical records would give accurate values from which to determine CO_2 levels but could provide clues to the weather in 1750.

19. Answers will vary. Considerations include: emissions from automobiles, which could be mitigated by using public transportation, carpooling, or cycling or walking on a regular basis; the use of household electrical products (which often rely on coal-fired electrical plants to generate the electricity), which can be mitigated by replacing worn-out appliances with energy-efficient versions, line-drying laundry rather than using a dryer, and turning off and unplugging electrical devices when they are not in use.

20. Answers will vary. These actions would contribute to a low-carbon footprint menu: (1) purchasing locally grown food, thus reducing the consumption of fossil fuels required to transport food long distances from the site of production to the site of consumption (different parts of the country would rely on different foods, depending on what is produced locally; (2) eating foods that do not require cooking, to reduce use of electricity or gas); (3) using a solar oven rather than a gas or electric oven to cook food

Chapter 24

1. Some advances discussed in the chapter are the development of agriculture and the use of antibiotics and other advances in public health.

2. An ecological footprint is a tool used to calculate how much of the earth's resources a population's lifestyle requires. It calculates how much biologically productive land and water area a human population needs to produce the resources it consumes and to absorb the waste it produces.

3. d

4. The factors contributing to the ecological footprint of each area are healthy food, energy for mobility and heat, fiber for paper, clothing and shelter, fresh air and clean water.

City	Footprint	Factors Contributing to Footprint
Dongtan, China	Relatively low	Designed to be sustainable: no cars are necessary because everything is within walking distance; energy is derived from solar and wind sources; buildings are designed to remain cool without reliance on air conditioning; extensive recycling to minimize waste and pollution.
Rural village in China	Relatively low	Less developed, therefore less use of fossil fuels for cars; food is produced locally, reducing transportation costs.
Calgary, Canada	Relatively high	Calgary is considering its ecological footprint in development decisions, and residents can use a light rail system that is powered through the use of wind turbines. Thus its footprint is probably smaller than would be expected for a developed city in North America.
Houston, Texas	Relatively high	Large area and highways necessitate travel by automobile, increasing fossil fuel use. Hot summers require high use of electricity for air conditioning; city acts as a heat trap.
New York	Relativity low	As noted in the chapter, the high population density and relatively small area in New York allows New Yorkers to use public transportation, rather than driving cars for their daily commutes.

5. The addition of 20 families will increase the population of the town. If the children remain in the town and raise families, the town's population will likely continue to increase in the future. Generally speaking, urban populations have a higher footprint than rural populations. As the new condominiums are on the outskirts of the town, the families will need to drive into town for school, work, and shopping. If each family has two cars, this represents 40 new vehicles, more than what was likely used on the farm. Similarly, 20 families will produce a variety of waste that will likely be in excess of that generated by farming a crop. There are now 20 households to be heated and cooled rather than the original farmhouse. It is thus likely that this population will use more energy and generate more waste than the farm, thereby enlarging the ecological footprint.
6. The developer could use solar or wind power to replace traditional fossil fuels for energy sources. The developer could also take into consideration public transportation and provide alternatives to driving, as well as making housing, workplaces, and school within walking and biking distance.
7. e
8. freshwater (R); coal (NR); codfish populations in the North Atlantic (R); wind (R); sunlight (R)
9. Coal, oil, and natural gas are fossil fuels that take millions of years to form from organic material in dead organisms and therefore these resources are not renewable on a useful timescale.
10. Renewable resources like food and water may not have the time to renew themselves when the population is growing so quickly and demand for them exceeds the space available to grow and harvest them. For example, at one point the North Atlantic codfish industry was in danger of collapsing because of high demand and overfishing–taking cod faster than they could reproduce was making the cod a renewable resource. Governments chose to limit the catch of cod, and the numbers of cod are gradually increasing. Similarly, while water is technically renewable, we can choose how to use water, and how to maintain clean water for drinking, even at the expense of recreation. And while we are unlikely to run out of wind or sunlight, we can make choices to make these resources more accessible. For example, we can agree to look at perhaps unattractive wind farms for the sake of taking advantage of this resource.
11. Answers will vary, depending on locale. Challenges might be space (solar panels and wind mills) and the expense of new technology.
12. a: Schools and shops located near residences will increase walking, biking, or the use of public transportation. These measures would decrease use of fossil fuels (because they reduce the use of fossil-fuel-powered cars), save energy, decrease CO_2 production, and reduce air pollution. b: If the buildings have green roofs, they have vegetation to absorb and filter rainwater, and provide growing space for food. This measure would save energy needed to bring food into the city from farms and recycles water and help keep buildings cool. b: Solar panels are mounted throughout a city would provide renewable energy with little pollution or waste.
13. Waste in the form of methane collected from landfills can be used to power dump trucks; batch reactors can digest human waste to use as fertilizer.
14. Examples: reducing home water use would help maintain water availability; installing compact fluorescent lightbulbs would help save electricity and decrease the amount of CO_2 released into the atmosphere; eating less meat–meat takes a great deal of grain and water to produce, affecting energy consumption; unplug electronics not in use would reduce vampire energy flow–energy flow from devices that are turned off; driving less would reduce fossil fuel burning and CO_2 emissions; recycling would reduce the amount of waste produced.
15. Cities in the desert must draw water from underground aquifers or rely on water from rivers. As the desert is typically hot (especially in the summer), the energetic costs of home cooling in the summer are substantial: a great deal of electricity is required to cool homes in the summer. Many southwestern cities are spread out, creating a demand for fossil fuels.
16. When water is taken out of the river to irrigate crops, water levels beyond this point will be much lower. This means that there is less water in the river (or reservoirs) for fishing, boating, and swimming. On the other hand, the water is being used to irrigate crops that represent food for the population. In addition to negative impacts on recreation downstream, the withdrawal of water for irrigation means that there is not enough water in the river to support an endangered species of minnow. This puts human needs (irrigation of crops) in direct conflict with the needs of another species. Different people will have different opinions about this issue.
17. In general, the United States has a much higher footprint than Bangladesh, primarily because the United States is a highly developed country. As countries develop, their footprints increase. This will lead to an overall increase in the global footprint. As the global footprint increases, the number of people that the earth can support will decrease (that is, carrying capacity will decrease). Different people will have different opinions about whether or not this is a shared responsibility–whether or not developed countries should reduce their footprints to accommodate the development of other countries.

Glossary

abiotic Refers to nonliving components of the environment such as temperature and precipitation.

acid A substance that increases the hydrogen ion concentration of solutions, making them more acidic.

activation energy The energy required for a chemical reaction to proceed. Enzymes accelerate reactions by reducing their activation energy.

active site The part of the enzyme that binds to substrates.

active transport The energy-requiring process by which solutes are pumped from an area of lower concentration to an area of higher concentration with the help of transport proteins.

adaptation The response of a population to environmental pressure, so that advantageous traits become more common in the population over time.

adaptive radiation The spreading and diversification of organisms that occur when they colonize a new habitat.

adenosine triphosphate (ATP) The molecule that cells use to power energy-requiring functions; the cell's energy "currency."

adult stem cells (somatic stem cells) Stem cells located in tissues that help maintain and regenerate those tissues.

aerobic respiration A series of reactions that occurs in the presence of oxygen and converts energy stored in food into ATP.

alga (plural: algae) A uni- or multicellular photosynthetic protist.

alleles Alternative versions of the same gene that have different nucleotide sequences.

allele frequency The relative proportion of an allele in a population.

allopatry Speciation that occurs because of geographic or climatic barriers to gene flow.

amino acids The building blocks of proteins. There are 20 different amino acids found in proteins.

amniocentesis A procedure that removes fluid surrounding a fetus to obtain and analyze fetal cells to diagnose genetic disorders.

anabolic reaction Any chemical reaction that combines simple molecules to build more-complex molecules.

anecdotal evidence An informal observation that has not been systematically tested.

aneuploidy An abnormal number of one or more chromosomes (either extra or missing copies).

angiosperm A seed-bearing flowering plant with seeds typically contained within a fruit.

animal A eukaryotic, usually multicellular, organism that obtains nutrients by ingesting other organisms or molecules produced by other organisms.

annelid A segmented worm, such as an earthworm.

antibiotic A chemical that can slow or stop the growth of bacteria; many antibiotics are produced by living organisms.

anticodon The part of a tRNA molecule that binds to a complementary mRNA codon.

apoptosis Programmed cell death; often referred to as cellular suicide.

aquifer Underground layers of porous rock from which water can be drawn.

archaea One of the two domains of prokaryotic life, the other is Bacteria.

arthropod An invertebrate having a segmented body, a hard exoskeleton, and jointed appendages.

atom The smallest unit of an element that cannot be chemically broken down into smaller units.

autosomes Paired chromosomes present in both males and females; all chromosomes except the X and Y chromosomes.

autotrophs Organisms such as plants, algae, and certain bacteria that capture the energy of sunlight by photosynthesis.

bacteria One of the two domains of prokaryotic life; the other is Archaea.

base A substance that reduces the hydrogen ion concentration of solutions, making them more basic.

bilateral symmetry The pattern exhibited by a body plan with clear right and left halves that are mirror images of each other

binary fission A type of asexual reproduction in which one parental cell divides into two.

biocapacity The amount of the earth's biologically productive area–cropland, pasture, forest, and fisheries–that is available to provide resources to support life.

biogeography The study of how organisms are distributed in geographical space.

biological species concept The definition of a species as a population whose members can interbreed to produce fertile offspring.

biome A large geographic area defined by its characteristic plant life, which in turn is determined by temperature and levels of moisture.

biotic Refers to living components of the environment.

blastocyst The stage of embryonic development in which the embryo is a hollow ball of cells. Researchers can derive embryonic stem cell lines from cells of a blastocyst stage embryo.

body mass index (BMI) An estimate of body fat based on height and weight.

bottleneck effect A type of genetic drift that occurs when a population is suddenly reduced to a small number of individuals, and alleles are lost from the population as a result.

bryophyte A nonvascular plant that does not produce seeds.

calorie The amount of energy required to raise the temperature of 1 gram of water by 1° Celsius.

Calorie 1,000 calories or 1 kilocalorie (kcal); the capital "C" in "Calorie" indicates "kilocalorie." The Calorie is the common unit of energy used in food nutrition labels.

cancer A disease of unregulated cell division: cells divide inappropriately and accumulate, in some instances forming a tumor.

capsule A sticky coating surrounding some bacterial cells used to adhere to surfaces.

carbohydrate An organic molecule made up of one or more sugars. A one-sugar carbohydrate is called a monosaccharide; a carbohydrate with multiple linked sugars is called a polysaccharide.

carbon cycle The movement of carbon atoms between organic and inorganic molecules in the environment.

carbon fixation The conversion of inorganic carbon (for example, CO_2) into organic forms (for example, sugars).

carbon footprint A measure of the total greenhouse gases we produce by our activities.

carcinogen Any chemical agent that causes cancer. Many carcinogens are mutagens.

carrier An individual who is heterozygous for a particular gene of interest, and therefore can pass on the recessive allele without showing any of its effects.

carrying capacity The maximum population size that a given environment or habitat can support, given its food supply and other natural resources.

catabolic reaction Any chemical reaction that breaks down complex molecules into simpler molecules.

cell The basic structural unit of living organisms.

cell cycle An ordered sequence of stages that a cell progresses through in order to divide during its life; the stages includes preparatory phases (G_1, S, G_2) and division phases (mitosis and cytokinesis).

cell cycle checkpoint A cellular mechanism that ensures that each stage of the cell cycle is completed accurately.

cell division The process by which a cell reproduces itself; cell division is important for normal growth, development, and repair of an organism.

cell membrane A phospholipid bilayer with embedded proteins that forms the boundary of all cells.

cell theory The concept that all living organisms are made of cells and that cells are formed by the reproduction of existing cells.

cell wall A rigid structure enclosing the cell membrane of some cells that helps the cell maintain its shape.

cellular differentiation The process by which a cell specializes to carry out a specific role.

centromere The specialized region of a chromosome where the sister chromatids are joined. This site is critical for proper alignment and separation of sister chromatids during mitosis.

chemotherapy The treatment of disease, specifically cancer, by the use of chemicals.

chlorophyll The pigment present in the green parts of plants that absorbs photons of light energy during the light reactions of photosynthesis.

chloroplast The organelle in plant and algal cells that is the site of photosynthesis.

chromosome A single, large DNA molecule wrapped around proteins. Chromosomes are located in the nuclei of most eukaryotic cells.

citric acid cycle A set of reactions that takes place in mitochondria and helps extract energy (in the form of high-energy electrons) from food; the second step of aerobic respiration.

coding regions Sequences of DNA that serve as instructions for making proteins.

coding sequence The part of a gene that specifies the amino acid sequence of a protein. Coding sequences determine the identity, shape, and function of proteins.

codominance A form of inheritance in which both alleles contribute equally to the phenotype.

codon A sequence of three mRNA nucleotides that specifies a particular amino acid.

coenzyme A small organic molecule, such as a vitamin, required for enzyme activity.

cofactor An inorganic substance, such as a metal ion, required for enzyme activity.

commensalism A type of symbiotic relationship in which one member benefits and the other is unharmed.

community A group of interacting populations of different species living together in the same area.

competitive exclusion principle The concept that when two species compete for resources in an identical niche, one is inevitably driven to extinction.

complementary Two strands of DNA are said to be complementary in that A always pairs with T, and G always pairs with C.

complex carbohydrate (polysaccharide) A carbohydrate made of many simple sugars linked together, that is, a polymer of monosaccharides; examples are starch and glycogen.

consumers Heterotrophs that eat other organisms or the organic molecules produced by organisms to obtain energy.

control group The group in an experiment that experiences no experimental intervention or manipulation.

convergent evolution The process by which organisms that are not closely related evolve similar adaptations as a result of independent episodes of natural selection.

correlation A consistent relationship between two variables.

covalent bond A strong chemical bond resulting from the sharing of a pair of electrons between two atoms.

cytokinesis The physical division of a cell into two daughter cells.

cytoplasm The gelatinous, aqueous interior of all cells.

cytoskeleton A network of protein fibers in eukaryotic cells that provides structure and facilitates cell movement.

decomposer An organism such as a fungus or bacterium that digests and uses the organic molecules in dead organisms as sources of nutrients and energy.

density-dependent factor A factor whose influence on population size and growth depends on the number and crowding of individuals in the population (for example, predation).

density-independent factor A factor that can influence population size and growth, regardless of the numbers and crowding within a population (for example, weather).

deoxyribonucleic acid (DNA) The molecule of heredity, common to all life forms, that is passed from parents to offspring.

dependent variable The measured result of an experiment, analyzed in both the experimental and control groups.

descent with modification Darwin's term for evolution, combining the ideas that all living things are related and that organisms have changed over time.

diabetes A disease characterized by abnormally high blood-sugar levels.

differential gene expression The process by which different genes are "turned on" (that is, expressed) in different cell types.

diploid Having two copies of every chromosome.

directional selection A type of natural selection in which organisms with phenotypes at one end of a spectrum are favored by the environment.

distribution pattern The way that organisms are distributed in geographic space, depending on resources and interactions with other members of the population.

diversifying selection A type of natural selection in which organisms with phenotypes at both extremes of the phenotypic range are favored by the environment.

DNA polymerase An enzyme that "reads" the sequence of a DNA strand and helps to add complementary nucleotides to form a new strand during DNA replication.

DNA profile A visual representation of a person's unique DNA sequence.

DNA replication The natural process by which cells make an identical copy of a DNA molecule.

domain The highest category in the modern system of classification; there are three domains–Bacteria, Archaea, and Eukarya.

dominant allele An allele that can mask the presence of a recessive allele.

double helix The spiral structure formed by two strands of DNA nucleotides bound together.

ecology The study of the interactions between organisms, and between organisms and their environment.

ecological footprint A measure of how much land and water area is required to supply the resources a person or population consumes and to absorb the wastes they produce.

ecosystem The living and nonliving components of an environment, including the communities of organisms present and the physical environment with which they interact.

electron A negatively charged subatomic particle with negligible mass.

electron transport chain A process that takes place in mitochondria and produces the bulk of ATP during aerobic respiration; the third step of aerobic respiration.

element A chemically pure substance that cannot be chemically broken down; each element is made up of and defined by a single type of atom.

embryo An early stage of development reached when a zygote undergoes cell division to form a multicellular structure.

embryonic stem cells Stem cells that make up an early embryo, which can differentiate into nearly every cell in the body.

endoplasmic reticulum A membrane-enclosed series of passages in eukaryotic cells in which proteins and lipids are synthesized.

endoskeleton A solid internal skeleton found in many animals, including humans.

endosymbiosis The theory that free-living prokaryotic cells engulfed other free-living prokaryotic cells billions of years ago, forming eukaryotic organelles such as mitochondria and chloroplasts.

energy The ability to do work. Cellular work includes processes such as building complex molecules and moving substances in and out of the cell.

enzyme A protein that speeds up the rate of a chemical reaction.

epidemiology The study of patterns of disease in populations, including risk factors.

essential amino acids Eight amino acids the human body cannot synthesize and must obtain from food.

essential nutrient A substance that cannot be synthesized by the body and must be obtained preassembled from the diet; certain amino acids and fatty acids, vitamins, and minerals are essential nutrients.

eukaryote Any organism of the domain Eukarya; eukaryotic cells are characterized by the presence of a membrane-enclosed nucleus and organelles.

evolution Change in allele frequencies in a population over time.

exoskeleton A hard external skeleton covering the body of many animals, such as arthropods.

experiment A carefully designed test, the results of which will either support or rule out a hypothesis.

experimental group The group in an experiment that experiences the experimental intervention or manipulation.

exponential growth The unrestricted growth of a population growing at a constant growth rate.

extinction The elimination of all individuals in a species; extinction may occur over time or in a sudden mass die-off.

facilitated diffusion The process by which large or hydrophilic solutes move across a membrane from an area of higher concentration to an area of lower concentration with the help of transport proteins.

falsifiable Describes a hypothesis that can be ruled out by data that show that the hypothesis does not explain the observation.

fermentation A series of chemical reactions that takes place in the absence of oxygen and converts some of the energy stored in food into ATP. Fermentation produces far less ATP than does aerobic respiration.

fern The first true vascular plants; ferns do not produce seeds.

fiber A complex plant carbohydrate that is not digestible by humans.

fitness The relative ability of an organism to survive and reproduce in a particular environment.

flagella (singular: flagellum) Whiplike appendages extending from the surface of some bacteria, used in movement of the cell.

folate A B vitamin also known as folic acid. Folate is an essential nutrient, necessary for basic bodily processes such as DNA replication and cell division.

food chain A linked series of feeding relationships in a community in which organisms further up the chain feed on ones below.

food web A complex interconnection of feeding relationships in a community.

fossils The preserved remains or impressions of once-living organisms.

fossil fuel Carbon-rich energy source, such as coal, petroleum, or natural gas, formed from the compressed, fossilized remains of once-living organisms.

fossil record An assemblage of fossils arranged in order of age, providing evidence of changes in species over time.

fungus (plural: fungi) A single-cell or multicellular eukaryotic organism that obtains nutrients by secreting digestive enzymes onto organic matter and absorbing the digested product.

gametes Specialized reproductive cells that carry one copy of each chromosome (that is, they are haploid). Sperm are male gametes; eggs are female gametes.

gel electrophoresis A laboratory technique that separates fragments of DNA by size.

gene A sequence of DNA that contains the information to make at least one protein.

gene expression The process of using DNA instructions to make proteins.

gene flow The movement of alleles from one population to another, which may increase the genetic diversity of a population.

gene pool The total collection of alleles in a population.

gene therapy A type of treatment that aims to cure disease by replacing defective genes with functional ones.

genetically modified organism (GMO) An organism that has been genetically altered by humans.

genetic code The particular amino acids specified by particular mRNA codons.

genetic drift Random changes in the allele frequency of a population between generations; genetic drift tends to have more dramatic effects in smaller populations than in larger ones.

genome One complete set of genetic instructions encoded in the DNA of an organism.

genotype The genetic makeup of an organism.

global hectare The unit of measurement of the ecological footprint, representing the biological productivity of an average hectare of land.

global warming An increase in the earth's average temperature.

glycogen A complex animal carbohydrate made of linked chains of glucose molecules; a source of stored energy.

glycolysis A series of reactions that breaks down sugar into smaller units; glycolysis takes place in the cytoplasm and is the first step of both aerobic respiration and fermentation.

Golgi apparatus An organelle made up of stacked membrane-enclosed discs that packages proteins and prepares them for transport.

Gram-negative Refers to bacteria with a cell wall that includes a thin layer of peptidoglycan surrounded by an outer lipid membrane that does not retain the Gram stain.

Gram-positive Refers to bacteria with a cell wall that includes a thick layer of peptidoglycan that retains the Gram stain.

greenhouse effect The normal process by which heat is radiated from earth's surface and trapped by gases in the atmosphere, helping to maintain the earth at a temperature that can support life.

greenhouse gas Any of the gases in earth's atmosphere that absorb heat radiated from the earth's surface and contribute to the greenhouse effect; for instance, carbon dioxide and methane.

growth rate The difference between the birth rate and the death rate of a given population; also known as the rate of natural increase

gymnosperm A seed-bearing plant with "naked" seeds typically held in cones.

habitat The physical environment where an organism lives and to which it is adapted.

half-life The time it takes for one-half of a substance to decay.

haploid Having only one copy of every chromosome.

Hardy-Weinberg equilibrium The principle that, in a nonevolving population, both allele and genotype frequencies remain constant from one generation to the next.

heat The kinetic energy generated by random movements of molecules or atoms.

herbivory Predation on plants, which may or may not kill the plant.

heterotrophs Organisms, such as humans and other animals, that obtain energy by eating other organisms or molecules produced by other organisms.

heterozygous Having two different alleles of a given gene.

homeostasis The maintenance of a relatively constant internal environment.

homologous chromosomes The two copies of each chromosome in a diploid cell. One chromosome in the pair is inherited from the mother, the other is inherited from the father.

homology Anatomical, genetic, or developmental similarity among organisms due to common ancestry.

hominid Any living or extinct member of the family Hominidae, the great apes–humans, orangutans, chimpanzees, and gorillas.

homozygous Having two identical alleles of a given gene.

hydrogen bond A weak electrical attraction between a partially positive hydrogen atom and another atom with a partial negative charge.

hydrophobic "Water-fearing"; hydrophobic molecules will not dissolve in water.

hydrophilic "Water-loving"; hydrophilic molecules dissolve in water.

hypha (plural: hyphae) A long, threadlike structure through which fungi absorb nutrients.

hypothesis A testable and falsifiable explanation for a scientific observation or question.

inbreeding Mating between closely related individuals. Inbreeding does not change the allele frequency within a population, but it does increase the proportion of homozygous individuals to heterozygotes.

inbreeding depression The negative reproductive consequences for a population associated with having a high frequency of homozygous individuals possessing harmful recessive alleles.

incomplete dominance A form of inheritance in which heterozygotes have a phenotype that is intermediate between homozygous dominant and homozygous recessive.

independent assortment The principle that alleles of different genes are distributed independently of one another during meiosis.

independent variable The variable, or factor, being deliberately changed in the experimental group.

induced pluripotent stem cell A pluripotent stem cell that was generated by manipulation of a differentiated somatic cell

inorganic molecule A molecule that lacks a carbon-based backbone and C-H bonds.

insect A six-legged arthropod with three body segments: head, thorax, and abdomen.

insulin A hormone secreted by the pancreas that regulates blood sugar.

interphase The stage of the cell cycle in which cells spend most of their time, preparing for cell division. There are three distinct phases within interphase (G_1, S, and G_2).

invertebrate An animal lacking a backbone.

ion An electrically charged atom, the charge resulting from the loss or gain of electrons.

ionic bond A strong electrical attraction between oppositely charged ions.

karyotype The chromosomal makeup of cells. Karyotype analysis can be used to detect trisomy 21 prenatally.

keystone species A species on which other species depend, and whose removal has a dramatic impact on the community.

kinetic energy The energy of motion or movement.

light energy The energy of the electromagnetic spectrum of radiation.

lipids Organic molecules that generally repel water.

logistic growth A pattern of growth that starts off fast and then levels off as the population reaches the carrying capacity of the environment.

lysosome An organelle in eukaryotic cells filled with enzymes that can degrade worn-out cellular structures.

macromolecules Large organic molecules that make up living organisms; they include carbohydrates, proteins, and nucleic acids.

macronutrients Nutrients, including proteins, carbohydrates, and fats, that organisms must ingest in large amounts to maintain health.

mammals Members of the class Mammalia; all members of this class have mammary glands and a fur-covered body.

matter Anything that takes up space and has mass.

meiosis A specialized type of cell division that generates genetically unique haploid gametes.

melanin A pigment, produced by a specific type of skin cell, that gives skin color.

messenger RNA (mRNA) The RNA copy of an original DNA sequence formed during transcription.

metabolism All the chemical reactions taking place in the cells of a living organism that allow it to obtain and use energy, including breaking down food molecules and building new molecules.

metastasis The spread of cancer cells from one location in the body to another.

micronutrients Nutrients, including vitamins and minerals, that organisms must ingest in small amounts to maintain health.

mineral An inorganic chemical element required by organisms for normal growth, reproduction, and tissue maintenance; examples are calcium, iron, potassium, and zinc.

mitochondria Membrane-bound organelles responsible for important energy-conversion reactions in eukaryotes.

mitochondrial DNA (mtDNA) The DNA in mitochondria that is inherited solely from the mother.

mitosis The segregation and separation of duplicated chromosomes during cell division.

molecule Atoms linked by covalent bonds.

mollusk A soft-bodied invertebrate, generally with a hard shell (which may be tiny, internal, or absent in some mollusks).

monomer One chemical subunit of a polymer.

monosaccharide The building block, or monomer, of a carbohydrate.

multifactorial inheritance An interaction between genes and the environment that contributes to a phenotype or trait.

multipotent Describes a cell with the ability to differentiate into a limited number of cell types in the body.

mutagen Any chemical or physical agent that can damage DNA by changing its nucleotide sequence.

mutation A change in the nucleotide sequence of DNA.

mutualism A type of symbiotic relationship in which both members benefit; a "win-win" relationship.

mycelium (plural: mycelia) A spreading mass of interwoven hyphae that forms the often subterranean body of multicellular fungi.

natural resources Raw materials that are obtained from the earth and are considered valuable even in their relatively unmodified, natural form.

natural selection Differential survival and reproduction of individuals in response to environmental pressure that leads to change in allele frequencies in a population over time.

neutron An electrically uncharged subatomic particle found in the nucleus of an atom.

niche The space, environmental conditions, and resources that a species needs in order to survive and reproduce.

nitrogen fixation The process of converting atmospheric nitrogen into a form that plants can use to grow.

nonadaptive evolution Any change in allele frequency that does not by itself lead a population to become more adapted to its environment; the causes of nonadaptive evolution are mutation, genetic drift, and gene flow.

noncoding regions DNA sequences that do not hold instructions to make proteins.

nondisjunction Failure of chromosomes to separate accurately during cell division; nondisjunction in meiosis leads to aneuploid gametes.

nonrenewable resources Natural resources that cannot be replaced.

nuclear envelope The double membrane surrounding the nucleus of a eukaryotic cell.

nucleic acids Organic molecules made up of linked nucleotide subunits; DNA and RNA are examples of nucleic acids.

nucleotides The building blocks of DNA. Each nucleotide consists of a sugar, a phosphate, and a base.

nucleus (atomic) The dense core of an atom.

nucleus (eukaryotic) The organelle in eukaryotic cells that contains the genetic material.

nutrients Components in food that the body needs to grow, develop, and repair itself.

obese Having 20% more body fat than is recommended for one's height, as measured by a body mass index greater than 30.

oncogene A mutated and overactive form of a proto-oncogene. Oncogenes drive cells to divide continually.

organelles The membrane-bound compartments of eukaryotic cells that carry out specific functions.

organic molecule A molecule with a carbon-based backbone and at least one C-H bond.

osmosis The diffusion of water across a semipermeable membrane from an area of lower solute concentration to an area of higher solute concentration.

osteoporosis A disease characterized by thinning bones.

paleontologist A scientist who studies ancient life by means of the fossil record.

parasitism A type of symbiotic relationship in which one member benefits at the expense of the other.

pathogen A disease-causing agent, usually an organism.

peer review A process in which independent scientific experts read scientific studies before their publication to ensure that the authors have appropriately designed and interpreted their study.

peptidoglycan A macromolecule that forms all bacterial cell walls and provides rigidity to the cell wall.

pH A measure of the concentration of H^+ in a solution.

phenotype The visible or measurable traits of an individual.

phospholipid A type of lipid that forms biological membranes.

phospholipid bilayer A double layer of phospholipid molecules that characterizes all biological membranes.

photons Packets of light energy, each with a specific wavelength and quantity of energy.

photosynthesis The process by which plants and other autotrophs use the energy of sunlight to make energy-rich molecules using carbon dioxide and water.

phylogenetic tree A branching tree of relationships showing common ancestry.

phylogeny The evolutionary history of a group of organisms.

pili (singular: pilus) Short, hairlike appendages extending from the surface of some bacteria, used to adhere to surfaces

pistil The female reproductive structure of a flower, made up of a stigma, style, and ovary.

placebo A fake treatment given to control groups to mimic the experience of the experimental groups.

placebo effect The effect observed when members of a control group display a measurable response to a placebo because they think that they are receiving a "real" treatment.

plant A multicellular eukaryote that has cell walls, carries out photosynthesis, and is adapted to living on land.

plate tectonics The movement of the earth's upper mantle and crust, which influences the geographical distribution of landmasses and organisms.

pluripotent Describes a cell with the ability to differentiate into nearly any cell type in the body.

polar molecule A molecule in which electrons are not shared equally between atoms, causing a partial negative charge at one end and a partial positive charge at the other; for example, water.

pollen Small, thick-walled structures that contain cells that will develop into sperm.

pollination The transfer of pollen from male to female plant structures so that fertilization can occur.

polygenic trait A trait whose phenotype is determined by the interaction between alleles of more than one gene.

polymer A molecule made up of individual subunits, called monomers, linked together in a chain.

polymerase chain reaction (PCR) A laboratory technique used to replicate, and thus amplify, a specific DNA segment.

population A group of organisms of the same species living and interacting in a particular geographic area.

population density The number of organisms per given area.

potential energy Stored energy.

predation An interaction between two organisms in which one organism (the predator) feeds on the other (the prey).

prion A protein-only infectious agent.

producers Autotrophs (photosynthetic organisms) that form the base of every food chain.

prokaryote A usually single-cell organism whose cell lacks internal membrane-bound organelles and whose DNA is not contained within a nucleus.

prokaryotic cells Cells that lack internal membrane-bound organelles.

protein An organic molecule made up of linked amino acid subunits. Proteins play many critical roles in living organisms.

protist A eukaryote that cannot be classified as a plant, animal, or fungus; usually unicellular.

proton A positively charged subatomic particle found in the nucleus of an atom.

proto-oncogene A gene that codes for a protein that helps cells divide normally.

punctuated equilibrium The theory that most species change occurs in periodic bursts as a result of sudden environmental change.

Punnett square A diagram used to determine probabilities of offspring having particular genotypes, given the genotypes of the parents.

radial symmetry The pattern exhibited by a body plan that is circular, with no clear left and right sides.

radiation therapy The use of ionizing (high-energy) radiation to treat cancer.

radioactive isotope An unstable form of an element that decays into another element by radiation, that is, by emitting energetic particles.

radiometric dating The use of radioactive isotopes as a measure for determining the age of a rock or fossil.

randomized clinical trial A controlled medical experiment in which subjects are randomly chosen to receive either an experimental treatment or a standard treatment (or placebo).

recessive allele An allele that reveals itself in the phenotype only if the organism has two copies of that allele.

recombination The stage of meiosis in which maternal and paternal chromosomes pair and physically exchange DNA segments.

regulatory sequence The part of a gene that determines the timing, amount, and location of protein produced.

relative dating Determining the age of a fossil on the basis of its position relative to layers of rock or fossils of known age.

renewable resources Natural resources that are replenished after use as long as the rate of consumption does not exceed the rate of replacement.

reproductive isolation Mechanisms that prevent mating (and therefore gene flow) between members of different species.

ribosome The cellular machinery that assembles proteins during the process of translation.

RNA polymerase The enzyme that accomplishes transcription. RNA polymerase copies a strand of DNA into a complementary strand of mRNA.

sample size The number of experimental subjects or the number of times an experiment is repeated. In human studies, sample size is the number of subjects.

saturated fat An animal fat, such as that found in butter; saturated fats are solid at room temperature.

science The process of using observations and experiments to draw evidence-based conclusions.

scientific theory A hypothesis that is supported by many years of rigorous testing and thousands of experiments.

seed The embryo of a plant, along with a starting supply of food, encased in a protective covering.

semi-conservative DNA replication is said to be semi-conservative because each newly made DNA molecule has one original and one new strand of DNA.

sex chromosomes Paired chromosomes that differ between males and females, XX in females, XY in males

short tandem repeats (STRs) Sections of a chromosome in which DNA sequences are repeated.

simple diffusion The movement of small, hydrophobic molecules across a membrane from an area of higher concentration to an area of lower concentration; simple diffusion does not require an input of energy.

simple sugar (monosaccharide) A carbohydrate made up of a single sugar subunit; an example is glucose.

sister chromatid One of the two identical DNA molecules that make up a duplicated chromosome following DNA replication.

solute A dissolved substance.

solution A mixture of solutes dissolved in a solvent.

solvent A substance in which other substances can dissolve; for example, water.

speciation The genetic divergence of populations owing to a barrier to gene flow between them, leading over time to reproductive isolation and the formation of new species.

stabilizing selection A type of natural selection in which organisms near the middle of the phenotypic range of variation are favored.

stamen The male reproductive structure of a flower, made up of a filament and an anther.

starch A complex plant carbohydrate made of linked chains of glucose molecules; a source of stored energy.

statistical significance A measure of confidence that the results obtained are "real," rather than due to random chance.

stem cells Immature cells that can divide and differentiate into specialized cell types.

substrate A compound or molecule that an enzyme binds to and on which it acts.

sustainability Using the earth's resources in a way that will not permanently destroy or deplete them; living within the limits of earth's biocapacity.

symbiosis A situation in which two different organisms live together, often interdependently

taxonomy The process of identifying, naming, and classifying organisms on the basis of shared traits.

testable A hypothesis is testable if it can be supported or rejected by carefully designed experiments or nonexperimental studies.

tetrapod An organism with four true limbs, that is, bony appendages with jointed wrists, ankles, and digits; mammals, amphibians, birds, and reptiles are tetrapods.

tissue An organized group of different cell types that work together to carry out a particular function.

totipotent Describes a cell with the ability to differentiate into any cell type in the body.

trans fat A type of vegetable fat which has been hydrogenated, that is, hydrogen atoms have been added, making it solid at room temperature.

transcription The first stage of gene expression, during which cells produce molecules of messenger RNA (mRNA) from the instructions encoded within genes.

transfer RNA (tRNA) A type of RNA that helps ribosomes assemble chains of amino acids during translation.

transgenic Refers to an organism that carries one or more genes from a different species.

translation The second stage of gene expression. Translation "reads" mRNA sequences and assembles the corresponding amino acids to make a protein.

transport proteins Proteins involved in the movement of molecules across the cell membrane.

triglyceride A type of lipid found in fat cells that stores excess energy for long-term use.

trisomy 21 Carrying an extra copy of chromosome 21; also known as Down syndrome.

trophic levels Feeding levels, based on positions in a food chain.

tumor suppressor genes Genes that code for proteins that monitor and check cell cycle progression. When these genes mutate, tumor suppressor proteins lose normal function.

unsaturated fat A plant fat, such as olive oil; unsaturated fats are liquid at room temperature.

vascular plant A plant with tissues that transport water and nutrients through the plant body.

vertebrate An animal with a bony or cartilaginous backbone.

vestigial structure A structure inherited from an ancestor that no longer serves a clear function in the organism that possesses it.

virus An infectious agent made up of a protein shell that encloses genetic information.

vitamin An organic molecule required in small amounts for normal growth, reproduction, and tissue maintenance.

vitamin D A fat-soluble vitamin necessary to maintain a healthy immune system and build healthy bones and teeth. The human body produces vitamin D when skin is exposed to UV light.

X chromosome One of the two sex chromosomes in humans.

X-linked trait A phenotype determined by an allele on an X chromosome.

Y chromosome One of two sex chromosomes in humans. The presence of a Y chromosome signals the male developmental pathway during fetal development.

zygote A cell that is capable of developing into an adult organism. The zygote is formed when an egg is fertilized by a sperm.

Photo Credits

Chapter 1

p. 1: Courtesy of Ann Warren. **p. 3:** Steve Bronstein/Getty Images. **p. 4:** Courtesy of Ann Warren. **p. 5:** *Infographic 1-1* (TL) Used by permission of *North Carolina Medical Journal* (Morrisville, NC: North Carolina Institute of Medicine), November/December 2003, www.ncmedicaljournal.org, (TR) Courtesy of *British Medical Journal,* 28 May 2009, Vol. 338, Issue 7706, (B) Aleksej Vasic/iStockphoto. **p. 6:** *Infographic 1.2* (TL) *Science,* Vol. 324, no. 5935, 26 June 2009. Reprinted with permission from AAAS, (TR) PLoS Med 2 (2), cover image, Krista Steinke. **p. 7:** Micha Pawlitzki/Zefa/Corbis: (TL) *Cell Metabolism,* March 4, 2009, cover illustration by Chris Lange, © Elsevier, 2009, (TC) © Elsevier, (TR) *Science,* Vol. 320, no. 5882, 13 June 2008. Reprinted with permission from AAAS; (B) Rick Wilson/The Florida Times-Union. **p. 10:** *Infographic 1.5* (L) Jessica Peterson/Photolibrary, (TR) Nancy Nehring/ iStockphoto, (CR) Interfoto/Alamy, (BL) Mark Moffett/Getty Images, (BR) G. Lasley/VIREO. **p. 12:** Vicki Wagner/Alamy. **p. 15:** *Infographic 1.8* Tony West/Alamy.

Chapter 2

p. 19: NASA/JPL. **p. 21:** Calvin J. Hamilton. **p. 22:** (T) NASA/JP; (B) NASA/JSC. **p. 23:** *Infographic 2.1* (T) Bildarchiv/ AgeFotostock, (C) globestock/iStockphoto, (B) Kazuo Ogawa/ AgeFotostock. **p. 27:** NASA/JSC. **p. 33:** *Infographic 2.6* (T) Mixa/Superstock, (C) B. Runk/S. Schoenberger/Grant Heilman, (B) AgeFotostock/Superstock. **p. 34:** NASA/Time Life Pictures/Getty Images. **p. 35:** GSFC/NASA.

Chapter 3

p. 39: Joe Raedle/Getty Images. **p. 41:** Pictorial Press Ltd/ Alamy. **p. 42:** Fleming, Alexander. 1929. On the Antibacterial Action of Cultures of a Penicillium, with Special Reference to Their Use in the Isolation of B. Influenzae. *British Journal of Experimental Pathology,* Vol. 10, pp. 226-236, Fig. 2; *Infographic 3.1* (L) The British Library/Photolibrary, (C and R) Biophoto Associates/Photo Researchers. **p. 43:** *Infographic 3.2* (TL) Scenics & Science/Alamy, (TC) Roland Birke/Photolibrary, (TR) Dennis Kunkel/Visuals Unlimited, (BL) Ed Reschke/ Photolibrary, (inset) David Toase/Photolibrary, (BC) Ed Reschke/Photolibrary, (inset) Christian Fischer/WIKI, Creative Commons, (BR) Michael Abbey/Photo Researchers, (inset) A. & F. Michler/Photolibrary. **p. 45:** SPL/Photo Researchers. **p. 47:** (T) Daily Herald Archive/SSPL/Getty Images; (B) Research and Development Division, Schenley Laboratories, Inc., Lawrenceburg, Indiana. **p. 54:** Joe Raedle/ Getty Images. **p. 55:** NIH.

Chapter 4

p. 59: ma-k/iStockphoto. **p. 61:** ma-k/iStockphoto. **p. 62:** Courtesy of Nestlé SA. **p. 63:** *Infographic 4.1* (TL) Denis Pepin/ Featurpics, (TC) Shadow216/Dreamstime, (TR) Fresh Food Images/Photolibrary, (CL) Jeffrey Coolidge/Getty Images, (CC) AgeFotostock/Superstock, (CR) Juanmonino/iStockphoto, (BL) AgeFotostock/Superstock, (BC) adlifemarketing/iStockphoto, (BR) Creative Commons: <http://www.flickr.com/photos/ tellumo/232317103/sizes/o/in/photostream/>. **p. 65:** (T) BostjanT/iStockphoto; (BL and BR) The Photo Works. **p. 66:** The Photo Works. **p. 67:** Boissonnet/AgeFotostock. **p. 71:** *Infographic 4.6* (TL) Rob Owen/Whal/StockXchng, (TR) ktphotog/iStockphoto, (C) Dr. Michael Klein/Peter Arnold, (B) shironosov/iStockphoto. **p. 73:** *Infographic 4.7* (L) Reprinted with permission © 2008 Southwest Research Institute. All rights reserved. (R) David M. Phillips/Photo Researchers. **p. 75:** *Infographic 4.8* Copyright © 2008. For more information about The Healthy Eating Pyramid, please see The Nutrition Source, Department of Nutrition, Harvard School of Public Health, http://www.thenutritionsource.org, and *Eat, Drink, and Be Healthy* by Walter C. Willett, M.D. **p. 76:** (L) Bengt-Göran Carlsson/TIOFOTO/Nordic Photos; (R) Alison Wright/ Photo Researchers

Chapter 5

p. 81: Philip Hart. **p. 83:** (T) Courtesy of Sapphire Energy, Inc., (B) Philip Hart. **p. 84:** egdigital/iStockphoto. **p. 85:** Courtesy of Sapphire Energy, Inc. **p. 86:** *Infographic 5.2* (T, from left) Kimberly Deprey/iStockphoto, mihtiander/FeaturePics, Bios/ Photolibrary, (B) AP Photo/Arthur Max, (CL and CR) Visuals Unlimited/Corbis, (R, from top) Gudella/FeaturePics, moori/ FeaturePics, Yobro10/Dreamstime.com, Photo168/ Dreamstime.com. **p. 88:** *Infographic 5.3* Corbis/SuperStock. **p. 89:** *Infographic 5.4* (from left) Pixelgnome/Dreamstime. com, Vasily Smirnov/Dreamstime.com, Rocky Reston/ Dreamstime.com, Shevelartur/Dreamstime.com, Aleksandr Lazarev/iStockphoto. **p. 91:** Ashley Cooper/Alamy; *Infographic 5.5* (top panel) (L) Visuals Unlimited/Corbis, (TR) Mark Hamblin/AgeFotostock, (BR) Visuals Unlimited/Corbis; (bottom panel) (L) 2ndLookGraphics/iStockphoto, (C) Jfybel/ Dreamstime, (R) NNehring/iStockphoto. **p. 93:** Michael Macor/San Francisco Chronicle/Corbis. **p. 94:** *Infographic 5.7* (L) Brand X Pictures, (R) Maxrale/iStockphoto. **p. 97:** Courtesy of Jim Sears/A2BE Carbon Capture LLC.

Chapter 6

p. 101: Leaf/Dreamstime. **p. 103:** (L) Jose Luis Pelaez Inc./ Getty Images; (TR) Tim Platt/Getty Images; (BR) Sian Kennedy/Getty Images. **p. 105:** (B) Courtesy of Paul Rozin; *Infographic 6.2* (L) Christian Handl/Photolibrary, (C, inset) Michael Gray/Dreamstime.com, (CT) iperl/Featurepics, (CB) Stephen Bonk/Dreamstime.com, (R) Photodisc. **p. 106:** (L) Steve Stock/Alamy; (R) Hartmann Christian/SIPA; *Infographic 6.3* (T) Royalty-Free/Corbis, (B) Robert Fried/Alamy. **p. 107:** *Infographic 6.4* Ilena Eisseva/Featurepics. **p. 110:** Tips Italia/ Photolibrary. **p. 111:** *Infographic 6.6* (T) apcuk/iStockphoto, (C) EricGerrard/iStockphoto, (B) bluestocking/iStockphoto, (R) DNY59/iStockphoto. **p. 115:** (L) Leaf/Dreamstime; (R) Konstik/ Dreamstime. **p. 116:** AP Photo/Harry Cabluck.

Chapter 7

p. 121: The Innocence Project. **p. 122:** (logo) The Innocence Project. **p. 123:** JUPITERIMAGES/Brand X/Alamy. **p. 124:** *Infographic 7.1* ISM/Phototake. **p. 125:** Uli Holz, Yeshiva

University. **p. 127:** Courtesy of Kary Mullis. **p. 128:** The Innocence Project. **p. 132:** *Infographic 7.6* (L) Biophoto Associates/Photo Researchers, (C) Dr. Gopal Murti/ Photo Researchers, (R) Biophoto Associates/Photo Researchers. **p. 133:** Kevin Rivoli/The New York Times/Redux.

Milestones in Biology: The Model Makers

p. 137: SSPL/Getty Images. **pp. 138–139:** James D. Watson Collection, Cold Spring Harbor Laboratory Archives. **p. 140:** *Rosalind Franklin and the Shape of DNA* (L) National Portrait Gallery, London, (R) Omikron/Photo Researchers. **p. 141:** *Erwin Chargaff's Work Provided a Clue to Base Pairing* National Library of Medicine. **p. 142:** SSPL/Getty Images.

Chapter 8

p. 143: Edwin Remsberg/Alamy. **p. 145:** Sebastian Knight/ Dreamstime.com. **p. 146:** Nigel Cattlin/Visuals Unlimited. **p. 148:** *Infographic 8.3* Eye of Science/Photo Researchers. **p. 151:** *Infographic 8.6* (from left) MedicalRF.com/ AgeFotostock, Lauritzasoare/Dreamstime.com, Courtesy of GTC Biotherapeutics, Courtesy of Lundbeck Inc. **p. 152:** Edwin Remsberg/Alamy. **p. 156:** Courtesy of GTC Biotherapeutics, Inc. **p. 159:** Dan Reynolds/CartoonStock.

Milestones in Biology: Sequence Sprint

p. 161: Mario Tama/Getty Images. **pp. 162–163:** Alex Wong/ Newsmakers/Getty Images. **p. 166:** Mario Tama/Getty Images. **p. 167:** Sinclair Stammers/Photo Researchers.

Chapter 9

p. 169: Stefano Lunardi/AgeFotostock. **p. 171:** Emilio Ereza/ Alamy. **p. 173:** *Infographic 9.1* (TL) NIH, (TR) Courtesy of Bristol-Myers Squibb Company, (BL) Steffen Hauser/ Botanikfoto/Alamy, (BR) Stefano Lunardi/AgeFotostock. **p. 174:** *Infographic 9.2* Targeting Connexin43 Expression Accelerates the Rate of Wound Repair. (2003). Cindy Qiu, Petula Coutinho, Stefanie Frank, Susanne Franke, Lee-yong Law, Paul Martin, Colin R. Green and David L. Becker. *Current Biology,* 13(19), 1697-1703. **p. 177:** Michael Abbey/Photo Researchers. **p. 180:** *Infographic 9.7* (L) Jim West/ AgeFotostock, (R) Peter Arnold, Inc./Alamy. **p. 181:** Steve Gschmeissner/Photo Researchers. **p. 182:** Alex Segre/ Photographers Direct.

Chapter 10

p. 187: Du Cane Medical Imaging Ltd./Photo Researchers. **p. 189:** UPI/BIll Greenblatt/Newscom. **p. 190:** *Infographic 10.1* (L) Stockbyte, (R) Ingram Publishing/Photolibrary. **p. 191:** Du Cane Medical Imaging Ltd./Photo Researchers. **p. 194:** *Infographic 10.4* (TL) Lepas/Dreamstime, (TC) Larry Jordan/ FeaturePics, (TR) Photodisc, (BL and BC) Royalty-Free/Corbis, (BR) Mary Lane/FeaturePics. **p. 197:** Glow Wellness/ SuperStock. **p. 199:** Courtesy of Lorene Ahern.

Chapter 11

p. 203: AP Photo/Carlos Osorio. **p. 205:** Courtesy of Emily Schaller. **p. 207:** *Infographic 11.2* ISM/Phototake. **p. 211:** Jeffrey Sauger. **p. 212:** *Infographic 11.6* Simon Fraser/Photo Researchers. **p. 217:** AP Photo/Carlos Osorio.

Milestones in Biology: Mendel's Garden

p. 221: Authenticated News/Getty Images. **pp. 222-223:** Malcolm Gutter/Visuals Unlimited. **p. 224:** *Concepts of Inheritance before Mendel* (L) Preformation, drawn by N. Hartsoecker, 1695, (R) Wellcome Library, London. **p. 227:** (T) Authenticated News/Getty Images, (B) Garden World Images/ AgeFotostock.

Chapter 12

p. 229: Courtesy of Peter Morenus. **p. 230:** Sebastian Kaulitzki/Alamy. **p. 232:** *Infographic 12.1* (L and R)) ISM/ Phototake. **p. 234:** Biophoto Associates/Photo Researchers. **p. 237:** AP Photo/Leslie Close. **p. 240:** (T, from left) Ludo Kuipers Photography; photographersdirect.com; Yuri Arcurs/ Fotolia; ImageSource/AgeFotostock; Tatiana Morozova/ FeaturePics; (C, from left) Denis Pepin/FeaturePics; PhotosIndia/Alamy; Jacob Langvad/Getty Images; Thomas Cockrem/Alamy; (BL) leaf/FeaturePics; (BR) Dmitriy Shironosov/Dreamstime.com. **p. 241:** Susumu Nishinaga/ Photo Researchers; *Infographic 12.6* (L) ImageSource/ AgeFotostock, (C) leaf/FeaturePics, (R) Ryan McVay/Getty Images. **p. 243:** Courtesy of Peter Morenus. **p. 245:** Jason Sitt/ Fotolia. **p. 246:** *Infographic 12.10* National Institute of Mental Health. **p. 247:** Markus Moellenberg/Corbis. **p. 249:** *Infographic 12.13* ISM/Phototake.

Chapter 13

pp. 253 and 254: AP Photo/PA. **p. 255:** Deco Images II/Alamy. **p. 256:** *Infographic 13.1* (T) Ed Reschke/Photolibrary, (C) Robert Knauft/Biology Pics/Photo Researchers, (B) Phototake/ Alamy. **p. 257:** Courtesy of Robert Langer, photo by Stu Rosner. **p. 258:** *Infographic 13.3* AP Photo/Brian Walker. **p. 259:** Wake Forest University Health Sciences/Center for Regenerative Medicine/Urology. **p. 264:** *Infographic 13.7* (T) James King-Holmes/Science Photo Library, (B) AP Photo/PA/ Files. **p. 266:** *Infographic 13.8* NIH.

Chapter 14

p. 271: ISM/Phototake. **p. 273:** CDC/ Janice Carr; Jeff Hageman. **p. 274:** *Infographic 14.1* (L) USDA/ARS, (TR) Paulo Cruz/Dreamstime.com, (CR, inset) DNY59/iStockphoto, (CR) Offscreen/Dreamstime.com, (BR) Genevieve Astrelli/ iStockphoto. **p. 281:** *Infographic 14.7* (T) CDC/Janice Haney Carr/Jeff Hageman, M.H.S., (C) Binh Tran/iStockphoto, (B) W. Lane/Minden Pictures. **p. 282:** ISM/Phototake. **p. 284:** *Infographic 14.8* (TL) Gallo Images/Getty Images, (TR) alandj/ iStockphoto, (CL) walik/iStockphoto, (CR) Steve Shepard/ iStockphoto, (BL) Ewa Walicka/Dreamstime.com, (BR) Christine Schuhbeck/AgeFotostock. **p. 287:** Nick D. Kim/ CartoonStock.

Milestones in Biology: Adventures in Evolution

p. 289: London Stereoscopic Company/Getty Images. **pp. 290-291:** HMS *Beagle* in the Galápagos by John Chancellor (1925-984). Courtesy of Gordon Chancellor. **p. 292:** *Lamarckianism* Bettmann/Corbis. **p. 293** *The Evolution of Darwin's Thought* (TC) public domain, (TR) Reproduced with permission from John van Wyhe, ed., The Complete Work of Charles Darwin Online (http://darwin-online.org.uk/), (BL) Classic Image/Alamy, (BC) Reproduced with permission from

John van Wyhe, ed., The Complete Work of Charles Darwin Online (http://darwin-online.org.uk/), (inset) World History Archive/Alamy, (BR) Charles Lyell, *Principles of Geology,* 11th ed. London: John Murray, 1872. Leith Storage P DG L. © Cambridge University. **p. 294:** HMS *Beagle* in the Galápagos by John Chancellor (1925–1984). Courtesy of Gordon Chancellor. **p. 296:** *The Evolution of Wallace's Thought* (TC) London Stereoscopic Company/Getty Images, (TR) © Cambridge University Library, (BL) Wallace, Alfred Russel, *The Geographical Distribution of Animals: With a Study of the Relations of Living and Extinct Faunas As Elucidating the Past Changes of the Earth's Surface,* 1876. (BC) © Natural History Museum, London. **p. 298:** University of South Carolina, Rare Books and Special Collections.

Chapter 15

p. 299: U.S. Geological Survey, Southeast Ecological Science Center, Sirenia Project. **p. 301:** Perrine Doug/Photolibrary. **p. 302:** (L) Courtesy of Sally W. Cushmore; (R) Mark Conlin, VW Pics/Superstock. **p. 306:** U.S. Geological Survey, Southeast Ecological Science Center, Sirenia Project. **p. 310:** Pat Canova/AgeFotostock. **p. 313:** *Infographic 15.5* Row 1: (L) Dmitry Deshevykh/iStockphoto, (R) Nico Smit/Dreamstime; Row 2: (L) Michelle Gilders/Alamy, (R) Gustav Verderber/Photolibrary; Row 3: (L) Rinusbaak/Dreamstime.com, (R) Henno Robert/AgeFotostock; Row 4: (L) Steve Byland/Dreamstime.com; (R) Nick Layton/Alamy; Row 5: (L) cynoclub/FeaturePics.com; (R) marilna/FeaturePics.com; Row 6: (L) Dimitar Marinov/Dreamstime.com, (R) Ziutograf/iStockphoto; Row 7: Frank W. Lane/FLPA/Minden Pictures. **p. 315:** *Infographic 15.7* Donald Fawcett, I. B. David, D. R. Wolsterholm/Visuals Unlimited. **p. 316:** *Infographic 15.8* (TL) G. Armistead/VIREO, (TR) J. Dunning/VIREO, (CL) Mark Jones/Photolibrary, (BL) Tim Laman/National Geographic/Getty Images.

Chapter 16

p. 321: Neil Shubin. **p. 323:** (T) Tyler Keillor/University of Chicago Fossil Lab; (B) Ted Daeschler/Academy of Natural Sciences/VIREO. **p. 324:** *Infographic 16.1* (T) choicegraphx/iStockphoto, (C) Colin Keates/Dorling Kindersley/Getty Images, (BL) Arpad Benedek/iStockphoto, (BR) Grafissimo/iStockphoto. **p. 325:** Ted Daeschler/Academy of Natural Sciences/VIREO. **p. 327:** *Infographic 16.2* Courtesy of Doug Shore. **p. 328:** *Infographic 16.3* Ted Daeschler/Academy of Natural Sciences/VIREO. **p. 330:** Neil Shubin. **p. 333:** *Infographic 16.6* (T, from left) Anatomical Travelogue/Photo Researchers, Eye of Science/Photo Researchers, Courtesy of Rachel M. Warga, Courtesy of Olivier Pourquie, Oxford Scientific/Photolibrary, (B, from left) Claudia Dewald/iStockphoto, Frank Wiechens/Fotolia, Vladimir Kozieiev/Dreamstime.com, Dolan Halbrook/iStockphoto, Tamara Murray/iStockphoto. **p. 334:** *Infographic 16.7* (L) Steven Hunt/Getty Images, (CL and CR) SuperStock, (R) PhotoAlto/Alamy. **p. 337:** Ruud de Man/iStockphoto.

Chapter 17

p. 339: Michael Melford/National Geographic/Getty Images. **p. 340:** Martin Shields/Alamy. **p. 341:** NASA/JSC. **p. 343:** *Infographic 17.2* James S. Kuwabara/U.S. Geolological Survey. **p. 344:** Roger Ressmeyer/Corbis. **p. 347:** Michael Melford/National Geographic/Getty Images. **p. 348:** *Infographic 17.4* (L) Jamie Carroll/iStockphoto, (R) Sylvie Bouchard/iStockphoto; (map) NASA Goddard Space Flight Center Image by Reto Stöckli (land surface, shallow water, clouds). Enhancements by Robert Simmon (ocean color, compositing, 3D globes, animation). Data and technical support: MODIS Land Group, MODIS Science Data Support Team, MODIS Atmosphere Group, MODIS Ocean Group. Additional data: USGS EROS Data Center (topography), USGS Terrestrial Remote Sensing Flagstaff Field Center (Antarctica), Defense Meteorological Satellite Program (city lights). **p. 352:** *Infographic 17.8* (from left) Eu Jin Chew/Dreamstime.com, Gert Vrey/Dreamstime.com, Musk/Alamy, Danita Delimont/Alamy. **p. 353:** WIKI, Creative Commons, Árbol de la vida según Haeckel, E. H. P. A. (1866). Generelle Morphologie der Organismen: allgemeine Grundzüge der organischen Formen-Wissenschaft, mechanisch begründet durch die von C. Darwin reformirte Decenden. **p. 357**: Bill Abbott/Cartoonstock

Chapter 18

pp. 359 and 361: Courtesy of the University of Washington, Lost City Science Team, IFE, URI-IAO, and NOAA. **p. 362:** (L) Courtesy of Gretchen Früh-Green, (R) Courtesy of NOAA. **p. 363:** *Infographic 18.1* (L) Courtesy of D. Kelley, University of Washington, (R) Image courtesy of Matt Schrenk, University of Washington. **p. 363:** Amy Nevala, Woods Hole Oceanographic Institution. **p. 364:** *Infographic 18.2* (TL) Courtesy of the University of Washington, (TR) Courtesy of the University of Washington, Lost City Science Team, IFE, URI-IAO, and NOAA, (BL and BR) University of Washington, School of Oceanography. **p. 365:** (L) Woods Hole Oceanographic Institute; (R) Courtesy of The Lost City 2005 Exploration; *Infographic 18.3* Per Ivar Somby. **p. 366:** *Infographic 18.4* (TL) Courtesy of D. Kelley, University of Washington, (T, inset) Courtesy of Matt Schrenk, University of Washington, (B, inset) Courtesy of Matt Schrenk, University of Washington, (TR) Woods Hole Oceanographic Institution, (inset) Julie Huber/Marine Biological Laboratory, (BL) Extremophiles. 2010 Jan, 14(1):61-9. Epub 2009 Nov 4. Novel ultramicrobacterial isolates from a deep Greenland ice core represent a proposed new species, Chryseobacterium greenlandense sp. nov. Loveland-Curtze J, Miteva V, Brenchley J. Department of Biochemistry and Molecular Biology, The Pennsylvania State University, (BCL) Courtesy of Mike Dyall-Smith, (BCR) Dennis Kunkel/Visuals Unlimited, (BR) Courtesy of Brookhaven National Laboratory. **p. 370:** *Infographic 18.6* (TL) E. Nelson and L. Sycuro, courtesy of the *Vibrio fischeri* Genome Project, (inset) William Ormerod/courtesy of Margaret McFall-Ngai, (TC) Dr. Kari Lounatmaa/Photo Researchers, (TR) Science Photo Library/Photolibrary, (inset) The Photo Works, (CL) MedicalRF/Visuals Unlimited, (inset) Scott Bodell/Photolibrary, (CC) Dennis Kunkel/Phototake, (CR) Biodisc/Visuals Unlimited/Alamy, (inset) Courtesy of Symbyos, Louisville, CO, (B) Sherman Thomson/Visuals Unlimited. **p. 371:** *Infographic 18.7* (TL) National Science Foundation, (inset) K.O. Stetter and R. Rachel, Univ. Regensburg, Germany, (TR) Courtesy of Jerry Ting, (inset) Eye of Science/Photo Researchers, (B) Corbis Premium RF/Alamy, (inset) Kenneth M. Stedman, Ph.D., NASA Astrobiology Institute-Center for

Life in Extreme Environments, Portland State University. **p. 374:** Courtesy of the University of Washington, Lost City Science Team, IFE, URI-IAO, and NOAA.

Chapter 19

p. 377: Courtesy of Northwest Trek. **p. 378:** (L and R) National Park Service, Olympic National Park, photo by Janis Burger. **p. 379:** Georgette Douwma/Getty Images. **p. 381** *Infographic 19.2* (TL) Patrick Robbins/Dreamstime.com, (TR) Courtesy of Bob Wightman, (BL) National Park Service, (BC) Marcopolo/FeaturePics, (BR) Courtesy of Peter Wigmore. **p. 382:** *Infographic 19.3* (L) Fotogal/FeaturePics, (inset) Mariya Bibikova/iStockphoto, (CL) Ferns at Muir Woods, CA, Sanjay ach/http://en.wikipedia.org/wiki/Fern, (inset) George Bailey/Dreamstime.com, (CR) Michael P. Gadomski/Photo Researchers, (inset) Ray Roper/iStockphoto, (R) Mark Turner/Photolibrary, (inset) Courtesy of Greg Rabourn. **p. 383** Courtesy of Northwest Trek. **p. 384:** *Infographic 19.4* (from left) Courtesy of Brooke et al., NOAA-OE, HBOI; Anky10/Dreamstime.com; Ed Reschke/Photolibrary; Photolibrary/Alamy; Manipulateur/Fotolia; London Scientific Films/Photolibrary; Stock.xchng; U.S. National Park Service; Karen Arnold/Dreamstime.com. **p. 385:** David Gomez/iStockphoto. **p. 387:** (TL) Jan Gottwald/iStockphoto, (TC) Outdoorsman/Dreamtime.com, (TR) Mark Conlin/Alamy, (BL) © Gary Nafis, (BC) Lon E. Lauber/Photolibrary, (BR) Chris Mattison/Alamy. **p. 389:** *Infographic 19.5* (C) Ed Reschke/Photolibrary, (TL) London Scientific Films/Photolibrary, (CL) Eye of Science/Photo Researchers, (BL) Steve Gschmeissner/Photo Researchers, (TR) Mike Norton/Dreamstime.com, (BR) Alexander Makarov/iStockphoto. **p. 390:** *Infographic 19.6* (L) Gary Retherford/Photo Researchers, (CL) Roland Birke/Photolibrary, (CR) http://en.wikipedia.org/wiki/File:Dog_vomit_slime_mold.jpg, (R) Oxford Scientific /Photolibrary. **p. 391:** *Infographic 19.7* (L) Wim van Egmond/Visuals Unlimited, (CL) Stock.xchng, (C) Zefiryn/Fotolia, (CR) Ximinez/Fotolia, (R) Wim van Egmond/Visuals Unlimited.

Chapter 20

p. 395: AP Photo/Rick Bowmer. **p. 397:** Janine Wiedel Photolibrary. **p. 398:** Mark Wilson/Getty Images; *Infographic 20.1* (TL) Library of Congress Prints and Photographs Division [LC-USW3-037939-E], (TR) Bachmann/AgeFotostock, (BL) AP Photo/Rick Bowmer, (BR) Courtesy of www.worldmap.com. **p. 399:** *Infographic 20.2* Carolina Biological Supply Company/Phototake. **p. 400:** *Infographic 20.3* (L) Elena Rostunova/Alamy, (R) Living Art Enterprises, LLC/Photo Researchers. **p. 401:** Nina Jablonski. **p. 402:** *Infographic 20.4* Adapted from Chaplin G., Geographic Distribution of Environmental Factors Influencing Human Skin Coloration, *American Journal of Physical Anthropology* 125:292-302, 2004; map updated in 2007. Designer: Emmanuelle Bournay, UNEP/GRID-Arendal. **p. 404:** *Infographic 20.6* (T) Dennis Kunkel/Visuals Unlimited, (B) CNRI/Photo Researchers. **p. 405:** *Infographic 20.7* (L) Anthropological Skull Model-KNM-ER 406, Omo L.7a 125, 3B Scientific®, (R) © 2001 David L. Brill/Brill Atlanta. **p. 410:** *Infographic 20.10* Adapted from Chaplin G., Geographic Distribution of Environmental Factors Influencing Human Skin Coloration, *American Journal of Physical Anthropology* 125:292-302, 2004; map updated in 2007. Designer: Emmanuelle Bournay, UNEP/GRID-Arendal.

Chapter 21

p. 413: Tom Ulrich/Visuals Unlimited. **p. 415:** John Vucetich. **p. 416:** *Infographic 21.1* (T) U.S. Fish and Wildlife Service, (TC) AP Photo/Michigan Technological University, John Vucetich, (BC) John Vucetich, (B) sherwoodimagery/iStockphoto. **p. 418:** *Infographic 21.2* (TL) Flirt/SuperStock, (CL) Tom Hansch/Dreamstime.com, (BL and BC) John Vucetich, (TR and CR) Ecological Studies of Wolves on Isle Royale Annual Report 2009-10 by John A. Vucetich and Rolf O. Peterson, School of Forest Resources and Environmental Science, Michigan Technological University, Figs. 6 and 11, courtesy of John Vucetich. **p. 419:** Russell Burden/Photolibrary; *Infographic 21.3* (L) Les Cunliffe/Dreamstime.com, (C) Melvinlee/Dreamstime.com, (R) Marcel Krol/Dreamstime.com. **p. 420:** *Infographic 21.4* Oksana Churakova/Dreamstime.com. **p. 421:** John Vucetich. **p. 422:** *Infographic 21.6* (T) Tom Ulrich/Visuals Unlimited, (C) Steve Kazlowski/DanitaDelimont.com, (B) Andrey Rozov/Dreamstime.com. **p. 423:** *Infographic 21.7* (T) Imagebroker/Alamy, (TC) Mark Duffy/Alamy, (BC) Cliff Keeler/Alamy, (B) John Vucetich. **p. 425:** *Infographic 21.8* (TL) Photobac/Dreamstime.com, (TR) Terry Morris/iStockphoto.com, (C) James Mattil/AgeFotostock, (BL) Oksana Churakova/Dreamstime.com, (BC) Jim Kruger/iStockphoto, (BR) John Vucetich. **p. 426:** *Infographic 21.9* (L) John Vucetich, (R) Courtesy of Sandy Updyke. **p. 427:** Ann & John Mahan.

Chapter 22

p. 431: OJO Images Ltd/Alamy. **p. 433:** Danish Ismail/Reuters/Landov. **p. 434:** OJO Images Ltd/Alamy. **p. 435:** *Infographic 22.2* (T) Dennis MacDonald/Photolibrary, (B) Ragnar/FeaturePics. **p. 436:** *Infographic 22.3* Olga Demchishina/iStockphoto. **p. 437:** Accent Alaska.com/Alamy. **p. 438:** *Infographic 22.4* (T) David Kay/Dreamstime.com, (C) Rui Miguel da Costa Neves Saraiva/iStockphoto, (B) Kyu Oh/iStockphoto. **p. 439:** *Infographic 22.5* Row 1: (L) Blend Images/Superstock, (C) Michael Sewell/Photolibrary, (R) Benny Rytter/iStockphoto; Row 2: (L) James Phelps Jr/Dreamstime.com, (CL) Abdolhamid Ebrahim/iStockphoto, (C) Lunamarina/Dreamstime.com, (CR) Rui Miguel da Costa Neves Saraiva/iStockphoto, (R) Image by Larry D. Moore, used under a Creative Commons ShareAlike License, http://en.wikipedia.org/wiki/File:Rabbit_in_montana.jpg; Row 3: (L) James Urbach/Photolibrary, (C) ElementalImaging/iStockphoto, (R) Lukrecja/FeaturePics; Row 4: (L) Kyu Oh/iStockphoto, (C) brytta/iStockphoto, (R) Roy T. Free/AgeFotostock. **p. 440:** *Infographic 22.6* (TL) Jon Yuschock/Fotolia, (TR) Courtesy of Donald Stahly, (CL) D. Harms/Photolibrary, (CR) Crown Copyright courtesy of Central Science Laboratory/Photo Researchers, (B) Harry Rogers/Photo Researchers. **p. 442:** *Infographic 22.7* (TL) Willi Schmitz/iStockphoto, (TR) Jim McKinley/Getty Images, (BL) Tim Martin/Dreamstime.com, (BC) Joanne Green/iStockphoto (BR) Steve Byland/iStockphoto. **p. 443:** *Infographic 22.8* (T) P-59 Photos/Alamy, (C) ElementalImaging/iStockphoto, (B) Nic Bothma/epa/Corbis. **p. 444:** (TL) Maigi/Dreamstime.com, (TC) Curt Pickens/iStockphoto, (TR) mrolands/Featurepics.com, (BL) Valentyn75/Dreamstime.com, (BR) Courtesy of Häagen-Dazs. **p. 445:** *Infographic 22.9* (TL) Custom Life Science Images/photographersdirect.com, BL) Custom Life Science Images/photographersdirect.com, (R, from top) Ann

Johansson Photography, Du an Kosti /iStockphoto, Courtesy of Mariano Higes and Raquel Martin-Hernandez, Courtesy of Beeologics and Professor Ilan Sela (Maori, et al., IAPV, a bee-affecting virus associated with Colony Collapse Disorder can be silenced by dsRNA ingestion, *Insect Molecular Biology* (2009) 18(1), 55-60).

Chapter 23

p. 449: WorldFoto/Alamy. **p. 450:** Lauri Patterson/iStockphoto. **p. 451:** Jeff Lepore/Panoramic Images. **p. 454:** Ned Therrien/Visuals Unlimited. **p. 455:** (L, from top) Richard Walters/iStockphoto, malerapaso/iStockphoto, Art33art/Dreamstime.com, (R, from top) Andoni Canel/Photolibrary, gsk/FeaturePics, Tom Bean/Alamy, Krzysztof Odziomek/iStockphoto, Jacka/Dreamstime.com. **p. 459:** *Infographic 23.6* (from left) Jim Zipp/Photo Researchers; Jack Thomas/Alamy; Robert L. Anderson, USDA Forest Service, Bugwood.org.; Evgeny Dubinchuk/Dreamstime.com. **p. 461:** *Infographic 23.8* (L) Goddard/NASA, (R) WorldFoto/Alamy. **p. 465:** *Infographic 23.10* (T) U.S. Dept. of Commerce, NOAA, Earth System Research Laboratory, (BL) Vin Morgan/AFP/Getty Images, (BR) Courtesy of Karin Kirk/Science Education Resource Center at Carleton College. **p. 467:** *Infographic 23.11* (TL) Anna Lubovedskaya/iStockphoto, (TC) Matt Meadows/Photolibrary, (TR) Stringer/epa/Corbis, (BL) olyniteowl/iStockphoto, (BR) Justin Kase Ztwoz/Alamy.

Chapter 24

p. 471: AP Photo/Chicago Department of Environment, Mark Farina. **p. 473:** Artist's impression of Dongtan, designed by Arup, © Arup. **p. 474:** Nir Elias/Reuters/Corbis. **p. 475:** *Infographic 24.2* (T, from left) Capricornis/Dreamstime.com, tank_bmb/FeaturePics, Photong/Dreamstime.com, Dimaberkut/Dreamstime.com, hfng/FeaturePics, (B, from left) Gary Whitton/Dreamstime.com, surpasspro/FeaturePics, Borut Trdina/iStockphotos, stu99/FeaturePics, Tommy Schultz/Dreamstime.com. **p. 476:** *Infographic 24.3* (map) http://commons.wikimedia.org/wiki/File:World_map_of_countries_by_ecological_footprint.png, (graph) © WWF, 2006. Living Plant Report 2006. WWF, Gland, Switzerland. **p. 478:** *Infographic 24.5* (TL) Frank Roeder/Dreamstime.com, (TR) Airwolf01/Dreamstime.com, (CL) Dennis Macdonald/Photolibrary, (CR) Lukasz Koszyk/iStockphoto, (BL) Hanhanpeggy/Dreamstime.com, (BC) Shaun Lowe/iStockphoto, (BR) Dmitro Tolokonov/iStockphoto. **p. 479:** (from left) Enrique Garcia Medina/Archivolatino, © 2009 Tyler Rush Photography and Water Taxi, LLC., Don Nichols/iStockphotos, Chine Nouvelle/Sipa/Newscom. **p. 480:** (T, from left) Rigucci/FeaturePics, (C) Jim West/Alamy, (R) AP Photo/Chicago Department of Environment, Mark Farina, (B, from left) photoneer/FeaturePics, Ashley Cooper/Visuals Unlimited, Frances Roberts/Alamy. **p. 481:** *Infographic 24.6* Artist's impression of Dongtan, designed by Arup, © Arup. **p. 483:** *Infographic 24.7* (from top) Aerial Archives/Alamy, Ron and Patty Thomas Photography/iStockphoto, Petr Nad/iStockphoto, LIU XIN/Xinhua/Landov
Pavle Marjanovic/Dreamstime.com, Rob Broek/iStockphoto. **p. 485:** *Infographic 24.9* (TL and BL) Courtesy of John C. Dohrenwend/USGS, (R) Jim Wark/Photolibrary. **p. 486:** *Infographic 24.10* (map) © 2003 World Resources Institute. **p. 487:** *Infographic 24.11* (T) AP Photo/Rick Bowmer, (BL) Chris Fourie/Dreamstime.com, (BC) http://en.wikipedia.org/wiki/Loggerhead_sea_turtle, (BR) Mark Conlin/Alamy. **p. 488:** *Infographic 24.12* (from top) stuartbur/iStockphoto, rockphoto/FeaturePics, okea /FeaturePics, Don Nichols/iStockphoto, Dimitri Vervitsiotis/Digital Vision/Getty Images, (inset) AP Photo/Damian Dovarganes, Yuliyan Velchev/Dreamstime.com, (inset) Dmitro Tolokonov/iStockphoto.

Index

Note: page numbers followed by f indicate figures; those followed by t indicate tables.